March 24–29, 2013
Fukuoka, Japan

# Association for Computing Machinery

*Advancing Computing as a Science & Profession*

# AOSD'13

Proceedings of the 2013 ACM on
## Aspect-Oriented Software Development

*Sponsored by:*
## AOSA

*In cooperation with:*
## ACM SIGPLAN and ACM SIGSOFT

*Supported by:*
## Microsoft Research, Oracle, Kyushu University, Ratuten, VMWare, Kayamori Foundation of Informational Science Advancement, FCVB, AOSD-Europe, and Cybozu

**Association for Computing Machinery**

*Advancing Computing as a Science & Profession*

**The Association for Computing Machinery**
2 Penn Plaza, Suite 701
New York, New York 10121-0701

**Notice to Past Authors of ACM-Published Articles**
ACM intends to create a complete electronic archive of all articles and/or other material previously published by ACM. If you have written a work that has been previously published by ACM in any journal or conference proceedings prior to 1978, or any SIG Newsletter at any time, and you do NOT want this work to appear in the ACM Digital Library, please inform permissions@acm.org, stating the title of the work, the author(s), and where and when published.

**ISBN:** 978-1-4503-1766-5 (Digital)

**ISBN:** 978-1-4503-2097-9 (Print)

Additional copies may be ordered prepaid from:

**ACM Order Department**
PO Box 30777
New York, NY 10087-0777, USA

Phone: 1-800-342-6626 (USA and Canada)
+1-212-626-0500 (Global)
Fax: +1-212-944-1318
E-mail: acmhelp@acm.org
Hours of Operation: 8:30 am – 4:30 pm ET

Printed in the USA

# MODULARITY:aosd.13 – Chairs' Welcome

It is our great pleasure to welcome you to MODULARITY: aosd.13, the premiere international research conference on modularity in software and software-intensive systems. MODULARITY: aosd.13 is the 12th annual international conference on Aspect-Oriented Software Development (AOSD).

This year's conference continues to broaden of the scope of the field to address all aspects of modularity, abstraction, and separation of concerns as they pertain to software, including new forms, uses, and analysis of modularity, along with the costs and benefits, and tradeoffs involved in their application. Modularity provides the international computer science research community and its many sub-disciplines (including software engineering, languages, and computer systems) with unique opportunities to come together to share and discuss perspectives, results, and visions with others interested in modularity as well as in the languages, development methods, architectures, algorithms, and other technologies organized around this fundamental concept.

The MODULARITY: aosd.13 conference comprises two main technical tracks: Research Results and Modularity Visions. Both tracks invited full, scholarly papers of the highest quality on results and new ideas in areas that include but are not limited to complex systems, software design and engineering, programming languages, cyber-physical systems, and other areas across the whole system life cycle.

Papers submitted to the Research Results track were reviewed in accordance with the highest established standards of scientific rigor applied in peer review of putative research results. Reviewers assessed works in terms of research problem formulations, novelty and sophistication of proposed solutions, clarity and significance of hypotheses, proper design and execution of experimental or analytical assessments, sound interpretation of data, and correct characterization of work in relation to existing knowledge.

The Research Results program committee accepted papers in three rounds. In each round, each paper was accepted, rejected, or (except in the last round) invited for revision and a second review.

Papers submitted to the Modularity Visions track were reviewed in accordance with the highest established standards of scientific rigor applied in peer review of scientific research proposals. Reviewers assessed works in terms of research problem formulations, novelty and sophistication of proposed solutions, clarity and significance of hypotheses, compelling preliminary results, proposals for sound future experimental or analytical assessments and interpretation of data, and correct characterization of work in relation to existing knowledge.

The Modularity Visions program committee comprising ten international experts from different areas of software modularity selected one technical paper after a rigorous peer review of the five submissions to this track. Each paper was assigned at least three reviewers. The program committee discussed each paper in detail assessing their respective novel vision and contribution to advancing the state-of-the-art in modularity.

Together, the Research Results (RR) and Modularity Visions (MV) tracks received 67 submissions (62 to the RR track; 5 to the MV track). Out of these submissions, 54 were distinct papers (13 papers were invited resubmissions from one round to the next, of which 11 were eventually accepted). In total, out of the 54 distinct submissions, 18 have been accepted (17 to the RR track; 1 to the MV track), which reflects an acceptance ratio of 33%.

The MODULARITY: aosd.13 conference features three keynote speakers on modularity: Takahiro Fujimoto, Kyo Chul Kang, and Steven P. Reiss. Takahiro will talk about the spectrum of architectural modularity and integrality from the perspectives of manufacturing management; Kang will talk about modularity in the context of product line variability; and Steven will talk about modularity in modern applications and tools to support it.

We thank all the authors who submitted papers to the program, all members of our organizing and program committees and all external reviewers.

We are looking forward to an interesting and inspiring MODULARITY: aosd.13.

**Hidehiko Masuhara**
*General Chair*
*The University of Tokyo, Japan*

**Shigeru Chiba**
*Organizing Co-Chair*
*The University of Tokyo, Japan*

**Naoyasu Ubayashi**
*Organizing Co-Chair*
*Kyushu University, Japan*

**Jörg Kienzle**
*Research Results Chair*
*McGill University, Canada*

**Elisa Baniassad**
*Modularity Visions Co-Chair*
*Australian National University, Australia*

**David H. Lorenz**
*Modularity Visions Co-Chair*
*The Open University of Israel, Israel*

# Table of Contents

## Session 6: Formal Reasoning about Modularity

## Session 7: Advances in Language Design

# MODULARITY:aosd.13 – Organization

**General Chair:** Hidehiko Masuhara *(The University of Tokyo, Japan)*

**Organizing Co-Chairs:** Shigeru Chiba *(The University of Tokyo, Japan)*
Naoyasu Ubayashi *(Kyushu University, Japan)*

**Research Results Chair:** Jörg Kienzle *(McGill University, Canada)*

**Modularity Visions Co-Chairs:** Elisa Baniassad *(Australian National University, Australia)*
David Lorenz *(The Open University of Israel, Israel)*

**Industry Track Co-Chairs:** Andrew Eisenberg *(VMware, Canada)*
Tomoji Kishi *(Waseda University, Japan)*

**Workshop Co-Chairs:** Tomoyuki Aotani *(Japan Advanced Institute of Science and Technology, Japan)*
Phil Greenwood *(Lancaster University, UK)*

**Demonstrations & BoFs Co-Chairs:** Walter Binder *(University of Lugano, Switzerland)*
Charles Zhang *(Hong Kong University of Science and Technology, China)*

**Student Event Co-Chairs:** Christoph Bockisch *(University of Twente, The Netherlands)*
Atsushi Igarashi *(Kyoto University, Japan)*

**Publicity Chair:** Eric Bodden *(EC SPRIDE / Technische Universitat Darmstadt, Germany)*

**Student Volunteer and Web Chair:** Yasutaka Kamei *(Kyushu University, Japan)*

**Design:** Fuminobu Takeyama *(Tokyo Institute of Technology, Japan)*

**Organizing members:** Tetsuo Kamina *(The University of Tokyo, Japan)*
Yoshiki Sato *(The University of Tokyo, Japan)*

# MODULARITY:aosd.13 – Sponsor & Supporters

Sponsor:

In cooperation with:

Supporters:

# Reify Your Collection Queries
# for Modularity and Speed!

Paolo G. Giarrusso
Philipps University Marburg

Klaus Ostermann
Philipps University Marburg

Michael Eichberg
Software Technology Group,
Technische Universität
Darmstadt

Ralf Mitschke
Software Technology Group,
Technische Universität
Darmstadt

Tillmann Rendel
Philipps University Marburg

Christian Kästner
Carnegie Mellon University

## ABSTRACT

Modularity and efficiency are often contradicting requirements, such that programers have to trade one for the other. We analyze this dilemma in the context of programs operating on collections. Performance-critical code using collections need often to be hand-optimized, leading to non-modular, brittle, and redundant code. In principle, this dilemma could be avoided by automatic collection-specific optimizations, such as fusion of collection traversals, usage of indexing, or reordering of filters. Unfortunately, it is not obvious how to encode such optimizations in terms of ordinary collection APIs, because the program operating on the collections is not reified and hence cannot be analyzed.

We propose SQuOpt, the Scala Query Optimizer—a *deep embedding* of the Scala collections API that allows such analyses and optimizations to be defined and executed within Scala, without relying on external tools or compiler extensions. SQuOpt provides the same "look and feel" (syntax and static typing guarantees) as the standard collections API. We evaluate SQuOpt by re-implementing several code analyses of the FindBugs tool using SQuOpt, show average speedups of 12x with a maximum of 12800x and hence demonstrate that SQuOpt can reconcile modularity and efficiency in real-world applications.

## Categories and Subject Descriptors

H.2.3 [**Database Management**]: Languages—*Query languages*; D.1.1 [**Programming Techniques**]: Applicative (Functional) Programming; D.1.5 [**Programming Techniques**]: Object-oriented Programming

## Keywords

Deep embedding; query languages; optimization; modularity

## 1. INTRODUCTION

In-memory collections of data often need efficient processing. For on-disk data, efficient processing is already provided by database management systems (DBMS) thanks to their query optimizers, which support many optimizations specific to the domain of collections. Moving in-memory data to DBMSs, however, typically does not improve performance [30], and query optimizers cannot be reused separately since DBMS are typically monolithic and their optimizers deeply integrated. A few collection-specific optimizations, such as shortcut fusion [11], are supported by compilers for purely functional languages such as Haskell. However, the implementation techniques for those optimizations do not generalize to many other ones, such as support for indexes. In general, collection-specific optimizations are not supported by the general-purpose optimizers used by typical (JIT) compilers.

Therefore programmers, when needing collection-related optimizations, perform them manually. To allow that, they are often forced to perform manual inlining [24]. But manual inlining modifies source code by combining distinct functions together, while often distinct functions should remain distinct, because they deal with different concerns, or because one function need to be reused in a different context. In either case, manual inlining reduces modularity — defined here as the ability to abstract behavior in a separate function (possibly part of a different module) to enable reuse and improve understandability.

For these reasons, currently developers need to choose between modularity and performance, as also highlighted by Kiczales et al. [18] on a similar example. Instead, we envision that they should rely on an automatic optimizer performing inlining and collection-specific optimizations. They would then achieve both performance and modularity.[1]

One way to implement such an optimizer would be to extend the compiler of the language with a collection-specific optimizer, or to add some kind of external preprocessor to the language. However, such solutions would be rather

---

[1]In the terminology of Kiczales et al. [18], our goal is to be able to decompose different *generalized procedures* of a program according to its primary decomposition, while separating the handling of some performance concerns. To this end, we are modularizing these performance concerns into a metaprogramming-based optimization module, which we believe could be called, in that terminology, *aspect*.

brittle (for instance, they lack composability with other language extensions) and they would preclude optimization opportunities that arise only at runtime.

For this reason, our approach is implemented as an embedded domain-specific language, that is, as a regular library. We call this library SQuOpt, the Scala QUery OPTimizer. SQuOpt consists of a domain-specific language (DSL) for queries on collections based on the Scala collections API. This DSL is implemented as an embedded DSL (EDSL) for Scala. An expression in this EDSL produces at run time an *expression tree* in the host language: a data structure which represents the query to execute, similar to an abstract syntax tree (AST) or a query plan. Thanks to the extensibility of Scala, expressions in this language look almost identical to expressions with the same meaning in Scala. When executing the query, SQuOpt optimizes and compiles these expression trees for more efficient execution. Doing optimization at run time, instead of compile-time, avoids the need for control-flow analyses to determine which code will be actually executed [3], as we will see later.

We have choosen Scala [23] to implement our library for two reasons: (i) Scala is a good meta-language for embedded DSLs, because it is syntactically flexible and has a powerful type system, and (ii) Scala has a sophisticated collections library with an attractive syntax (for-comprehensions) to specify queries.

To evaluate SQuOpt, we study queries of the FindBugs tool [17]. We rewrote a set of queries to use the Scala collections API and show that modularization incurs significant performance overhead. Subsequently, we consider versions of the same queries using SQuOpt. We demonstrate that the automatic optimization can reconcile modularity and performance in many cases. Adding advanced optimizations such as indexing can even improve the performance of the analyses beyond the original non-modular analyses.

Overall, our main contributions are the following:

- We illustrate the tradeoff between modularity and performance when manipulating collections, caused by the lack of domain-specific optimizations (Sec. 2). Conversely, we illustrate how domain-specific optimizations lead to more readable and more modular code (Sec. 3).

- We present the design and implementation of SQuOpt, an embedded DSL for queries on collections in Scala (Sec. 4).

- We evaluate SQuOpt to show that it supports writing queries that are at the same time modular and fast. We do so by re-implementing several code analyses of the FindBugs tool. The resulting code is more modular and/or more efficient, in some cases by orders of magnitude. In these case studies, we measured average speedups of 12x with a maximum of 12800x (Sec. 5).

## 2. MOTIVATION

In this section, we show how the absense of collection-specific optimizations forces programmers to trade modularity against performance, which motivates our design of SQuOpt to resolve this conflict.

As our running example through the paper, we consider representing and querying a simple in-memory bibliography. A book has, in our schema, a title, a publisher and a list of authors. Each author, in turn, has a first and last name. We

```scala
package schema
case class Author(firstName: String, lastName: String)
case class Book(title: String, publisher: String,
  authors: Seq[Author])

val books: Set[Book] = Set(
  new Book("Compilers: Principles, Techniques and Tools",
      "Pearson Education",
      Seq(new Author("Alfred V.", "Aho"),
          new Author("Monica S.", "Lam"),
          new Author("Ravi", "Sethi"),
          new Author("Jeffrey D.", "Ullman"))))
/* other books ... */)
```

**Figure 1: Definition of the schema and of some content.**

```scala
case class BookData(title: String, authorName: String,
  coauthors: Int)

val records =
  for {
    book ← books
    if book.publisher == "Pearson Education"
    author ← book.authors
  } yield new BookData(book.title,
                author.firstName + " " +
                author.lastName,
                book.authors.size - 1)

def titleFilter(records: Set[BookData],
  keyword: String) =
  for {
    record ← records
    if record.title.contains(keyword)
  } yield (record.title, record.authorName)

val res = titleFilter(records, "Principles")
```

**Figure 2: Our example query on the schema in Fig. 1, and a function which postprocesses its result.**

represent authors and books as instances of the Scala classes `Author` and `Book` shown in Fig. 1. The class declarations list the type of each field: Titles, publishers, and first and last names are all stored in fields of type `String`. The list of authors is stored in a field of type `Seq[Author]`, that is, a sequence of authors – something that would be more complex to model in a relational database. The code fragment also defines a collection of books named `books`.

As a common idiom to query such collections, Scala provides *for-comprehensions*. For instance, the for-comprehension computing `records` in Fig. 2 finds all books published by Pearson Education and yields, for each of those books, and for each of its authors, a record containing the book title, the full name of that author and the number of additional coauthors. The *generator* `book ← books` functions like a loop header: The remainder of the for-comprehension is executed once per book in the collection. Consequently, the *generator* `author ← book.authors` starts a nested loop. The return value of the for-comprehension is a collection of all yielded records. Note that if a book has multiple authors, this for-comprehensions will return multiple records relative to this book, one for each author.

We can further process this collection with another for-comprehension, possibly in a different module. For example, still in Fig. 2, the function `titleFilter` filters book titles containing the word "Principles", and drops from each record the number of additional coauthors.

In Scala, the implementation of for-comprehensions is not fixed. Instead, the compiler desugars a for-comprehension to a series of API calls, and different collection classes can implement this API differently. Later, we will use this flexibility to provide an optimizing implementation of for-comprehensions, but in this section, we focus on the behavior of the standard Scala collections, which implement for-comprehensions as loops that create intermediate collections.

## 2.1 Optimizing by Hand

In the naive implementation in Fig. 2 different concerns are separated, hence it is modular. However, it is also inefficient. To execute this code, we first build the original collection and only later we perform further processing to build the new result; creating the intermediate collection at the interface between these functions is costly. Moreover, the same book can appear in `records` more than once if the book has more than one author, but all of these duplicates have the same title. Nevertheless, we test each duplicate title separately whether it contains the searched `keyword`. If books have 4 authors on average, this means a slowdown of a factor of 4 for the filtering step.

In general, one can only resolve these inefficiencies by manually optimizing the query; however, we will observe that these manual optimizations produce less modular code.[2]

To address the first problem above, that is, to avoid creating intermediate collections, we can manually inline `titleFilter` and `records`; we obtain two nested for-comprehensions. Furthermore, we can *unnest* the inner one [6].

To address the second problem above, that is, to avoid testing the same title multiple times, we *hoist* the filtering step, that is, we change the order of the processing steps in the query to first look for `keyword` within `book.title` and then iterate over the set of authors. This does not change the overall semantics of the query because the filter only accesses the title but does not depend on the author. In the end, we obtain the code in Fig. 3. The resulting query processes the title of each book only once. Since filtering in Scala is done lazily, the resulting query avoids building an intermediate collection.

This second optimization is only possible after inlining and thereby reducing the modularity of the code, because it mixes together processing steps from `titleFilter` and from the definition of `records`. Therefore, reusing the code creating records would now be harder.

To make `titleFilterHandOpt` more reusable, we could turn the publisher name into a parameter. However, the new versions of `titleFilter` cannot be reused as-is if some details of the inlined code change; for instance, we might need to filter publishers differently or not at all. On the other hand, if we express queries modularly, we might lose some opportunities for optimization. The design of the collections API, both in Scala and in typical languages, forces us to manually optimize our code by repeated inlining and subsequent application of query optimization rules, which leads to a loss of modularity.

---

[2]The existing Scala collections API supports optimization, for instance through non-strict variants of the query operators (called 'views' in Scala), but they can only be used for a limited set of optimizations, as we discuss in Sec. 6.

```scala
def titleFilterHandOpt(books: Set[Book],
                       publisher: String,
                       keyword: String) =
  for {
    book ← books
    if book.publisher == publisher &&
  book.title.contains(keyword)
    author ← book.authors
  } yield (book.title, author.firstName + " " +
          author.lastName)
val res = titleFilterHandOpt(books,
  "Pearson Education", "Principles")
```

**Figure 3: Composition of queries in Fig. 2, after inlining, query unnesting and hoisting.**

```scala
import squopt._
import schema.squopt._

val recordsQuery =
  for {
    book ← books.asSquopt
    if book.publisher ==# "Pearson Education"
    author ← book.authors
  } yield new BookData(book.title,
    author.firstName + " " + author.lastName,
    book.authors.size - 1)

// ...
val records = recordsQuery.eval

def titleFilterQuery(records: Exp[Set[BookData]],
                     keyword: Exp[String]) = for {
  record ← records
  if record.title.contains(keyword)
} yield (record.title, record.authorName)
val resQuery = titleFilterQuery(recordsQuery, "Principles")
val res = resQuery.optimize.eval
```

**Figure 4: SQuOpt version of Fig. 2; `recordQuery` contains a reification of the query, `records` its result.**

## 3. AUTOMATIC OPTIMIZATION WITH SQUOPT

The goal of SQuOpt is to let programmers write queries modularly and at a high level of abstraction and deal with optimization by a dedicated domain-specific optimizer. In our concrete example, programmers should be able to write queries similar to the one in Fig. 2, but get the efficiency of the one in Fig. 3. To allow this, SQuOpt overloads for-comprehensions and other constructs, such as string concatenation with + and field access `book.author`. Our overloads of these constructs reify the query as an expression tree. SQuOpt can then optimize this expression tree and execute the resulting optimized query. Programmers explicitly trigger processing by SQuOpt, by adapting their queries as we describe in next subsection.

### 3.1 Adapting a Query

To use SQuOpt instead of native Scala queries, we first assume that the query does not use side effects and is thus *purely functional*. We argue that purely functional queries are more declarative. Side effects are used to improve performance, but SQuOpt makes that unnecessary through automatic optimizations. In fact, the lack of side effects enables more optimizations.

In Fig. 4 we show a version of our running example adapted to use SQuOpt. We first discuss changes to `records`. To enable SQuOpt, a programmer needs to (a) import the SQuOpt library, (b) import some wrapper code specific to the types the collection operates on, in this case `Book` and `Author` (more about that later), (c) convert explicitly the native Scala collections involved to collections of our framework by a call to `asSquopt`, (d) rename a few operators such as `==` to `==#` (this is necessary due to some Scala limitations), and (e) add a separate step where the query is evaluated (possibly after optimization). All these changes are lightweight and mostly of a syntactic nature.

For parameterized queries like `titleFilter`, we need to also adapt type annotations. The ones in `titleFilterQuery` reveal some details of our implementation: Expressions that are reified have type `Exp[T]` instead of `T`. As the code shows, `resQuery` is optimized before compilation. This call will perform the optimizations that we previously did by hand and will return a query equivalent to that in Fig. 3, after verifying their safety conditions. For instance, after inlining, the filter `if book.title.contains(keyword)` does not reference `author`; hence, it is safe to hoist. Note that checking this safety condition would not be possible without reifying the predicate. For instance, it would not be sufficient to only reify the calls to the collection API, because the predicate is represented as a boolean function parameter. In general, our automatic optimizer inspects the whole reification of the query implementation to check that optimizations do not introduce changes in the overall result of the query and are therefore safe.

## 3.2 Indexing

SQuOpt also supports the transparent usage of indexes. Indexes can further improve the efficiency of queries, sometimes by orders of magnitude. In our running example, the query scans all books to look for the ones having the right publisher. To speed up this query, we can preprocess `books` to build an index, that is, a dictionary mapping, from each publisher to a collection of all the books it published. This index can then be used to answer the original query without scanning all books.

We construct a *query* representing the desired dictionary, and inform the optimizer that it should use this index where appropriate:

```
val idxByPublisher =
  books.asSquopt.indexBy(_.publisher)
Optimization.addIndex(idxByPublisher)
```

The `indexBy` collection method accepts a function that maps a collection element to a key; `coll.indexBy(key)` returns a dictionary mapping each key to the collection of all elements of `coll` having that key. Missing keys are mapped to an empty collection.[3] `Optimization.addIndex` simply preevaluates the index and updates a dictionary mapping the index to its preevaluated result.

A call to `optimize` on a query will then take this index into account and rewrite the query to perform index lookup instead of scanning, if possible. For instance, the code in Fig. 4 would be transparently rewritten by the optimizer to a query similar to the following:

```
val indexedQuery =
  for {
    book ← idxByPublisher("Pearson Education")
```

---

[3]For readers familiar with the Scala collection API, we remark that the only difference with the standard `groupBy` method is the handling of missing keys.

```
    author ← book.authors
  } yield new BookData(book.title, author.firstName
    + " " + author.lastName, book.authors.size - 1)
```

Since dictionaries in Scala are functions, in the above code, dictionary lookup on `idxByPublisher` is represented simply as function application. The above code iterates over books having the desired publisher, instead of scanning the whole library, and performs the remaining computation from the original query. Although the index use in the listing above is notated as `idxByPublisher("Pearson Education")`, only the cached result of evaluating the index is used when the query is executed, not the reified index definition.

This optimization could also be performed manually, of course, but the queries are on a higher abstraction level and more maintainable if indexing is defined separately and applied automatically. Manual application of indexing is a crosscutting concern because adding or removing an index affects potentially many queries. SQuOpt does not free the developer from the task of assessing which index will 'pay off' (we have not considered automatic index creation yet), but at least it becomes simple to add or remove an index, since the application of the indexes is modularized in the optimizer.

## 4. IMPLEMENTATION

After describing how to use SQuOpt, we explain how SQuOpt represents queries internally and optimizes them. We give only a brief overview of our implementation technique; it is described in more detail in a technical report that accompanies this paper [10].

### 4.1 Expression Trees

In order to analyze and optimize collection queries at runtime, SQuOpt reifies their syntactic structure as *expression trees*. The expression tree reflects the syntax of the query after desugaring, that is, after for-comprehensions have been replaced by API calls. For instance, `recordsQuery` from Fig. 4 points to the following expression tree (with some boilerplate omitted for clarity):

```
new FlatMap(
  new Filter(
    new Const(books),
    v2 ⇒ new Eq(new Book_publisher(v2),
            new Const("Pearson Education"))),
    v3 ⇒ new MapNode(
          new Book_authors(v3),
          v4 ⇒ new BookData(
                new Book_title(v3),
                new StringConcat(
                  new StringConcat(
                    new Author_firstName(v4),
                    new Const(" ")),
                  new Author_lastName(v4)),
                new Plus(new Size(new Book_authors(v3)),
                  new Negate(new Const(1))))))
```

The structure of the for-comprehension is encoded with the `FlatMap`, `Filter` and `MapNode` instances. These classes correspond to the API methods that for-comprehensions get desugared to. SQuOpt arranges for the implementation of `flatMap` to construct a `FlatMap` instance, etc. The instances of the other classes encode the rest of the structure of the collection query, that is, which methods are called on which arguments. On the one hand, SQuOpt defines classes such as `Const` or `Eq` that are generic and applicable to all queries. On the other hand, classes such as `Book_publisher` cannot be predefined, because they are specific to the user-defined types used in a query. SQuOpt provides a small code generator, which creates a case class for each method and field of a user-defined type. Functions in the query are represented by functions that create expression trees; representing functions in this way is frequently called higher-order abstract syntax [25].

We can see that the reification of this code corresponds closely to an abstract syntax tree for the code which is executed; however, many calls to specific methods, like `map`, are represented by special nodes, like `MapNode`, rather than as method calls. For the optimizer it becomes easier to match and transform those nodes than with a generic abstract syntax tree.

Nodes for collection operations are carefully defined by hand to provide them highly generic type signatures and make them reusable for all collection types. In Scala, collection operations are highly polymorphic; for instance, `map` has a single implementation working on all collection types, like `List`, `Set`, and we similarly want to represent all usages of `map` through instances of a single node type, namely `MapNode`. Having separate nodes `ListMapNode`, `SetMapNode` and so on would be inconvenient, for instance when writing the optimizer. However, `map` on a `List[Int]` will produce another `List`, while on a `Set` it will produce another `Set`, and so on for each specific collection type (in first approximation); moreover, this is guaranteed statically by the type of `map`. Yet, thanks to advanced typesystem features, `map` is defined only once avoiding redundancy, but has a type polymorphic enough to guarantee statically that the correct return value is produced. Since our tree representation is strongly typed, we need to have a similar level of polymorphism in `MapNode`. We achieved this by extending the techniques described by Odersky and Moors [22], as detailed in our technical report [10].

We get these expression trees by using Scala implicit conversions in a particular style, which we adopted from Rompf and Odersky [26]. Implicit conversions allow to add, for each method `A.foo(B)`, an overload of `Exp[A].foo(Exp[B])`. Where a value of type `Exp[T]` is expected, a value of type `T` can be used thanks to other implicit conversions, which wrap it in a `Const` node. The initial call of `asSquopt` triggers the application of the implicit conversions by converting the collection to the leaf of an expression tree.

It is also possible to call methods that do not return expression trees; however, such method calls would then only be represented by an opaque `MethodCall` node in the expression tree, which means that the code of the method cannot be considered in optimizations.

Crucially, these expression trees are generated at runtime. For instance, the first `Const` contains a reference to the actual collection of books to which `books` refers. If a query uses another query, such as `records` in Fig. 4, then the subquery is effectively *inlined*. The same holds for method calls inside queries: If these methods return an expression tree (such as the `titleFilterQuery` method in Fig. 4), then these expression trees are inlined into the composite query. Since the reification happens at runtime, it is not necessary to predict the targets of dynamically bound method calls: A new (and possibly different) expression tree is created each time a block of code containing queries is executed.

Hence, we can say that expression trees represent the computation which is going to be executed after inlining; control flow or virtual calls in the original code typically disappear—

especially if they manipulate the query as a whole. This is typical of deeply embedded DSLs like ours, where code instead of performing computations produces a representation of the computation to perform [5, 3].

This inlining can duplicate computations; for instance, in this code:

```
val num: Exp[Int] = 10
val square = num * num
val sum = square + square
```

evaluating `sum` will evaluate `square` twice. Elliott et al. [5] and we avoid this using common-subexpression elimination.

## 4.2 Optimizations

Our optimizer currently supports several algebraic optimizations. Any query and in fact every reified expression can be optimized by calling the `optimize` function on it. The ability to optimize reified expressions that are not queries is useful; for instance, optimizing a function that produces a query is similar to a "prepared statement" in relational databases.

The optimizations we implemented are mostly standard in compilers [21] or databases:

- *Query unnesting* merges a nested query into the containing one [6, 14], replacing for instance

  ```
  for {val1 ← (for {val2 ← coll} yield f(val2))}
    yield g(val1)
  ```

  with

  ```
  for {val2 ← coll; val1 = f(val2)} yield g(val1)
  ```

- *Bulk operation fusion* fuses higher-order operators on collections.

- *Filter hoisting* tries to apply filters as early as possible; in database query optimization, it is known as selection pushdown. For filter hoisting, it is important that the full query is reified, because otherwise the dependencies of the filter condition cannot be determined.

- We reduce during optimization tuple/case class accesses: For instance, `(a, b)._1` is simplified to `a`. This is important because the produced expression does not depend on `b`; removing this false dependency can allow, for instance, a filter containing this expression to be hoisted to a context where `b` is not bound.

- *Indexing* tries to apply one or more of the available indexes to speed up the query.

- *Common subexpression elimination (CSE)* avoids that the same computation is performed multiple times; we use techniques similar to Rompf and Odersky [26].

- Smaller optimizations include constant folding, reassociation of associative operators and removal of identity maps (`coll.map(x ⇒ x)`), typically generated by the translation of for-comprehensions.

Each optimization is applied recursively bottom-up until it does not trigger anymore; different optimizations are composed in a fixed pipeline.

Optimizations are only guaranteed to be semantics-preserving if queries obey the restrictions we mentioned: for instance, queries should not involve side-effects such as assignments or I/O, and all collections used in queries should implement the specifications stated in the collections API. Obviously the choice of optimizations involves many tradeoffs; for that reason we believe that it is all the more important that the optimizer is not hard-wired into the compiler but implemented as a library, with potentially many different implementations.

To make changes to the optimizer more practical, we designed our query representation so that optimizations are easy to express; restricting to pure queries also helps. For instance, filter fusion can be implemented simply as:[4]

```
val mergeFilters = ExpTransformer {
  case Sym(Filter(Sym(Filter(collection, pred2)), pred1)) ⇒
    coll.filter(x ⇒ pred2(x) && pred1(x))
}
```

The above code matches on reified expression of form `coll.filter(pred2).filter(pred1)` and rewrites it. A more complex optimization such as filter hoisting requires only 20 lines of code.

We have implemented a prototype of the optimizer with the mentioned optimizations. Many additional algebraic optimizations can be added in future work by us or others; a candidate would be loop hoisting, which moves out of loops arbitrary computations not depending on the loop variable (and not just filters). With some changes to the optimizer's architecture, it would also be possible to perform cost-based and dynamic optimizations.

## 4.3 Query Execution

Calling the `eval` method on a query will convert it to executable bytecode; this bytecode will be loaded and invoked by using Java reflection. We produce a thunk that, when evaluated, will execute the generated code.

In our prototype we produce bytecode by converting expression trees to Scala code and invoking on the result the Scala compiler, `scalac`. Invoking `scalac` is typically quite slow, and we currently use caching to limit this concern; however, we believe it is merely an engineering problem to produce bytecode directly from expression trees, just as compilers do.

Our expression trees contain native Scala values wrapped in `Const` nodes, and in many cases one cannot produce Scala program text evaluating to the same value. To allow executing such expression trees we need to implement cross-stage persistence (CSP): the generated code will be a function, accepting the actual values as arguments [26]. This allows sharing the compiled code for expressions which differ only in the embedded values.

More in detail, our compilation algorithm is as follows. (a) We implement CSP by replacing embedded Scala values by references to the function arguments; so for instance `List(1, 2, 3).map(x ⇒ x + 1)` becomes the function `(s1: List[Int], s2: Int) ⇒ s1.map(x ⇒ x + s2)`. (b) We look up the produced expression tree, together with the types of the constants we just removed, in a cache mapping to the generated classes. If the lookup fails we update the cache with the result of the next steps. (c) We apply CSE on the expression. (d) We convert the tree to code, compile it and load the generated code.

**Preventing errors in generated code** Compiler errors in generated code are typically a concern; with SQuOpt,

however, they can only arise due to implementation bugs in SQuOpt (for instance in pretty-printing, which cannot be checked statically), so they do not concern users. Since our query language and tree representation are statically typed, type-incorrect queries will be rejected statically. For instance, consider again `idxByPublisher`, described previously:

```
val idxByPublisher =
  books.asSquopt.indexBy(_.publisher)
```

Since `Book.publisher` returns a `String`, `idxByPublisher` has type `Exp[Map[String, Book]]`. Looking up a key of the wrong type, for instance by writing `idxByPublisher(book)` where `book: Book`, will make `scalac` emit a static type error.

## 5. EVALUATION

The key goals of SQuOpt are to reconcile *modularity* and *efficiency*. To evaluate this claim, we perform a rigorous performance evaluation of queries with and without SQuOpt. We also analyze modularization potential of these queries and evaluate how modularization affects performance (with and without SQuOpt).

We show that modularization introduces a significant slowdown. The overhead of using SQuOpt is usually moderate, and optimizations can compensate this overhead, remove the modularization slowdown and improve performance of some queries by orders of magnitude, especially when indexes are used.

### 5.1 Study Setup

Throughout the paper, we have already shown several compact queries for which our optimizations increase performance significantly compared to a naive execution. Since some optimizations change the complexity class of the query (e.g. by using an index), so the speedups grow with the size of the data. However, to get a more realistic evaluation of SQuOpt, we decided to perform an experiment with existing real-world queries.

As we are interested in both performance and modularization, we have a specification and three different implementations of each query that we need to compare:

(0) **Query specification:** We selected a set of existing real-world queries specified and implemented independently from our work and prior to it. We used only the specification of these queries.

(1) **Modularized Scala implementation:** We reimplemented each query as an expression on Scala collections — our baseline implementation. For modularity, we separated reusable domain abstractions into subqueries. We confirmed the abstractions with a domain expert and will later illustrate them to emphasize their general nature.

(2) **Hand-optimized Scala implementation:** Next, we asked a domain expert to performed manual optimizations on the modularized queries. The expert should perform optimizations, such as inlining and filter hoisting, where he could find performance improvements.

(3) **SQuOpt implementation:** Finally, we rewrote the modularized Scala queries from (1) as SQuOpt queries. The rewrites are of purely syntactic nature to use our library (as described in Sec. 3.1) and preserve the modularity of the queries.

---

[4]`Sym` nodes are part of the boilerplate we omitted earlier.

Since SQuOpt supports executing queries with and without optimizations and indexes, we measured actually three different execution modes of the SQuOpt implementation:

($3^-$) **SQuOpt without optimizer:** First, we execute the SQuOpt queries without performing optimization first, which should show the SQuOpt overhead compared to the modular Scala implementation (1). However, common-subexpression elimination is still used here, since it is part of the compilation pipeline. This is appropriate to counter the effects of excessive inlining due to using a deep embedding, as explained in Sec. 4.1.

($3^o$) **SQuOpt with optimizer:** Next, we execute SQuOpt queries after optimization.

($3^x$) **SQuOpt with optimizer and indexes:** Finally, we execute the queries after providing a set of indexes that the optimizer can consider.

In all cases, we measure query execution time for the generated code, excluding compilation: we consider this appropriate because the results of compilations are cached aggressively and can be reused when the underlying data is changed, potentially even across executions (even though this is not yet implemented), as the data is not part of the compiled code.

We use additional indexes in ($3^x$), but not in the hand-optimized Scala implementation (2). We argue that indexes are less likely to be applied manually, because index application is a crosscutting concern and makes the whole query implementation more complicated and less abstract. Still, we offer measurement ($3^o$) to compare the speedup without additional indexes.

This gives us a total of five settings to measure and compare (1, 2, $3^-$, $3^o$, and $3^x$). Between them, we want to observe the following interesting performance ratios (speedups or slowdowns, computed through the indicated divisions):

(**M**) Modularization overhead (the relative performance difference between the modularized and the hand-optimized Scala implementation: 1/2).

(**S**) SQuOpt overhead (the overhead of executing unoptimized SQuOpt queries: $1/3^-$; smaller is better).

(**H**) Hand-optimization challenge (the performance overhead of our optimizer against hand-optimizations of a domain expert: $2/3^o$; bigger is better). This overhead is partly due to the SQuOpt overhead (S) and partly to optimizations which have not been automated or have not been effective enough. This comparison excludes the effects of indexing, since this is an optimization we did not perform by hand; we also report (**H'**) = $2/3^x$, which includes indexing.

(**O**) Optimization potential (the speedup by optimizing modularized queries: $1/3^o$; bigger is better).

(**X**) Index influence (the speedup gained by using indexes: $3^o/3^x$) (bigger is better).

(**T**) Total optimization potential with indexes ($1/3^x$; bigger is better), which is equal to $(O) \times (X)$.

In Figure 5, we provide an overview of the setup. We made our raw data available and our results reproducible [31].[5]

[5]Data available at: http://www.informatik.uni-marburg.de/~pgiarrusso/SQuOpt

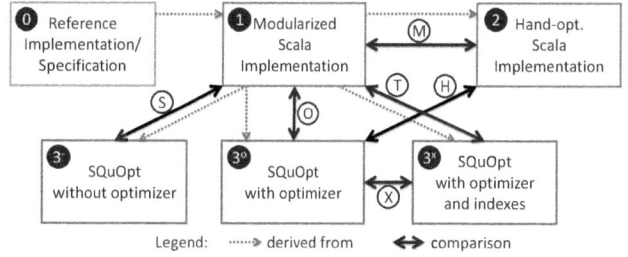

**Figure 5: Measurement Setup: Overview**

| Abstraction | Used |
|---|---|
| All fields in all class files | 4 |
| All methods in all class files | 3 |
| All method bodies in all class files | 3 |
| All instructions in all method bodies and their bytecode index | 5 |
| Sliding window (size $n$) over all instructions (and their index) | 3 |

**Table 1: Description of abstractions removed during hand-optimization and number of queries where the abstraction is used (and optimized away).**

## 5.2 Experimental Units

As experimental units, we sampled a set of queries on code structures from FindBugs 2.0 [17]. FindBugs is a popular bug-finding tool for Java Bytecode available as open source. To detect instances of bug patterns, it queries a structural in-memory representation of a code base (extracted from bytecode). Concretely, a single loop traverses each class and invokes all visitors (implemented as listeners) on each element of the class. Many visitors, in turn, perform activities concerning multiple bug detectors which are fused together. An extreme example is that, in FindBugs, query 4 is defined in class `DumbMethods` together with other 41 bug detectors for distinct types of bugs. Typically a bug detector is furthermore scattered across the different methods of the visitor, which handle different elements of the class. We believe this architecture has been chosen to achieve good performance; however, we do not consider such manual fusion of distinct bug detectors together as modular. We selected queries from FindBugs because they represent typical non-trivial queries on in-memory collections and because we believe our framework allows expressing them more modularly.

We sampled queries in two batches. First, we manually selected 8 queries (from approx. 400 queries in FindBugs), chosen mainly to evaluate the potential speedups of indexing (queries that primarily looked for declarations of classes, methods, or fields with specific properties, queries that inspect the type hierarchy, and queries that required analyzing methods implementation). Subsequently, we *randomly* selected a batch of 11 additional queries. The batch excluded queries that rely on control-/dataflow analyses (i.e., analyzing the effect of bytecode instructions on the stack), due to limitations of the bytecode tookit we use. In total, we have 19 queries as listed in Table 2 (the randomly selected queries are marked with the superscript *R*).

We implemented each query three times (see implementations (1)–(3) in Sec. 5.1) following the specifications given in

| | | Performance (ms) | | | | | Performance ratios | | |
|---|---|---|---|---|---|---|---|---|---|
| Id | Description | 1 | 2 | $3^-$ | $3^o$ | $3^x$ | M (1/2) | H (2/3$^o$) | T (1/3$^x$) |
| 1 | Covariant compareTo() defined | 1.1 | 1.3 | 0.85 | 0.26 | 0.26 | 0.9 | 5.0 | 4.4 |
| 2 | Explicit garbage collection call | 496 | 258 | 1176 | 1150 | 52 | 1.9 | 0.2 | 9.5 |
| 3 | Protected field in final class | 11 | 1.1 | 11 | 1.2 | 1.2 | 10.0 | 1.0 | 9.8 |
| 4 | Explicit runFinalizersOnExit() call | 509 | 262 | 1150 | 1123 | 10.0 | 1.9 | 0.2 | 51 |
| 5 | clone() defined in non-Cloneable class | 29 | 14 | 55 | 46 | 0.47 | 2.1 | 0.3 | 61 |
| 6 | Covariant equals() defined | 29 | 15 | 23 | 9.7 | 0.20 | 1.9 | 1.6 | 147 |
| 7 | Public finalizer defined | 29 | 12 | 28 | 8.0 | 0.03 | 2.3 | 1.5 | 1070 |
| 8 | Dubious catching of IllegalMonitorStateException | 82 | 72 | 110 | 28 | 0.01 | 1.1 | 2.6 | 12800 |
| $9^R$ | Uninit. field read during construction of super | 896 | 367 | 3017 | 960 | 960 | 2.4 | 0.4 | 0.9 |
| $10^R$ | Mutable static field declared public | 9527 | 9511 | 9115 | 9350 | 9350 | 1.0 | 1.0 | 1.0 |
| $11^R$ | Refactor anon. inner class to static | 8804 | 8767 | 8718 | 8700 | 8700 | 1.0 | 1.0 | 1.0 |
| $12^R$ | Inefficient use of toArray(Object[]) | 3714 | 1905 | 4046 | 3414 | 3414 | 2.0 | 0.6 | 1.1 |
| $13^R$ | Primitive boxed and unboxed for coercion | 3905 | 1672 | 5044 | 3224 | 3224 | 2.3 | 0.5 | 1.2 |
| $14^R$ | Double precision conversion from 32 bit | 3887 | 1796 | 5289 | 3010 | 3010 | 2.2 | 0.6 | 1.3 |
| $15^R$ | Privileged method used outside doPrivileged | 505 | 302 | 1319 | 337 | 337 | 1.7 | 0.9 | 1.5 |
| $16^R$ | Mutable public static field should be final | 13 | 6.2 | 12 | 7.0 | 7.0 | 2.0 | 0.9 | 1.8 |
| $17^R$ | Serializable class is member of non-ser. class | 12 | 0.77 | 0.94 | 1.8 | 1.8 | 16 | 0.4 | 6.9 |
| $18^R$ | Swing methods used outside Swing thread | 577 | 53 | 1163 | 45 | 45 | 11 | 1.2 | 13 |
| $19^R$ | Finalizer only calls super class finalize | 55 | 13 | 73 | 11 | 0.10 | 4.4 | 1.1 | 541 |

Table 2: Performance results. As in in Sec. 5.1, (1) denotes the modular Scala implementation, (2) the hand-optimized Scala one, and ($3^-$), ($3^o$), ($3^x$) refer to the SQuOpt implementation when run, respectively, without optimizations, with optimizations, with optimizations and indexing. Queries marked with the $R$ superscript were selected by random sampling.

| | M (1/2) | S (1/3$^-$) | H (2/3$^o$) | H' (2/3$^x$) | O (1/3$^o$) | X (3$^o$/3$^x$) | T (1/3$^x$) |
|---|---|---|---|---|---|---|---|
| Geometric means of performance ratios | 2.4x | 1.2x | 0.8x | 5.1x | 1.9x | 6.3x | 12x |

Table 3: Average performance ratios. This table summarizes all interesting performance ratios across all queries, using the geometric mean [7]. The meaning of speedups is discussed in Sec. 5.1.

```
for {
  classFile ← classFiles.asSquopt
  method ← classFile.methods
  if method.isAbstract && method.name ==# "equals" &&
  method.descriptor.returnType ==# BooleanType
  parameterTypes ← Let(method.descriptor.parameterTypes)
  if parameterTypes.length ==# 1 && parameterTypes(0) ==#
  classFile.thisClass
} yield (classFile, method)
```

Figure 6: Find covariant equals methods.

the FindBugs documentation (0). Instead of using a hierarchy of visitors as the original implementations of the queries in FindBugs, we wrote the queries as for-comprehensions in Scala on an in-memory representation created by the Scala toolkit BAT.[6] BAT in particular provides comprehensive support for writing queries against Java bytecode in an idiomatic way. We exemplify an analysis in Fig. 6: It detects all co-variant equals methods in a project by iterating over all class files (line 2) and all methods, searching for methods named "equals" that return a boolean value and define a single parameter of the type of the current class.

**Abstractions** In the reference implementations (1), we identified several reusable abstractions as shown in Table 1.

[6] http://github.com/Delors/BAT

The reference implementations of all queries except $17^R$ use exactly one of these abstractions, which encapsulate the main loops of the queries.

**Indexes** For executing ($3^x$) (SQuOpt with indexes), we have constructed three indexes to speed up navigation over the queried data of queries 1–8: Indexes for method name, exception handlers, and instruction types. We illustrate the implementation of the method-name index in Fig. 7: it produces a collection of all methods and then indexes them using indexBy; its argument extracts from an entry the key, that is the method name. We selected which indexes to implement using guidance from SQuOpt itself; during optimizations, SQuOpt reports which indexes it could have applied to the given query. Among those, we tried to select indexes giving a reasonable compromise between construction cost and optimization speedup. We first measured the construction cost of these indexes:

| Index | Elapsed time (ms) |
|---|---|
| Method name | 97.99±2.94 |
| Exception handlers | 179.29±3.21 |
| Instruction type | 4166.49±202.85 |

For our test data, index construction takes less than 200 ms for the first two indexes, which is moderate compared to the time for loading the bytecode in the BAT representa-

```
val methodNameIdx: Exp[Map[String, Seq[(ClassFile, Method)]]] =
  (for {
  classFile ← classFiles.asSquopt
  method ← classFile.methods
} yield (classFile, method)).indexBy(entry ⇒ entry._2.name)
```

**Figure 7: A simple index definition**

tion ($4755.32 \pm 141.66$). Building the instruction index took around 4 seconds, which we consider acceptable since this index maps each type of instruction (e.g. INSTANCEOF) to a collection of all bytecode instructions of that type.

## 5.3 Measurement Setup

To measure performance, we executed the queries on the preinstalled JDK class library (rt.jar), containing 58M of uncompressed Java bytecode. We also performed a preliminary evaluation by running queries on the much smaller ScalaTest library, getting comparable results that we hence do not discuss. Experiments were run on a 8-core Intel Core i7-2600, 3.40 GHz, with 8 GB of RAM, running Scientific Linux release 6.2. The benchmark code itself is single-threaded, so it uses only one core; however the JVM used also other cores to offload garbage collection. We used the preinstalled OpenJDK Java version 1.7.0_05-icedtea and Scala 2.10.0-M7.

We measure steady-state performance as recommended by Georges et al. [9]. We invoke the JVM $p = 15$ times; at the beginning of each JVM invocation, all the bytecode to analyze is loaded in memory and converted into BAT's representation. In each JVM invocation, we iterate each benchmark until the variations of results becomes low enough. We measure the variations of results through the coefficient of variation (CoV; standard deviation divided by the mean). Thus, we iterate each benchmark until the CoV in the last $k = 10$ iterations drops under the threshold $\theta = 0.1$, or until we complete $q = 50$ iterations. We report the arithmetic mean of these measurements (and also report the usually low standard deviation on our web page).

## 5.4 Results

**Correctness** We machine-checked that for each query, all variants in Table 2 agree.

**Modularization Overhead** We first observe that performance suffers significantly when using the abstractions we described in Table 1. These abstractions, while natural in the domain and in the setting of a declarative language, are not idiomatic in Java or Scala because, without optimization, they will obviously lead to bad performance. They are still useful abstractions from the point of view of modularity, though—as indicated by Table 1—and as such it would be desirable if one could use them without paying the performance penalty.

**Scala Implementations vs. FindBugs** Before actually comparing between the different Scala and SQuOpt implementations, we first ensured that the implementations are comparable to the original FindBugs implementation. A direct comparison between the FindBugs reference implementation and any of our implementations is not possible in a rigorous and fair manner. FindBugs bug detectors are not fully modularized, therefore we cannot reasonably isolate the implementation of the selected queries from support code. Furthermore, the architecture of the implementation has many differences that affect performance: among others,

FindBugs also uses multithreading. Moreover, while in our case each query loops over all classes, in FindBugs, as discussed above, a single loop considers each class and invokes all visitors (implemented as listeners) on it.

We measured *startup performance* [9], that is the performance of running the queries only once, to minimize the effect of compiler optimizations. We setup our SQuOpt-based analyses to only perform optimization and run the optimized query. To setup FindBugs, we manually disabled all unrelated bug detectors; we also made the modified FindBugs source code available. The result is that the performance of the Scala implementations of the queries ($3^-$) has performance of the same order of magnitude as the original FindBugs queries – in our tests, the SQuOpt implementation was about twice as fast. However, since the comparison cannot be made fair, we refrained from a more detailed investigation.

**SQuOpt Overhead and Optimization Potential** We present the results of our benchmarks in Table 2. Column names refer to a few of the definitions described above; for readability, we do not present all the ratios previously introduced for each query, but report the raw data. In Table 3, we report the geometric mean [7] of each ratio, computed with the same weight for each query.

We see that, in its current implementation, SQuOpt can cause a overhead S ($1/3^-$) up to 3.4x. On average SQuOpt queries are 1.2x faster. These differences are due to minor implementation details of certain collection operators. For query $18^R$, instead, we have that the the basic SQuOpt implementation is 12.9x faster and are investigating the reason; we suspect this might be related to the use of pattern matching in the original query.

As expected, not all queries benefit from optimizations; out of 19 queries, optimization affords for 15 of them significant speedups ranging from a 1.2x factor to a 12800x factor; 10 queries are faster by a factor of at least 5. Only queries $10^R$, $11^R$ and $12^R$ fail to recover any modularization overhead.

We have analyzed the behavior of a few queries after optimization, to understand why their performance has (or has not) improved.

Optimization makes query $17^R$ slower; we believe this is because optimization replaces filtering by lazy filtering, which is usually faster, but not here. Among queries where indexing succeeds, query 2 has the least speedup. After optimization, this query uses the instruction-type index to find all occurrences of invocation opcodes (INVOKESTATIC and INVOKEVIRTUAL); after this step the query looks, among those invocations, for ones targeting runFinalizersOnExit. Since invocation opcodes are quite frequent, the used index is not very specific, hence it allows for little speedup (9.5x). However no other index applies to this query; moreover, our framework does not maintain any selectivity statistics on indexes to predict these effects. Query $19^R$ benefits from indexing without any specific tuning on our part, because it looks for implementations of finalize with some characteristic, hence the highly selective method-name index applies. After optimization, query 8 becomes simply an index lookup on the index for exception handlers, looking for handlers of IllegalMonitorStateException; it is thus not surprising that its speedup is thus extremely high (12800x). This speedup relies on an index which is specific for this kind of query, and building this index is slower than executing the unoptimized query. On the other hand, building this index is entirely appropriate in a situation where similar queries are common enough. Similar considerations apply to usage of indexing in general, similarly to what happens in databases.

**Optimization Overhead** The current implementation of the optimizer is not yet optimized for speed (of the optimization algorithm). For instance, expression trees are traversed and rebuilt completely once for each transformation. However, the optimization overhead is usually not excessive and is $54.8 \pm 85.5$ ms, varying between 3.5 ms and 381.7 ms (mostly depending on the query size).

**Limitations** Although many speedups are encouraging, our optimizer is currently a proof-of-concept and we experienced some limitations:

- In a few cases hand-optimized queries are still faster than what the optimizer can produce. We believe these problems could be addressed by adding further optimizations.

- Our implementation of indexing is currently limited to immutable collections. For mutable collections, indexes must be maintained incrementally. Since indexes are defined as special queries in SQuOpt, incremental index maintenance becomes an instance of incremental maintenance of query results, that is, of incremental view maintenance. We plan to support incremental view maintenance as part of future work; however, indexing in the current form is already useful, as illustrated by our experimental results.

**Threats to Validity** With rigorous performance measurements and the chosen setup, our study was setup to maximize internal and construct validity. Although we did not involve an external domain expert and we did not compare the results of our queries with the ones from FindBugs (except while developing the queries), we believe that the queries adequately represent the modularity and performance characteristics of FindBugs and SQuOpt. However, since we selected only queries from a single project, external validity is limited. While we cannot generalize our results beyond FindBugs yet, we believe that the FindBugs queries are representative for complex in-memory queries performed by applications.

**Summary** We demonstrated on our real-world queries that relying on declarative abstractions in collection queries often causes a significant slowdown. As we have seen, using SQuOpt without optimization, or when no optimizations are possible, usually provides performance comparable to using standard Scala; however, SQuOpt optimizations can in most cases remove the slowdown due to declarative abstractions. Furthermore, relying on indexing allows to achieve even greater speedups while still using a declarative programming style. Some implementation limitations restrict the effectiveness of our optimizer, but since this is a preliminary implementation, we believe our evaluation shows the great potential of optimizing queries to in-memory collections.

# 6. RELATED WORK

This paper builds on prior work on language-integrated queries, query optimization, techniques for DSL embedding, and other works on code querying.

**Language-Integrated Queries** Microsoft's Language-Integrated Query technology (LINQ) [20, 2] is similar to our work in that it also reifies queries on collections to enable analysis and optimization. Such queries can be executed against a variety of backends (such as SQL databases or in-memory objects), and adding new back-ends is supported. Its implementation uses *expression trees*, a compiler-supported implicit conversion between expressions and their reification as a syntax tree. There are various major differences, though. First, the support for expression trees is hard-coded into the compiler. This means that the techniques are not applicable in languages that do not explicitly support expression trees. More importantly, the way expression trees are created in LINQ is generic and fixed. For instance, it is not possible to create different tree nodes for method calls that are relevant to an analysis (such as the map method) than for method calls that are irrelevant for the analysis (such as the toString method). For this reason, expression trees in LINQ cannot be customized to the task at hand and contain too much low-level information. It is well-known that this makes it quite hard to implement programs operating on expression trees [4].

LINQ queries can also not easily be decomposed and modularized. For instance, consider the task of refactoring the filter in the query from x in y where x.z == 1 select x into a function. Defining this function as bool comp(int v) { return v == 1; } would destroy the possibility of analyzing the filter for optimization, since the resulting expression tree would only contain a reference to an opaque function. The function could be declared as returning an expression tree instead, but then this function could not be used in the original query anymore, since the compiler expects an expression of type bool and not an expression tree of type bool. It could only be integrated if the expression tree of the original query is created by hand, without using the built-in support for expression trees.

Although queries against in-memory collections could theoretically also be optimized in LINQ, the standard implementation, LINQ2Objects, performs no optimizations.

A few optimized embedded DSLs allow executing queries or computations on distributed clusters. DryadLINQ [35], based on LINQ, optimizes queries for distributed execution. It inherits LINQ's limitations and thus does not support decomposing queries in different modules. Modularizing queries is supported instead by FlumeJava [3], another library (in Java) for distributed query execution. However, FlumeJava cannot express many optimizations because its representation of expressions is more limited; also, its query language is more cumbersome. Both problems are rooted in Java's limited support for embedded DSLs. Other embedded DSLs support parallel platforms such as GPUs or many-core CPUs, such as Delite [28].

Willis et al. [33, 34] add first-class queries to Java through a source-to-source translator and implement a few selected optimizations, including join order optimization and incremental maintenance of query results. They investigate how well their techniques apply to Java programs, and they suggest that programmers use manual optimizations to avoid expensive constructs like nested loops. While the goal of these works is similar to ours, their implementation as an external source-to-source-translator makes the adoption, extensibility, and composability of their technique difficult.

There have been many approaches for a closer integration of SQL queries into programs, such as HaskellDB [19] (which also inspired LINQ), or Ferry [15] (which moves part of a program execution to a database). In Scala, there are also

APIs which integrate SQL queries more closely such as Slick.[7] Its frontend allows to define and combine type-safe queries, similarly to ours (also in the way it is implemented). However, the language for defining queries maps to SQL, so it does not support nesting collections in other collections (a feature which simplified our example in Sec. 2), nor distinguishes statically between different kinds of collections, such as `Set` or `Seq`. Based on Ferry, ScalaQL [8] extends Scala with a compiler-plugin to integrate a query language on top of a relational database. The work by Spiewak and Zhao [29] is unrelated to [8] but also called ScalaQL. It is similar to our approach in that it also proposes to reify queries based on for-comprehensions, but it is not clear from the paper how the reification works.[8]

**Query Optimization** Query optimization on relational data is a long-standing issue in the database community, but there are also many works on query optimization on objects [6, 13]. Compared to these works, we have only implemented a few simple query optimizations, so there is potential for further improvement of our work by incorporating more advanced optimizations.

**Scala and DSL Embedding** Technically, our implementation of SQuOpt is a deep embedding of a part of the Scala collections API [22]. Deep embeddings were pionereed by Leijen and Meijer [19] and Elliott et al. [5]. The technical details of the embedding are not the main topic of this paper; we are using some of the Scala techniques presented by Rompf and Odersky [26] for using implicits and for adding infix operators to a type. Similar to Rompf and Odersky [26], we also use the Scala compiler on-the-fly. A plausible alternative backend for SQuOpt would have been to use Delite [27], a framework for building highly efficient DSLs in Scala. Using this framework, in concurrent work, Rompf et al. [28] also optimize collection queries; while their work allows for imperative programs, they do not support embedding arbitrary libraries in an automated way. On the other hand, they can reuse support for automatic parallelization and multiple platforms present in Delite. Ackermann et al. [1] present Jet, which also optimizes collection queries but targets MapReduce-style computations in a distributed environment. Moreover, both works do not apply typical database optimizations such as indexing or filter hoisting.

We regard the Scala collections API [22] as a shallowly embedded query DSL. Query operators immediately perform collection operations when called, so that it is not possible to optimize queries before execution. In addition to these eager query operators, the Scala collections API also provides *views* to create lazy collections. Views are somewhat similar to SQuOpt in that they reify query operators as data structures and interpret them later. However, views are not used for automatic query optimization, but for explicitly changing the evaluation order of collection processing. Unfortunately, views are not suited as a basis for the implementation of SQuOpt because they only reify the outermost pipeline of collection operators, whereas nested collection operators as well as other Scala code in queries, such as filter predicates or `map` and `flatMap` arguments, are only shallowly embedded. Deep embedding of the whole query is necessary for many optimizations, as discussed in Sec. 3.

**Code Querying** In our evaluation we explore the usage of SQuOpt to express queries on code and re-implement a subset of the FindBugs [17] analyses. There are various other specialized code query languages such as CodeQuest [16] or D-CUBED [32]. Since these are special-purpose query languages that are not embedded into a host language, they are not directly comparable to our approach.

## 7. FUTURE WORK

As part of future work we plan to add support for *incremental view maintenance* [12] to SQuOpt. This would allow, for instance, to update incrementally both indexes and query results.

To make our DSL more convenient to use, it would be useful to use the virtualized pattern matcher of Scala 2.10, when it will be more robust, to add support for pattern matching in our virtualized queries.

Finally, while our optimizations are type-safe, as they rewrite an expression tree to another of the same type, currently the Scala type-checker cannot verify this statically, because of its limited support for GADTs. Solving this problem conveniently would allow checking statically that transformations are safe and make developing them easier.

## 8. CONCLUSIONS

We have illustrated the tradeoff between performance and modularity for queries on in-memory collections. We have shown that it is possible to design a deep embedding of a version of the collections API which reifies queries and can optimize them at runtime. Writing queries using this framework is, except minor syntactic details, the same as writing queries using the collection library, hence the adoption barrier to using our optimizer is low.

Our evaluation shows that using abstractions in queries introduces a significant performance overhead with native Scala code, while SQuOpt, in most cases, makes the overhead much more tolerable or removes it completely. Optimizations are not sufficient on some queries, but since our optimizer is a proof-of-concept with many opportunities for improvement, we believe a more elaborate version will achieve even better performance and reduce these limitations.

**Acknowledgements** The authors thank Sebastian Erdweg for helpful discussions on this project, Katharina Haselhorst for help implementing the code generator, and the anonymous reviewers, Jacques Carette and Karl Klose for their helpful comments on this paper. This work is supported in part by the European Research Council, grant #203099 "ScalPL".

## References

[1] S. Ackermann, V. Jovanovic, T. Rompf, and M. Odersky. Jet: An embedded DSL for high performance big data processing. In *Int'l Workshop on End-to-end Management of Big Data (BigData)*, 2012.

[2] G. M. Bierman, E. Meijer, and M. Torgersen. Lost in translation: formalizing proposed extensions to C#. In *OOPSLA*, pages 479–498. ACM, 2007.

[3] C. Chambers, A. Raniwala, F. Perry, S. Adams, R. R. Henry, R. Bradshaw, and N. Weizenbaum. FlumeJava: easy, efficient data-parallel pipelines. In *PLDI*, pages 363–375. ACM, 2010.

---

[7] http://slick.typesafe.com/
[8] We contacted the authors; they were not willing to provide more details or the sources of their approach.

[4] O. Eini. The pain of implementing LINQ providers. *Commun. ACM*, 54(8):55–61, 2011.

[5] C. Elliott, S. Finne, and O. de Moor. Compiling embedded languages. *JFP*, 13(2):455–481, 2003.

[6] L. Fegaras and D. Maier. Optimizing object queries using an effective calculus. *ACM Trans. Database Systems (TODS)*, 25:457–516, 2000.

[7] P. J. Fleming and J. J. Wallace. How not to lie with statistics: the correct way to summarize benchmark results. *Commun. ACM*, 29(3):218–221, Mar. 1986.

[8] M. Garcia, A. Izmaylova, and S. Schupp. Extending Scala with database query capability. *Journal of Object Technology*, 9(4):45–68, 2010.

[9] A. Georges, D. Buytaert, and L. Eeckhout. Statistically rigorous Java performance evaluation. In *OOPSLA*, pages 57–76. ACM, 2007.

[10] P. G. Giarrusso, K. Ostermann, M. Eichberg, R. Mitschke, T. Rendel, and C. Kästner. Reify your collection queries for modularity and speed! *CoRR*, abs/1210.6284, 2012. URL http://arxiv.org/abs/1210.6284.

[11] A. Gill, J. Launchbury, and S. L. Peyton Jones. A short cut to deforestation. In *FPCA*, pages 223–232. ACM, 1993.

[12] D. Gluche, T. Grust, C. Mainberger, and M. Scholl. Incremental updates for materialized OQL views. In *Deductive and Object-Oriented Databases*, volume 1341 of *LNCS*, pages 52–66. Springer, 1997.

[13] T. Grust. *Comprehending queries*. PhD thesis, University of Konstanz, 1999.

[14] T. Grust and M. H. Scholl. How to comprehend queries functionally. *Journal of Intelligent Information Systems*, 12:191–218, 1999.

[15] T. Grust, M. Mayr, J. Rittinger, and T. Schreiber. FERRY: database-supported program execution. In *Proc. Int'l SIGMOD Conf. on Management of Data (SIGMOD)*, pages 1063–1066. ACM, 2009.

[16] E. Hajiyev, M. Verbaere, and O. de Moor. *CodeQuest*: Scalable source code queries with Datalog. In *ECOOP*, pages 2–27. Springer, 2006.

[17] D. Hovemeyer and W. Pugh. Finding bugs is easy. *SIGPLAN Notices*, 39(12):92–106, 2004.

[18] G. Kiczales, J. Lamping, A. Menhdhekar, C. Maeda, C. Lopes, J.-M. Loingtier, and J. Irwin. Aspect-oriented programming. In *ECOOP*, pages 220–242, 1997.

[19] D. Leijen and E. Meijer. Domain specific embedded compilers. In *DSL*, pages 109–122. ACM, 1999.

[20] E. Meijer, B. Beckman, and G. Bierman. LINQ: reconciling objects, relations and XML in the .NET framework. In *Proc. Int'l SIGMOD Conf. on Management of Data (SIGMOD)*, page 706. ACM, 2006.

[21] S. S. Muchnick. *Advanced Compiler Design and Implementation*. Morgan Kaufmann, 1997. ISBN 1-55860-320-4.

[22] M. Odersky and A. Moors. Fighting bit rot with types (experience report: Scala collections). In *IARCS Conf. Foundations of Software Technology and Theoretical Computer Science*, volume 4, pages 427–451, 2009.

[23] M. Odersky, L. Spoon, and B. Venners. *Programming in Scala*. Artima Inc, 2 edition, 2011.

[24] S. Peyton Jones and S. Marlow. Secrets of the Glasgow Haskell Compiler inliner. *JFP*, 12(4-5):393–434, 2002.

[25] F. Pfenning and C. Elliot. Higher-order abstract syntax. In *PLDI*, pages 199–208. ACM, 1988.

[26] T. Rompf and M. Odersky. Lightweight modular staging: a pragmatic approach to runtime code generation and compiled DSLs. In *GPCE*, pages 127–136. ACM, 2010.

[27] T. Rompf, A. K. Sujeeth, H. Lee, K. J. Brown, H. Chafi, M. Odersky, and K. Olukotun. Building-blocks for performance oriented DSLs. In *DSL*, pages 93–117, 2011.

[28] T. Rompf, A. K. Sujeeth, N. Amin, K. J. Brown, V. Jovanovic, H. Lee, M. Jonnalagedda, K. Olukotun, and M. Odersky. Optimizing data structures in high-level programs: new directions for extensible compilers based on staging. In *POPL*, pages 497–510. ACM, 2013.

[29] D. Spiewak and T. Zhao. ScalaQL: Language-integrated database queries for Scala. In *Proc. Conf. Software Language Engineering (SLE)*, 2009.

[30] M. Stonebraker, S. Madden, D. J. Abadi, S. Harizopoulos, N. Hachem, and P. Helland. The end of an architectural era: (it's time for a complete rewrite). In *Proc. Int'l Conf. Very Large Data Bases (VLDB)*, pages 1150–1160. VLDB Endowment, 2007.

[31] J. Vitek and T. Kalibera. Repeatability, reproducibility, and rigor in systems research. In *Proc. Int'l Conf. Embedded Software (EMSOFT)*, pages 33–38. ACM, 2011.

[32] P. Węgrzynowicz and K. Stencel. The good, the bad, and the ugly: three ways to use a semantic code query system. In *OOPSLA*, pages 821–822. ACM, 2009.

[33] D. Willis, D. Pearce, and J. Noble. Efficient object querying for Java. In *ECOOP*, pages 28–49. Springer, 2006.

[34] D. Willis, D. J. Pearce, and J. Noble. Caching and incrementalisation in the Java Query Language. In *OOPSLA*, pages 1–18. ACM, 2008.

[35] Y. Yu, M. Isard, D. Fetterly, M. Budiu, U. Erlingsson, P. K. Gunda, and J. Currey. DryadLINQ: a system for general-purpose distributed data-parallel computing using a high-level language. In *Proc. Conf. Operating systems design and implementation*, OSDI'08, pages 1–14. USENIX Association, 2008.

# Supporting Data Aspects in Pig Latin

Curtis E. Dyreson, Omar U. Florez, Akshay Thakre, and Vishal Sharma
Department of Computer Science
Utah State University
Logan, Utah, USA
curtis.dyreson@usu.edu,{omar.florez,akshay.thakre,vishal.sharma}@aggiemail.usu.edu

## ABSTRACT

In this paper we apply the aspect-oriented programming (AOP) paradigm to Pig Latin, a dataflow language for cloud computing, used primarily for the analysis of massive data sets. Missing from Pig Latin is support for cross-cutting data concerns. Data, like code, has cross-cutting concerns such as versioning, privacy, and reliability. AOP techniques can be used to weave metadata around Pig data. The metadata imbues the data with additional semantics that must be observed in the evaluation of Pig Latin programs. In this paper we show how to modify Pig Latin to process data woven together with metadata. The data weaver is a layer that maps a Pig Latin program to an augmented Pig Latin program using Pig Latin templates or patterns. We also show how to model additional levels of advice, i.e., meta-metadata.

## Categories and Subject Descriptors

H.2.3 [**Database Management**]: Metadata, cloud computing

## General Terms

Management, Languages

## Keywords

Aspect-oriented, Pig, cross-cutting concerns

## 1. INTRODUCTION

No matter whether data is stored in a database, flat file, spreadsheet, or as persistent objects, data has cross-cutting concerns. A *cross-cutting data concern* is a data need that is *universal* (potentially applicable to an entire database) and *widespread* (can be used to enhance many different databases). Many data collections have cross-cutting data concerns, and as a collection evolves, new concerns may arise. For instance, a new privacy policy is implemented to hide certain information in a Facebook page. A *privacy* cross-cutting concern could be added to the relevant Facebook data to hide it from the general public.

There are many data needs that are universal and widespread. Heretofore, these needs have not been seen as cross-cutting concerns. Data *security* and *privacy* policies govern every interaction with a datum, and have been researched for many years [5, 11, 16]. Security and privacy are of special concern in cloud computing since data is stored and processed in the cloud on potentially untrustworthy computers [7, 31, 39]. Data quality is another potential cross-cutting concern [2]. Data warehouses aggregate data from a variety of data sources of varying quality and queries that mix low and high quality data should provide a measure of quality along with a result. Data *provenance* [9, 10] and *lineage* [3, 8] track the data and/or processes that produce a query result, which aids in debugging and understanding complex queries. *Time* is another potential cross-cutting concern, both the time of the transaction that creates a datum and the time that it is valid in the real-world need to be tracked for many applications [30]. Each of these potential cross-cutting concerns has an individual, distinct semantics.

Currently, there does not exist a general framework to support cross-cutting data concerns (though systems often support individual concerns, e.g., security). Data management systems are large and complex, and are not designed to be easily configured or modified to support cross-cutting data concerns. Developers currently have to rely on ad-hoc techniques to add concerns to a data collection, or use a database management system that already supports a particular concern. To support cross-cutting data concerns a new paradigm is needed, one that looks to fields outside of databases for useful techniques and insights. Aspect-oriented programming (AOP) provides a framework that can be adapted to our needs. AOP was developed to extend existing programs with new functionality without having to reprogram.

### 1.1 Harnessing Aspects for Data Cross-cutting

Previously we used aspect-oriented techniques to create aspect-oriented data (AOD) for data stored in the relational model [12, 14]. AOD "tags" data with metadata from a cross-cutting data concern to create a data aspect. The aspect becomes active whenever the data is used. A data aspect weaver weaves behavior for the cross-cutting concern into the evaluation of a query, constraint, or object management operation. We showed how to weave behavior into the relational algebra [12]. There has also been other research in using AOP in databases. Research has addressed using aspect-oriented techniques to program databases [32], using a relational database to support AOP [33], and applying AOP to XML schema [15].

Figure 1 gives a broad classification of the space of cross-cutting data concerns using an AOP approach. In general, a data aspect has access to two things: *data* and *advice*, which is the metadata that annotates the data. A data aspect becomes active when the data is used in an operation in the sense that the aspect can *change* (insert, update, or modify) the data or make *no change*. The aspect

|  | Data | |
|---|---|---|
|  | change | no-change (noop) |
| **change** | temporal privacy security quality | lineage provenance probabilistic |
| **no-change (noop)** | vacuuming profiling | authored by language |

(Advice — left spanning label)

**Figure 1: The space of cross-cutting concerns**

could also change the advice. In general, "change" or "no change" are the only possible effects (ignoring side effects like computation time involved). In Figure 1 the concerns are partitioned into four categories based on whether the advice and/or data changes. For example, a temporal cross-cutting data concern constructs new timestamps during some query operations, such as a join operation. The new timestamps become advice for some data, e.g., a tuple in the join result. These timestamps may (logically) delete data since the constructed times may be shorter. As a second example, consider data lineage. Lineage keeps track of all of the data that contributes to a particular result, that is, it constructs advice (metadata) for data, but the constructed advice does not change the data. As a third example, a profiling cross-cutting concern generates statistics (new data) about the data usage, but the advice itself does not change.

AOP can successfully model the kinds of cross-cutting concerns already researched in databases (e.g., time, provenance) and new kinds not yet researched. For instance, *versioned security* where a magazine subscriber has access to articles at the time the subscription was current even after the subscription has ended. Versioned security can be modeled as a temporal aspect tagging a security aspect in our framework, i.e., as meta-metadata. Recursively higher levels of advice (meta-meta-metadata) can also be modeled.

## 1.2 Pig Latin

In this paper we propose adapting AOP to Pig Latin [19, 29] to support cross-cutting data concerns. Pig Latin is a dataflow language and cloud computing platform for the analysis of massive datasets. Developed by researchers at Yahoo, Pig Latin is one of the first, and is (in our opinion) the best, of the emerging cloud computing languages for data analysis. Though relatively new, Pig Latin already has a strong user and development community.[1] Pig Latin is a typical "NoSQL" language. A NoSQL language replaces SQL, the *de facto* query language for databases, with a language that is better suited to programmers. As a dataflow language, Pig Latin is more amenable than declarative languages, like SQL, to aspect-oriented techniques. A Pig Latin program is a sequence of statements. Each statement represents a transformation of some data.

Pig Latin currently has **no support** for cross-cutting data concerns. Users must resort to ad hoc techniques to support, for instance, temporal semantics for data. Snodgrass has pointed out the perils of relying on user good faith to correctly implement temporal semantics [36]. Often users will not know which cross-cutting concerns are present nor the semantics of each individual concern.

Pig Latin lacks many of the features found in other database languages, such as SQL. In the relational model, data is rigidly de-

[1]http://hadoop.apache.org/pig/

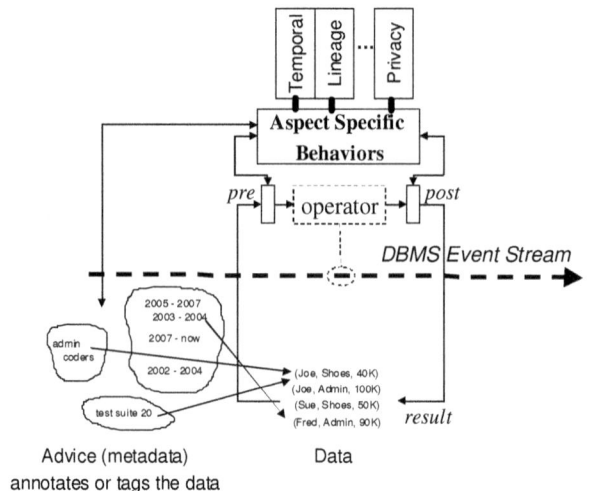

**Figure 2: Opening Pig Latin programs to AOD**

scribed by a fixed schema, Pig Latin, on the other hand, is schemaless. Pig Latin users dynamically load data into a query from text files or back-end databases. Pig Latin also lacks data modification operators, such as INSERT, DELETE, and UPDATE. Pig Latin data is created and maintained by other processes. Not surprisingly, Pig Latin also has no data constraint specifications. All constraints are maintained by other processes. Finally, the Pig Latin data model supports sets and bags, as well as tuples, i.e., it is a non-first normal form data model. So while sharing some commonalities with other database query languages, Pig Latin is different, over and above the cloud computing framework (Hadoop) that supports its back-end.

Figure 2 illustrates the role that a data aspect weaver has to fulfill in the evaluation of a Pig Latin program. In the figure, "Data" is annotated or tagged with "Advice," which is metadata from a cross-cutting concern such as privacy. Over time, a stream of Pig Latin "operators," e.g., a join, are evaluated. When the operator is executed, the advice becomes active. Associated with each kind of advice are "pre," "post," and "intra" advice operators that kick in before, during, and after evaluation of the operation. These operators are specified in a code module which is plugged into the Pig Latin evaluation engine. For example, suppose that Pig Latin evaluates a join. For each pair of tuples in the join with temporal advice, a pre-join operator is called in the temporal module prior to joining, the intra-join operator is called as part of the join, and a post-join operator is called after the join. For temporal advice the pre- and post-joins are no-ops (no action is taken) and the intra-join is temporal intersection. The operations are specific to each kind of cross-cutting concern. All of the advice-specific operators are specified in code that is called at the appropriate time during program evaluation. The plug-in module for each kind of advice consists of these operations.

Previously we described how to model data aspects in a relational database [14], and in the relational algebra [12]. This paper shares a common motivation with our previous work, but in this paper we focus on Pig Latin, which has a different data model and query language. The main contribution of this paper is an extension of each Pig Latin transformation to support data aspects.

This paper is organized as follows. The next section develops a motivating example. After that, data aspects are developed in greater detail. The paper then presents aspect-oriented Pig Latin. The final sections cover related work and summarize the paper.

| Subscribers | | | |
|---|---|---|---|
| (Name, | City, | Amt, | Id) |
| (Maya, | Logan, | $20, | 1) |
| (Jose, | Logan, | $15, | 2) |
| (Knut, | Ogden, | $20, | 3) |

**Table 1: Some data about subscribers to Magazine.com**

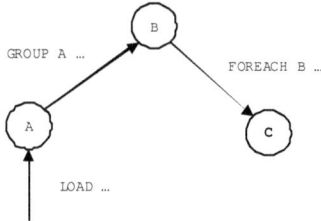

**Figure 3: Dataflow in the simple program**

# 2. MOTIVATION

Assume that Magazine.com stores data about its subscribers in a collection of Pig relations. A Pig relation is a bag of tuples, similar to a table in an SQL database. Each tuple is an ordered list of fields. Each field is a piece of data. Unlike an SQL table, not all tuples have to have the same number of fields. Moreover, Pig relations can have values that are themselves tuples, bags, or maps, something that is not allowed in a relational database. A portion of the data, the **Subscribers** relation, is shown in Table 1. Each tuple in Subscriber records, in order, a name (**Name**), city (**City**), subscription amount (**Amt**), and a tuple identifier (**Id**).

## 2.1 Pig Latin

Magazine.com would like to count the subscribers per city. The following Pig Latin program computes the desired count.

```
A = LOAD 'subscribers' USING PigStorage()
        AS (name: chararray, city: chararray,
            amount: int);
B = GROUP A BY city;
C = FOREACH B GENERATE city, COUNT(B.name);
DUMP C;
```

The program has four statements. The first statement loads the data, and gives a name and a type to each field within a tuple. The statement also establishes the **Subscribers** relation as the data node A. A grouping transformation is applied to the data in node A to produce node B. The data is grouped into bags by value as shown in Table 2. The data in node B is then processed to generate the name and count for each city as shown in Table 3. The final statement, DUMP, displays the data accumulated at node C.

This program has a very simple dataflow, with only three nodes. To evaluate the program, Pig Latin first constructs a representation of the dataflow as illustrated in Figure 3. Next it applies query optimization rules to optimize the data flow (for instance the FOREACH transformation could be combined with the GROUP transformation to generate only the needed fields while grouping). Only when the DUMP statement is parsed is the optimized dataflow program evaluated using Hadoop, that is, the program is transformed to map-reduce constructs and executed in parallel.

## 2.2 Cross-cutting Data Concerns

On-line magazines earn revenue by restricting content to paid subscribers. *Security* enforces the restriction. Each subscriber should

| B |
|---|
| (Logan, {(Maya, Logan, $20), (Jose, Logan, $15)}) |
| (Ogden, {(Knut, Ogden, $20)}) |

**Table 2: Subscribers grouped by city**

| C |
|---|
| (Logan, 2) |
| (Ogden, 1) |

**Table 3: The count of subscribers**

be able to see their own data, but not that of others. Subscribers complain that once their subscription ends, they are no longer able to see the content to which they once subscribed, but they should be able to do so. Magazine.com decides to support both security and *versioned security*, whereby subscribers still have access to content as of the time when they subscribed. To help the programmers implement the system, Magazine.com also decides that it is important to support *lineage* in query evaluation. Lineage keeps track of which facts were used to produce a result, thereby helping programmers understand how the query produced a particular result.

To accommodate the new requirements, which are all cross-cutting concerns, the designers need to add new data and functionality to their existing database and its applications. Ideally, the designers will be able to add without changing a line of existing Pig Latin programs.

## 2.3 Aspect-oriented Pig Latin Data

In an aspect-oriented approach, the database designers "tag" data in the database with advice, creating aspects. The tagging could be at different levels, i.e., in the Pig Latin data model, the tagged data could be an *attribute value*, a *tuple*, or a *relation*. We focus on *tuple-* and *relation-tagging* in this paper. The advice that tags a tuple is assumed to pertain to all of the attribute values within that tuple, and for a relation, the advice applies to all of the tuples in the relation. Relation-tagging is useful for establishing default advice for each tuple in the relation.

Though aspects are developed independently, more than one kind of advice can tag a tuple or relation, for instance a tuple could be tagged with both lineage and security advice. The advice can be combined into a single *perspective* [13], or remain independent. Finally, since the advice is data, it too can be advised by *meta-metadata*, i.e., metadata is to data as meta-metadata is to metadata.

Several data cuts are concretely represented in Table 4 and Table 5 which extend and refine the Magazine.com database example of the previous section. Each advice tuple is prefixed with a "perspective id" (**Per**), which is the first field in each advice tuple. The first tuple of **Security Advice** has a **Per** of A. A data cut is a pairing of a tuple id with an advice id. Table 4 shows the aspected **Subscribers** relation. The Data Cuts relation weaves advice to data identified by the **Id** column. This tagging scheme is repeated for the meta-metadata (**Temporal Meta Advice** and **Metadata Data Cuts**). Each subscriber is tagged by a security aspect that records the security on the tuple and a lineage aspect that denotes how the tuple was constructed. Initially, the lineage is just the identifier of the tuple itself. The security is a partial order from the lowest levels (Paid and Lapsed) to the top level (DBA). Only paid subscribers have access to the content. The meta-metadata records when the security advice is current. Jose was a paid subscriber from 2007 to 2008 at which time his subscription lapsed. If the data is rolled back to its state current in 2007, Jose should have access to the content of the site. Said differently, Jose paid for the 2007 to 2008 content and

15

therefore should have access to that data by setting his content perspective to some time in that range. An advice tuple shaded in grey denotes default advice, that is, data advised by relation-tagging. The default temporal advice starts in 2006 when the site began.

Table 5 extends the database with an aspected **Personal Info** relation that records personal information about each subscriber. By default, only the DBA has access to this data.

## 2.4 Aspect-Oriented Pig Latin Programs

Advice is involved whenever data is used in a query. For instance, suppose that two tuples are to be joined. Sequenced temporal semantics permits the tuples to be joined *only at the times they both existed*. For instance, the tuples with **Id** 1 and 2 in Table 4 can be joined only at times 2007-now since tuple 2 was not in the database in 2006.

An advice's behavior is woven into the evaluation of a Pig Latin program as shown in Figure 4. In Figure 4(a), the typical dataflow as depicted. Data at a node $R$ is transformed to that at node $X$. Figure 4(b) shows a Pig Latin pattern or template that will replace the transformation of Figure 4(a). In the aspected case, the relation at node $R$ consists of three components: a data relation, $R_D$, an advice relation, $R_A$, and a data cuts relation, $R_C$. Without loss of generality we focus on a single kind of advice, more generally there would be several advice relations. Each of these relations must be transformed to create the three components of the result data node: $X_D$, $X_A$, and relation, $X_C$.

Another way to conceptualize the weaving is to imagine a Pig Latin program evaluated simultaneously at two levels: the *data level* and the *advice level* (the cuts attach the advice to the data). At the data level, the Pig Latin program proceeds as written by evaluating each transformation on the data. The weaving needs to add transformations at the advice level to the dataflow. There are three basic patterns for the dataflow at the advice level.

1. Single tuple - Pig Latin transformations that process a single relation tuple-by-tuple have no special operations for advice. Security, privacy, data quality measures, etc. already annotate and describe the data. The advice sticks to the data through the transformation. For example a transformation to project the values in a column retains the advice for each value.

2. Pair of tuples - Two Pig Latin relations can be related by processing pairs of tuples, one chosen from each relation. For a pair of tuples, the advice that annotates each tuple must be processed together. For instance, a join operation will need to join the advice for a pair of tuples while joining the pair.

3. Multiple tuples - Some Pig Latin transformations relate many tuples in a single relation, for instance, when grouping a relation, the advice for all of the tuples in the group must be processed.

The weaving also has to *synchronize* the advice and data levels after each transformation since the advice level can impact the data level (and vice-versa). For example, in the temporal sequenced join discussed previously, tuples at the data level join only if they also join at the advice level.

In the next section we develop a specific template for each kind of Pig Latin transformation. But in general, the weaving is a Pig Latin program modification whereby each statement in a program is replaced by a sequence of statements constructed by instantiating a template for the transformation. This strategy can be extended

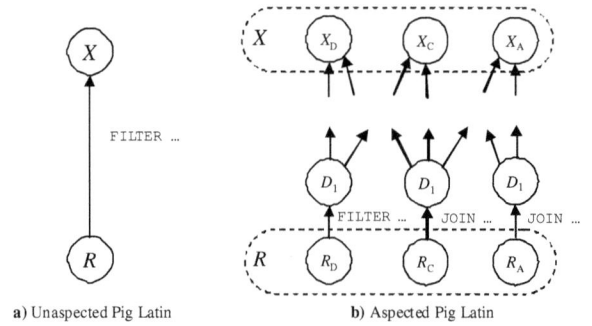

a) Unaspected Pig Latin          b) Aspected Pig Latin

**Figure 4: Dataflow in the simple program**

to additional levels of advice, e.g., meta-metadata. The pattern is repeatedly applied for each level.

## 3. ASPECT-ORIENTED PIG LATIN

This section describes changes to Pig Latin to support aspect-oriented data. Recall that each kind of aspect (e.g., security) enforces a semantics on the use of the data. All uses must obey that semantics. We model the bulk of Pig Latin transformations, showing how each is redefined to support data aspects. Each modification is described in terms of a pattern or template. The template is applied to rewrite the corresponding transformation in a Pig Latin program at the data and advice levels. Each transformation is redefined using (non-aspect-oriented) Pig Latin to illustrate that Pig itself can be used to become aspect-oriented. A key optimization to the basic strategy, which we call *advice inlining*, is presented in Section 3.5. Only data aspects are initially considered; in Section 3.6 program aspects are introduced.

We consider three broad categories of Pig Latin transformations: single, paired, and multiple tuple transformations. We first model single tuple transformations which involve only one tuple at a time and are generally simpler than the other cases.

### 3.1 Single Tuple

The single tuple transformations are FILTER, FOREACH, SPLIT, SAMPLE, LOAD, DUMP, and STORE. We discuss the FILTER in detail and only breifly present the other transformations.

#### 3.1.1 FILTER

We first describe FILTER and then present an detailed example of the transformation using aspected data.

The FILTER transformation selects tuples from a data node that meet some condition, $P$.

```
X = FILTER R ON P;
```

The aspected-oriented transformation, TFILTER$_{AO}$, first applies a FILTER at the data level. As the FILTER may remove some tuples, the data cuts and advice should then be synchronized with the data, removing extraneous advice, an operation that we call TRIM$_{AO}$.

Figure 5 illustrates the basic pattern for FILTER. The TRIM$_{AO}$ pattern to the right of the figure should be repeated for each level of advice. The Pig Latin code template for FILTER$_{AO}$ is given below, with comments enclosed within '/* */'.

```
/* Filter the data */
X_D = R_D ON P;
```

| Subscribers | Data Cuts | Security Advice | Lineage Advice | Metadata Data Cuts | Temporal Meta Advice |
|---|---|---|---|---|---|
| (Name, City, Amt, Id) | (Id, Per) | (Per, Sec) | (Per, Lin) | (Per, MetaPer) | (MetaPer, Start, End) |
| (Maya, Logan, $20, 1) | (1, A) | (A, Paid) | (A, {1}) | (B, X) | (X, 2007, 2008) |
| (Jose, Logan, $15, 2) | (2, B) | (B, Paid) | (B, {2}) | (C, Y) | (Y, 2009, now) |
| (Knut, Ogden, $20, 3) | (2, C) | (C, Lapsed) | (C, {2}) | (A, Z) | (Z, 2006, now) |
|  | (3, D) | (D, Lapsed) | (D, {3}) | (D, Z) |  |

**Table 4: Aspected Subscribers**

| Personal Info | Data Cuts | Security Advice | Lineage Advice |
|---|---|---|---|
| (Name, City, Amt, Id) | (Id, Per) | (Per, Sec) | (Per, Lin) |
| (Maya, maya@aol.com, 5) | (5, E) | (E, DBA) | (E, {5}) |
| (Jose, jose@aol.com, 6) | (6, F) | (F, DBA) | (F, {6}) |
| (Knut, knut@aol.com, 7) | (7, G) | (G, DBA) | (G, {7}) |

**Table 5: Aspected personal information about subscribers**

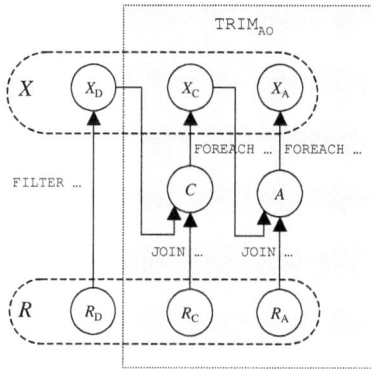

**Figure 5: The template for FILTER$_{AO}$**

```
/* TRIM the advice level */
/* Remove extraneous cuts*/
C = JOIN R_C BY id, X_D BY id;
X_C = FOREACH C GENERATE C.id, C.per;


/* Remove extraneous advice */
A = JOIN R_A BY per, X_C BY per;
X_A = FOREACH A GENERATE A.per, A.metadata;
```

As an example, consider a query to filter subscribers below $20. At the data level, Maya and Knut pass the filter, but Jose is filtered. The result is shown in Table 6. In the table, extraneous, inert advice and data cuts that could be trimmed are highlighted in gray. TRIM$_{AO}$ will clean up this extra advice and synchronize the advice level to the data level, but the extraneous advice is harmless (except for occupying space) and can be left in place leading to the alternative, cheaper plan shown in Figure 6.

### 3.1.2 FOREACH

The FOREACH projects only specified fields, $f_1, \ldots, f_n$, into the result.

$$X = \text{FOREACH } R \text{ GENERATE } f_1, \ldots, f_n;$$

As all the tuples are retained, the data cuts and advice are unchanged, and so the template for FOREACH$_{AO}$ is simple.

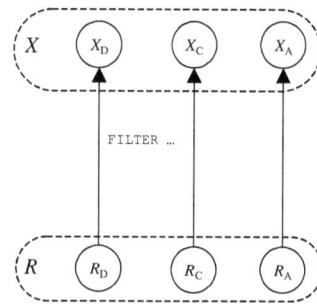

**Figure 6: An alternative, cheaper template for FILTER$_{AO}$**

```
/* GENERATE the data */
X_D = FOREACH R_D GENERATE f_1, ..., f_n;

/* Repeat rest of pattern for each level of advice */
X_C = R_C;
X_A = R_A;
```

The template is illustrated in Figure 7. As an example, consider generating subscriber cities. Each city is generated along with all of the advice and meta advice.

### 3.1.3 Split and Sample

A SPLIT transformation partitions a relation into $n$ relations for parallel processing. The split is based on conditions $c_1, \ldots, c_n$ where each condition is a predicate involving field values.

```
SPLIT R INTO X_1 IF c_1, ..., X_n IF c_n;
```

A SAMPLE transformation is chooses a random sampling of a relation. It is used to estimate results. The *sample_size* is a percentage of the size of the relation, e.g., 0.01 would represent 1%.

```
X = SAMPLE R sample_size;
```

The aspect-oriented versions of these transformations are similar to FILTER$_{AO}$. They apply the transformation at the data level and then remove extraneous data cuts and advice using TRIM$_{AO}$, or alternatively, leave the cuts and advice unchanged since extraneous cuts and advice are harmless.

17

| Subscribers | Data Cuts | Security Advice | Lineage Advice | Metadata Data Cuts | Temporal Meta Advice |
|---|---|---|---|---|---|
| (Name, City, Amt, Id) | (Id, Per) | (Per, Sec) | (Per, Lin) | (Per, MetaPer) | (MetaPer, Start, End) |
| (Maya, Logan, $20, 1) | (1, A) | (A, Paid) | (A, {1}) | (B, X) | (X, 2007, 2008) |
| (Knut, Ogden, $20, 3) | (2, B) | (B, Paid) | (B, {2}) | (C, Y) | (Y, 2009, now) |
| | (2, C) | (C, Lapsed) | (C, {2}) | (A, Z) | (Z, 2006, now) |
| | (3, D) | (D, Lapsed) | (D, {3}) | (D, Z) | |

**Table 6:** `Subscribers` **that paid $20 or more for their subscription, the cells shaded gray can be trimmed**

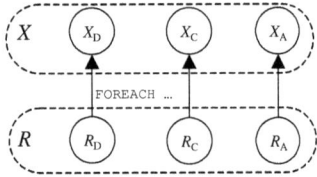

**Figure 7: The template for FOREACH**$_{AO}$

### 3.1.4 Load, Store, and Dump

The LOAD transformation loads a relation from disk into memory, STORE stores an in-memory relation to disk, and DUMP displays a relation. The aspect-oriented versions of these transformations must be trivially augmented to deal with three relations (the data, the data cuts, and the advice) rather than a single relation.

### 3.1.5 User-defined Functions and MapReduce

User-defined functions (UDFs) implemented in Java can be added to a Pig Latin dataflow. As of Pig version 0.10.0, a MAPREDUCE transformation can run a Map-Reduce job, also coded in Java. In both cases, each tuple in a relation is streamed through an arbitrary program, and output tuples are collected. Our current design only supports data aspects in Pig Latin, not in Java. Hence users must rewrite the UDF and Map-Reduce jobs to handle aspected data. For Aspect-oriented Pig Latin, we currently give the user the option to join the data relation to the cuts and advice relations prior to streaming the tuples, or to simply stream the data relation (potentially violating the semantics of the data concerns involved). We discuss alternative strategies in future work.

## 3.2 Pair of Tuple Transformations

Paired tuple transformations involve a pair of tuples and generally invoke an advice-specific operation to process the advice at the advice level.

### 3.2.1 Joins

Pig Latin has several kinds of joins: cross, replicated, inner, outer, skewed, and merge. Additionally Pig Latin has a COGROUP transformation that groups tuples that would join. Semantically they are all a variant of an equi-join, where two tables are joined on the values of one or more fields being equivalent.

$X = $ JOIN $R$ BY $f_R$, $S$ BY $f_S$;

For an aspect-oriented join, JOIN$_{AO}$, advice constrains the join. If two tuples potentially join at the data level, their advice also needs to "join" at the advice level (the meaning of a join depends on the kind of advice). For instance two tuples only join when their temporal advice overlaps (i.e., the tuples exist at the same time).

Figure 8 shows the template for JOIN$_{AO}$. First each relation involved in the join is merged by joining the data to the data cuts and then to advice (the joins must be repeated for each metadata level).

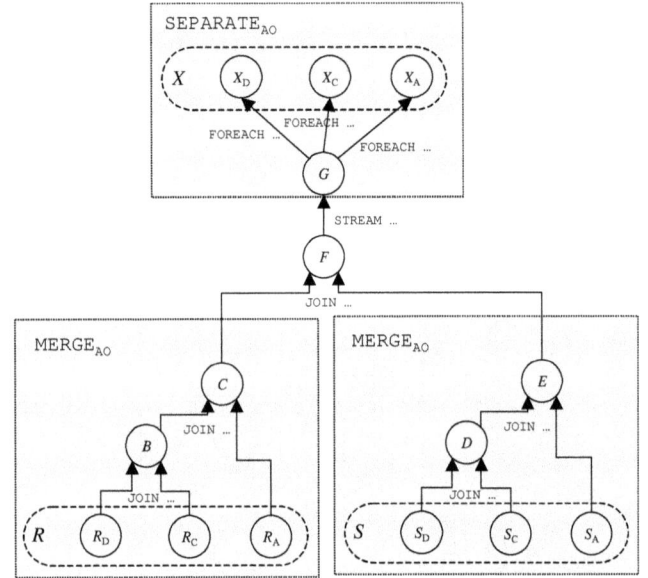

**Figure 8: A JOIN**$_{AO}$ **first merges both relations, then joins, and finally separates**

We call this operation MERGE$_{AO}$. Next, the two merged relations are joined, effectively joining at the data level. Then, the result is streamed through an advice-specific join operation (described in detail below). Finally, the merged relations are separated into data, cuts, and advice using the FOREACH transformation. We call the separation step, SEPARATE$_{AO}$.

At the advice level, advice is streamed through an advice-specific join (ADVICE-JOIN). This user-defined function "joins" the advice for a tuple using an advice-specific technique. Example advice-specific joins are listed below. These examples assume that the last four (or two) fields in a tuple are the advice pairs to be tested, and that the data, ids, and perspectives are called "*rest*."

- Temporal advice — Computes the temporal join for pairs of time periods, i.e., the time when the periods overlap.

  $$\textbf{temporal-join}((rest, t, u, v, w)) =$$
  $$\{(rest, \textbf{max}(t, v, ), \textbf{min}(u, w))\}$$

- Lineage advice — Lineage $x$ always joins with lineage $y$ and manufactures new advice that is the union of the previous lineage.

  $$\textbf{lineage-join}((rest, x, y)) = \{(rest, x \bigcup y))\}$$

- Security advice — A partial order join is performed by keeping the most private group.

$$\textbf{security-join}((rest, x, y)) = \{(rest, x), (rest, \textbf{lca}(x,y))\}$$

The `ADVICE-JOIN` also manufactures a new data cut identifier and advice reference.

As an example, consider the $\text{JOIN}_{AO}$ of **Subscribers** with **Personal Info**. First, the two relations are individually merged. Next, the join is performed at the data level, resulting in the relation shown in Table 7. At this point, the data level has been joined, but the advice level has not. After the advice-specific joins are performed and the relation is separated, Table 8 results. In this example, no tuples were removed from the join due to incompatible or mismatching advice, but some of the advice has been trimmed, For instance the security advice joins only at the level of the `DBA` which is the least common ancestor of each pair of security advice values. The meta-metadata (**Temporal Meta Advice**) joins only on the interesection of the pair of time intervals. In the final result, the data cuts identifiers (perspectives) are composed values, manufactured from the underlying identifiers (perspectives).

### 3.2.2 Union

In a `UNION` transformation, the tuples in each relation are put into a single bag. The union does not eliminate duplicates.

$$X = \text{UNION } R, S;$$

The aspect-oriented union, $\text{UNION}_{AO}$, similarly combines the advice and data cuts (assuming that the ids and references are disjoint).

```
X_D = UNION R_D, S_D;
/* Repeat for each level of advice */
X_C = UNION R_C, S_C;
X_A = UNION R_A, S_A;
```

## 3.3 Multiple Tuple

Multiple tuple transformations involve groups or collections of tuples. The advice for a group needs to be processed as a group.

### 3.3.1 Grouping

Grouping is important when computing aggregates. Pig Latin has a `GROUP` transformation that groups tuples on fields, $f_1, \ldots, f_n$.

$$X = \text{GROUP } R \text{ USING } f_1, \ldots, f_n;$$

Advice constrains the grouping. Two tuples potentially group only if their advice also groups. For instance two tuples are in the same group only when their temporal advice overlaps (i.e., the tuples exist at the same time). The template for $\text{GROUP}_{AO}$ is sketched in Figure 9. First, the data is merged with the cuts and advice, using the $\text{MERGE}_{AO}$ pattern. Next, at the data level, the data is grouped. Then, the data is streamed through an advice-specific grouping operator, `ADVICE-GROUP`, to compute the groups for the advice. The semantics of this operator depends on the kind of advice. The input to this operator is a set of advice values (the advice for all of the group members). The output is a refined set of advice.

- Temporal advice — Compute membership constant periods, that is those intervals of time for which group membership does not change.

  **temporal-group**$(T)$
  $$= \{(t,u) \mid (t,\_) \in T \wedge (\_,u) \in T \wedge$$
  $$\neg(\exists(w,\_) \in T) \vee \exists(\_,w) \in T[t < w < u])\}$$

- Lineage advice — Lineage forms a set of the ids of all of the tuples in the group.

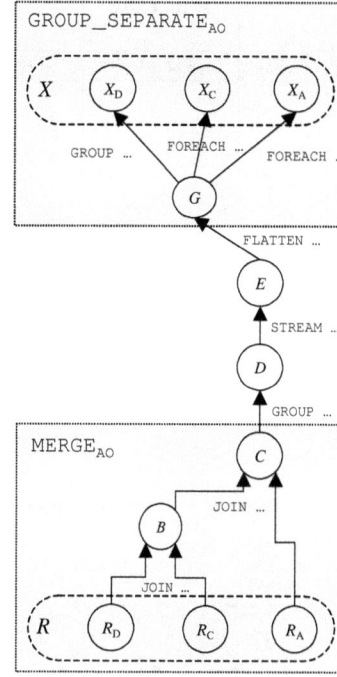

**Figure 9: The template for** $\text{GROUP}_{AO}$

$$\textbf{lineage-group}(T) = \{t.\text{id} \mid t \in T\}$$

- Security advice — Each level in the hierarchy is its own group.

  $$\textbf{security-group}(T) = T$$

Finally, the grouped data must be separated into cuts and advice by a $\text{GROUP\_SEPARATE}_{AO}$ pattern.

As an example, consider grouping the **Subscribers** relation using the `City` field. To simplify this example we assume a single kind of advice: security advice. The result is shown in Table 9. Each city will end up in a separate group. But because the cities have different advice, they will be further split into more groups. As part of the aspect-specific grouping, new advice corresponding to each group is manufactured.

### 3.3.2 Distinct

The `DISTINCT` transformation eliminates duplicate tuples from a relation.

$$X = \text{DISTINCT } R;$$

For aspect-oriented distinct, $\text{DISTINCT}_{AO}$, when duplicates of a tuple are eliminated, the data cuts to the duplicates must be changed to attach to the tuple that was not eliminated. The duplicate elimination does not *coalesce*, that is, it does not eliminate or reduce overlapping or redundant advice.

The template for $\text{DISTINCT}_{AO}$ is shown in Figure 10. The cuts are first merged with the data (the advice does not have to be since it the distinct is applied at the data level, not the advice level). Next, the tuples are grouped on all of the data fields, yielding distinct groups of tuples. Finally, the cuts and advice are separated and each data tuple gets a new identified. Since the pattern involves some complexities, we give a code template below.

19

| Data | Data Cuts | Security Advice | Lineage Advice | Metadata Data Cuts | Temporal Meta Advice |
|---|---|---|---|---|---|
| Name, ..., Id, ..., Id | Id, Per, Id, Per | Per, Sec, Per, Sec | Per, Lin, Per, Lin | Per, M, Per, M | M, Start, End, M, Start, End |
| Maya, ..., 1, ..., 5 | 1, A, 5, E | A, Paid, E, DBA | A, {1}, E, {5} | A, Z, E, Z | Z, 2006, now, Z, 2006, now |
| Jose, ..., 2, ..., 6 | 2, B, 6, F | B, Paid, F, DBA | B, {2}, F, {6} | B, Y, F, Z | Y, 2009, now, Z, 2006, now |
| Jose, ..., 2, ..., 6 | 2, C, 6, F | C, Lapsed, F, DBA | C, {2}, F, {6} | C, X, F, Z | X, 2007, 2008, Z, 2006, now |
| Knut, ..., 3, ..., 7 | 3, D, 7, G | D, Lapsed, G, DBA | D, {3}, G, {7} | D, Z, G, Z | Z, 2006, now, Z, 2006, now |

**Table 7: Subscribers joined with Personal Info prior to advice-specific joins**

| Data | Data Cuts | Security Advice | Lineage Advice | Metadata Data Cuts | Temporal Meta Advice |
|---|---|---|---|---|---|
| (Name, ..., Id) | (Id, Per) | (Per, Sec) | (Per, Lin) | (Per, MetaPer | (MetaPer, Start, End) |
| (Maya, ..., 1.5) | (1.5, A.E) | (A.E, DBA) | (A.E, {1,5}) | (A.E, Z) | (X, 2007, 2008) |
| (Jose, ..., 2.6) | (2.6, B.F) | (B.F, DBA) | (B.F, {2,6}) | (B.F, Y) | (Y, 2009, now) |
| (Jose, ..., 2.6) | (2.6, C.F) | (C.F, DBA) | (C.F, {2,6}) | (C.F, X) | (X, 2007, 2008) |
| (Knut, ..., 3.7) | (3.7, D.G) | (D.G, DBA) | (D.G, {3,7}) | (D.G, Z) | (Z, 2006, now) |

**Table 8: Subscribers joined with Personal Info after advice-specific join and SEPARATE$_{AO}$**

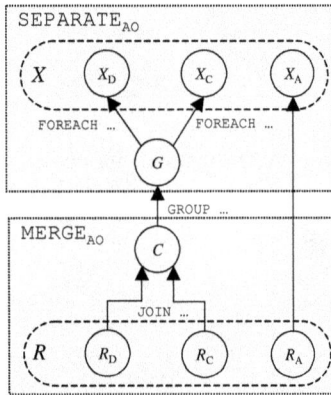

**Figure 10: The template for DISTINCT$_{AO}$**

```
/* Merge the data with the data cuts */
C = JOIN R_D BY id, R_C BY id;

/* Group the duplicates */
G = GROUP C on all data fields;

/* Generate the data with a minimum cut Id */
X_D = FOREACH G GENERATE data fields, min(R_D.id);

/* Generate the data cuts */
X_C = FOREACH G GENERATE , min(R_D.id), R_C.ref;

/* Advice is not changed */
X_A = R_A;
```

As an example, consider computing distinct cities. First the subscriber names are generated yielding three tuples. Next the DISTINCT transformation is applied, yielding two tuples in the result (Logan and Ogden). The advice for the Logan tuples remains as three distinct perspectives. The result is shown in Table 10.

## 3.4 The Example Revisited

We return to the example query of Section 2.1. Assume that we have a single temporal aspect. The aspect-oriented version of the program is given below.

```
A = LOAD_AO 'subscribers' USING PigStorage()
      AS (name:chararray, city:chararray,
          amount:int, id:chararray);
B = GROUP_AO A BY dept;
C = FOREACH B GENERATE_AO dept, COUNT(B.name);
DUMP_AO C;
```

The aspect-oriented behavior is woven into the program using the templates described in this section, yielding the following Pig Latin program.

```
AD = LOAD 'subscribers.data' USING PigStorage()
          AS (name:chararray, city:chararray,
              amount:int, id:chararray);
AC = LOAD 'subscribers.cuts' USING PigStorage()
          AS (id:chararray, per:chararray);
AA = LOAD 'subscribers.advice' USING PigStorage()
          AS (per:chararray, start:int, end:int);

/* Merge the data with the cuts and advice */
B = JOIN AD BY id, AC BY id;
C = JOIN B BY per, AA BY per;

/* Group on the data values */
D = GROUP C USING dept;

/* Stream through aspect-specific grouping */
E = STREAM D THROUGH TEMPORAL-GROUP;

/* Flatten it and regroup using the data and id */
F = FLATTEN E;
BD = GROUP F USING dept, id;

/* Generate the data cuts */
CC = FOREACH F GENERATE id, per;

/* Generate the advice */
CA = FOREACH F GENERATE per, start, end;

/* Generate the result */
CD = FOREACH BD GENERATE dept, COUNT(BD.name);
DUMP CD;
```

| Subscribers | Data Cuts | Security Advice |
|---|---|---|
| (Logan, 11 {(Maya, Logan, \$20, 1)}) (Jose, Logan, \$15, 2)}) | (11, J) | (J, Paid) |
| (Logan, 12, {(Jose, Logan, \$15, 2)}) | (12, H) | (H, Lapsed) |
| (Ogden, 13, {(Knut, Ogden, \$20, 3)}) | (13, I) | (I, Paid) |

**Table 9: Grouped subscribers**

| Cities | Data Cuts | Security Advice | Lineage Advice | Metadata Data Cuts | Temporal Meta Advice |
|---|---|---|---|---|---|
| (City, Id) | (Id, Per) | (Per, Sec) | (Per, Lin) | (Per, MetaPer) | (MetaPer, Start, End) |
| (Logan, 1) | (1, A) | (A, Paid) | (A, 1) | (A, X) | (X, 2007, 2008) |
| (Ogden, 3) | (1, B) | (B, Paid) | (B, 2) | (B, Y) | (Y, 2009, now) |
|  | (1, C) | (C, Lapsed) | (C, 3) | (C, Z) | (Z, 2006, now) |
|  | (3, D) | (D, Lapsed) | (D, 4) | (D, Z) |  |

**Table 10: Distinct cities**

## 3.5 Optimizing by Advice In-lining

Maintaining separate data, cuts, and advice relations throughout the evaluation of a Pig Latin program can incur many invocations of $\text{TRIM}_{AO}$, $\text{MERGE}_{AO}$, and $\text{SEPARATE}_{AO}$. Since they involve JOINs, which are expensive, a more optimal strategy is to *in-line* the advice at the start of a query, and separate at the end. The in-lining removes data cuts by attaching the advice directly to the data. The following template can be used to in-line advice for an aspected relation. Let $R_D$ be the data relation, $R_C$ by the cuts relation, and $R_{A_1}, \ldots, R_{A_n}$ be $n$ advice relations. The template should be applied for each level of advice.

```
I₁ = JOIN R_{A₁} by per, R_{A₂} by per;
I₂ = JOIN I₁ by per, R_{A₃} by per;
...
Iₙ = JOIN I_{n-1} by per, R_{Aₙ} by per;
I = JOIN Iₙ by per, R_C by per;
R_I = COGROUP R_D by id, I by id;
```

The first $n$ steps join each kind of advice to form a combined perspective. Next, the perspective is joined to the data cuts. Finally, the data is co-grouped with individual perspectives (all the perspectives that would join with the advice).

An example of advice in-lining in given in Table 11.

The benefit of in-lining is that in the templates presented in Sections 3.1 (Figure 5) to 3.3 (Figure 10), only the operations **not** shaded in gray need to be performed on the in-lined data and advice. The cost of in-lining is incurred once for the query. Prior to a DUMP or STORE the data must be separated as well.

## 3.6 Program Aspects

A Pig Latin program can also be aspected. A program aspect represents a constraint on the relations that are evaluated. Suppose that a program involves a relation, $[R_D, R_C, R_A]$, and is aspected by a perspective consisting solely of advice, $P_A$. Then the program aspect transformation constrains $R_D$ to tuples that have advice consistent with the perspective prior to evaluating the program. It does so as follows.

```
/* First relate all of the advice, CROSS is
   the Cartesian product transformation. */
B = CROSS R_A, P_A;
```

```
/* Generate new advice */
X_A = STREAM B THROUGH ADVICE-PROGRAM-ASPECT;

/* Use new advice to remove extraneous data
   cuts */
C = COGROUP R_C BY ref, X_A BY ref;
X_C = FOREACH C GENERATE C.id, C.ref;

/* Use new data cuts to remove extran. tuples */
D = COGROUP R_D BY id, X_C BY id;
X_D = FOREACH D GENERATE D.data, D.id;
```

Each advice tuple is passed through the advice-specific stream, which leaves the tuple unchanged, trims the tuple, or removes it.

## 3.7 Complexity Analysis

The increased modeling power of aspect-oriented data comes with an increased cost. In this section we analyze the worst-case time complexity, assuming that all of the aspect-oriented transformations are implemented in Pig Latin (some advice-specific behaviors are implemented as user-defined functions). Let $D$ be the size of each data relation, $C$ be the size of a data cuts relation, and $A$ be the size of an advice relation. Typically $A$ will be much smaller than $D$, and if there is a lot of default advice, $C$ will also be much smaller than $D$. Finally, let $z$ be the number of different kinds of advice, e.g., $z$ is three for the examples in this paper, and let $k$ be the number of levels of advice, e.g., in our examples, $k$ is 2. We assume that all binary operations, i.e., the various kinds of join, cost $O(n * m)$ where $n$ and $m$ are the size of the operands, and unary operations, i.e., filtering, sampling, generating, and grouping, cost $O(n)$. Though this assumption overestimates the join cost, which in practice (e.g., in a good hash join) can be nearly linear, the assumption is appropriate for our complexity analysis.

To determine the cost of each aspect-oriented transformation, we summed the cost of every Pig Latin transformation in a template. For example a $\text{FILTER}_{AO}$, costs 1 FILTER $+k$(2JOINs and 2 FOREACHs), which yields a total cost of $O(D) + O(kDC + kzAC) + O(kC + kzA)$, or $O(kDC + kzAC)$ by simplifying the equation. The analysis states that the cost of $\text{FILTER}_{AO}$ is dominated by the cost of the $k2$ JOINs, which concurs with the general rule of thumb, that the cost of joins dominates query cost.

The analysis is summarized in Table 12. Most of the opera-

| Subscribers | | | Advice | | | | |
|---|---|---|---|---|---|---|---|
| Name, City, Amt, Id | | | Id, Per, Sec, Lin, {Per, MetaPer, Start, End} | | | | |
| (Maya, Logan, $20, 1) | | | {(1, A, Paid, {1}, {(A, Z, 2006, now)})} | | | | |
| (Jose, Logan, $15, 2) | | | {(2, B, Paid, {2}, {(B, X, 2007, 2008)}), | | | | |
| | | | (2, C, Lapsed, {2}, {(C, Y, 2009, now)})} | | | | |
| (Knut, Ogden, $20, 3) | | | {(3, D, Lapsed, {3}, I, {(D, Z, 2006, now)})} | | | | |

**Table 11: Advice in-lined `Subscribers`**

| Transform | Pig Latin | Aspect-oriented Pig Latin |
|---|---|---|
| FILTER | $O(D)$ | $O(D+ \text{TRIM}_{AO})$ |
| FOREACH | $O(D)$ | $O(D + kA + kC)$ |
| DISTINCT | $O(D)$ | $O(D+ \text{MERGE}_{AO} + \text{SEPARATE}_{AO})$ |
| JOIN | $O(D^2)$ | $O(D^2+ \text{MERGE}_{AO} + \text{SEPARATE}_{AO})$ |
| GROUP | $O(D)$ | $O(kD+ \text{MERGE}_{AO} + \text{SEPARATE}_{AO})$ |
| UNION | $O(D^2)$ | $O(D^2 + kC^2 + kzA^2)$ |
| SPLIT/SAMPLE | $O(D)$ | $O(k(D+ \text{TRIM}_{AO}))$ |
| LOAD/STORE | $O(D)$ | $O(D + kz(C + A))$ |
| $\text{TRIM}_{AO}$ | - | $O(D(z(CA^k)))$ |
| $\text{MERGE}_{AO}$ | - | $O(D(z(CA^k)))$ |
| $\text{SEPARATE}_{AO}$ | - | $O(D + z(kCA))$ |
| Program aspects | - | $O(kDC + kzAC)$ |

**Table 12: Complexity Analysis**

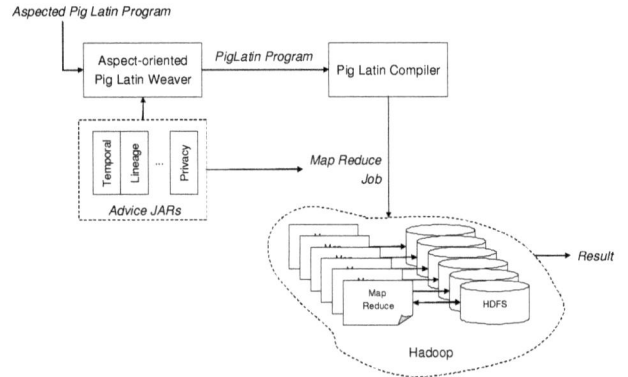

Figure 11: An overview of the implementation architecture

tions include only the additional cost of processing the cuts and advice, but five operations are much more expensive: $\text{FILTER}_{AO}$, $\text{JOIN}_{AO}$, $\text{GROUP}_{AO}$, $\text{SPLIT}_{AO}$, and $\text{SAMPLE}_{AO}$. We consider each in turn. $\text{FILTER}_{AO}$, increases the cost by trimming extraneous cuts and advice (those that have been filtered from the relation). But as we pointed out in Section 3.1.1 the extraneous cuts and advice are harmless and do not need to be removed (other than for consistency), lowering the cost to $O(D)$. A similar speedup applies to $\text{SPLIT}_{AO}$ and $\text{SAMPLE}_{AO}$. $\text{JOIN}_{AO}$, is expensive because the data has to be joined to the advice and the data cuts for each join. This adds the cost of a merge and separate phase. Finally $\text{GROUP}_{AO}$ is inherently more expensive than GROUP, because data must be grouped by both data and metadata, increasing both the number of groups and the cost of computing each group.

Advice in-lining can lower the cost of many of the operations since the relations do not have to be merged, separated, or trimmed. The cost of in-lining is an additional merge and separate phase.

## 4. IMPLEMENTATION

Apache Pig is an open source, Java implementation of Pig Latin.[2] We modified the source code to implement Aspect-oriented Pig Latin. Our modifications and an experimental reproducibility package can be found at the project's website.[3] The modified architecture is given in Figure 11. A Pig Latin program is input to the Aspect-oriented Pig Latin Weaver. The weaver translates an aspected Pig Latin to a Pig Latin program by weaving the aspects into the code using the transformations discussed in Section 3. JARs that contain the advice-specific operations are also part of the transformation. That program is then compiled into a Map-Reduce job, which can be run on Hadoop (or on a local machine). The input data resides on the Hadoop Distributed File System (HDFS). As the Map-Reduce job evaluates it also calls advice-specific behavior in the JARs.

We implemented the data weaver by rewriting the Pig Latin parser (which is specified using ANTLR[4]). We also had to modify some run-time libraries to pass schema information within Hadoop. Normally a Map-Reduce job does not know the schema of the data, but since we need to sometimes find the advice data in a tuple each tuple needs to know the schema of the relation to which it belongs. So we modified the system to retain schema information.

The cost of weaving is trivial; the weaver takes a fraction of a second since most Pig Latin programs are tens of lines in length. To get an idea of the cost of evaluating a data analysis program that enforces metadata semantics, we designed an experiment to measure the cost of single-tuple and tuple-pair operations. We chose $\text{FILTER}_{AO}$ as a representative single-tuple operation, and $\text{JOIN}_{AO}$ for the tuple-pair operation. We generated three datasets of 5 million, 10 million, and 15 million tuples, respectively. We then aspected the tuples with "best-case" advice (each tuple is aspected by the same advice, so the advice is a single tuple) and "worst-case" advice (each tuple has different advice, so there are 5 million advice tuples for 5 million data tuples). We used only one kind of advice: temporal. For both cases the cuts relation has the same number of tuples as the data relation (so 5 million data cuts for 5 million data tuples). To mitigate the impact of parallel evaluation, we ran Hadoop on a single Linux machine with two Intel 686 2.66 GHZ chips, 3.5 GB of RAM, and mirrored 500 GB disks (RAID level 1). We tested using Java 1.6, Pig 0.10.0, and Hadoop 1.0.1. We ran each test five times and took the average cost. Times were measured using the "real" system time captured by Linux's time command. We measured the cost of loading and storing each dataset together with the operation. We then subtracted from the cost, the I/O time. Aspect-oriented I/O costs are approximately triple the non-aspected case in the experiment since advice and cuts relations must also be read and stored. In a long program with several transformations the I/O cost would be spread out over several operations.

Table 13 gives the result of the experiment. The times are given

---

[2]http://pig.apache.org
[3]http://cs.usu.edu/~cdyreson/AOD
[4]http://www.antlr.org

| Transform | Pig Latin | Aspect-oriented Pig Latin | | Advice In-lined |
|---|---|---|---|---|
| | | best | worst | |
| LOAD/STORE | | | | |
| 5 million | 22s (4.4) | 48s (9.7) | 67s (13.5) | - |
| 10 million | 37s (3.7) | 88s (8.8) | 114s (11.4) | - |
| 15 million | 52s (3.5) | 112s (7.4) | 178s (11.8) | - |
| FILTER - Filters on the condition "true" | | | | |
| 5 million | 5s (1.0) | 120s (24.0) | 132s (26.4) | 5s (1.0) |
| 10 million | 10s (1.0) | 220s (22.0) | 250s (25.0) | 11s (1.1) |
| 15 million | 13s (0.9) | 310s (20.6) | 371s (24.7) | 15s (1.0) |
| JOIN - Equi-join on data fields | | | | |
| 5 million | 112s (22.4) | 236s (47.2) | 353s (70.6) | 150s (30.0) |
| 10 million | 230s (23.0) | 437s (43.7) | 587s (58.7) | 271s (27.1) |
| 15 million | 335s (22.3) | 678s (45.2) | 825s (55.0) | 370s (24.6) |

**Table 13: Evaluation Experiment**

in seconds and in parentheses, the throughput, which is measured as the number of seconds per million tuples (lower is better). While aspect-oriented Pig Latin programs cost more, they also do more; they enforce metadata semantics in data processing. Un-aspected Pig Latin programs do not observe such semantics. The big increase in cost is for $FILTER_{AO}$. This is because $FILTER_{AO}$ adds two joins to the filtering. But as we observed in Section 3.1.1, the joins can be deferred since they only remove "inert" data.

The table also shows the cost for the advice in-lining optimization on the "best" data case. For $FILTER_{AO}$, advice in-lining makes the tuples slightly larger, but performs the same filtering as the non-aspected case. For $JOIN_{AO}$, the non-aspected JOIN is coupled with a STREAM transformation to join advice. The advice in-lining operations do not show the cost of the in-lining.

## 5. RELATED WORK

There is a little previous research on support for manifold kinds of metadata in database management systems, though descriptive metadata has been studied in detail (c.f., [6, 10, 20]). Most closely related to this paper is the AUCQL language for querying different kinds of metadata in a semi-structured data model [13], which was later developed into a query language, MetaXQuery, for XML data [23, 24]. This paper in contrast focuses on Pig Latin.

The database research community has researched models and support for specific kinds of metadata, or in our terminology, specific kinds of aspects. One of the most important and most widely researched kinds is temporal. Temporal extensions of every data model exist, for instance, relational [35], object-oriented [34], and XML [17]. This paper generalizes the work in relational temporal databases by proposing an infrastructure that supports many kinds of advice, not just temporal advice. More specifically we extend tuple-timestamped models [22], whereby the temporal metadata modifies the entire tuple. Other tuple-level, relational model extensions to support security, privacy, probabilities, uncertainty, and reliability have been researched, but no general framework or infrastructure exists which can support all the disparate varieties, other than our own work [12, 14]. This paper makes two important novel contributions. First, we develop aspect-oriented programming for Pig Latin, which shares some operations in common with relational algebra, but has different transformations, such as grouping, distinct, sample, split, load, and store. Additionally, Pig Latin has a non-1NF data model, though in this paper we focused only on

the 1NF aspects. The second contribution is extending the framework to meta-metadata as advice tagging advice.

There are several systems that have aspect-like support for combining different kinds of metadata. Mihaila et al. suggest annotating data with quality and reliability metadata and discuss how to query the data and metadata in combination [27]. The SPARCE system wraps or super-imposes a data model with a layer of metadata [28]. The metadata is active during queries to direct and constrain the search for desired information. Systems that provide mappings between metadata (schema) models are also becoming popular [4, 26]. Our approach differs from these systems by focusing on Pig Latin to support AOP, and by building a framework whereby the behavior of individual data aspects can be specified as "plug-in" components.

The information retrieval community has been very active in researching descriptive metadata, in particular metadata that is used to classify knowledge [37]. The Dublin Core is a commonly used classification standard [38]. Commercial and research systems [1] to manage (descriptive) metadata collections have been developed, as well as methods to automatically extract content-related metadata [21, 25]. The focus of the information retrieval research is on how to best use, manage, and collect metadata to describe data to improve search [18]. In contrast, our focus is on modeling data aspects which *impose* a semantics on the *use* of the data, i.e., they go beyond the simple, descriptive tagging of data.

Finally, our goal in this paper, consistent with AOP and unlike many of the above approaches, is to maximally reuse existing languages and systems. Hence we focus on using Pig Latin itself to support aspect-oriented data by weaving the support for cross-cutting concerns, expressed in Pig Latin, into Pig Latin programs.

## 6. CONCLUSIONS

Cloud computing data analysis platforms like Hadoop do a poor job of supporting cross-cutting data concerns. Data has a wide variety of cross-cutting concerns: time, security, reliability, privacy, quality, summaries, rankings, and uncertainty. In this paper we adapted techniques from aspect-oriented programming (AOP) to Pig Latin. Pig Latin is a dataflow language for analyzing data in the cloud. We proposed annotating Pig data using data aspects. A data aspect binds advice (metadata) to data. The advice also has semantics that must be observed when the data is used in a query. We showed how to weave Pig Latin into a Pig Latin program to support cross-cutting concern data concerns.

In future we plan to address optimization, and management of data cuts, in particular mechanisms for tagging data with advice. Pig Latin does not have any data management role, so this would be prior to analysis by Pig Latin. We also plan to investigate new optimization rules for Aspect-oriented Pig Latin. We anticipate that several optimizations are possible, such as keeping a relation joined to its advice throughout evaluation of a program rather than splitting it into data, cuts, and advice relations after each transformation. We also need better support for UDFs. Each UDF must currently be written to correctly implement aspects. This places a great burden on UDF coders. To avoid requiring a user to modify each UDF, we need to instead partition the data into sets of tuples that have the same advice, and evaluate the UDF for only the given set. To the best of our knowledge there is no research yet on such partitioning for cloud computing. An alternative solution is to re-engineer a programming language like Java to support data aspects. Consider for instance an IF statement that compares two values tagged with temporal metadata. For some time periods the IF condition may be true, yet false for others, so each branch might need to be evaluated (for different time periods), resulting in a very different IF than currently exists in Java. This is also an open problem to the best of our knowledge.

# 7. ACKNOWLEDGMENTS

This material is based upon work supported by the National Science Foundation under Grant No. 1144404 entitled "III: EAGER: Aspect-oriented Data Weaving." Any opinions, findings, and conclusions or recommendations expressed in this material are those of the authors and do not necessarily reflect the views of the National Science Foundation.

# 8. REFERENCES

[1] M. Q. W. Baldonado, K. C.-C. Chang, L. Gravano, and A. Paepcke. Metadata for Digital Libraries: Architecture and Design Rationale. In *ACM DL*, pages 47–56, 1997.

[2] C. Batini and M. Scannapieco. *Data Quality: Concepts, Methodologies and Techniques*. Data-Centric Systems and Applications. Springer, 2006.

[3] O. Benjelloun, A. D. Sarma, A. Y. Halevy, M. Theobald, and J. Widom. Databases with Uncertainty and Lineage. *VLDB J.*, 17(2):243–264, 2008.

[4] P. A. Bernstein. Applying Model Management to Classical Meta Data Problems. In *CIDR*, 2003.

[5] E. Bertino, G. Ghinita, and A. Kamra. Access Control for Databases: Concepts and Systems. *Foundations and Trends in Databases*, 3(1-2):1–148, 2011.

[6] D. Bhagwat, L. Chiticariu, W. C. Tan, and G. Vijayvargiya. An Annotation Management System for Relational Databases. In *VLDB*, pages 900–911, 2004.

[7] K. Birman, G. Chockler, and R. van Renesse. Toward a Cloud Computing Research Agenda. *SIGACT News*, 40(2):68–80, June 2009.

[8] R. Bose and J. Frew. Lineage Retrieval for Scientific Data Processing: A Survey. *ACM Computing Surveys*, 37(1):1–28, 2005.

[9] P. Buneman and W. C. Tan. Provenance in Databases. In *SIGMOD Conference*, pages 1171–1173, 2007.

[10] L. Chiticariu, W. C. Tan, and G. Vijayvargiya. DBNotes: A Post-It System for Relational Databases Based on Provenance. In *SIGMOD Conference*, pages 942–944, 2005.

[11] D. E. Denning and P. J. Denning. Data Security. *ACM Comput. Surv.*, 11(3):227–249, Sept. 1979.

[12] C. E. Dyreson. Aspect-Oriented Relational Algebra. In *EDBT*, pages 377–388, 2011.

[13] C. E. Dyreson, M. H. Böhlen, and C. S. Jensen. Capturing and Querying Multiple Aspects of Semistructured Data. In *VLDB*, pages 290–301, 1999.

[14] C. E. Dyreson and O. U. Florez. Data Aspects in a Relational Database. In *CIKM*, pages 1373–1376, 2010.

[15] C. E. Dyreson, R. T. Snodgrass, F. Currim, S. Currim, and S. Joshi. Weaving Temporal and Reliability Aspects into a Schema Tapestry. *Data Knowl. Eng.*, 63(3):752–773, 2007.

[16] B. Fung, K. Wang, R. Chen, and P. S. Yu. Privacy-preserving Data Publishing: A Survey of Recent Developments. *ACM Comput. Surv.*, 42(4):14:1–14:53, June 2010.

[17] D. Gao and R. T. Snodgrass. Temporal Slicing in the Evaluation of XML Queries. In *VLDB*, pages 632–643, 2003.

[18] H. Garcia-Molina, D. Hillmann, C. Lagoze, E. D. Liddy, and S. Weibel. How Important is Metadata? In *JCDL*, page 369, 2002.

[19] A. Gates, O. Natkovich, S. Chopra, P. Kamath, S. Narayanam, C. Olston, B. Reed, S. Srinivasan, and U. Srivastava. Building a Highlevel Dataflow System on top of MapReduce: The Pig Experience. *PVLDB*, 2(2):1414–1425, 2009.

[20] F. Geerts, A. Kementsietsidis, and D. Milano. MONDRIAN: Annotating and Querying Databases through Colors and Blocks. In *ICDE*, page 82, 2006.

[21] L. Gravano, P. G. Ipeirotis, and M. Sahami. QProber: A System for Automatic Classification of Hidden-Web Databases. *ACM Trans. Inf. Syst.*, 21(1):1–41, 2003.

[22] C. S. Jensen, M. D. Soo, and R. T. Snodgrass. Unification of Temporal Data Models. In *ICDE*, pages 262–271, 1993.

[23] H. Jin and C. E. Dyreson. Sanitizing using Metadata in MetaXQuery. In *SAC*, pages 1732–1736, 2005.

[24] H. Jin and C. E. Dyreson. Supporting Proscriptive Metadata in an XML DBMS. In *DEXA*, pages 479–492, 2008.

[25] D. Lee and Y. Hwang. Extracting Semantic Metadata and its Visualization. *ACM Crossroads*, 7(3):19–27, Mar. 2001.

[26] S. Melnik, E. Rahm, and P. A. Bernstein. Rondo: A Programming Platform for Generic Model Management. In *SIGMOD Conference*, pages 193–204, 2003.

[27] G. A. Mihaila, L. Raschid, and M.-E. Vidal. Using Quality of Data Metadata for Source Selection and Ranking. In *WebDB (Informal Proceedings)*, pages 93–98, 2000.

[28] S. Murthy, D. Maier, L. M. L. Delcambre, and S. Bowers. Superimposed Applications using SPARCE. In *ICDE*, page 861, 2004.

[29] C. Olston, B. Reed, U. Srivastava, R. Kumar, and A. Tomkins. Pig Latin: A Not-so-Foreign Language for Data Processing. In *SIGMOD Conference*, pages 1099–1110, 2008.

[30] G. Özsoyoglu and R. T. Snodgrass. Temporal and Real-Time Databases: A Survey. *IEEE Trans. Knowl. Data Eng.*, 7(4):513–532, 1995.

[31] S. Pearson and A. Benameur. Privacy, Security and Trust Issues Arising from Cloud Computing. In *IEEE Cloud Computing (CloudCom)*, pages 693–702, Dec. 2010.

[32] A. Rashid. *Aspect-Oriented Database Systems*. Springer, 2003.

[33] A. Rashid and N. Loughran. Relational Database Support for Aspect-Oriented Programming. In *NetObjectDays*, pages 233–247, 2002.

[34] R. T. Snodgrass. Temporal Object-Oriented Databases: A Critical Comparison. In *Modern Database Systems*, pages 386–408. 1995.

[35] R. T. Snodgrass, editor. *The TSQL2 Temporal Query Language*. Kluwer, 1995.

[36] R. T. Snodgrass. *Developing Time-Oriented Database Applications in SQL*. Morgan Kaufmann, 1999.

[37] A. Tannenbaum, editor. *Metadata Solutions: Using Metamodels, Repositories, XML, and Enterprise Portals to Generate Information on Demand*. Addison-Wesley, 2001.

[38] H. Wagner and S. Weibel. The Dublin Core Metadata Registry: Requirements, Implementation, and Experience. *J. Digit. Inf.*, 6(2), 2005.

[39] M. Zhou, R. Zhang, W. Xie, W. Qian, and A. Zhou. Security and Privacy in Cloud Computing: A Survey. In *International Conference on Semantics Knowledge and Grid (SKG)*, pages 105–112, Nov. 2010.

# KFusion: Optimizing Data Flow without Compromising Modularity

Liam Kiemele
University of Victoria
liamk@cs.uvic.ca

Celina Berg
University of Victoria
celina@cs.uvic.ca

Aaron Gulliver
University of Victoria
agullive@ece.uvic.ca

Yvonne Coady
University of Victoria
ycoady@cs.uvic.ca

## ABSTRACT

Programming language support for multi-core architectures introduces a fundamentally new mechanism for modularity—a *kernel*. Though it can be used as a means to separate concerns, a kernel is given a clean slate of memory at execution time. As a consequence, application developers attempting to leverage libraries of kernels often incur substantial unanticipated performance penalties. Currently, the only recourse is to compromise modularity for the sake of optimizing data flow on an application-specific basis.

KFusion is our prototype tool for optimizing libraries of kernels according to application-specific needs. Our goal is to shield application developers from loop fusion and deforestation in compositions of low level kernels that share data. Libraries, augmented by domain experts with annotations to ensure correct compositions of kernels, provide application developers with the opportunity to supply hints according to customized data flow needs—keeping modularity intact. In the worst case, an inaccurate hint incurs no penalty. Case studies of applications using general-purpose libraries for linear algebra, image manipulation and physics engines show that KFusion can substantially improve performance associated memory bandwidth bottlenecks.

## Categories and Subject Descriptors

D.1.3 [**PROGRAMMING TECHNIQUES**]: Concurrent Programming—*ParalletProgramming*

## General Terms

Design, Performance

## Keywords

OpenCL, Parallelism, Performance, Modularity

## 1. INTRODUCTION

With the shift to multi-core hardware, the speed of computation is often not the performance bottleneck for many applications—instead, memory bandwidth is. Previously, well modularized conceptual representations of concerns could reap the benefits of compiler optimizations and good cache utilization to mitigate costs of accessing memory. Now, new languages for new paradigms have new separation of concerns mechanisms, with entirely different consequences with respect to these costs.

For example, a language such as OpenCL [14], commonly used for Graphical Processing Units (GPUs), has a fundamentally different way to separate concerns: *kernels*. By design, the execution of a kernel is isolated. Kernels do not share memory, and the cache is effectively reset every time a kernel is launched. Consequently, functionality that is separated into different kernels cannot amortize costs of moving data to and from memory.

For domain experts in multi-core programming, this is a well-known problem addressed by kernel *fusion*. Recently, tools to help predict cost savings of classes of optimizations have been proposed to help developers tackle these challenges [29, 1]. Application developers relying on libraries that use kernels, however, should be insulated from these complexities. Our goal is to try to achieve a balance by separating the details of these concerns from an application perspective, while providing a means to semi-automate fusion in low level libraries.

The contribution of this work is twofold. First, we present *KFusion*, a prototype for explicitly managing kernel compositions from a high-level within an OpenCL codebase. The compositions are optimized using two well-known techniques: (1) *loop fusion* to amortize memory access costs by combining operations performed on the same dataset, and (2) *deforestation* to eliminate intermediate values. Second, we provide an evaluation of KFusion in terms of its usability and performance in general-purpose libraries for linear algebra, image processing and a physics engine.

## 2. BACKGROUND AND RELATED WORK

To provide some context with respect to the trade-offs between data flow optimizations and modularity, we first provide an overview of OpenCL and performance in general. We then consider related work on the tension between modularity versus performance, and examine related tools with similar goals to ours.

## 2.1 GPU Programming

Modern programming paradigms for current multi-core architectures necessarily involve concurrency [30]. The era where the clock speed of processors continued to increase is over, and instead performance gains will be in terms of core utilization and careful treatment of memory-bandwidth bottlenecks.

Parallel hardware is now ubiquitous and of particular interest are Graphical Processing Units (GPUs). The two major standards for GPU programming are CUDA [23] and OpenCL [14]. This work focuses on OpenCL as it is an open standard which has potential to be expanded to many platforms and architectures, whereas CUDA is vendor specific. In this section, we provide an overview of the essential elements of an OpenCL application. These elements are expected to translate directly to CUDA, though in some cases with different names.

Functionality is generally organized into compute kernels which are executed by threads operating over a *global work space*. An example of an extremely simple kernel is shown in Listing 1. Though this appears to just add two values, this code is designed to sum each corresponding element in vectors $v1$ and $v2$ in parallel. In this case, the global work space consists of the vectors. Each output value is computed in parallel—each kernel is launched as a thread which is part of a *work group* with a unique *global id*. The id is used to carry out computation on the correct element of data in the work space. In this example, this is stored as the index $i$, and this code operates much like a classic parallel *for* loop.

```
1  __kernel void vectorAdd(__global float* v1,
                           __global float* v2,
3                          __global float* v3)
   {
5      int i = get_global_id(0);
       v3[i] = v2[i] + v1[i];
7  }
```

**Listing 1: A vector addition kernel. These instructions will be executed for each element $i$ in vectors v1,v2 and v3**

Kernels provide the essential building blocks for parallelizing OpenCL computation and also effectively act as units for modularity. Each kernel is given a clean slate of memory to operate within. When a kernel finishes executing, none of its cached data can be reused. This makes each kernel execution a relatively heavy-weight event, but the advantage is that this isolation prevents unintentional interaction with other kernels. This clean slate means that there is much less to manage, but also resets any beneficial cached data—two consecutive kernels cannot take advantage of shared memory or access patterns.

## 2.2 Modularity versus Performance

Initial work on modularity in the systems domain includes Dijkstra's *THE multiprogramming system* [10], which introduced a layered approach to separation of concerns. Ultimately, boundaries have been proven to be difficult to achieve in modern operating systems, in particular due to optimizations aimed at improving resource utilization.

Similarly, Parnas reasoned that a system should be broken down by design decisions which were likely to change as opposed to an execution path [25]. He also established that separation should be enforced by information hiding. These criteria were designed to assist with program comprehension and prevent cascading changes, and were considered key to addressing fundamental flaws associated with programming in the 70s [9], where performance considerations dominated modularity concerns. Explicit support for information hiding and data abstraction later began to be provided by languages such as Simula [7] and CLU [20].

Aspects [15] established that with the right language constructs, the benefits of modularity can be extended to include concerns that inherently span several modules in a system, including optimizations. As aspects, *crosscutting* concerns that are inherently challenging to structure with conventional mechanisms can be untangled from the codebase.

Well-known techniques for optimizations associated with data flow include loop fusion and deforestation. Loop fusion is used in imperative as well as parallel applications to improve memory locality [13], consolidating otherwise separate traversals of data structures to amortize costs. Deforestation involves the elimination of intermediate data structures [31]. Fusion often results in deforestation, as combining operations on traversals of data structures typically removes the need for intermediate results. Automated application of these techniques is advantageous as they can preserve modularity but still significantly improve performance.

## 2.3 Frameworks for Performance

The body of work from parallel compilers [16, 3] includes formalisms for the automatic loop fusion and deforestation of serial code [5]. KFusion leverages these approaches to optimize for application-specific data flow needs, and though our prototype uses a naive approach, it does not preclude experimentation with other contenders in this domain.

Leung et. al have produced a parallel compiler with effective automatic parallelization of Java code and offloading of computation to the GPU when there is a performance benefit [18]. Baskaran et al. use a polyhedral model to convert affine programs written in C to CUDA [2]. For the range of programs they considered, they were able produce an implementation with performance close to hand optimized CUDA code. Others have converted OpenMP [6] to CUDA [17], allowing for incremental and rapid development of GPU code, though somewhat limited to the applications that OpenMP parallelizes well.

BONES [21], is based on the idea of *skeleton* algorithms. These are highly parameterized general routines to which sections of source code can be mapped. They cover parallel operations such as *map* and *reduce*, but can be expanded to cover a large variety of problems. BONES also leverages a fusion optimizer from well-known techniques. SkePU [8] also relies on skeleton algorithms, and enables lazy copying of data from host to device. One downside when using a set of skeletons is that application-specific needs may not be met, as developers may have to add to the skeleton, confronting low-level optimizations instead of focusing on application level logic.

Within the Domain Specific Language (DSL) community, the Delite framework [4] has achieved significant compile time optimizations. The framework allows for the development of a DSL which can then be compiled down to Scala [24], C++ or OpenCL. The DSL can take advantage of several optimizations such as replacing higher order functions with their first order equivalent, and support fusing operations in order to take advantage of data locality. The end result is the creation of a series of domain specific libraries, which may have tradeoffs in more general cases depending on the ease in which they can be expanded. While Delite creates DSLs, KFusion is capable of working with any given general purpose library.

Yang et al. [32] takes naive GPU kernels and optimizes them in two major respects: memory use and parallelism. The naive kernel is optimized in terms of memory access patterns, memory coalescing, vectorization and loop unrolling. They achieve performance close to, and in some cases superior to, finely tuned hand optimized code. This shows that kernels themselves can be optimized and improved automatically, but it does not handle optimization which may exist between kernels.

Though not specific to GPUs, yet relevant to this work is the idea of *active libraries* such as the SPIRAL framework [26]. The general idea behind active libraries leverages a series of tests executed at compile time to inform the end compilation. The best example is FFTW [11], which has many implementations of various components and tests to identify the best implementation for a given hardware configuration. While our work does not use this approach, we feel it could be leveraged to accomplish a hardware specific implementation.

The key difference between KFusion and all of these approaches that it is designed to explore the ability to offer optimizations from the application perspective, without requiring the application developer to compromise modularity and tangle the application with lower level library code to get performance. While other frameworks convert small sections of code—typically for loops—into parallel code implementations, KFusion endeavours to customize existing OpenCL libraries according to application-specific data flow needs at compile time. As a result, KFusion could complement any of these existing approaches. A parallelizing compiler or skeleton algorithm approach could be used to create an OpenCL enabled library, which could be further optimized according to application-specific data flow by KFusion. Finally, existing optimizing compilers could be leveraged before or after KFusion.

# 3. OPENCL TO OPENCL TRANSFORMER

Optimizations associated with memory access patterns are hard to modularize, in particular because these patterns are often dictated by application-specific needs. For example, consider a library of operations that each of which modifies the same data structure. Inherently, in OpenCL, this library would support these individual operations through separate kernels (and in some operations, potentially several kernels). This modularity is important as it ensures the library can be used in a general case. Unfortunately, modularity also hinders efficient memory access—especially due to the fact cache is reset between kernels. Though application-specific combinations of these operations might be able to amortize costs through fusion and deforestation, this information is only available at the time the application is compiled.

For example, leveraging a general-purpose linear algebra library as shown in Algorithm 1, an application can benefit from parallelism offered by the operations in the library. Within the library, each operation is computed by a separate kernel, executed in order. The API protects the application

---

**Algorithm 1** An application-specific use of a general-purpose library, where each variable $x,y,c$ represents a vector or array of length $n$.

1: square(x) - square the values of array x
2: square(y) - square the values of array y
3: add(c,x,y) - add x and y and store in c
4: sqrt(c) - obtain the square root of each value in c

| Kernel | Operations | Global Memory | Cycles |
|--------|-----------|---------------|--------|
| square(x) | 1 | 1 load and 1 store | 804 |
| square(y) | 1 | 1 load and 1 store | 804 |
| add(c,x,y) | 1 | 2 loads and 1 store | 804 |
| sqrt(c) | 1 | 1 load and 1 store | 804 |
| **total** | **4** | **9** | **3216** |

Table 1: **List of kernels and the load and store operation costs.**

developer using this library from having to deal with the lower level implementation details. Though each kernel may be individually optimized, the precise combination of kernels dictated by the application would typically not have been anticipated in advance.

## 3.1 Costs of Modularity

It is possible to reason about the costs both in terms of computation and memory accesses. Equation 1 provides a simplified formula for obtaining performance estimation: $C$ is the number of computation instructions and $\alpha$ is the average number of clock cycles required to execute each instruction. $M_G$ and $M_L$ each represent the number of memory operations at each level: global and local. $\beta_G, \beta_C, \beta_L$ and $\beta_P$ represent the costs. These will change depending on hardware, we will assume a standard GPU. There exists more complex models [27], but this works for our motivating example.

$$T(c) = C\alpha + M_G\beta_G + M_L\beta_L \qquad (1)$$

Using the NVIDIA optimization guidelines [22], the costs can be expressed in terms of clock cycles. The minimum latency of a global memory optimization is 400 clock cycles. Local memory on the other hand has a latency of 5 clock cycles. The maximum number of clock cycles required for an arithmetic operation is 4. The equation then becomes simplified to be Equation 2.

$$T(c) = 4C + 400M_G + 5M_L \qquad (2)$$

When these operations are in separate kernels, the summary of the global load and store costs can be seen in Table 1. If instead all of these kernels were fused, only a minimum number of loads and store operations are required. The amortized costs can be seen in Table 2.

The end result is a reduced number of cycles required to execute the kernel—to about a third. Also, as opposed to 4 mathematical operations to 9 memory operations, there are now 4 arithmetic instructions for 3 memory operations. This will improve OpenCL's ability to hide latency by overlapping communication and computation. The actual performance results with different vector sizes can be seen in Figure 1. While there is not a 3 times performance increase because OpenCL has the ability to hide some latency through parallelism, there is still a significant performance boost.

| Kernel | Operations | Global Memory | Cycles |
|--------|:----------:|:-------------:|:------:|
| square(x) | 1 | 1 load | 404 |
| square(y) | 1 | 1 load | 404 |
| add(c,x,y) | 1 | 0 | 4 |
| sqrt(c) | 1 | 1 store | 404 |
| total | 4 | 3 | 1216 |

**Table 2: The fused kernel costs which reduce the cost to approximately one third of the unfused version.**

**Figure 1: A performance comparison between unfused and fused kernels for different vector sizes. There is a significant gain primarily do the amortization is memory access costs. In both cases, OpenCL can mitigate some latency associated with memory access, but fusion greatly helps with this.**

The key concept here is data flow. If operations are combined, data is reused while still in cache or ideally hardware registers. But as systems increase in size and complexity, manually fusing kernels to obtain better performance not only becomes unsustainable, but also suggests deep compromises in information hiding and separation of concerns if application developers must fuse functions below a library's API.

## 3.2   The Prototyope Tool: KFusion

KFusion uses a semi-automated approach to accomplish low-level kernel fusion that maintains original source semantics. This allows for independent, low-level, kernels to be transformed into high-performance monolithic kernels on an application-specific basis.

KFusion leverages two types of annotations in order to guide source code transformations:

**Application Level** - At this level there is a single annotation used by application developers to provide hints as to which regions of the code could possibly be fused according to data flow.

**Library Level** - Designed for domain experts, these annotations augment general-purpose library files that in turn use OpenCL kernel files.

This separation allows the library developer to handle the low level details associated with fusion, allowing application developer to reason at a higher level about data flow.

### 3.2.1   Application Annotations

At the application level, only one annotation is required, and it is a hint with no penalty if not used:

**kfuse(param){ ... }** - The data flow of the parameters is such that the enclosed set of function calls could be fused for better performance.

An example of the annotation is shown in Listing 2. Kfuse will attempt to combine functions on synchronization annotations detailed at the library level. By default, intermediate results are not kept according to optimizations of deforestation. So only the final mutation to the vector $c$ is propagated from the fused kernels.

```
1    kfuse(x,y,c) {
         square(x)
3        square(y)
         add(c,x,y)
5        squareroot(c)
     }
```

**Listing 2: The resulting fused implementation will make the operations non-destructive for all vectors except $c$ which will output the correct result**

### 3.2.2   Library Function Annotations

Library functions are annotated with synchronization information through a simple *pragma* before the function definition. They determine if a function and any of the kernels it contains can be fused. Most functions will not require this annotation, but they allow for restrictions which can be used to ensure safe fusions.

**#pragma sync in** - This function requires a synchronized input. It will not be fused with any previous functions in the control flow path.

**#pragma sync out** - This function requires a synchronized output. It will not be fused with following functions in the control flow path.

These situations will arise when either a kernel requires the entire result of a previous operation or when a kernel ends in a reduction type operation. Examples of these are shown in Listing 3. Another good example is a resize operation which will end in a result which will be incompatible with the following kernels. The synchronization *pragmas* are currently very coarse grained and could be improved by specifying which specific inputs and outputs must be synchronized. The stronger limitations do prove to be safe, but may be unnecessary.

```
    #pragma sync in
2   void matrixMult(Vector *b, Matrix *A,
      Vector *x);

    #pragma sync out
6   void dotProduct(Vector *a, double &result);
```

**Listing 3: In example of synchronization annotations for a matrix multiplication and dot product operation. The matrix multiplication requires the entirety of x and the dot product reduces to a single value.**

```
   __kernel void add(__global double * c,  const double
      alpha, __global double * a, const double beta,
      __global double * b) {
2  kload{
      int gid = get_global_id(0);
4     double aVal = a[gid];
      double bVal = b[gid];
6     double cVal;
   }
8
   cVal = alpha*aVal + beta*bVal;
10
   kstore{
12    c[gid] = cVal;
   }
14 }
```

**Listing 4: An example of annotating load and store operations with kload and kstore**

### 3.2.3 Kernel Annotations

The library developer, or perhaps a domain expert in multi-core programming, will also use annotations with OpenCL kernels which are executed on the GPU or other target device:

**kload { ... }** - regions of kernel operations where data is loaded from global memory

**kstore { ... }** - regions of kernel operations where data is stored

**#pragma immovable** - for asynchronous instructions *prefetch* and *async_work_group_copy* that should not be rearranged by KFusion

An example of these annotations can be seen in Listing 4. These annotations detail the load and store operations.

*kload* is used to outline which private or local variables are assigned to from global memory. These effectively identify which variables are assigned values as opposed to kernel parameters. Kload is also used to mark any variables which will be carried through and used to store data at the end of execution. Kload must occur in the root scope of a given kernel and cannot be within control structures such as *loops*. A good rule of thumb is that all initial declarations in a function are Kload instructions while temporary variables are not.

*kstore* is used to annotate any output statements which are assigned to global memory. Generally this is a kernel parameter. A good rule of thumb is that these are the last few or single last assignment statement in a kernel.

*#pragma immovable* is used to annotate that the following statement should not be moved during optimization. An example of its use is shown in Listing 5. This relates to OpenCL asynchronous instructions dealing with memory transfer and prefetching. In the event kernels are combined, any asynchronous instructions at the root scope will be moved earlier in execution in order hide latency and maximize the use of available bandwidth. In most cases, moving these functions forward in the code should not cause problems, but in the even data is transferred to and from a location multiple times, some of these statements will have to labelled as immovable.

```
   __kernel void example(__global double * A,  __global
      double * x, __global double * b) {
2  kload{
4     __local double[32] var;
   event_t copy = async_work_group_copy(var,x,32,0);
6  }
   ...
8  #pragma immovable
   copy = async_work_group_copy(var,&x[32],32,0);
10 ...
   #pragma immovable
12 copy = async_work_group_copy(var,&x[64],32,0);
   ...
```

**Listing 5: An example of #pragma immovable. When values are asynchronously copied to the same variable multiple times, the later operations will have to be labelled immovable in order to ensure correctness.**

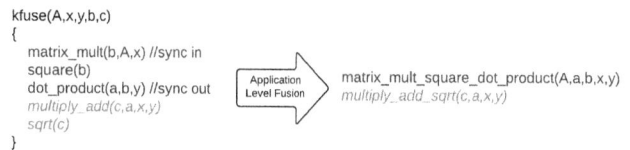

**Figure 2: Creation of a new function call at the end of the first synthesis stage. This example takes into account the synchronization information given in the library file. In this example a dot product produces a synchronized result and the matrix multiplication requires a synchronized input. When fused, this will create two separate functions. This will ensure correctness.**

## 3.3 KFusion Transformation

KFusion has two major phases, analysis and synthesis. This subsection will briefly discuss analysis and examine synthesis in more detail.

Analysis is straightforward and involves collecting information given through annotations which will later be used to inform fusion. From application files, it collects sets of functions as specified by *kfuse*. From each function in the library files it collects which kernels are called with which arguments as well as any synchronization information specified by *#pragma sync [in][out]*. From kernel files it collects the arguments to each kernel, the inputs and outputs as specified by *kload* and *kstore* as well as any immovable statements as specified by *#pragma immovable*.

The synthesis process of is more complex. It leverages the given annotations to create new functions and kernels.

### 3.3.1 Application Files

The first step in synthesis takes a sets of functions defined by the *kfuse* annotation. It then uses synchronization information to break each set into subsets. Each subset will be fused independently. KFusion then outputs a new function call for each subset. An example of this with synchronization is shown in Figure 2. This is what the application developer will see should they view the source code after the transformation. This provides feedback on what was successfully fused.

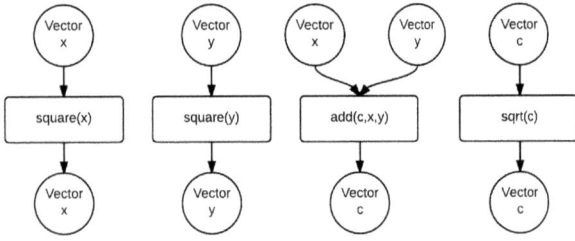

Figure 3: An overview of the inputs and outputs to each kernel. These inputs and outputs will later be used to perform kernel fusion accomplishing both loop fusion and deforestation

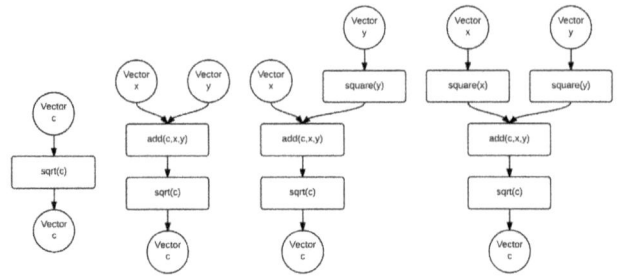

Figure 4: An example of constructing the dependency graph given the set of inputs and outputs as shown in Figure 3. Inputs are recursively matched with outputs to produce a dependency graph. The inputs are represented by circles and it is relatively easy to see how the data can be pipelined through each operation.

### 3.3.2  Library Files

Synthesis of the library files is a somewhat more complex process. For each function call, KFusion duplicates the called function and replaces its parameter names with generated ones which are guaranteed to be unique for each argument. The new functions are then combined and duplicate instructions are removed.

A new *kernel invocation* is created by KFusion. We define a kernel invocation as the set of functions which launch a kernel with a set of arguments. For each library function, the kernel invocations are extracted and combined to generate the new invocation. Currently we assume one kernel per library function. This new invocation contains all the arguments of the child invocations and removes any duplicates or redundancies. The component kernels followed by the new kernel can be seen Table 3.

| Kernel | Arguments |
|---|---|
| square_kernel | x:arg_1 |
| square_kernel | y:arg_2 |
| add_kernel | c:arg_3, y:arg_2, x:arg_1 |
| sqrt_kernel | x:arg_1 |
| **square_square _add_sqrt_kernel** | **y:arg_2, x:arg_1 c:arg_3** |

Table 3: The list of kernels with arguments—each forming a kernel invocation—along with the new kernel in bold.

This part of the KFusion process results in a single kernel with all available arguments and only one set of pre-instructions and post instructions. The list of kernel executions and arguments is collected and will be used to generate a new kernel in the next KFusion stage.

### 3.3.3  Kernel Files

The last stage of the KFusion synthesis is entered with a list of kernels, each called with a unique set of arguments. The analysis phase provided a clearly defined set of inputs and outputs for each kernel. An example of this list can be shown in Figure 3. The synthesis of the kernel files is similar to that of the library functions in that all kernel parameters are replaced with unique identifiers and local variables are validated to keep track of data flow and to ensure they have unique names. This provides a method of dependency propagation and identifies which outputs are tied to which inputs. A set of unique kernels is created.

Using the additional data flow information added through dependency propagation, KFusion builds a dependency graph from the bottom up starting at the last kernel executed and moving back to the first. This is a recursive algorithm performing a depth first search in order to match outputs with inputs. A node is a parent of another node if and only if one of its outputs satisfies one or more of the child's inputs.

An example of the construction can be seen in Figure 4. By adding the kernels in reverse order and always adding them to the newest or topmost possible node in the graph, a dependency graph which accurately represents the correct execution order is created.

In the event a kernel does not fill any dependency in the graph, it is effectively independent. In this case it is combined with bottom most node. This is accomplished by combining the loads, stores and core operations of each kernel into one. This will add to the number of inputs and outputs of the bottommost node and make it amenable to fusion later when KFusion adds more nodes to the graph.

The dependency graph is then used to correctly accomplish kernel fusion. The algorithm for this process can be seen in Algorithm 2 and an overview of this process is illustrated in Figure 5. KFusion scans the input operations of the root kernel and looks for a parent node in the graph with an output that satisfies the given input. If a match is found, the kernel is fused. This starts by replacing any global load operations—which are inputs—with the matching final results of the parent kernel—which are the parent's outputs. This replaces global load operations with copy operations from private memory or local memory. These are much less expensive and avoid many problems which can arise from global memory access. This reduction significantly cuts down on the I/O involved.

The parent kernel's load and core operations are then appended to the start of the root kernels load operations. This effectively adds another set of loads as well as dependencies to the root kernel. At this point the parent node's parent are assigned to the root node, then the parent node is discarded.

Some bookkeeping is required. The transformer uses scope to mitigate any similarly named local variables from the parent kernel, and a single assignment maybe necessary to move

1: given a dependency graph $G$ and root node $R$
2: **while** r.parents $\neq 0$ **do**
3:    **for** parent $p$ **do**
4:       **if** $p$.outputs match at least one of $R$.inputs **then**
5:          **for** each assignment $s$ in $R$.loads **do**
6:             **if** right hand side of $s$ is in $p$.outputs **then**
7:                replace right hand side of $s$ with output
8:             **end if**
9:          **end for**
10:          $R$.load = $p$.loads + $\{p.\text{core}\}$ + $R$.load
11:          remove $p$ from parents
12:          add $p$.parents to parents
13:       **else**
14:          return an error.
15:       **end if**
16:    **end for**
17: **end while**
18: move all root level declarations to the top of the scope
19: move all asynchronous operations to below declarations
20: output new kernel

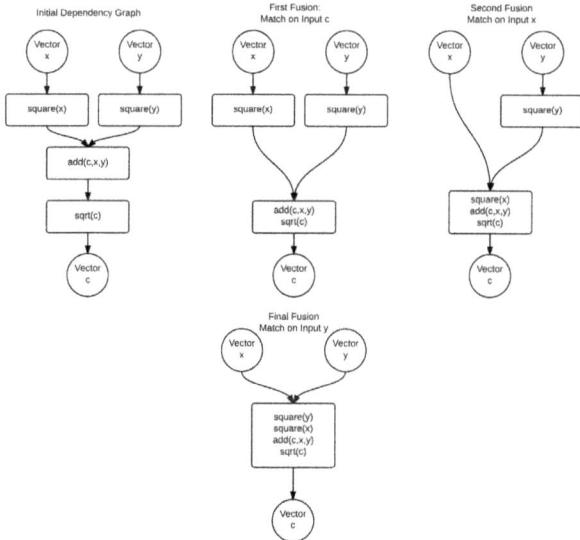

**Figure 5: An overview of fusing the dependency graph. At each stage KFusion eliminates a parent node of the root node and then reassigns the parent nodes to the root. This collapses the graph creating a fused kernel. While it may initially appear that the inputs and outputs are unchanged, each kernel has to reload the inputs its inputs and store the outputs. When reduced to a mono-kernel, everything is loaded and stored the minimum number of times.**

between scopes. The fusion process is repeated until the dependency graph is reduced to a single node.

Once reduced to a single node, KFusion can move statements around in order to improve the execution. Declarations at the kernels root scope are moved to the top of the scope in order to improve readability and ensure variables are declared before they are used. Asynchronous instructions at the base scope are rearranged and moved just after the declarations. These instructions are typically used to move data from global to local memory, executing them early will hide latency associated with loading values. Moving asynchronous instructions earlier in the execution is an optimization that could not occur with smaller independent kernels.

### 3.3.4 End Result

Three artifacts are generated for each successful fusion:

**New function call** - exists in the application file which leverages the newly created function. This passes the required arguments into the library function and eliminates several function calls and kernel executions.

**New function** - in the library file a new function which performs the necessary operations to execute a single monolithic kernel. The new function takes care of any and all bookkeeping required by the new underlying kernel. This accomplishes deforestation at a library level as any redundant computation is removed.

**New kernel** - built from a dependency based fusion. This new kernel accomplishes both deforestation and loop fusion as well as reorders asynchronous instructions. It removes redundant computation and allows for improved data flow. Values stay in cache and local memory longer and throughput is dramatically increased. Data can be loaded asynchronously hiding latency and improving throughput. In ideal cases this kernel can accomplish the work of its $N$ component kernels in the time it would take to execute a single one.

These new structures are created via a source-to-source transformation method, and the application developer can read the end result and perhaps make further changes and optimizations. This is a desirable feature especially when considering the possibility of daisy chaining further optimization tools. It also ensures portability and most devices will have their own OpenCL compiler invoked at runtime in order to best compile code for the target platform.

## 3.4 Limitations

KFusion is still a prototype and has many limitations. The first is that is can only travel down one level of abstraction. This means it is unable to handle functions which call other functions which call kernels. While it will not fail in these instances, it cannot fuse kernels which have two levels of indirection. Expanding KFusion to handle several levels of abstraction is possible, but the analysis and fusion would be much more considerably more complex than our current proof-of-concept.

KFusion could be improved if the synchronization was more fine grained and did not limit fusion possibilities. The current synchronization primitives should be expanded upon to allow for the specific declaration of which variables need to be synchronized in and out. KFusion only performs a static analysis. A dynamic analysis may be more powerful in determining data flow and measure performance characteristics in order to determine what should be fused.

KFusion requires the convention of having an application file, a library file and kernel file. Deviating from this format or mixing kernel code in with library code will cause significant problems for KFusion. There is an OpenCL style which stores the kernel files as raw strings embedded within library source code. This is not supported currently and most likely will not be, as the two formats should be interchangeable and support for embedded kernels will not impact the findings of this work.

When we take a step back and look at the kernel fusion technique, we notice some general limitations. We cannot perform kernel fusion on instructions within loops and is limited to more straightforward transformations. This is because a loop will inherently create a series of dependencies on the previous operation which cannot be passed forward in private memory. This means that while KFusion can effectively accomplish loop fusion by combining kernels, it cannot accomplish fusion of loops within the kernels itself.

There are also limitations with the GPU's local memory. Local memory is much faster than global, but limited in size. Each work group only has access to a specific amount of local memory. Combining several kernels which all use an independent space in local memory may cause requested memory to exceed the available amount.

## 4. CASE STUDIES

This section describes KFusion in the context of applications for linear algebra, image manipulation and physics simulation. Image manipulation provides the best performance case and shows and how to improve operations using KFusion which are driven by data flow from the application's perspective. Linear algebra provides another good performance case, but also highlights how correctness has been maintained. The final case is a pool game which uses OpenCL to handle all the game play mechanics and physics. While somewhat simple, this shows how KFusion can operate at the heart of a larger system with many components and still achieve performance improvements.

### 4.1 Linear Algebra

Building an OpenCL BLAS library allows for the common operations and data types present in linear algebra to be abstracted so the programmer does not have to deal with the internal details of every operation. It also allows for the OpenCL details to be abstracted away making it easy for the application programmer to combine operations.

We consider three examples in increasing complexity and memory access requirements. The first is a simple equation involving a few vectors, the second involves a series of vectors and the third involves a matrix multiplication and a set of vectors.

**Application 1: Simple Equation**
The first example uses a minimum number of equations to show how fusion is possible within the realm of linear algebra. The algorithm is shown in Algorithm 3. The microbenchmark solves the Pythagorean Theorem for thousands of values in parallel.

While there are only a few operations present here, it is possible to see how they chain together and how fusing the

---

**Algorithm 3** Operations present in c = sqrt($a^2 + b^2$) where $a$ $b$ and $c$ are vectors of length $n$

1: vector_mult(a2,a,a) - $a2 = a * a$
2: vector_mult(b2,b,b) - $b2 = b * b$
3: vector_add(c2 ,1,a2,1,b2) - $c2 = a2 + b2$
4: vector_sqrt(c,c2) - $c = sqrt(c2)$

---

kernels might improve performance. Each element of the vectors will be operated on independently in parallel.

**Application 2: Complex Equations: Distance Calculations**
The next test uses a larger problem—calculating in the distance between a series of $x, y$ coordinates. The algorithm itself can be seen in Algorithm 4. This has significantly more data, intermediate results and reduces to a set of final outcomes. There are still a small number of dependencies between operations and each element of the result vector can be computed in parallel.

---

**Algorithm 4** Operations present in the distance calculations between sets of points $x1$ $y1$ and $x2$ $y2$

1: vector_add(dx,1,x1,-1,x2)
2: vector_add(dy,1,y1,-1,y2)
3: vector_mult(dx2,dx,dx)
4: vector_mult(dy2,dy,dy)
5: vector_add(c2,1,dx2,1,dy2)
6: vector_sqrt(c,c2);

---

In this case there are a series of intermediate products and a larger set of operations accessing more memory. This provides a more complicated but well suited problem for fusion.

**Application 3: Matrix Equations: Initial Steps of Conjugate Gradient Solver**
The final example is the initial few operations of the conjugate gradient algorithm. The other major components of the conjugate gradient algorithm can also be fused. This subset provides a representative example of several operations, as shown in Algorithm 5.

---

**Algorithm 5** Initial set of operations for conjugate gradient algorithm

1: matrix_vector_mult(b,A,x);
2: vector_add(r,1,b2,-1,b);
3: vector_add(p,1,r,0,r);
4: vector_mult(r,1,p,1,p);

---

In this case, there is a matrix vector multiplication and the memory I/O for this operation is difficult to mitigate, but KFusion can fuse the following operations resulting in one large mono-kernel. This is effectively the worst case for fusion as the majority of I/O is not mitigated. Likewise, because a matrix vector multiplication requires some synchronization, it is always going to be difficult to mitigate the I/O present. Finally, because most of the memory access associated with the matrix occurs within a loop, it becomes impossible to mitigate the I/O.

### 4.2 Image Manipulation

OpenCL libraries, such as ViennaCL [28], are bringing GPU performance to various domains, without requiring domain experts to become OpenCL experts. This case study considers four different application scenarios, which differ according to their inherent dependencies and data flow.

**Application 1: Low Data Dependency**

The best case scenario in terms of possible gains is one that has few dependencies other than the preceding operations. The operations are all single pixel operations.

**Application 2: Medium Data Dependency**

The average case scenario is one that involves more dependencies, but still allows for reordering to ensure efficient combination of operations. This case involves a convolution as well as a binary operation between two images. This represents a problem with dependencies and it becomes harder to effectively combine operations in order to improve performance. It is still possible to combine them, but there will be less of an improvement. Primarily because even if kernels are combined, there will still be a significant amount of global load operations.

**Application 3: High Dependency**

The high dependency case involves two convolutions and very little overlap in computation leading to a smaller amount of fusion. This provides a case where fusion should not make a significant performance improvement as there are a large number of operations which do not lend themselves to deforestation. In this case there are not a large number of operations to chain together. There are two disjoint operations which are brought together to create a single output value. Nonetheless KFusion can produce a reasonable fusion scenario which may have a slight improvement over the original.

**Application 4: High Dependency-Low Fusion**

In order to complete the spectrum, the highest dependency case represents a high amount of dependencies leading to very little fusion opportunity. It is a bare-bones case where overlap is minimum, but the kernel needs to access a large number of values.

## 4.3 Physics Application

The physics simulation is a working game of pool—simplified in order to reduce development time. It deals primarily with rigid body physics and collisions in a 2D space. Whereas the previous two case studies involve specialized code for solving specific problems, this case study was designed to show a more general case where a small section of a larger program is sped up using OpenCL and Kernel Fusion.

When developing the application it is possible to order the operations so they are even more amenable to optimization. One such ordering can be seen in algorithm 6. This is because *moveBalls* requires synchronization after execution and the collision functions will work best if there is a synchronization just before all of them. When applying fusion, it makes sense to ignore impulse as it is only executed once. As a result, fusing the contents of the while loop provides the best potential benefit.

---

**Algorithm 6** Initial set of operations for conjugate gradient algorithm

---

1: impulse(balls,clue)
2: **while** balls are moving **do**
3:     collideBalls(balls)
4:     collidePockets(balls, table)
5:     collideWalls(balls, table)
6:     moveBalls(balls)
7: **end while**

---

```
1  kfuse(image)
   {
3      resize(image, image->width*1.1,
          image->height*1.1);
5      toHSV(image);
       colorize(image,color);
7      toRGB(image);
       RGBinvert(image);
9  }
```

**Listing 6: The use of kfuse in our first image manipulation application scenario, LDD (best case).**

```
1  kfuse(image1, image2)
   {
3      convolve(image1,blur);
       convolve(image2,sharpen);
5      binOp(image1,image2,SUB);
   }
```

**Listing 7: The use of kfuse in our fourth image manipulation application scenario, HDDLF (worst case).**

## 5. RESULTS

Costs and benefits of KFusion from the perspective of the application programmer can be framed in terms of programming efforts and performance rewards. We first consider these in terms of lines of code and annotations involved, followed by a performance assessment.

## 5.1 Lines of Code and Annotations

KFusion achieves application-specific performance benefits but eliminates the need to write a substantial amount of code. Table 4 demonstrates the magnitude of the lines of code automatically generated in terms of the case studies presented here, with up to a 75% increase. Writing this code would require the application developer to break the API and create new kernels and library functions. This goes against basic principles of modularity such as information hiding.

In addition to this extra code inside the kernels and libraries, KFusion automatically generates the configuration code required for each new kernel including: 1) code to compile the kernel at run time, 2) code to assign arguments and execute the kernel and 3) code to transfer data to and from the GPU.

The automatic code generation of the KFusion process allows the application developer to be completely insulated from the generated code. This eases evolution at the application level, allowing developers to make changes to their applications and regenerate the fused code as opposed to evolving both the application and the manually written kernel and library code.

Examples of application level annotations required in our case studies are shown in Listings 6 and 7. The *kfuse* mechanism demonstrated here, wraps a set of library function calls that are related by data flow. This approach does require the application level programmer to have some knowledge of the underlying library as only library functions that contain OpenCL kernels can be fused.

The annotations at the kernel level require knowledge of the underlying library and associated kernels but as de-

| Case Study | Library | Kernel | Fused Library | Fused Kernel | % Increase |
|---|---|---|---|---|---|
| Image Manip. | 815 | 322 | 1034 | 759 | 34.5% |
| Linear Algebra | 422 | 159 | 626 | 393 | 75.3% |
| Physics | 593 | 232 | 624 | 377 | 26.9% |

Table 4: Lines of code for fused and libraries and kernels.

| Case Study | Kernel & Library Annotations | Application Annotations |
|---|---|---|
| Image Manipulation | 20 | 4 |
| Linear Algebra | 12 | 3 |
| Physics | 8 | 1 |

Table 5: Lines of code for fused and libraries and kernels.

scribed above, we envision a library specialist would work at this level. For example, it is necessary to mark *load* and *store* operations as well a certain asynchronous instructions which should not be optimized by KFusion. Even these points are well defined and distinguishable by a library developer who would identify a load as any assignment from a global value and a store as any assignment to a global value.

Table 5 provides an overview of the number of annotations that were required for three of the case studies described in this paper.

## 5.2 Performance

In this subsection, we present the performance results for each case study. Each was executed 1000 times and the results where averaged. We used two sets of hardware. The first was an NVidia Tesla c2075 GPU with 448 cores and 6GB of memory [19]. The second was an Intel i7 2600k CPU with 4 cores and eight hardware threads [12]. This provides two major points on the spectrum. When compiling, the device code us given to a just in time OpenCL compiler which optimizes for the specific hardware.

The results of the linear algebra case study can be seen in Figure 6 and Figure 7. The image manipulation results are represented in Figures 8 and 9. Figure 10 shows an example of how KFusion compares to hand fused code. The GPU results are followed by the CPU results. The physics are shown in Figure 11. Lastly, Tables 6 and 7 show the difference in performance between CPU execution and fused GPU execution for the linear algebra and image manipulation case studies respectively.

Each case study has very similar performance characteristics. In applications with low dependencies it was possible to fuse kernels and achieve a significant performance increase. As operations had more and more dependencies, fusion produced worse results. KFusion works best when it can mitigate as much I/O as possible while loading few values. This reflects the hardware's capabilities. In the best case, the low data dependency example in the image manipulation case study, KFusion produces approximately a 4 times speedup. In the worst cases, KFusion does not degrade performance. This occurs when there is a large number of I/O operations which KFusion cannot mitigate—such as in a matrix vector multiplication.

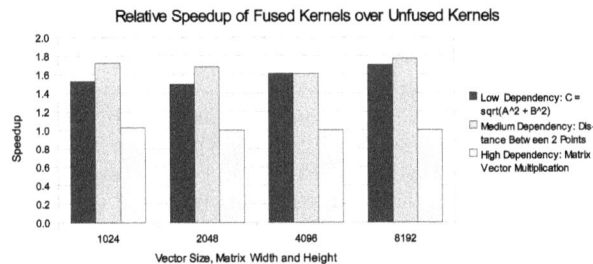

Figure 6: GPU Results—Linear Algebra: The relative speedups of fusing in each application scenario. We notice that lower dependencies provide significantly higher performance boosts. In the final case, KFusion cannot produce a measurable performance increase, but does not hinder performance.

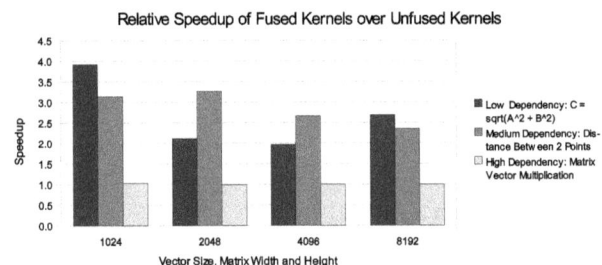

Figure 7: CPU Results—Linear Algebra: The relative speedups of fusing in each application scenario.

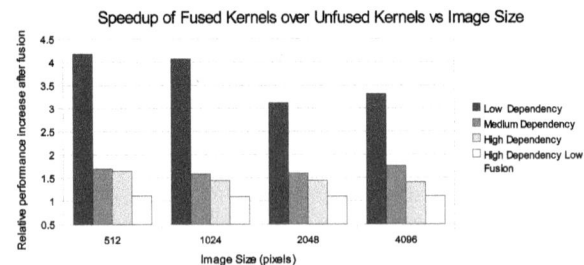

Figure 8: GPU Results—Image Manipulation: The relative speedups of fusing in each application scenario. As with linear algebra, we see that the low dependency cases produce the largest performance increases. This is due to the fact a large amount of redundant I/O an be eliminated.

34

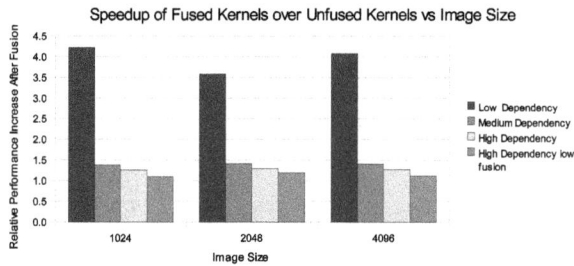

Figure 9: CPU Results—Image Manipulation: The relative speedups of fusing in each application scenario.

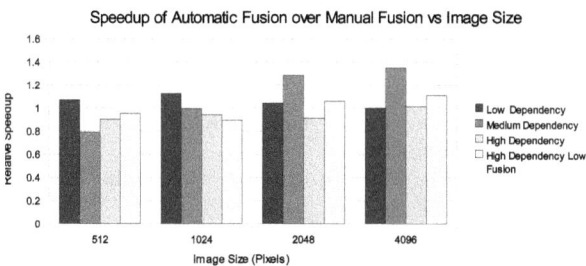

Figure 10: GPU Results—Image Manipulation: The relative speedup of KFusion over manually fused code. They key here is that KFusion is approximately as good, and sometimes better than, hand fused code.

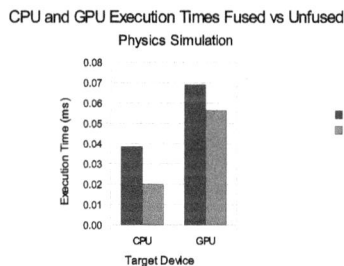

Figure 11: OpenCL Physics: The performance of Fused vs Unfused for both CPU and GPU. We can see that KFusion can significantly increase the performance of application with regards to both CPU and GPU performance.

|  | Execution Time (ms) | |
| Case | Unfused CPU | Fused GPU |
| --- | --- | --- |
| Low Dependency | 0.12 | 0.04 |
| Medium Dependency | 0.13 | 0.03 |
| High Dependency | 102.35 | 3.52 |

Table 6: Linear Algebra Results: A comparison between the CPU execution vs the fused GPU execution time with a 4096 element vectors and 4096x4096 matrices. In the best case, moving computation to the GPU improves performance by two orders of magnitude.

|  | Execution Time (ms) | |
| Case | Unfused CPU | Fused GPU |
| --- | --- | --- |
| Low Dependency | 3155 | 29 |
| Medium Dependency | 3236 | 65 |
| High Dependency | 4873 | 202 |
| High Dependency Low Fusion | 4241 | 213 |

Table 7: Image Manipulation Results: A comparison between the CPU execution vs the fused GPU execution time with a 4096x4096 image. In the best case performance increases by over two orders of magnitude.

## 6. CONCLUSION AND FUTURE WORK

The KFusion prototype takes preliminary steps to explore how simple annotations can result in optimizations based on application-specific data flow. Relative to previous approaches, KFusion allows for improved decoupling of application and library level concerns. Application developers merely provide hints for possible data flow optimizations, from which libraries annotated by domain experts can be customized. KFusion allows a developer to use a modular library and obtain monolithic performance without having to understand the internals of the library. Having established the initial costs and benefits of this approach, much future work remains.

First, we plan to determine specific ways in which our approach is complementary to other code generators and optimizing compilers, further improving the state of the art. For example, combining this work with an optimizing source-to-source compiler that operates at an intra-kernel level. As mentioned previously a process involving (1) optimization, (2) fusion and (3) re-optimization is most likely ideal. Second, we plan to explore fusing kernels to allow for concurrent execution of multiple kernels even if they do not share data. This should solve another inherently tangled concern on GPUs: under-utilization of available processors. This also opens the idea of recursively fusing kernels both vertically and horizontally. Most likely this could only go so far, as OpenCL has no global synchronization mechanisms within kernels. Finally, associated with these avenues of future work would be the development of tool support. Augmenting existing IDEs to help developers navigate and trace from high to low-level, and most effectively establish the right balance of optimizations in their code.

## 7. REFERENCES

[1] D. F. Bacon, S. L. Graham, Oliver, and J. Sharp. Compiler Transformations for High-Performance Computing. *ACM Computing Surveys*, 26:345–420, 1994.

[2] M. Baskaran, J. Ramanujam, and P. Sadayappan. Automatic C-to-CUDA Code Generation for Affine Programs. In R. Gupta, editor, *Compiler Construction*, volume 6011 of *Lecture Notes in Computer Science*, chapter 14, pages 244–263. Springer Berlin / Heidelberg, Berlin, Heidelberg, 2010.

[3] M. G. Burke and R. K. Cytron. Interprocedural Dependence Analysis and Parallelization. *SIGPLAN Not.*, 39(4):139–154, Apr. 2004.

[4] H. Chafi, A. K. Sujeeth, K. J. Brown, H. Lee, A. R.

Atreya, and K. Olukotun. A Domain-specific Approach to Heterogeneous Parallelism. In *Proceedings of the 16th ACM symposium on Principles and Practice of Parallel Programming*, PPoPP '11, pages 35–46, New York, NY, USA, 2011. ACM.

[5] L. Correnson, E. Duris, D. Parigot, and G. Roussel. Declarative Program Transformation: a Deforestation Case-study, 1999.

[6] L. Dagum and R. Menon. OpenMP: An Industry-Standard API for Shared-Memory Programming. *IEEE Comput. Sci. Eng.*, 5:46–55, January 1998.

[7] O.-J. Dahl and K. Nygaard. SIMULA. In *Encyclopedia of Computer Science*, pages 1576–1578. John Wiley and Sons Ltd., Chichester, UK, 2003.

[8] U. Dastgeer, J. Enmyren, and C. W. Kessler. Auto-tuning SkePU: a Multi-backend Skeleton Programming Framework for Multi-GPU Systems. In *Proceedings of the 4th International Workshop on Multicore Software Engineering*, IWMSE '11, pages 25–32, New York, NY, USA, 2011. ACM.

[9] E. Dijkstra. *A Discipline Of Programming.* Prentice-Hall series in automatic computation. Prentice-Hall, 1976.

[10] E. W. Dijkstra. The structure of the "THE" multiprogramming system. In *Proceedings of the first ACM symposium on Operating System Principles*, SOSP '67, pages 10.1–10.6, New York, NY, USA, 1967. ACM.

[11] M. Frigo, Steven, and G. Johnson. The Design and Implementation of FFTW3. In *Proceedings of the IEEE*, volume 93, pages 216–231, 2005.

[12] Intel. Intel Core i7-2600K Processor (8M Cache, 3.40 GHz). http://ark.intel.com/products/52214/Intel-Core-i7-2600K-Processor-%288M-Cache-3_40-GHz%29, 2011. date accessed: July 2011.

[13] K. Kennedy and K. McKinley. Maximizing Loop Parallelism and Improving Data Locality via Loop Fusion and Distribution. In U. Banerjee, D. Gelernter, A. Nicolau, and D. Padua, editors, *Languages and Compilers for Parallel Computing*, volume 768 of *Lecture Notes in Computer Science*, pages 301–320. Springer Berlin / Heidelberg, 1994. 10.1007/3-540-57659-2_18.

[14] Khronos. OpenCL. http://www.khronos.org/opencl/, 2011.

[15] G. Kiczales, J. Lamping, A. Mendhekar, C. Maeda, C. V. Lopes, J.-M. Loingtier, and J. Irwin. Aspect-Oriented Programming. In *ECOOP*, pages 220–242, 1997.

[16] L. Lamport. The Parallel Execution of DO Loops. *Commun. ACM*, 17(2):83–93, Feb. 1974.

[17] S. Lee, S.-J. Min, and R. Eigenmann. OpenMP to GPGPU: a Compiler Framework for Automatic Translation and Optimization. In *Proceedings of the 14th ACM SIGPLAN symposium on Principles and practice of parallel programming*, PPoPP '09, pages 101–110, New York, NY, USA, 2009. ACM.

[18] Leung, Alan and Lhoták, Ondřej and Lashari, Ghulam. Automatic parallelization for Graphics Processing Units. In *Proceedings of the 7th International Conference on Principles and Practice of Programming in Java*, PPPJ '09, pages 91–100, New York, NY, USA, 2009. ACM.

[19] E. Lindholm, J. Nickolls, S. Oberman, and J. Montrym. NVIDIA Tesla: A Unified Graphics and Computing Architecture. *IEEE Micro*, 28(2):39–55, Mar. 2008.

[20] B. Liskov. In *The Second ACM SIGPLAN Conference on History of Programming Languages April 20 – 23, 1993, Cambridge, United States*, New York.

[21] C. Nugteren and H. Corporaal. Introducing 'Bones': A Parallelizing Source-to-source Compiler Based on Algorithmic Skeletons. In *Proceedings of the 5th Annual Workshop on General Purpose Processing with Graphics Processing Units*, GPGPU-5, pages 1–10, New York, NY, USA, 2012. ACM.

[22] NVIDIA. *NVIDIA CUDA Programming Guide 2.0.* 2008.

[23] NVIDIA. CUDA Zone. www.nvidia.com/object/cuda_home.html, 2010. date accessed: March 2009.

[24] M. Odersky and al. An Overview of the Scala Programming Language. Technical Report IC/2004/64, EPFL Lausanne, Switzerland, 2004.

[25] D. L. Parnas. On the Criteria To Be Used in Decomposing Systems into Modules. *Communications of the ACM*, 15:1053–1058, 1972.

[26] M. Püschel, J. M. F. Moura, B. Singer, J. Xiong, J. Johnson, D. Padua, M. Veloso, and R. W. Johnson. Spiral: A Generator for Platform-Adapted Libraries of Signal Processing Algorithms. *Int. J. High Perform. Comput. Appl.*, 18.

[27] A. Resios. GPU Performance Prediction using Parametrized Models. Master's thesis, Utrecht University, The Netherlands, 2011.

[28] K. Rupp, J. Weinbub, and F. Rudolf. Automatic Performance Optimization in ViennaCL for GPUs. In *Proceedings of the 9th Workshop on Parallel/High-Performance Object-Oriented Scientific Computing*, POOSC '10, pages 6:1–6:6, New York, NY, USA, 2010. ACM.

[29] E.-M. Sha, C. Lang, and N. Passos. Polynomial-Time Nested Loop Fusion with Full Parallelism. In *Parallel Processing, 1996. Vol.3. Software., Proceedings of the 1996 International Conference on*, volume 3, pages 9 –16 vol.3, aug 1996.

[30] H. Sutter. The free lunch is over: A fundamental turn toward concurrency in software. *Dr. Dobb's Journal*, 30(3):202–210, 2005.

[31] P. Wadler. Deforestation: Transforming Programs to Eliminate Trees. In *Proceedings of the 2nd European Symposium on Programming*, ESOP '88, pages 344–358, London, UK, UK, 1988. Springer-Verlag.

[32] Y. Yang, P. Xiang, J. Kong, and H. Zhou. A GPGPU compiler for Memory Optimization and Parallelism Management. In *Proceedings of the 2010 ACM SIGPLAN conference on Programming language design and implementation*, PLDI '10, pages 86–97, New York, NY, USA, 2010. ACM.

# Reactive Behavior in Object-oriented Applications: An Analysis and a Research Roadmap*

Guido Salvaneschi, Mira Mezini
Software Technology Group
Technische Universität Darmstadt
{salvaneschi, mezini}@informatik.tu-darmstadt.de

## ABSTRACT

Reactive applications are difficult to implement. Traditional solutions based on event systems and the Observer pattern have a number of inconveniences, but programmers bear them in return for the benefits of OO design. On the other hand, reactive approaches based on automatic updates of dependencies – like functional reactive programming and dataflow languages – provide undoubted advantages but do not fit well with mutable objects.

In this paper, we provide a research roadmap to overcome the limitations of the current approaches and to support reactive applications in the OO setting. To establish a solid background for our investigation, we propose a conceptual framework to model the design space of reactive applications and we study the flaws of the existing solutions. Then we highlight how reactive languages have the potential to address those issues and we formulate our research plan.

## Categories and Subject Descriptors

D.1 [**Software**]: Programming Techniques—*Object-oriented Programming*; D.3.3 [**Programming Languages**]: Language Constructs and Features

## Keywords

Reactive Programming; Functional-reactive Programming; Object-oriented Programming; Incremental Computation

## 1. INTRODUCTION

Most contemporary software systems are reactive: Graphical user interfaces need to respond to the commands of the user, embedded software needs to react to the signals of the hardware and control it, and a distributed system needs to react to the requests coming over the network. While a simple batch application just needs to describe the algorithm for computing outputs from inputs, a reactive system must also react to the changes of the inputs and update the outputs correspondingly. Moreover, there are more tight

---

*This work is supported in part by the European Research Council, grant No. 321217

constraints on computation time, because reactive systems work in real-time and need to react quickly – within seconds or even milliseconds. When the reactive behavior involves non-trivial computations or large amounts of data, various optimization strategies, such as caching and incremental updating, need to be employed.

Object-oriented programming does not provide specific mechanisms for implementing reactive behavior, with two consequences. First, reactive behavior is usually encoded by using the Observer design pattern, whose drawbacks have been extensively highlighted in literature [10, 31, 30]. For example, the code responsible for update of outputs is usually tangled with the code changing the inputs. As a result, it becomes difficult to understand the computational relations between inputs and outputs and, thus, the intended behavior of the system. Second, the update functionality with the necessary strategies to achieve the desired performances must be implemented manually for each application. Such optimizations, however, introduce a lot of additional complexity, so that it becomes an act of balance between complexity and efficiency.

Various approaches aim to address different aspects of these issues. Event-driven programming (EDP) creates inversion of control to enable modularization of the update code [38, 22, 18]. Aspect-oriented programming (AOP) enables complete separation of the update concern, by specifying in the aspects the points where the update needs to be triggered [24, 42, 37, 5]. The above approaches fit well with mutable objects, but retain some of the problems related to a programming style based on inversion of control, similar to the well-discussed problems of the Observer design pattern.

Declarative reactive approaches, most notably functional-reactive programming (FRP) [17] and reactive languages, like FrTime [10], Flapjax [31] and Scala.React [30], completely automate the update process. The developer specifies only how a changing value is computed from other values, and the framework ensures that the computed value is automatically updated whenever the inputs are changed. It is, however, not clear whether they can obsolete manual implementation of update code. FRP and reactive languages deal with update of primitive values, and may be too inefficient since they do not provide incremental update of complex structures. Incremental update is provided by other approaches such as LiveLinq [29], but they are limited to specific data structures. Also, declarative approaches based on the functional paradigm do not offer the advantages of typical object-oriented designs, including modularization and component reuse.

In summary, the current state of the affairs is rather disappointing: Developers implement reactive applications in the *comfortable* world of objects, at the cost of relying on programming models whose limitations have been known for a long time. On the other hand, alternatives based on reactive programming offer an appealing solution, but do not succeed because they do not provide the necessary flexibility and do not integrate with the OO design.

In this paper, we propose a research roadmap to fill the gap between OO design and reactive approaches. Our vision is that the concepts developed by FRP and dataflow programming can be integrated with object-orientation to provide dedicate support for reactive applications in mainstream languages. This goal is challenging because reactive abstractions have been explored mainly in the functional setting or in special domains, like reactive data structures. However, the analysis presented in this paper provides a solid background and the first steps in our research plans are already ongoing. Our initial effort is the integration of events and behaviors *a la* FRP into our prototype language RESCALA [40], and results are promising. In summary, we provide the following contributions:

- We characterize the design space of reactive applications and discuss the strategies that can be applied to implement reactivity.

- To understand the practical impact of each update strategy, we analyze the implementation of reactive behavior in several real-world OO applications. Our analysis highlights the drawbacks of traditional abstractions.

- We analyze the existing language solutions for reactive systems. We underline their limitations, and the key achievements to take into account in further research.

- We propose a research roadmap which addresses the issues found in the current approaches and has the ultimate goal of combining objects and reactive abstractions in a flexible and efficient language.

The paper is structured as follows. In Section 2 we analyze the design space of reactive software. In Section 3 we present an empirical evaluation of real-world OO reactive applications. Section 4 outlines a possible alternative and discusses other research solutions. Section 5 presents our research roadmap.

## 2. DESIGN SPACE IN OO LANGUAGES

In reactive systems, the outputs of the program need to be updated based on changes of inputs and time. The ways of achieving this goal are however very diverse. In this section, we overview the possible update strategies and discuss the rationale of choosing them.

To make the discussion more clear, in Figure 1 we show a model of a reactive software: The outputs provided to the clients (the objects $a$ and $e$) must reflect the current state of the inputs of the application (objects $b$, $c$, and $d$) according to certain transformations $f$ and $g$. Objects can be composed: For example, the object $a$ contains the references $a_1$ and $a_2$ to other objects (e.g. by storing them in fields). Dashed arrows model dependencies: Output objects are computed from certain input objects. To clarify how the model applies, consider a weighted graph used to compute a *derived* graph. The derived graph is composed of references to the edges exceeding a MIN weight value. In the model of Figure 1 the basic graph can be represented by the $b$ object that contains the references $b_1$, $b_2$ and $b_3$ to the edges. The derived graph can be represented by $a$ and contains the references $a_1$ and $a_2$ to the edges. Finally, $f$ is the transformation that produces the derived graph $a$ from the basic graph $b$ by filtering the edges according to the MIN value, modeled by the $c$ object. For each update strategy, we show in a Java-like language how the dependent graph is obtained.

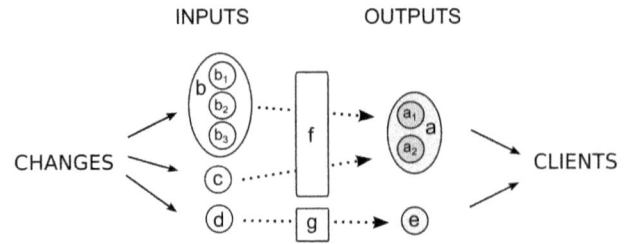

Figure 1: A model of reactive behavior among objects.

### 2.1 On-demand Recomputation

The most straightforward approach is to recompute the output values each time they are needed. For example, when they are requested by the user to generate a report, or when she refreshes a view. Similarly, in real-time computer games and simulation applications, it is common to recompute outputs automatically at certain intervals of time or simply as often as possible. In OO design, the typical example of this approach are methods returning values computed from the state of the object each time these methods are called.

The distinguishing aspect of on-demand recomputation is that, after the evaluation, the output is discarded. For example, in Figure 1, every time a client requests $a$, $a$ is recomputed and $b$ is evaluated to calculate $a$. Figure 2(a) shows the on-demand recomputation strategy applied to the graph example: Every time the getDerivedGraph method is called, the dependent graph is computed from scratch by filtering all the edges (Line 6) and it is returned to the client.

The advantage of this approach is that it is simple to carry out, because the developer just needs to implement procedures computing the outputs from inputs. It also guarantees that the values are always computed from the current state of the program, and thus are always consistent with their inputs. The approach is also memory efficient, because only the inputs need to be stored, but not the outputs or any intermediate computation.

### 2.2 Caching

Recomputing outputs every time they are requested may be too inefficient, especially in the cases when the computations are expensive or need to deal with large amounts of data. Caching a computed result is a general optimization strategy that avoids repeating the computation. In the model of Figure 1, caching is obtained by saving $a$ and $e$, letting them available for more than one client access. A typical design is to introduce a field for storing the computation results. The method that computes the dependent value is modified to return the value of the field if it is valid, and to compute and save the result otherwise. Figure 2(b) shows an implementation of the caching strategy: The derivedGraph is maintained in a field (Line 3) and returned only if valid, otherwise, the dependent graph is recomputed. When an edge is added to the base graph, the derived graph is invalidated (Line 17).

The cached values are valid only as long as the inputs of the computation do not change. When the inputs change, the cached value must be either recomputed, or invalidated and recomputed at the next request. The latter approach is more efficient when the computed value is used not so frequently, but it is also slightly more complicated. A major issue is to detect changes of the inputs and decide which cached values need to be invalidated. A straightforward approach is to invalidate all cached values after the change of every input. An efficient solution, however, is to analyze the actual

```
1  class Graph {
2    Edge [] edges;
3
4    getDerivedGraph(){
5      Graph g = new Graph();
6      for (Edge edge : edges){
7        if (edge.weight > MIN)
8          g.add(edge);
9      }
10     return g;
11   }
12   ...
13 }
14
15
16
17
18
19
```

```
1  class Graph {
2    Edge [] edges;
3    Graph derivedGraph;
4    boolean valid;
5
6    getDerivedGraph(){
7      if (!valid){
8        derivedGraph=new Graph();
9        for (Edge edge : edges){
10         if (edge.weight > MIN)
11           derivedGraph.add(edge);
12   } }
13     return derivedGraph;
14   }
15   addEdge(Edge e){
16     edges.add(e);
17     valid = false;
18   } ...
19 }
```

(a)                              (b)

**Figure 2: On-demand recomputation (a) and caching with in-validation (b).**

```
1  class Graph {
2    Edge [] edges;
3    Graph derivedGraph;
4
5    getDerivedGraph(){
6      return derivedGraph;
7    }
8    addEdge(Edge e){
9      edges.add(e);
10     if (e.weight > MIN)
11       derivedGraph.add(edge);
12   } ...
13 }
```

```
1  class Graph {
2    List<Edge> orderedEdges;
3
4    getDerivedGraph(){
5      derivedGraph = new Graph();
6      for(Edge edge:orderedEdges){
7        if(edge.weight > MIN)
8          derivedGraph.add(edge);
9        else break;
10     }
11     return derivedGraph;
12   } ...
13 }
```

(a)                              (b)

**Figure 3: Tracking dependencies with caching (a) and tracking with on-demand recomputation (b).**

dependencies between inputs and outputs, and, after a change to an input, update only the outputs that depend on it – as we explain hereafter.

## 2.3 Tracking Dependencies

Instead of updating all outputs after a change to an input, the programmer can rather update only the outputs that actually depend on the changed input. For example, in Figure 1, a change in $c$ requires an update of $a$, but $e$ is still valid and should not be recomputed. A finer-grained tracking of dependencies can take into account that $a$ only depends on the elements among $b_1$, $b_2$ and $b_3$ that exceed MIN. Figure 3(a) shows an implementation of dependency tracking with caching: The dependent graph is maintained in a field, and it is updated only when one of the edges with weight greater than MIN is added, i.e. the logic keeps track of the edges on which the derived graph depends. Figure 3(b) shows an implementation of dependency tracking with on-demand recomputation. In this case, the dependent graph is recomputed on every client request. The knowledge about dependencies is maintained by keeping edges in an ordered list (Line 2). In this way, the computation of the dependent graph can be performed by evaluating only a subset of the edges of the base graph. The evaluation is interrupted when an edge not part of the dependencies is encountered (Line 9).

Although tracking dependencies may seem straightforward, implementing this strategy in practice is usually not easy. The programmer needs a precise knowledge of computational relations between outputs and inputs. The dataflow of an application is usually not explicit in imperative code, and a careful code analysis is required to reconstruct it. Moreover, the actual dataflow of an application may depend on dynamic conditions (e.g. dynamic type of a variable in case of subtype polymorphism) and thus may be statically not determinable. Developers must implement the update functionality that corresponds to the detected computational dependencies: After a change of each different input, the update of the corresponding outputs must be called. This may introduce a substantial amount of additional code. The update functionality may also cause modularity problems, because, when implemented in a straightforward way, it may introduce undesired dependencies from inputs to outputs. To avoid such dependencies, the programmer may employ various callback mechanisms (e.g. the Observer

pattern), but this further increases the complexity of the implementation.

## 2.4 Update Incrementalization

Completely recomputing a cached value each time it is invalidated may be too expensive, especially if this value is a complex data structure, such as an array or a graph. A common optimization, in that case, is to update the cached value incrementally depending on the changes to the input. In the model of Figure 1, update incrementalization is an optimization of the functions $f$ and $g$. This kind of optimization applies in presence of caching: to make updates incremental, the entity to update must be available to receive the changes. Figure 4(a) shows an example of update incrementalization: When a change occurs, the derived graph is not recomputed from scratch but it is modified gradually by adding only the edges that satisfy the condition (Line 13). Note that there is no dependency tracking like in Figure 3(a), i.e. the dependent graph is *always* notified of the change, regardless of the weight of the added edge (Figure 4(a), Line 7).

Incremental update requires more fine-grained analysis of the changes to the inputs. It is not sufficient to detect that a certain input has changed, but it is also necessary to get precise information about the change. In addition, the programmer must design algorithms to update the value incrementally after different kinds of changes to the inputs. For example, in case the derived graph is the Minimum Spanning Tree of the original graph, specific domain knowledge in graph theory is required to implement the update algorithm incrementally.

## 2.5 Accumulating Changes

Accumulating changes is an optimization of the computation of outputs from inputs (i.e. an optimization of $f$ and $g$ in the model of Figure 1). Changes are stored and applied to a cached output, so caching is subsumed by this strategy.

Accumulating changes also implies the incrementalization of the update. Indeed, incrementalization is required to combine the existing object with the incoming changes. Accumulation allows one to arbitrarily choose when to apply the stored changes. One extreme is every time a change occurs (no accumulation), the other is every time the client requests the output. If the update of a value is postponed until the client request, this strategy avoids redundant updates of rarely requested values. However, combining a lot of accumulated changes is more expensive and the response time increases. In some cases, like in databases, the update is postponed until the end of some logical transaction in the inputs. Figure 4(b) shows an example of change accumulation. The changes to the

39

```
1  class Graph {
2    Edge [] edges;
3    DerivedGraph derivedGraph;
4
5    addEdge(Edge e){
6      edges.add(e);
7      derivedGraph.add(e);
8    } ...
9  }
10 class DerivedGraph
11      extends Graph {
12   addEdge(Edge e){
13     if (e.weight > MIN)
14       this.add(e);
15   } ...
16 }
17
18
19
```

```
1  class Graph {
2    Edge [] edges;
3    Graph derivedGraph;
4    Changes [] changes;
5
6    getDerivedGraph(){
7      applyChanges();
8      return derivedGraph;
9    }
10   applyChanges(){
11     /* Update derivedGraph
12     based on changes.
13     Clean added and removed */
14   }
15   addEdge(Edge e){
16     edges.add(e);
17     changes.add(new Add(e));
18   } ...
19 }
```

(a)                              (b)

**Figure 4: Update incrementalization (a) and change accumulation (b).**

base graph are accumulated in the `changes` array (Line 4). When the client requests the derived graph, the changes are applied and the derived graph is returned (Lines 6–8).

Updating a value after accumulating changes is usually more complicated than updating a value after each primitive change, because it requires more sophisticated data structures to describe the accumulated changes and more sophisticated algorithms to implement the update. As a result, this strategy increases memory consumption. However, accumulating changes also offers opportunities for optimization. For example, some changes can cancel each other. Nevertheless, a complex a domain-specific logic is usually required to take advantage of such cases.

## 3. CASE STUDIES

To analyze the design issues of OO reactive software, we inspected four reactive Java applications. Our goal is not to develop a systematic empirical study on OO reactive software. Instead, we want to provide a solid background for our research by surveying concrete examples of how reactive features impact OO software design. Due to space reasons, we show only a summary of our analysis. The interested reader can find more details in a technical report [39].

The case studies are of different sizes and cover different kinds of software (two desktop applications, a mobile application, and a library) as well as a variety of external sources of reactive behavior, like network messages, data sampling, values from sensors and user input. Figure 5 summarizes the main metrics of each application.

The **SWT Text Editor** (the `StyledText` widget) implements a text editor in the popular SWT library used by the Eclipse IDE [15]. The application reacts to the insertion of characters and to formatting commands by the user.

The **FreeCol Game** [20] is an open-source turn-based strategy game. The AI of the game controls the opponent players, so the application reacts to the user and the AI. Updates concern the game model and the map in the GUI.

**Apache Jmeter** [26] supports the performance assessment of several server types (e.g. HTTP). The user specifies a test plan by adding graphical elements to a panel. First, the application must react to changes in the test plan. Additionally, the application is reactive to network events, as the results of the test are visualized in real-time.

The **AccelerometerPlay** Android application is one of the example applications provided by the Android platform [2]. It displays a set of particles rolling on the screen. The inclination of the device is detected by the accelerometer and the particles are updated accordingly.

### 3.1 Design Choices in the Case Studies

Different design choices concerning reactive behavior are motivated in the case studies by the design, size and kind of software. Our analysis clearly indicates that different applications are better "served" by different points in the design space depicted in Section 2.

The SWT text editor adopts caching and dependency tracking to achieve good performance. It is a typical example of an object with an internal state, which changes in the process of interacting with the user. The editor is implemented in the conventional event-driven style and it is highly optimized to ensure low reaction time. To this end, a lot of fields are used to cache intermediate values, and a complex logic takes care of updating values only when needed.

The FreeCol game lies at the opposite side of the design space and mostly adopts the on-demand recomputation strategy. Since the game behavior is inherently complex, the major issue is managing complexity to keep the development time and the stability of the game within reasonable limits. So, design decisions reducing or at least limiting complexity are favored. A substantial part of the code implements computations of values that are used in the user interface or for AI decision making. Almost all of these values are computed every time they are requested by the user or by the AI.

In the AccelerometerPlay application particle positions are recomputed on-demand every time the screen is refreshed, and incrementalization is applied to efficiently update the positions after conflict resolution. The logic is summarized in Figure 7: Position update comes first, then conflicts with the screen border and conflicts among particles are resolved. Finally, the particles are displayed. Since the application must react quickly, the update logic is based on imperative changes and an iterative algorithm is used for conflict resolution.

JMeter is optimized to make the GUI fast and reactive, changing it in response both to the user and the test events. Due to the different nature of these sources of change, the optimization strategies slightly differ. Caching is used for graphic widgets when the same graphic interface is associated to multiple elements in the test plan and can be reused when the user switches from one element to the other. Incrementalization is applied to optimize updates of graphs and statistics displaying the results of an ongoing test.

### 3.2 Problem Statement

Our analysis revealed several design issues. We argue that these problems – commented hereafter – are not due to bad choices by the programmers. Instead, as we explain in Section 3.3, they are the consequence of the design limitations and the trade-offs imposed by OO language abstractions.

#### 3.2.1 Code Complexity

Manually caching intermediate values requires an accurate logic that is responsible to perform the updates and maintain consistency. Performances increase, but the application becomes more complex. In the SWT text editor, the presence of a lot of fields in each class (70 mutable fields in the worst case) makes reasoning on the behavior of the application really hard, since computations depend on previous state. In addition, the update logic for enabling reactiv-

| Case study | LOC | Types | Cycl. Compl. | LOC/Method | Methods/Type | Fields/Type |
|---|---|---|---|---|---|---|
| SWT Text Editor | 9,227 | 48 | 4.77 | 17.39 | 10.64 | 4.75 |
| FreeCol Game | 170,597 | 1,175 | 2.60 | 10.67 | 5.77 | 2.11 |
| Apache JMeter | 90,704 | 1,081 | 1.84 | 8.39 | 7.19 | 2.13 |
| AccelerometerPlay | 460 | 4 | 2.00 | 10.73 | 4.00 | 8.00 |

**Figure 5: Main metrics for the case studies.**

| Time% | Call/Value | Avg.Change | Calls | Time |
|---|---|---|---|---|
| 99.57% | 5.1 | 0.2650 | 14,798 | 76,705 |
| 97.98% | 5.0 | 0.2672 | 14,732 | 75,821 |
| 92.13% | 1.4 | 3.8125 | 29 | $3.6 \cdot 10^6$ |
| 62.20% | 85.0 | 0.0009 | 8,507,562 | 83 |
| 40.35% | 176.5 | 0.0038 | 5,437,321 | 84 |

**Figure 6: Redundancy analysis of the FreeCol game.**

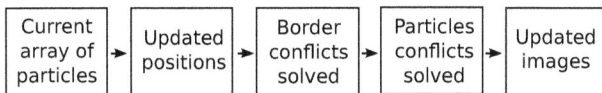

**Figure 7: The logic of the AccelerometerPlay application.**

ity pervades a considerable part of the application. For example, the `StyledText` class includes 11 "addListener" methods, 11 "removeListener" methods and 18 "handleEvent" methods. Moreover, the event-handling code includes anonymous classes created on the fly, which also expose callback methods. Finally, since values are separated from their update logic, local reasoning is impossible and understanding the application behavior requires inspecting a lot of code.

On the other hand, on-demand recomputation, like in the FreeCol game, clearly simplifies the logic of the application: As values are generated only when required, the behavior is not hidden by the code that maintains the dependencies.

### 3.2.2 Hidden Design Intent

The AccelerometerPlay application is an example of how reactive functionalities can hide the design intent of the developer. Although it is quite simple (less than 500 LOC), the reactive logic is spread all over the code and a conceptual model like the one in Figure 7 must be harvested from the system of callbacks and events. The origin of this complexity is that the design intent of each update strategy is not explicit in the implementation. For example, certain values are functionally dependent on other values but the design does not express this aspect. Only a careful analysis of the code reveals that a field is never changed directly, but updated after changes of other fields. To reconstruct the intended computational dependencies, the developer must analyze all the update code scattered across the application. Our analysis revealed that computational dependencies are quite common in complex applications. For example, we determined that in the `StyledText` class of the SWT text editor, about half of the 70 mutable fields are not freely changeable, but store values functionally dependent on other fields. Despite that, all fields are declared in the same way and identifying dependent fields requires to reverse engineer the logic of the application, which is lost in the callbacks.

### 3.2.3 Redundant Computations

The major advantage of on-demand computation is to keep the design simple. However, the overhead that is observed when this design choice is prevalent can be relevant.

We used a profiler and instrumentation via AspectJ to count the potentially redundant calls of the most time-consuming methods in the case studies. The computations after which the returned value does not change (for the same parameters) are potentially redundant. The impact of the design choices on redundancy can be seen in Figure 8. It compares the level of redundancy (calls per different observed values) in the 20 most expensive methods in the FreeCol game and in the SWT text editor, which lie at the opposite positions in the design space. The SWT text editor (right), thanks to its complex logic, shows values of redundancy which are substantially lower. We devoted further investigation to the potential optimizations in the FreeCol game, which largely adopts on-demand recomputation. For example, we discovered that the most expensive method (99.57% of the time) has 80% of potentially redundant calls, i.e. the computed value changes only in ~26% of the method calls (Figure 6). Further details are in [39].

### 3.2.4 Scattering and Tangling of Update Code

When values are intermediately cached, they must be updated in every point of the application where the inputs of the computation are changed. This leads to scattering of the update code.

To evaluate code scattering in the case studies, we considered the places where a field is directly written except the initialization. Since in many cases fields are not modified directly, but a setter method is used, we also included setter methods in the analysis. With the exception of the AccelerometerPlay application which is too small to suffer from scattering and tangling issues, we found that update scattering is extremely common. For example, in the FreeCol game, in JMeter and in the SWT text editor, respectively, only 38.4%, 46.0% and 30.7% of the fields is updated in just one place. In the worst case, a field was updated in 96 places!

### 3.2.5 Error-proneness and Code Repetitions

When the updates of the dependencies are managed manually, it is often hard for developers to understand when to trigger an update. For this reason, programmers code in a defensive way and introduce an update even when not necessary. For example, in the JMeter application we found cases in which selecting a GUI with the cursor triggers a sequence of four updates of the interface even before the user changes any value. However, update errors are not the only inconvenience that manual updates can cause. When update functionalities are complex, managing consistency by hand can easily lead to code repetitions because the same update pattern is cloned in many places. We evaluated the occurrence of code similarities in the case studies; the results are in Figure 9. The numbers in the table show that for applications of significant size, even when the update functionalities are carefully designed, like in the case of the SWT text editor, it is hard to keep the code clean.

Calls / Value

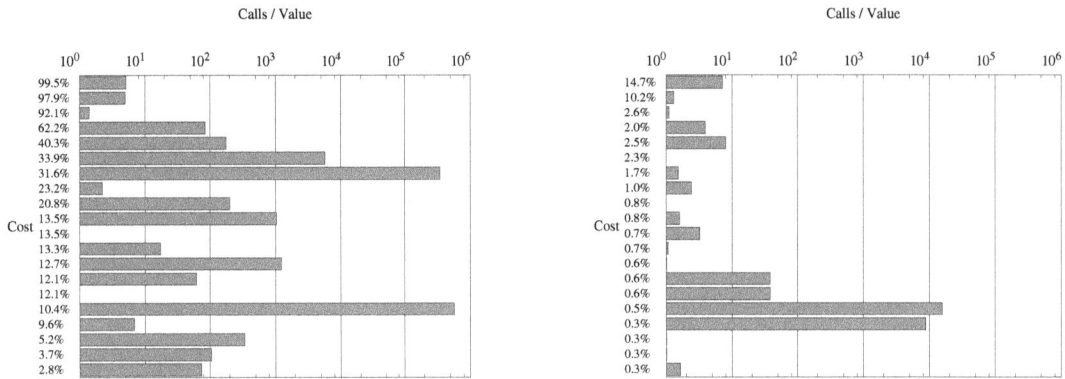

Figure 8: Redundancy of the most expensive methods in the FreeCol game and the SWT text editor.

| Application | Similarities | Similarities / LOC |
|---|---|---|
| SWT Text Editor | 29 | 0.00314 |
| FreeCol Game | 281 | 0.00106 |
| Apache Jmeter | 381 | 0.00420 |
| AccelerometerPlay | 0 | 0 |

Figure 9: Code repetitions.

## 3.3 Lesson Learned from the Case Studies

In this section, we summarize the major results from the case studies. A crucial observation is that, in OO applications, reactive entities are separated from the code responsible to keep them updated. This has two bad consequences. First, the dependencies are not explicit, so the design rationale of the application is hard to grasp even for trivial cases. Second, updates are scattered across the application and tangle the rest of the code.

Unfortunately, modularization of update code is hard to achieve in the OO style, because dependencies must be imperatively updated every time an input value is changed – which can occur in several places of the application. Furthermore, manually analyzing dependencies and writing corresponding update code is error-prone; certain dependencies may be overlooked and consequently the programmers can fail to update all functionally dependent values. Therefore, values are often updated defensively without precise knowledge of whether it is actually necessary.

Manually written update code also produces a maintenance problem, because there are no automatic checks ensuring the consistency of the update code with the actual dependencies of the computation. In addition, each time the computational dependencies are changed, the developer must correctly *update the update functionality* to reflect the current state of the dependencies! Errors of such a manual maintenance activity may remain undetected for a long time; forgetting to update a certain value usually does not lead to a crash, and a redundant update may even not cause any visible effects at all, only inefficiency.

Importantly, in our analysis, we observe a clear trade-off between efficiency and complexity. To keep the design simple, programmers accept the cost of on-demand recomputation and potential redundancy. For example, intermediate caching via object fields highly complicates the application because the update logic must be implemented manually. In conclusion, keeping the design simple has a high cost in performance. In some cases, like the FreeCol game, there is a wide space for potential optimization.

However, this is not easy to achieve, because the computations involve complicated algorithms and depend on various different inputs.

## 4. ANALYSIS OF ADVANCED LANGUAGES

In this section, we discuss some advanced language concepts to support reactive applications. For each approach, we analyze the problems it addresses and its limitations. We start the discussion with functional reactive approaches, as they provide interesting insights as how to overcome the problems discussed in the previous section and are the source of inspiration for our planned research, a roadmap of which is presented in the following section.

### 4.1 Functional Reactive Languages

Functional-reactive programming (FRP) was introduced by Elliott [17] to model time-changing values as dedicated language abstractions. Further approaches refined and extended the concept, mostly focusing on the formal semantics of continuous time [35]. More contemporary incarnations of the concepts are integrated in recent languages such as FrTime [10], Flapjax [31], and Scala.React [30]. To make the argumentation more concrete, in Figure 10, we show an example of Flapjax, a dataflow language that overcomes several limitations of reactive design based on inversion of control. Our considerations can be substantially generalized to FrTime and Scala.React. The functionality presented in Figure 10 consists of displaying the elapsed time since the user clicked on a button in a Web page.

Flapjax supports behaviors, i.e. reactive abstractions that model time-changing values (named with a final "B" in the code snippet). For example, the value `nowB` is a behavior that represents the current time updated every second. Behaviors can be sampled to obtain "traditional" values via the `valueNow` function: `startTm` is the initial instant of the simulation. In addition, behaviors can be combined with events: In Line 3, the `snapshot` function captures the instant value of the `nowB` behavior every time the `click` event of the `reset` button occurs. As a result, `clickTmsB` always contains the time of the previous click (or `startTm` before any click event). `elapsedsB` keeps the value of the time elapsed from the last click, and `insertValueB` updates the value in the graphic every time the `elapsedsB` value changes (Line 6). The crucial aspect of the reactive semantics is that a declaration like the one in Line 5 expresses a *constraint* rather than a *statement*. The example shows how the language creates implicit dependencies among time-changing values. The general idea is that when the programmer defines a constraint `a=f(b)` and `b` is a time-changing value, the framework automatically detects the dependency of `a` on `b` and is responsible for performing the updates automatically. In Line 5, when either `nowB` or `clickTmsB` changes, the value of `elapsedB` is automatically updated. So, whenever the programmer accesses `elapsedB`, she sees the updated value. Reactive languages provide abstractions to compose time-changing values and combine them with event streams. Eventually, time-changing values are bound to the GUI which automatically reflects the changes. The reader interested in more details can refer to [31].

Automatic dependency tracking addresses several issues highlighted in the case studies. The application is simplified, because the programmer does not shoulder the burden of keeping dependent values consistent (Section 3.2.1). As a consequence, the errors that can derive from forgetting the updates are automatically avoided (Section 3.2.5). The update code, which captures the behavior of a program entity, is modularized with the entity, allowing local reasoning and avoiding scattering and tangling with the rest of the application (Section 3.2.4). In contrast to callbacks, which return void, reactive behaviors can be easily composed. As a result, software is much more readable because the design intention of the programmer is explicit and direct modeling of relations among objects enforces a more declarative style (Section 3.2.2). Finally, reactive languages automatically derive dependencies and perform only necessary updates (Section 3.2.3).

In summary, reactive languages are an appealing solution to the issues identified for OO languages. In particular, since updates are performed by the runtime and do not add complexity to the application logic, they have the potential of solving the trade-off between efficiency and simplicity described in Section 3.3. However, there are some crucial issues that prevent their broader adoption.

*Functional flavor and immutability.*

Reactive languages impose a functional style, while OO programming features an imperative style. When a lot of code already exists, a functional refactoring of the entire application is in general not acceptable. Some computations are cumbersome to express functionally, while retaining acceptable performance and algorithmic clarity. For example, the conflict resolution algorithm in the AccelerometerPlay application is expressed in an imperative style using for loops, sequences of imperative statements to detect the conflicts, and imperative updates of the particles positions. The AccelerometerPlay application is also an example of performance-critical software. Conflicts among potentially hundred of particles must be solved in a sufficiently short time that the movement appears fluid to the user. Expressing the resolution algorithm in functional style with somewhat acceptable performance, would involve accumulative recursion: Functions in this style are rather hard to understand and, yet, probably not as efficient as encodings based on loops and imperative updates.

A consequence of the functional flavor of reactive languages is that they are effective with primitive values, but do not fit well with mutable objects. Strategies for incremental computation are highly application-specific and a framework can hardly address the problem in a domain-agnostic way. As a result, there is no way to automatically incrementalize object updates. So, reactive languages recompute the dependent object every time the base object changes. To clarify this point, consider the expression `list2 = list1.filter(x>10)`, an instance of the case `a=f(b)` – discussed previously – where `a` and `b` are `list2`, respectively `list1` and `f` is the `filter` function for the given predicate. The expression establishes a dependency between `list1` and `list2` via `filter`. In our case, each time object `list1` changes, the `filter` operator produces a new `list2` object. Instead, imperative approaches update the mutable dependent object in-place, which is more efficient and

```
1  var nowB = timerB(1000);
2  var startTm = nowB.valueNow();
3  var clickTmsB = $E("reset", "click").snapshotE(nowB)
4                  .startsWith(startTm);
5  var elapsedB = nowB − clickTmsB;
6  insertValueB(elapsedB, "curTime", "innerHTML");
7
8  <body onload="loader()">
9    <input id="reset" type="button" value="Reset"/>
10   <div id="curTime"> </div>
11 </body>
```

**Figure 10: Automatic dependency tracking in Flapjax.**

preserves the object identity. For example, the SWT text editor employs mutable data structures to efficiently store the inserted text.

These observations are symptoms of a more general problem in reactive languages: The update strategy is hardcoded in the reactive framework, so only a point in the design space described in Section 2 is available to the programmer. As a result, efficiency might be an issue even for trivial cases.

*Design of complex systems.*

So far, we defended the mutability of OO. Another, even more important reason, why we do not want to abandon the OO style is that it has established itself as the paradigm of choice for complex applications, for reasons related to design clarity and evolvability. A first remark is about modeling: Objects are effective in modeling complex systems because they reproduce the interaction of real-world entities. For example, the hundreds of simulation elements used by the FreeCol game are conveniently represented by objects.

Objects enable the development of large applications by modularizing rather large pieces of functionalities while abstracting over implementation details. For example, all the SWT widgets are ready-to-use components, but at the same time open for future modifications via subtype polymorphism: E.g., the SWT text editor uses a default implementation for the text container, but clients can provide a custom container by implementing a proper interface.

OO also supports reuse by class inheritance. This aspect is crucial in libraries which are developed incrementally, e.g., the `StyledText` class is part of an inheritance hierarchy of depth 7. Another key requirement of complex systems is runtime variability. Objects address this issue via dynamic polymorphism. For example, the simulation elements in JMeter are treated uniformly and late bound depending on the user decisions. In the SWT library, a text editor can be used wherever a generic widget is expected.

## 4.2  Observer and Event-Driven Programming

The Observer pattern enables decoupling the code that changes a value from the code that updates the values depending on it. The Observer pattern has a number of drawbacks that have been extensively analyzed by researchers. The main points of criticism are summarized hereafter; the interested reader can refer to [31, 30] for a detailed discussion. First, with the Observer pattern, applications are harder to understand because the natural order of dependencies is inverted. A lot of boilerplate code is introduced to correctly implement the pattern. Another problem is that, since callbacks do not return a type but perform imperative updates on the objects state, reactions do not compose. Finally, the notification to the observers is triggered in an imperative fashion and can be easily scattered and tangled with the rest of the application. Interestingly, many of these points of criticism clearly emerged in our case studies.

Event-based languages like C# [11], Ptolemy [38], EScala [22] and EventJava [18] directly support events and event composition as language constructs. These languages reduce the boilerplate code introduced by the Observer pattern and provide advanced features like quantification over events, event combination and implicit events. Event-based languages integrate well with the imperative paradigm: The callback defines all the operations required for the update of the object state. As a result, this approach preserves object identity – opposed to functional solutions which necessarily compute a different object – and supports efficient fine-grained changes of the updated object.

The main drawback of event-based programming is that a significant amount of modeling reactive dependencies is still encoded in a programmatic way, rather than being supported by the language. The update functionality must be designed and implemented explicitly in the callback method. Caching must be managed manually, deciding a proper policy and coding it in the callback as well. Similar considerations apply for any optimization strategy. For example, accumulation of changes must be entirely implemented by the programmer. Another issue is that, since all the update functionality must be coded manually, consistency is not automatically guaranteed. Instead, the developer must take care of correctly notifying all the entities which are functionally dependent. This leads to error-prone code with the risk of notifying *too rarely*, breaking functional dependency, or *too often*, defensively inserting unnecessary updates. Finally, non-functional design choices are hardcoded in the callback implementation: The developer has no option to choose among different non-functional trade-offs discussed in Section 2, such as caching or incremental updates.

### 4.3 Aspect-oriented Programming

In the context of reactive applications, AOP can be used to intercept objects modifications and keep dependent entities updated. Since AOP supports proper modularization of crosscutting concerns, the update functionalities are separated from the code of the object. For example, the Observer pattern can be implemented in a modular way by using AOP techniques [24]. Other researchers proposed AOP languages to modularize complex relations among objects [42, 37]. AOP integrates well with the imperative style and the mutability of object's state. A point of criticism is that AspectJ-like pointcut-advice models, which dominate AOP design, can potentially break OO modularity. However, it has been shown that pointcuts can be integrated with an event system preserving OO-style modular reasoning [22].

Finally, many limitations of event-based programming hold for AOP, too. Most noticeably, updates must be performed explicitly in the aspect code and dependencies are not automatically tracked as in reactive languages. In addition, composition of reactive behaviors is not easy to obtain, since aspects interactions are complex to master compared to expression composition in functional programming.

### 4.4 Reactive Collections

Reactive collections define functional dependencies among data structures (often expressed via SQL-like queries). The crucial point is that the framework keeps the dependent structures automatically updated when the basic ones change. Efficient incremental updates have been investigated by the database research community for a long time in the context of the view maintenance problem [8]. More recently, researchers introduced these solutions into programming languages, intercepting the updates via AspectJ [44] or using code generation techniques [41] to trigger the updates on the dependencies. Off-the-shelf libraries include LiveLinq [29] and Glazed Lists [23].

Reactive data structures share some design advantages with re-

active languages. Computational dependencies are expressed explicitly, so the application is easier to read: *Queries* describe the functional relations between program entities in a declarative manner. Update functionalities are automatically derived – with little or no additional complexity. The framework is in charge of keeping the functional dependencies specified by the query constantly up to dated. Finally, reactive collections implement efficient update of various reusable operators by using incremental changes and caching to avoid redundant computation.

The main limitation of reactive data structures it that the approach is restricted to a specific domain – changing collections. These frameworks provide out-of-the-box reactive data structures, but they do not support automatic update of other types of objects. Another issue is that reactive collections do not automatically detect all computational dependencies, but only exploit a set of predefined ones. In general, it is not possible to specify a generic expression and leave the framework the responsibility of tracking all the dependencies. For example, the predicate in a `filter` operator (Section 4.1) is not part of the dependencies mechanism, so any change to the predicate remains undetected. Apart from providing custom indexes to speedup certain queries, reactive collections rely on hard-coded non-functional design choices depending on the internal implementation. The programmer cannot fine-tune the update strategy or customize the caching behavior. Finally, most of these frameworks come in a very relational flavor. This further limits their integration with OO languages in which they are usually embedded.

### 4.5 Other Approaches

Constraint-based languages, like Kaleidoscope [21], support constraints that the framework attempts to enforce according to a priority ranking. One-way constraints have proved effective in the scope of graphical interfaces: The Lisp-based Garnet [33] and Amulet [34] graphical toolkits support automatic constraints resolution to relive the programmer from manual updates of the view.

In synchronous dataflow languages like Esterel [3] and LUS-TRE [7], the program defines a network in which a synchronous signal propagates and triggers the computations in the nodes. These languages are usually compiled to finite state machines and – at the cost of limiting the language expressively – provide guarantees of real-time and memory-bound execution.

Self-adjusting computation [1] studies the automatic derivation of incremental programs from traditional ones. This solution adopts an algorithmic approach, focusing on complexity boundaries of the incremental computation.

Complex event processing (CEP) frameworks, e.g. TelegraphCQ [9] and SASE [45] perform queries over time-changing streams of data. Typical scenarios are monitoring of environmental data and trading applications. Compared to database software, in which the user triggers the evaluation of the queries, in CEP, the queries must be reactively evaluated upon data arrival. However, usually, queries are specified in a separate language which does not integrate with the in-language values and with the other language abstractions.

## 5. A RESEARCH ROADMAP

In this section, we present a research roadmap for embedding direct support of reactive applications in object-oriented programming languages. The milestones in the roadmap are ordered from the basic ones – which are ongoing, like the integration of reactive abstractions into an event model, to more elaborate ones – which address complex systems that require automatic adaptation. Beside language design, we plan to work on the improvement of the performance of reactive languages. Previous work mostly focused on language abstractions, with less attention to optimization or performance assessment. However, the high overhead of current reactive languages is also among the factors that limit the spreading of this technology.

## 5.1 Integration with Event-based Programming

Languages with support for event-based programming do make an important step forward in more directly supporting reactive behavior in an imperative object model. As such, they are an ideal starting point for our research. Yet, as argued, they lack the capability of declaratively expressing reactive computations dependent on changing values. In summary, both events and reactive expressions are needed. Events support fine-grained updates of mutable objects. Reactive abstractions capture reactive computations in a compact and declarative way.

Hence, a first step in our plan is to seamlessly integrate reactive abstractions into object-oriented event systems. This goal requires the design of the interface between the reactive abstractions and the abstractions for imperative events, such that they can be treated uniformly in computations and become composable. Such interface is fundamental to support a mixed programming approach and gradual migration of existing software to a more functional and declarative style.

A further step is to integrate reactive abstractions in other aspects of the language. For example, collections must react to changes of the contained elements. It has been shown that changes can be suitably provided to the clients via an event-based interface[1]. Similarly, reactive abstractions can be conveniently used in data structures to model properties which are functionally dependent on other values (e.g. the size or the head of a list, both functionally dependent on the content of the list). The result is a library of *reactivity-enabled* data structures, which expose certain values as reactive abstractions.

This work is currently ongoing in the incarnation of RESCALA [40] (*Reactive*EScala), a language which integrates the advanced event system of EScala [22] and reactive abstractions in the style of Scala.React [30].

## 5.2 Integration with Object-oriented Design

Reactive languages provide abstractions to represent time-changing values. For simplicity, we assume the Flapjax terminology established in Section 4.1 and we refer to these abstractions as behaviors. Behavior values are bound to expressions that capture the dependencies over other values. For example, in Figure 10, Line 5, the `elapsedB` value is bound to the behavior expression `nowB-clickTmsB`. It is unclear, however, how behaviors should integrate with OO design.

We believe that behaviors should be part of the interface of an object and clients should attach to public behaviors to build complex reactive expressions. Private behaviors should instead model functionally dependent values that are consumed only inside the object. Clearly, object encapsulation should be supported to hide implementation details from clients. These results can be trivially achieved by applying visibility modifies to behaviors. However, the next steps in the integration with OO design require more investigation.

An open question is whether behavior expressions can be reassigned. A negative answer leads to a design more similar to method bodies in most OO languages: They are statically defined at devel-

opment time and at runtime can only be executed. On the contrary, modifiable behaviors imply a design similar to fields that can be accessed by getter and setter methods. In that case, behavior expressions would be changeable. Since behavior expressions capture the dependencies over other application entities, allowing their reassignment introduces a potentially excessive degree of dynamicity, especially if behavior expressions can be reassigned from outside the object. However, in a large application, it can happen that the dependencies of a long-lived component are not known when the component is instantiated and, depending on the evolution of the system, must be assigned during its lifetime.

While in the existing literature behaviors are usually assigned and not modified later, this mostly seems due to accidental circumstances rather than justified by design considerations. First, the use cases provided in literature are mostly small examples in which reassignment is not really needed. Second, the functional flavor of the existing solutions presumably favors single assignment. Third, reassignment complicates the reactive model both from an implementation and a semantic standpoint, so non reassignment has been favored also for the sake of simplicity. As a result, we still lack a broad discussion of these issues that concretely justifies the preference for a model or the other.

Another open issue concerns inheritance. Should it be possible to override a reactive value with a new dependency expression or refer to the overridden one via *super*? Intuitively, this seems desirable, but the consequences on the propagation model need careful investigation as well as the expected benefits. A final consideration is about polymorphism. We envisage a scenario in which reactive entities are late bound – like objects – and the dynamic type of the reactive value captures the dependencies over the other entities of the application.

In summary, while reactive abstractions have been applied in the context of OO languages before, previous approaches focused on reactive fragments that only superficially challenge OO design. As a result, we still lack a systematic investigation of the interaction between OO features and reactive abstractions. A starting point for our work is [25] which focuses on the specific scenario of an application in a functional reactive style interfacing with an OO graphic library.

## 5.3 Efficient Reactive Expressions

Reactive languages enable to define arbitrary constraints on dependencies between objects and leave the framework shouldering the recomputation of the dependent objects. However, as shown in Section 4.1, current approaches enforce immutability and recompute dependent objects every time, which negatively impacts efficiency. On the other hand, reactive collections (Section 4.4) overcome this problem by applying advanced strategies such as update incrementalization for a predefined set of operators.

Unfortunately, optimizations are provided out-of-the-box for built-in operators and are not at the fingertips of end developers. As discussed in Section 2.4, in certain circumstances, only domain knowledge enables the developer to provide a mechanism that supports incremental updates. Hence, a predefined set of operators is not sufficient.

Motivated by the above observations, a fundamental step in our research roadmap is to design a framework that combines the open approach of reactive languages, which support arbitrary reactive computations, with the efficiency of built-in reactive data structures. This solution overcomes the frustrating state of the art, where efficient reactive data structures and reactive languages are separate worlds. We will follow two lines of research.

As a first step towards this goal, we aim at bringing efficient

---

[1]For example, the .NET framework provides the `System.Collections.ObjectModel.ObservableCollection<T>` class which exposes to the clients the `PropertyChanged` and the `CollectionChanged` events.

built-in operators to reactive abstractions. We will provide a variety of efficient operators that seamlessly operate on reactive collections and reactive abstractions. Highly efficient libraries will be designed along the lines of [44, 41], but integrated with the abstractions of existing reactive languages. For example, by allowing behavior-like expressions in the predicate of a filter operator or by modeling the result of the reactive operators as behaviors. Predefined operators must cover the most common applicative scenarios such as collections and relational operations.

The second step of research will aim at reconciling the openness of reactive languages with efficient reactive operators. This is only possible if the optimization of generic computations is available to application developers. Optimizing a particular step of a reactive computation process must be handy: Providing a faster version of a reactive computation must be as easy as – say – overriding an existing method.

Our idea is to separate the *creation* of dependent objects, which is performed from scratch, form the *maintenance* of objects. In the default case, both creation and maintenance are accomplished by complete recomputation, i.e. by applying the function that relates basic and derived objects, for example `filter` in the `list2 = list1.filter(x>10)` expression. However, the programmer can provide refinements for the maintenance case. Those refinements can be implemented by imperative algorithms or by taking advantage of domain-specific knowledge to efficiently update dependent objects. In this way, efficient event-based computations that apply the optimization principles well-known from the OO context can be conveniently hidden behind high level operators expressed in a functional style. Reactive objects are then connected by those operators to compose constraints. To further open the framework, the programmer should be able to refine existing operators with a more efficient version when better performance is needed. Finally, late binding can be leveraged to obtain the dynamic selection of the best operator (i.e. the best refinement for a set of types).

A second line of research concerns optimizations that must take into account a broader scope than single operators. For example, considerable performance improvement in relational expressions comes from reordering by anticipating selections and deferring joins. In addition, those optimizations must be performed, at least partially, at runtime, to allow cross-module analysis. Deeply-embedded DSLs come with powerful interfaces to support custom optimizations for DSL expressions embedded into the host languages. For example, in LINQ, the developer is provided with the raw compiler output in the form of an – internally untyped – expression tree. Scala-virtualized [32] employs a similar approach, but fosters more typing guarantees. We will investigate the applicability of these techniques to optimizing reactive expressions. However, these mechanisms are quite low-level. As a result, optimizations are hard to perform and require highly specialized skills. It has been reported that building a LINQ provider for the RavenDB database took more time than building the database [16]. Also, they do not support dynamic optimizations. We will opt for hiding the complexity of those techniques behind higher-level abstractions.

## 5.4 Propagation Model

To enforce the constraints defined by the user, reactive languages keep a runtime model of the dependencies in the application. Usually, this model is a directed graph in which a change in a node triggers an update over the transitive closure of the dependency relation. Reactive languages mostly enforce a push propagation model in which changes are proactively applied to dependents in the graph [10, 31]. However, also lazy models with invalidation of the cached values and on-demand recomputation have been pro-

posed [30]. The propagation of the changes along the graph has a considerable performance impact.

Optimization techniques regarding the propagation model have been already proposed. For example, *lowering* is a technique that applies static analysis to collapse several reactive nodes in the graph into one [6]. As a result, the computation is moved from the reactive model to the usual (and more efficient) call-by-value system. The Yampa FRP framework employs a similar approach but merges the computations at runtime [35].

Based on the above observations, one research direction that we plan to follow concerns optimizations related to the propagation model.

First, alternative graph constructions can be performed that lead to observationally equivalent reactive models. As a result, performance considerations can guide the choice. For example, as noted in [6], always collapsing the computations can lead to poor performance in certain cases. Consider the following code snippet from [6]. A time consuming operation depends on an operation whose output rarely changes. The second operation, instead, depends on a frequently-changing value:

```
(time-consuming-op
   (infrequently-changing-op frequent-emitter))
```

Consider the case in which this code results in three nodes: A source node `frequent-emitter`, an `infrequently-changing-op` node which depends on the first one, and a `time-consuming-op` node, which depends on `infrequently-changing-op`. Every time `frequent-emitter` emits a new value, `infrequently-changing-op` is executed. Since the outcome of `infrequently-changing-op` rarely changes, `time-consuming-op` is executed just a few times. Instead, if `infrequently-changing-op` and `time-consuming-op` are collapsed into the same node, `time-consuming-op` is executed at the same rate `frequent-emitter` changes its value. This effect is even more significant in languages like Scala.React, which apply collapsing of computations as the principal composition mechanism. In summary, collapsing should not be accepted or refused in its entirety and code analysis or dynamic techniques must be applied to detect where each solution leads the best results.

A second aspect of the propagation model that needs further investigation is the choice of a push-based implementation that is adopted by most reactive languages. According to the design space presented in Section 2, this solution favors caching over on-demand recomputation. The choice is motivated by a constraint: A push-based solution is necessary to guarantee that possible side effects in the reaction are really performed. However, change propagation does not always involve side effects. An optimization should explore the space between caching and on-demand recomputation and provide a convenient compromise. For example, reactive data structures dynamically switch to a caching strategy when the requests exceed a threshold [44].

However, this solution is quite simple and does not consider factors like the current machine load. For example, if the load is low, it may be convenient to recompute cached reactive values even if they are rarely requested. On the other hand, with heavy load, this strategy can further degrade the performances without providing significant benefits. More advanced approaches relying on concepts from control systems need to be investigated. The latter have been e.g., explored for parallel data structures to provide the best performance adapting to different machines, configurations and workloads [14].

Finally, update propagation models are common to both reactive languages and event-based systems. The former use these models to propagate updates across dependencies, the latter to trigger dependent events [22]. Since these mechanisms are more and more used in programming languages, an obvious question is if those functionalities should be supported at the VM level. In previous work, starting from similar considerations, we investigated dedicated VM support for AOP [4]. Research work on runtime environments which natively support the concept of reactive memory was recently carried out at the OS level [12]. However, implementing a similar approach in a managed environment which specifically supports propagation of changes across reactive entities is still a research challenge.

A second research direction is optimization by design. This approach is enabled by making the performance implications of the language abstractions explicit and leaving the choice in the hands of the programmer. Current reactive languages focus on expressivity rather than performance. As a result, the programmer has no clear control of how performance is affected by the design choices. Unlike reactive languages, dataflow languages, like Esterel and LUSTRE, intentionally limit the expressive power of the available abstractions to achieve memory and time-bound execution. Attempts to limit expressivity to improve performances have already been done in the reactive languages community. For example, real-time FRP [43] is a time-bound and space-bound subset of FRP. However, real-time FRP is a *closed language* [28], i.e. it is not embedded in a larger general-purpose language, which considerably limits the applicability of this approach.

In summary, programmers face a black-or-white choice: Relinquish performance for expressivity or abdicating abstraction for efficiency. Instead, reactive languages should incorporate reactive primitives that, at the cost of a reduced expressive power, have a high-performance profile. Static analysis or a dedicated static type system can ensure that those primitives are not combined with the rest of the reactive system in a way that cancels the performance improvement.

A starting point is to implement lexically scoped dependencies. In current reactive approaches, reactive dependencies are established dynamically. When, during the evaluation of an expression, a reactive value is found, the value is inserted in the dependency graph. This approach introduces considerable overhead. In fact, the evaluation is slowed down by the process of double-liking dependent values with their dependencies. Similarly, when the value of a node changes, it must be unlinked from the nodes depending on it. This behavior is required to keep the graph updated, since dependencies can change dynamically. For example, the value of the expression `if(a) b else c` depends on either b or c on the bases of a. As a consequence, the structure of the graph is not fixed but must be continuously restructured to capture the current dependencies [30]. In contrast, lexically scoped dependencies are fixed. This results in regions in which the graph structure does not change, avoiding the computations required to keep the graph updated and improving performances.

## 5.5 Evaluation

To make sure that our research leads to concrete results, we plan to evaluate each progress. However, evaluating language design is not easy, because design quality is hard to capture with synthetic metrics and design choices have long term effects which are hard to predict. For example, the impact of programmers experience or the maintainability of large systems can be evaluated only when a considerable amount of projects have been developed. As a consequence, we believe that a priori reasoning and careful analysis of the available options remain fundamental steps [27]. Nevertheless, where applicable, we plan to perform objective evaluations of our results.

Performances can be evaluated effectively by running benchmarks. For example, former studies compared the performance

of OO programming (Java) with the mixed OO-functional style (Scala) in the context of parallel applications and multicore environments [36]. We believe that similar experiments can evaluate the performance impact of reactive abstractions compared to the traditional solutions for reactive applications.

Other aspects of the language design can be evaluated by using software metrics. Common metrics include coupling and cohesion, lines of code, number of operations and others [19]. We plan to adopt this approach to evaluate our design choices in the advanced state of our research, when studies can comprise several artifacts. Other researchers already used metrics to validate language design choices in reactive applications. Recently, the combined use of synthetic metrics and manual inspection (to investigate specific issues) was successfully applied to evaluate event quantification in software product lines [13].

## 6. REFERENCES

[1] U. A. Acar, G. E. Blelloch, and R. Harper. Adaptive functional programming. *ACM Trans. Program. Lang. Syst.*, 28(6):990–1034, Nov. 2006.

[2] Android developers site. http://developer.android.com/index.html.

[3] G. Berry and G. Gonthier. The Esterel synchronous programming language: design, semantics, implementation. *Science of Computer Programming*, 19(2):87 – 152, 1992.

[4] C. Bockisch, S. Kanthak, M. Haupt, M. Arnold, and M. Mezini. Efficient control flow quantification. In *Proceedings of the 21st annual ACM SIGPLAN conference on Object-oriented programming systems, languages, and applications*, OOPSLA '06, pages 125–138, New York, NY, USA, 2006. ACM.

[5] E. Bodden, R. Shaikh, and L. Hendren. Relational aspects as tracematches. In *Proceedings of the 7th international conference on Aspect-oriented software development*, AOSD '08, pages 84–95, 2008.

[6] K. Burchett, G. H. Cooper, and S. Krishnamurthi. Lowering: a static optimization technique for transparent functional reactivity. In *Proceedings of the 2007 ACM SIGPLAN symposium on Partial evaluation and semantics-based program manipulation*, PEPM '07, pages 71–80, 2007.

[7] P. Caspi, D. Pilaud, N. Halbwachs, and J. A. Plaice. LUSTRE: a declarative language for real-time programming. In *Proceedings of the 14th ACM SIGACT-SIGPLAN symposium on Principles of programming languages*, POPL '87, pages 178–188, 1987.

[8] S. Ceri and J. Widom. Deriving production rules for incremental view maintenance. In *Proceedings of the 17th International Conference on Very Large Data Bases*, VLDB '91, pages 577–589, San Francisco, CA, USA, 1991. Morgan Kaufmann Publishers Inc.

[9] S. Chandrasekaran, O. Cooper, A. Deshpande, M. J. Franklin, J. M. Hellerstein, W. Hong, S. Krishnamurthy, S. R. Madden, F. Reiss, and M. A. Shah. TelegraphCQ: continuous dataflow processing. In *Proceedings of the 2003 ACM SIGMOD international conference on Management of data*, SIGMOD '03, pages 668–668, 2003.

[10] G. H. Cooper and S. Krishnamurthi. Embedding dynamic dataflow in a call-by-value language. In *ESOP'06*, pages 294–308, 2006.

[11] Microsoft corporation. C# language specification. version 3.0.

http://msdn.microsoft.com/en-us/vcsharp/aa336809.aspx, 2007.

[12] C. Demetrescu, I. Finocchi, and A. Ribichini. Reactive imperative programming with dataflow constraints. In *Proceedings of the 2011 ACM international conference on Object oriented programming systems languages and applications*, OOPSLA '11, pages 407–426, 2011.

[13] R. Dyer, H. Rajan, and Y. Cai. An exploratory study of the design impact of language features for aspect-oriented interfaces. In *AOSD '12: 11th International Conference on Aspect-Oriented Software Development*, AOSD '12, March 2012.

[14] J. Eastep, D. Wingate, and A. Agarwal. Smart data structures: an online machine learning approach to multicore data structures. In *Proceedings of the 8th ACM international conference on Autonomic computing*, ICAC '11, pages 11–20, 2011.

[15] Eclipse site. http://www.eclipse.org/.

[16] O. Eini. The pain of implementing LINQ providers. *Queue*, 9(7):10:10–10:22, July 2011.

[17] C. Elliott and P. Hudak. Functional reactive animation. In *Proceedings of the second ACM SIGPLAN international conference on Functional programming*, ICFP '97, pages 263–273. ACM, 1997.

[18] P. Eugster and K. R. Jayaram. EventJava: An extension of Java for event correlation. In *Proceedings of the 23rd European Conference on ECOOP 2009 – Object-Oriented Programming*, ECOOP '09, pages 570–594, 2009.

[19] E. Figueiredo, N. Cacho, C. Sant'Anna, M. Monteiro, U. Kulesza, A. Garcia, S. Soares, F. Ferrari, S. Khan, F. Castor Filho, and F. Dantas. Evolving software product lines with aspects: an empirical study on design stability. In *Proceedings of the 30th international conference on Software engineering*, ICSE '08, pages 261–270, 2008.

[20] FreeCol game. http://www.freecol.org/.

[21] B. N. Freeman-Benson. Kaleidoscope: mixing objects, constraints, and imperative programming. OOPSLA/ECOOP '90, pages 77–88, 1990.

[22] V. Gasiunas, L. Satabin, M. Mezini, A. Núñez, and J. Noyé. EScala: modular event-driven object interactions in Scala. AOSD '11, pages 227–240. ACM, 2011.

[23] Glazed Lists site. http://www.glazedlists.com/.

[24] J. Hannemann and G. Kiczales. Design pattern implementation in Java and AspectJ. In *Proceedings of the 17th ACM SIGPLAN conference on Object-oriented programming, systems, languages, and applications*, OOPSLA '02, pages 161–173, 2002.

[25] D. Ignatoff, G. H. Cooper, and S. Krishnamurthi. Crossing state lines: Adapting object-oriented frameworks to functional reactive languages. In *FLOPS*, pages 259–276, 2006.

[26] Jmeter developers site. http://jakarta.apache.org/jmeter/index.html.

[27] G. Kiczales and M. Mezini. Separation of concerns with procedures, annotations, advice and pointcuts. In *Proceedings of the 19th European conference on Object-Oriented Programming*, ECOOP'05, pages 195–213, Berlin, Heidelberg, 2005. Springer-Verlag.

[28] R. B. Kieburtz. Implementing closed domain-specific languages. In *Proceedings of the International Workshop on Semantics, Applications, and Implementation of Program Generation*, SAIG '00, pages 1–2, London, UK, UK, 2000. Springer-Verlag.

[29] LiveLinq site. http://www.componentone.com/SuperProducts/LiveLinq/.

[30] I. Maier, T. Rompf, and M. Odersky. Deprecating the Observer Pattern. Technical report, 2010.

[31] L. A. Meyerovich, A. Guha, J. Baskin, G. H. Cooper, M. Greenberg, A. Bromfield, and S. Krishnamurthi. Flapjax: a programming language for Ajax applications. OOPSLA '09, pages 1–20, 2009.

[32] A. Moors, T. Rompf, P. Haller, and M. Odersky. Scala-virtualized. In *Proceedings of the ACM SIGPLAN 2012 workshop on Partial evaluation and program manipulation*, PEPM '12, pages 117–120, New York, NY, USA, 2012. ACM.

[33] B. A. Myers, D. A. Giuse, R. B. Dannenberg, D. S. Kosbie, E. Pervin, A. Mickish, B. V. Zanden, and P. Marchal. Garnet: Comprehensive support for graphical, highly interactive user interfaces. *Computer*, 23(11):71–85, Nov. 1990.

[34] B. A. Myers, R. G. McDaniel, R. C. Miller, A. S. Ferrency, A. Faulring, B. D. Kyle, A. Mickish, A. Klimovitski, and P. Doane. The Amulet environment: New models for effective user interface software development. *IEEE Trans. Softw. Eng.*, 23(6):347–365, 1997.

[35] H. Nilsson, A. Courtney, and J. Peterson. Functional reactive programming, continued. In *Proceedings of the 2002 ACM SIGPLAN workshop on Haskell*, Haskell '02, pages 51–64, 2002.

[36] V. Pankratius, F. Schmidt, and G. Garreton. Combining functional and imperative programming for multicore software: An empirical study evaluating Scala and Java. In *Software Engineering (ICSE), 2012 34th International Conference on*, ICSE '12, pages 123 –133, 2012.

[37] D. J. Pearce and J. Noble. Relationship aspects. In *Proceedings of the 5th international conference on Aspect-oriented software development*, AOSD '06, pages 75–86, New York, NY, USA, 2006. ACM.

[38] H. Rajan and G. T. Leavens. Ptolemy: A language with quantified, typed events. In J. Vitek, editor, *ECOOP 2008 Paphos, Cyprus*, volume 5142 of *LNCS*, pages 155–179. Springer-Verlag, July 2008.

[39] http://www.stg.tu-darmstadt.de/media/st/publications/oo_-reactive_report.pdf.

[40] REScala web site. http://www.stg.tu-darmstadt.de/research/escala/.

[41] T. Rothamel and Y. A. Liu. Generating incremental implementations of object-set queries. GPCE '08, pages 55–66, 2008.

[42] K. Sakurai, H. Masuhara, N. Ubayashi, S. Matsuura, and S. Komiya. Association aspects. In *Proceedings of the 3rd international conference on Aspect-oriented software development*, AOSD '04, pages 16–25, 2004.

[43] Z. Wan, W. Taha, and P. Hudak. Real-time FRP. In *International Conference on Functional Programming (ICFP'01)*, 2001.

[44] D. Willis, D. J. Pearce, and J. Noble. Caching and incrementalisation in the Java query language. *SIGPLAN Not.*, 43(10):1–18, Oct. 2008.

[45] E. Wu, Y. Diao, and S. Rizvi. High-performance complex event processing over streams. In *Proceedings of the 2006 ACM SIGMOD international conference on Management of data*, SIGMOD '06, pages 407–418, 2006.

# Enhancing Design Models with Composition Properties: A Software Maintenance Study

Francisco Dantas[1,2], Alessandro Garcia[1], Jon Whittle[3], João Araújo[4]

[1]Opus Research Group – Software Engineering Lab, Informatics Department, PUC-Rio, Brazil
[2]Department of Computer Science, State University of Rio Grande do Norte, Brazil
[3]School of Computing and Communications, Lancaster University –Lancaster, United Kingdom
[4]CITI/FCT, Universidade Nova de Lisboa, Portugal
{fneto, afgarcia}@inf.puc-rio.br, j.n.whittle@lancaster.ac.uk, joao.araujo@fct.unl.pt

## ABSTRACT

A considerable part of software design is dedicated for the composition of two or more modules. The implication is that changes made later in the implementation often require some reasoning about module composition properties. However, these properties are often not explicitly specified in design artefacts. Moreover, they cannot be easily inferred from the source code either. As a result, implicit composition properties may represent a major source of software maintenance complexity. This fact is particularly true with the advent of post object-oriented techniques, which are increasingly providing advanced mechanisms to enable flexible module composition. However, there is little empirical knowledge on how design models with explicitly-specified composition properties can improve software maintenance tasks. This paper reports an experiment that analyses the impact of design models enriched with composition properties on system maintenance. Explicit composition modelling was achieved in the experiment through a conservative approach, i.e., a specific set of additional UML stereotypes dedicated to model composition properties. The experiment involved 28 participants, who were asked to realize four maintenance tasks using UML design models with different levels of module composition details. Our findings suggested that explicit composition modelling contributed to better results in the realization of program change tasks, regardless of the developers' expertise. Users of composition-enhanced models consistently yielded better results when compared to users of plain UML models. The use of explicit composition specification also led to an average increase of 44.7% in the quality of changes produced by the participants.

## Categories and Subject Descriptors

D.2.10 [**Software Engineering**]: Design - Representation.

## General Terms

Design, Experimentation.

## Keywords

Composition Mechanisms, Software Maintenance, Design Models.

## 1. INTRODUCTION

Over the last decades, software design has increasingly become an incremental task. The use of powerful mechanisms for composing software modules is often required to support agile realization of software changes [1-4]. The composition part of a system defines how two or more modules are bound. Post object-oriented programming techniques (e.g., [1-4][35]) are increasingly providing composition mechanisms to enable the flexible definition of module composition. They are moving away from supporting only simple composition mechanisms, such as method calls [8]. Examples of advanced composition mechanisms range from pointcut-advice [1] to bindings [2].

A side effect of this trend is that the composition part of a system is increasingly playing a major role on the software maintenance [6][7][46][47][48]. This promotes a shift in the structure of software designs: while complexity is factored out of software modules, the composition specification is much more complex [1-5]. Because of this new complexity flavour, software changes might become harder to realize [6][7][18]. Almost inevitably, software developers need to reason about the properties of module composition in order to accomplish software changes. Developers may need to analyse design artefacts to understand the composition properties and implement the changes. The module composition has a number of properties [18], which are not directly supported by modelling languages.

For instance, the scope of a module composition is a property that refers to the set of modules taking part in such a composition. Let's consider a composition realized by a single method call `update(M1.setX())` in a given module `M2`. The scope of this composition is defined by three modules, namely `M1`, `M2`, and `M3`. The operation `update(M1.setX())` updates the value of the attribute x declared in module `M1`. However, let us consider that the original value of x is used by another module `M3`. Thus, other update operations over x cannot be ignored by `M3` as the correct execution of these two modules depends on the computation of the correct value of x. Then, the scope of the aforementioned composition is not limited to `M1` and `M2`; `M3` also implicitly joins the composition, but it is not directly visible in the composition statement. The situation is likely to get more complicated if `M3` is, for instance, an aspect. If the composition scope is not well understood, it can impact negatively on the program maintenance. Then, developers might need to rely on design models, such as sequence diagrams, to understand the composition scope.

Previous studies [8-15] state that Unified Modelling Language (UML) design models provide a means to manage the increasing complexity of programs, which in turn assist developers in understanding and managing them. However, nothing is said in the literature about the impact of making available UML design models that show different kinds of composition details, which allow us to express the nuances of its properties (e.g., composition scope). Unfortunately, the reasoning about composition properties is often hampered by the fact that most developers do not express critical composition properties in their models. However, in the vast majority of cases, detailed models are harmful for code change [16]. This happens because explicit information could actually create so much noise that the models become too complex [16][17]. Nevertheless, our assumption is that having a more detailed UML design [13], where composition properties are more explicitly specified, will successfully support developers in maintaining their programs, thus improving the quality of the realized changes.

Several assessments of advanced programming techniques have been carried out (e.g., [6][7]). They represent the first studies focusing on the analysis of the impact of composition code properties on software changes. According to them, the complexity of the composition code is directly associated with the quality of changes realized in a program. As a consequence, developers would potentially better realize program changes in evolving programs if module composition properties are explicitly specified somewhere. In this context, considering the promising UML benefits to manage the complexity of software systems, there is little empirical knowledge about the impact of design models with explicit composition details on change quality.

This paper evaluates the impact of using UML design models, enriched with composition details, on the quality of program change tasks. This goal is achieved through the design and execution of an Internet-based experiment (Section 3). This experiment involved 28 participants (intermediate to senior-level) that were required to perform individually the same four maintenance tasks to an evolving game application. All the participants had at least an intermediate expertise in the technologies explored in the experiment. In order to evaluate the impact of the specification of composition details on the realization of programming tasks, the subjects were divided in two subgroups of 14 participants each. In addition to the same source code, the first group had access to a complementary composition design in plain UML notation and the second one had access to a complementary design with an enriched UML notation (Section 2.2), where the composition properties are highlighted[1] (Section 2.1). Explicit composition modelling was achieved in the experiment through a conservative approach, i.e., a specific set of additional UML stereotypes dedicated to support the modelling of composition properties.

In particular, our main findings and contributions (Section 4) are the following. First, we provide findings that confirm that, in general, the programming tasks supported by UML+ models bring benefits in terms of software maintenance. Our investigation has shown that the use of UML+ models improves the quality of the changes in 44.7% of the cases. Second, we have also identified that availability of UML+ models brings benefits in terms of change quality regardless of the programming task complexity and participants' expertise. For instance, we have evidence that as the change quality increases, the number of changes decreases. Related work is discussed in Section 5. Finally, the threats to

validity are discussed in Section 6. Final considerations and future work are presented in Section 7.

## 2. BACKGROUND

This section presents the composition properties analysed in our study and the UML-based notation to specify composition properties. The example provided will be used along Sections 3.1 and 3.2.

## 2.1 Composition Properties

Composition code defines how two or more modules are bound and entails new dimensions of complexity in a program. Moreover, composition code refers to the code elements associated with the use of advanced composition mechanisms, such as intertype declaration and pointcut-advice [1]. The essence of composition code relies on the definition of common properties underlying the binding of the modules. Programs are built to have certain properties, including in the module composition, which may exert an impact on the quality attributes, such as software maintainability (Section 4.3). In our previous analysis [6][18], we observed that composition code is characterized by at least three basic properties: *diversity*, *scope* and *volatility*.

First, composition code might involve several modules, which may be of diverse types, such as classes and aspects. Therefore, the property *composition diversity* refers to the amount and types of different modules that comprise the composition code. The example of Figure 1 illustrates how diverse the composition code can be. In this example, there are different modules used to realize the composition, namely abstract aspects, aspects, classes and interfaces. The property *composition scope* refers to the extent of the enclosing context where the program elements involved in the composition are associated with. In the example illustrated in Figure 1, the composition scope is defined by the set of modules, which belongs to the composition code. For instance, the scope of the composition encompasses all modules that are affected by the modules which make up their code, either by changing code or influencing behaviour.. Finally, the property *composition volatility* refers to the extent that these dependencies are broken when a single change is performed in the code pertaining to the composition or the involved modules.

## 2.2 UML+ Notation

We chose UML as the design modelling language in our study due to its popularity. It is well-known and widely used in industry and academic research settings. Additionally, due to its wide adoption in software projects we did not need to provide any "artificial" training to the participants. For each programming task, class and sequence diagrams were provided. Both of them were presented with different kinds of details: UML and UML+, where the latter provides complementary information about the composition code. UML+ refers to those diagrams that make the composition properties more explicit to developers (Section 2.1). With UML+, aspect-oriented design models are represented using the notation proposed by Chavez *et al.* [19][20]. Table 1 describes the UML+ design model representation proposed by Chavez *et al.* [19][20] for Aspect-Oriented Programming (AOP) using the example provided in Figure 1.

---

[1] From herein, plain UML is referred just as UML, and the UML enriched with composition properties is simply referred as UML+.

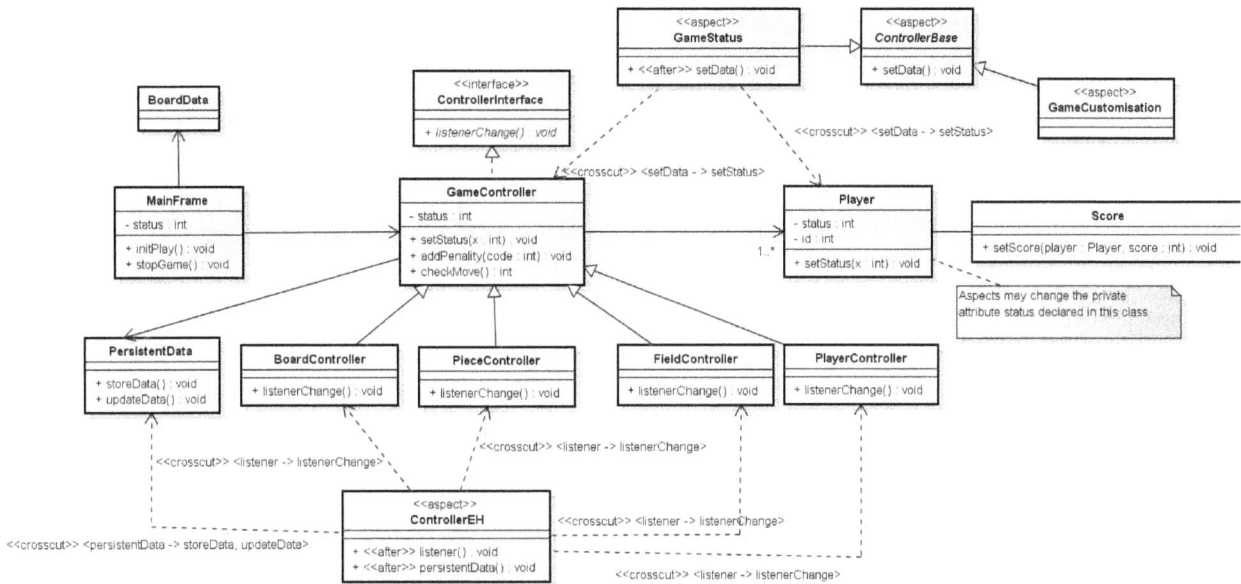

**Figure 1. Design model in UML+**

The additional composition information in UML+ models was introduced via existing UML notations, such as stereotypes and comment boxes. Moreover, UML abstractions are well aligned with the key abstractions of AspectJ. This language was chosen because it supports both (i) conventional programming mechanisms for module composition, such as inheritance, which are very directly supported by UML, and (ii) advanced composition mechanisms, such as pointcuts and intertype declaration, which are not directly represented by UML.

**Table 1. UML+ Notation**

| AOP Notation | Description |
|---|---|
| <<aspect>> | It defines an aspect through a special stereotype for UML classes. In Figure 1, examples of aspects are GameStatus, GameCustomization |
| <<precede>> | It defines an order between two aspects that affect the same joinpoint. The aspect on the start of the arrow executes before the aspect touching the arrow's head. |
| <<crosscut>> | It defines the crosscutting relationship between the composition modules and the other modules. In this case, the aspect GameStatus crosscuts the classes GameController |
| <<require>> | It provides an explicit dependency between two aspects; this means that a given aspect requires the presence of another. |
| <pointcutname → operations> | It defines the operations that are affected by an advice. In this case, we have an advice that is executed after the operation click (<setData → setStatus>) declared in the GameController. |

## 3. EXPERIMENTAL DESIGN

The experiment follows the standard Between-Subject design as each group performed the experiment with exactly one treatment: UML and UML+ models. We chose this design for several reasons: (i) we aim at comparing the effectiveness of using UML and UML+ for realizing a set of programming tasks (Section 3.3) in the same application, (ii) as the application was the same, participants could not be part of both groups using UML and UML+ due to learning-related side effects, (iii) to reduce the threat on the use of this design, the separate groups were equally created, treated and composed of participants with equivalent expertise (see Section 3.4) and (iv) due to practical reasons, such as execution time restrictions and availability of suitable applications. Figure 2 illustrates in detail the design in five steps: (1) we picked participants for the experiment; (2) after we assigned them to the UML and UML+ groups; (3) we treated the experiment's independent variables (Section 3.6); (4) we compared the experiment results in terms of stability and finally, (5) we interpreted the results statistically in order to test our hypotheses (Section 3.1). The main elements of our experiment design encompass: the research question and hypotheses (Section 3.1); the object of our experiment (Section 3.2); the design of the change tasks (Section 3.3); the participants involved in the experiment (Section 3.4), the experimental procedures (Section 3.5); and the variables and analysis (Section 3.6).

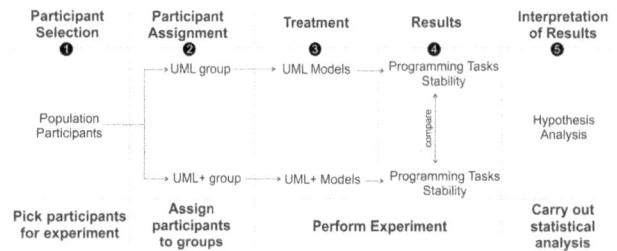

**Figure 2. Experiment Design**

51

## 3.1 Research Question and Hypotheses

Our experiment aims at analysing whether the influence of design models enriched with composition details affects the quality of changes made in software maintenance tasks. Based on that, we define the following research question: *"Do design models with a richer composition specification lead to better code maintenance than plain design models?"* In order to answer this question, we compare the quality of code changes made by maintainers when using either plain UML models (named UML group) or enriched UML models (named UML+ group).

The refinement of our research question consists of terms of a null hypothesis, denoted H1, which can be formulated as follows:

**H1:** Code change quality with UML is the same as code change quality with UML+

The hypothesis is defined to test the impact of realizing change tasks on a program when either UML or UML+ design models are available. Through this hypothesis, it is possible to evaluate whether the variation in the *"change quality"* is influenced (or not) by the use of UML+ or UML models. The notion of change quality measured in our study is presented in Section 3.6.

## 3.2 Object

We selected four code change tasks for our experiment. The tasks (Section 3.3) mimic recurring maintenance scenarios that emerged in the project history of four evolving applications: MobileMedia (4KLOC), iBatis (16KLOC), HealthWatcher (5KLOC) and GameUP (2KLOC). The change tasks were extracted from these projects. These applications were chosen as the objects of our analysis as they were, in fact, implemented in AspectJ.

We designed the programming tasks in the experiment using only the GameUP application [45] because its functionalities are self-explanatory. This game application is from a well-known domain and enabled us to embrace the aforementioned maintenance tasks. In addition, its design follows the MVC decomposition, a well-known architecture in industry and academy [21]. This means that the participants will not spend much of their time trying to become familiar with the overall design decomposition. Therefore, they can mostly concentrate their effort on the realization of the code change tasks (Section 3.3). Still, the design models can support them on learning the global design decisions, including the constraints governing module composition.

GameUP is a board game application, which is intended to support three other board games: Shogi, JHess and Checkers. Checkers is an American checker game whereas Shogi and JHess are chess games. GameUP provides functionalities to manage various functionalities for customizing its board (e.g., indicating moveable pieces) and the matches between players (e.g., indicating player turns). The main GameUP modules are presented in Figure 3, which is a simplified version of its architecture. Its architecture is divided into three layers by following the MVC (model-view-controller) structure. The MVC architecture provides a low coupling between the modules present in each layer. In this way, the modules that implement the GameUP interface are completely decoupled from the modules that implement GameUp controllers. Similarly, the controller modules are architecturally decoupled from data modules. Consequently, the MVC architecture facilitates the process of changing the existing source code of modules in each layer by separating the different responsibilities of the game application.

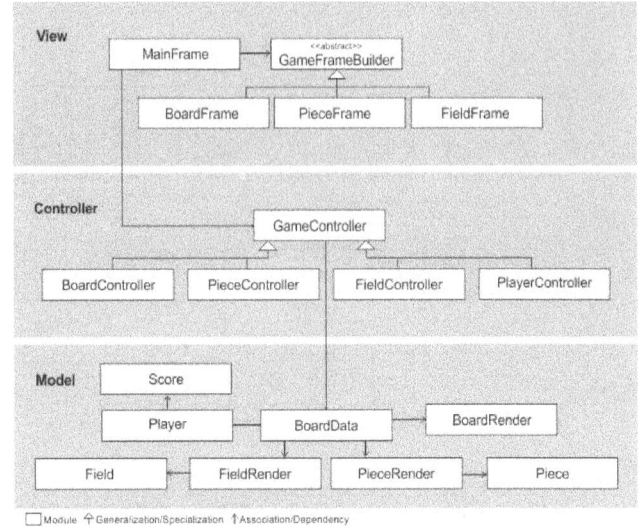

**Figure 3. GameUp Architecture**

The interaction between GameUP and users is realized by the modules that implement the GameUP screens, such as `MainFrame` and `BoardFrame` (Figure 3). The access to the game functionalities is performed through the controllers, such as `GameController` and `BoardController`. Data are persisted via classes that belong to the Model layer (Figure 3).

## 3.3 Task Design

In order to define the code change tasks, we followed three important criteria: (i) they should be representative of recurring program maintenance scenarios that require the use of composition mechanisms, (ii) they should be diverse enough to require reasoning on module composition properties (Section 2.1), (iii) it should be doable to complete the set of tasks in the limit of one hour, and (iv) they should be representative of programming scenarios where the understanding of composition properties is required. To this end, we defined four tasks on program maintenance. All the tasks require reasoning about the modules and their composition properties. They also require different types of changes, such as refactorings and functionality increments [26][34][39]. Therefore, these are maintenance tasks where the use of UML+ models could play a role. We also focused on this limited, albeit significant, set of tasks in order to follow the design guidelines for controlled experiments, i.e., we could not demand more than one hour of work.

In addition, in order to fulfil our fourth criterion, we revisited research work, independently performed by several researchers in the community (e.g., [3][4][31][33][35][38]), who observed that changes made in the composition code are the cause of maintenance problems. Katz *et al.* [33], for instance, discuss the role of 'invasive aspects' in software maintenance (explored in task 3). The problem of 'invasive aspects', although realized through different implementation mechanisms, have their counterparts in hybrid FOP languages [32] (e.g., CaesarJ), composition filters [4], and delta-oriented programming [35]. We had to focus on AOP in our experiment as both (i) it does not require artificial training (see Figure 4), and (ii) it has been used in development of industrial applications given its popularity [40][41].

**Table 2. Description of the Tasks**

| Task | Description |
|------|-------------|
| T1 | The GameUP developers must evolve the games with a new functionality, which consists of displaying a new screen before the match starts, where the player can specify whether the game will be played on the network or not. This new functionality must be implemented through a method named `setupGame( )` using AspectJ mechanisms. The existing method `startGame()` is in charge of initializing the game. |
| T2 | The player's status is indicated by a colored button. When the button is green (status=0) it indicates that the player can play. If the button is red (status=1) the player cannot play due to some restrictions and another match must be started. Having said this, you are required to add a new functionality which aims to change the button color to yellow (status=2) when a player suffers a penalty and passes your next turn on to your opponent. This new functionality must be implemented in a separate aspect using an after returning advice. |
| T3 | The current version of Game UP allows both saving the board configuration in JPG format and informing the player about the result of such an operation by a message on the screen. The GameUP developers need your help to evolve the source code in order to allow saving the game in an XML format in addition to the JPG format. |
| T4 | Help GameUP developers to add the method `int saveScore( )` to the class `PersistentData` using AspectJ's mechanisms. This method aims at providing a new functionality, which is to save the player's score at the end of each game match. |

**Table 3. Tasks vs. Code Change**

| Task | Manipulating Code Elements - Necessary Modifications |
|------|------------------------------------------------------|
| T1 | T1 requires at least the addition of one module and the modification of two other modules, which represent 30% of the code involved in the composition. |
| T2 | This task requires modification in at least two modules. To be more specific this task requires the change of operations implicitly related to other modules. Approximately, 60% of the code involved in the composition is modified. |
| T3 | This task requires the addition of one new module and modification in two other ones. T3 requires modification in 50% of the modules involved in the composition. |
| T4 | T4 requires the addition of at least one new module, the modification of another one. This task is directly associated with the expressiveness power of composition mechanisms. In this task, at least 50% of the modules are modified. |

For each task (Table 2), we provided both a class and a sequence diagram with the relevant design elements. Regarding the tasks, we have chosen to work with open questions instead of multiple-choice ones in order to avoid resorting to simply guessing and to make them more representative under the perspective of real maintenance scenarios. Each task consisted of a request to change slices of code involving module composition. The expected answer in each task is made up by a number of required steps (operations). For each answer in a task, the subject can earn from 0 to 1 point, depending on the number of steps performed correctly (Section 3.5).

To successfully realize the tasks, it is required that the developers (Section 3.4) understand and observe the composition properties underlying the composition code. For instance, in the UML+ class diagram of task T2 (Figure 1) the UML note in class Player helps (as a recall mechanism) the programmer to reason about the program elements affected by the use of the *after returning* mechanism. This UML note highlights important information about the implicit and wide scope (Section 2.1) of *after returning* advice over the affected modules. An overview of the expected modifications is presented in Table 3.

## 3.4 Participants

The Internet-based experiment involved 28 participants: 22 Ph.D. students, 4 M.Sc. students, 2 postdocs in Computer Science from different Brazilian Universities and companies. All the students had experience with software development in industry. In particular, 71.42% of the participants have worked for more than two years in the industry. We tried to diversify the subject's profiles as we understand that a group made up entirely by

students might not represent developer population as stated by Di Penta et al. [22]. All the participants were invited to participate without any obligation to accept the invitation. All of them have a good level of experience (at least two years) in working with AspectJ, the chosen programming language (see Figure 4).

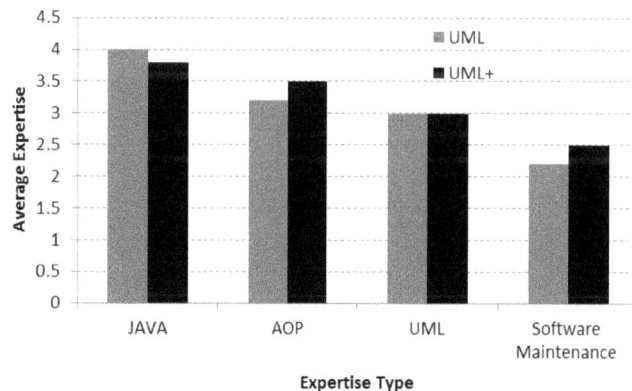

**Figure 4. Average expertise of the subject groups**

The participants were asked to answer a questionnaire in order to assess their level of expertise in areas that are essential in the experiment. For instance, they were questioned about their knowledge about Java, AOP, UML and software maintenance. Using a five-point Likert scale, we quantified the subject's level of expertise regarding Java, AOP, UML and software maintenance, varying from 0 (no knowledge) to 4 (expert). In particular, we required a minimum score of 2 for all fields. The result is illustrated in Figure 4. The participants of each group (UML or UML+) were chosen randomly. The random distribution of the participants is presented in Table 4 in terms of expertise level and groups.

**Table 4: Participant distribution: expertise vs. groups**

| Group | Participants Expertise Level | | |
|---|---|---|---|
| | 2 | 3 | 4 |
| UML | 5 | 5 | 4 |
| UML+ | 6 | 5 | 3 |

2 = good expertise / 3 = very good expertise / 4 = expert

## 3.5 Procedures

The participants were required to fill in an Internet-based questionnaire about their technical skills given their different locations. Based on their answers, we designed the experiment to be performed in 60-minute Internet-based sessions. The total execution time was determined by running a pilot experiment. Each session had four scenarios on the maintenance of composition code. Considering the high level of the subject's knowledge and in order to avoid bias towards the tasks, no tutorial or training on the required technologies were needed. We divided the participants into two groups of 14 participants each: one group using UML design diagrams and the other one using UML+ design diagrams. It is important to highlight that the participants were not familiar with the experimental design goal.

For the first group, the maintenance tasks (Section 3.3) were given to them, along with the plain UML diagrams and the corresponding slice of source code, which refers to the existing code – available to downloading – relevant to the task. The second group received the same tasks given to the first group. In addition, we provided UML+ design diagrams, where information about composition properties is explicitly represented. These models were enriched with the use of stereotypes and UML notes to explicitly represent information related to composition properties. The diagrams contained all the information about module and composition properties required to answer each task. The premise behind the separation of each group in two subsets aims at answering our research question (Section 3.1).

At the end of the experiment they were required to give their opinion on points such as time pressure and difficulty of the tasks. The participants were required to realize the experiment within 60 minutes. We decided to establish this time limit in order to prevent the participants from interrupting the experiment and restarting it later. We just estimated a time of 60 minutes in order to not get the participants tired and, thus, affect their performance. Fortunately, the adjustments in the complexity of the tasks promoted a desired effect as none of the participants reached the execution time estimate (60 minutes) to finish the experiment. The minimum and maximum execution time was 50 and 57 from UML+ subject and 51 and 55 minutes from UML subject respectively, except one UML+ participant who exceeded this time and needed 5 additional minutes. As a consequence, time was not taken into consideration in this experiment.

The experiment procedures can be summarized in three steps: (i) filling the Internet-based technical skill questionnaire, (ii) performing the code change tasks, and (iii) filling in the feedback questionnaire. The latter helped us to conduct a qualitative analysis and justify the quantitative differences between the UML and UML+ groups. In order to know if the participants took advantage of the provided UML-based design models, a set of questions was posed to them at the end of the experiment and also on the feedback questionnaire. These questions are summarized in Table 8.

We can summarize the procedures in four steps as illustrated in Figure 5: (i) the participants answered a technical skill questionnaire to capture their background and expertise; (ii) a set of tasks was designed and realized by two groups of participants: those who used UML models and those who used UML+; (iii) the answers were analysed; (iv) at the end, a feedback questionnaire was filled in by the participants. It is important to highlight that a set of tasks was designed and realized by a small group of participants in a pilot experiment. Based on the pilot experiment feedback, we adjusted the experiment in different ways, such as number of tasks and the required total execution time.

## 3.6 Variable and Analysis

The independent variables in our experiment are the UML-based design models. Based on them we consider one dependent variable in our experiment: the change quality (or solution quality) provided in each programming task. There is more than one solution available and they are associated with varying degrees of change. The given answer is evaluated considering the number of acceptable steps to realize each task. The change quality of each answer is measured as follows. Firstly, we transform the answers into quantitative values.

An entire correctly answered task is worth 1 point. Wrong answers count negatively. Thus, if the task contains 5 operations, each correct operation is worth 0.2 point. If the final score received in the task is 1 point, we classify it as correct. If it is between 0 and 1, not inclusive, we classify it as partially correct. Finally, if the final score is equal to 0 we classify it as incorrect.

```
01  class MainFrame {
02      int status;
03      GameController controller;
04      void startGame(){
05          ...
06          Controller.setStatus(status);
07      }
08      void stopGame(){ //stop the game  }
09  }
10  aspect ExecutionOrder {
11      declare precedence: BoardStartup,
                            NickNameDefinition
12  }
```

**Figure 6. Slice of code of task 1 using AspectJ**

The slice of code provided in Figure 6 was given to the participants in task 1. In this task they were required to add a new method called setupGame() using AspectJ mechanisms (see code slices provided in Figure 6). In order to realize this operation, we expected each solution to take at least 3 operations: (1) add a new aspect; (2) introduce a pointcut expression to advise the method startGame() (line 04 – Figure 6) before its call, and (3) include the new aspect in the declare precedence statement within the aspect ExecutionOrder (lines

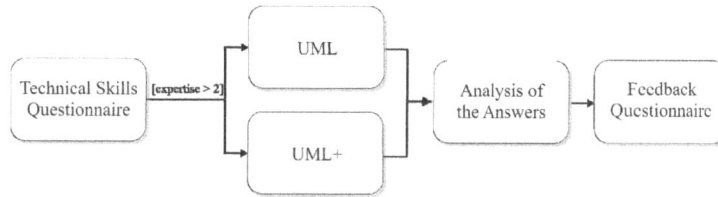

**Figure 5. The Experimental Overview**

**Table 5. Descriptive Statistics of the Experimental Results**

| | Group | Min | Max | Avg | Mean | % Diff | Shapiro p-value | Mann-Whitney p-value |
|---|---|---|---|---|---|---|---|---|
| Solution Quality | UML | 0.57 | 4.00 | 2.39 | 2.28 | 44.70 | 0.242 | 0.021 |
| | UML+ | 2.62 | 4.00 | 3.46 | 3.75 | | 0.006 | |

**Table 6. Participants' Performance per Programming Task**

| ORDER (*) | Task T1 | | Task T2 | | Task T3 | | Task T4 | |
|---|---|---|---|---|---|---|---|---|
| | UML | UML+ | UML | UML+ | UML | UML+ | UML | UML+ |
| 01 | 1.00 | 1.00 | 1.00 | 1.00 | 1.00 | 1.00 | 1.00 | 1.00 |
| 02 | 1.00 | 1.00 | 1.00 | 1.00 | 1.00 | 1.00 | 1.00 | 1.00 |
| 03 | 1.00 | 1.00 | 0.80 | 1.00 | 1.00 | 1.00 | 1.00 | 1.00 |
| 04 | 1.00 | 1.00 | 0.40 | 0.75 | 1.00 | 1.00 | 1.00 | 1.00 |
| 05 | 0.75 | 1.00 | 0.50 | 0.75 | 0.80 | 1.00 | 0.75 | 1.00 |
| 06 | 0.75 | 1.00 | 0.60 | 1.00 | 0.75 | 1.00 | 0.66 | 1.00 |
| 07 | 0.75 | 0.75 | 0.66 | 1.00 | 0.75 | 1.00 | 1.00 | 1.00 |
| 08 | 0.50 | 1.00 | 0.40 | 1.00 | 0.50 | 1.00 | 0.32 | 1.00 |
| 09 | 0.50 | 0.80 | 0.40 | 0.75 | 0.32 | 0.75 | 0.32 | 0.75 |
| 10 | 0.32 | 0.66 | 0.66 | 0.75 | 0.32 | 0.75 | 0.50 | 0.75 |
| 11 | 0.25 | 0.80 | 0.00 | 0.50 | 0.00 | 0.66 | 0.32 | 0.75 |
| 12 | 0.32 | 0.66 | 0.40 | 0.75 | 0.32 | 0.75 | 0.50 | 0.80 |
| 13 | 0.32 | 0.60 | 0.25 | 0.75 | 0.25 | 0.75 | 0.50 | 0.80 |
| 14 | 0.25 | 0.50 | 0.25 | 0.66 | 0.25 | 0.66 | 0.25 | 0.80 |
| **AVG** | **0.62** | **0.84** | **0.52** | **0.83** | **0.59** | **0.88** | **0.65** | **0.90** |

(*) Order numbers do not identify participants. They are used for reinforce the idea that each line in the table refers to a different participant

10 to 12). In this case each operation is worth approximately 0.33, which is the resulting value for the math operation 1 divided by 3 (total of operations). After analysing the participants' answers, a feedback questionnaire was applied. The goal was to understand both the level of satisfaction regarding the experiment execution and the relevance of the UML+ and UML models during the tasks.

## 4. DISCUSSION

This section presents and discusses the experimental results. First, we present the statistical test used in our analysis (Section 4.1). A qualitative discussion follows in Section 4.2. Finally, we summarize the main findings of our experiment (Section 4.3). A complete data analysis can be found at the website of this study [28]. There, the reader can also find additional results of UML-based models on code change quality.

## 4.1 Statistical Test

The statistical tests were used to evaluate the hypotheses listed in Section 3.1. To this end, we used the R language and environment[2]. First, we verified whether the sample was normalized by applying the Shapiro test. After that, we applied t-

test and Mann-Whitney; this latter is used when the sample distribution was not normal. Given that samples were not normalized, we selected the non-parametric Mann-Whitney test to analyse our hypothesis (Section 3.1). We used a confidence level of 95% ($\alpha$=0.05). The statistical results are summarized in Table 5. In summary, the UML+ group had 44.70% more acceptable solutions than the UML group. This means that our hypothesis can be accepted (Section 3.1).

**UML+ vs. UML: Change Quality.** Tables 6 and 7 summarize respectively the performance of all subjects and the statistical analysis per task. The values of each task vary from zero to 1, in which case the subject had a fully-acceptable solution to a given task. Figure 7 illustrates the quality distribution of the 56 answers from each group (14 subjects answering 4 questions). As illustrated in this figure, there is a significant advantage in favour of the group that worked with UML+ design models. For instance, none of the UML+ participants had a hit rate less than 25%. More importantly, the majority of UML+ participants provided answers with hit rate superior to 50%, but with a trend towards the range of 75-100%. The same consistent performance was not achieved by the UML group, where: (i) the hit rate tended to vary from poor (0-25%) to very good (76-100%), and (ii) most of the participants achieved a hit rate in the range of 26-50%.

---

[2] http://www.r-project.org/ (18/04/2012)

In Table 7 we present statistical data of our analysis per task. The average values presented in the second and third columns refer to the quality of the change given by all the participants of each group (UML and UML+). The most significant difference is 59.61%, which was identified for task T2 with a p-value of 0.003. We observed that most of the participants in the UML group were not conscious about the modules affected by the new advice required by the change. In this task, a new advice with an after returning semantics was added to the existing code in order to advise join point shadows placed in the class GameController. Figure 1 illustrates this example. The advice changes the attribute status by reference and its new value is incorrectly manipulated by the class Player. The subject did not realize that the composition scope also includes the class Player. Similarly to tasks T2, T3 and T4 are equally statistically relevant with p-values equal to 0.036 and 0.025, respectively.

In these tasks, the use of stereotypes to highlight the advised operations helps the participants recall the role of each module on the composition. In task T1 the scenario was different. Our statistical test indicated that in this task the availability of UML+ models (with enhanced composition information) did not make any difference in terms of the change quality. This happens because the information associated with the composition makes explicit the required execution order between aspects. As this information could be inferred from the UML models as well, the difference was not significant. Thus, the conclusion is that the composition models bring benefits in the realization of tasks regardless of their degree of complexity. Corroborating with our answer, the participants' feedback attested this. When asked about the tasks design, participants from both groups pointed out that tasks T2 and T4 presented a higher complexity when compared to the others. This consideration strengthens our perception that the explicit composition details in design models exert a positive influence regardless of the tasks complexity.

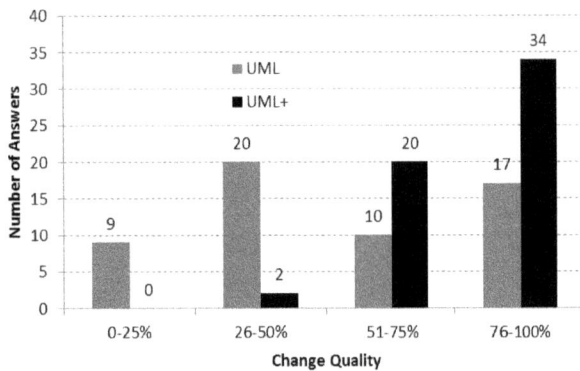

**Figure 7. Average percentage of change quality vs. number of answers provided**

**Table 7. Descriptive Statistics per Task**

| Task | UML | UML+ | % Diff | Mean(UML+) - Mean(UML) | Mann-Whitney |
|------|-----|------|--------|------------------------|--------------|
| T1 | 0.62 | 0.84 | 35.48 | 0.22 | 0.051 |
| T2 | 0.52 | 0.83 | 59.61 | 0.31 | 0.003 |
| T3 | 0.59 | 0.88 | 49.15 | 0.29 | 0.036 |
| T4 | 0.65 | 0.90 | 38.46 | 0.25 | 0.025 |

## 4.2 Qualitative Discussion

In terms of the change quality, the highest gain with UML+ was observed in task T2. This task requires that the subject takes into consideration that there are multiple modules affected by a single pointcut-advice pair. However, this broadly-scoped impact of the composition properties is only implicit in the code and in the UML models. The pointcut-advice pair added to the task T2 affected the value of the attribute status in the class Player. Thus, the gains reached in task T2 are associated with the composition details provided in the UML+ design models. These details make evident the *scope* associated with the use of the composition mechanism *after returning*, which was previously explained in Section 4.1.

**UML+ group succeed in the realization of tasks.** The benefits associated with the availability of UML+ design diagrams are attested by the participants. When asked *"which were the characteristics present in the source code and also in the UML-based diagrams that make the changes more difficult to be performed?"* 35.7% of the participants from the UML group answered both *"the lack of information associated with the aspects"* and *"the dependency 1:n between aspects and classes"*. This claim about the lack of information was not mentioned by any subject of the UML+ group, which makes us infer that this lack of information is associated with the need for specification of composition properties (Section 2.1).

In addition, it should be noted that the tasks T2 and T4 were pointed out by the participants as the most difficult ones. Interestingly, these same tasks were the ones that the UML+ group performed much better than the UML group. These results might suggest that the additional composition information available in UML+ design models became more useful as the task complexity increased. From our quantitative results we point out two main findings: (i) the use of UML+ was always favourable in terms of requiring less changes involving module composition, and (ii) the use of UML+ promotes gains in terms of the change quality regardless of subject expertise, which varies from intermediate to advanced (see Figure 4). We did not explicitly analyse the data with respect to the experience factor. However, we could assert that that expertise did not affect the key finding as the participants of both groups have approximately the same level of expertise and the UML+ users always yielded better results than UML participants. Finally, UML+ and UML groups consisted of participants with heterogeneous expertise, with different combinations of skills on UML, Java, AOP and maintenance (Figure 4).

**Implicit composition properties were detrimental to change quality.** The qualitative analysis derived from both groups (i.e., UML and UML+ design models) is presented in Table 8. The questions in this table appeared on the feedback Internet-based questionnaire (Section 3.5). The number of participants who answered each question positively is indicated in the second (UML group) and third (UML+ group) columns. For both groups, the majority of participants answered that overall the UML design models were useful for the realization of tasks (Q6, Q7 and Q8 – Table 8). When answering the question *"Which UML models were used?"*, the majority of the participants (around 70%) in UML and UML+ groups confirmed they used both class and sequence diagrams. Only two participants for both groups said that UML diagrams were not useful. The participants, who did not take advantage of UML design models, explained why they have not used the models: they were very concerned on spending more time by analysing them and, therefore, to not being able to realize all the tasks. Finally, a total of 9 and 17 participants used either the sequence diagrams or class diagrams only, respectively. Regarding the relevance of these diagrams in the execution of the tasks, the most frequent answer was "*the UML models were useful to understand how different modules interact with each other*". In this case, different modules are related to the involvement of both aspects and classes in a composition, which is associated with the composition diversity (Section 2.1).

Finally, most participants indicated that the two main factors that led them to change the existing code were: (i) the dependencies among classes and aspects are highly based on the program language syntax, and (ii) similar pattern of method signatures (see Table 8 – Q4 e Q5). These answers indicate that both UML and UML+ groups were aware of changes violating existing composition properties (Section 2.1). However, participants who used UML+ in their tasks also pointed out the scope of the composition (Q1) as a detrimental factor to generate a solution with better quality. This means that even though the participants do not have a deep understanding about the effect of these composition properties in evolving programs, they were able to identify them as detrimental factors to provide acceptable solution to the program maintenance tasks.

## 4.3 Solution Quality meets Stability

The UML+ design models were consistently beneficial in terms of the change quality. However, we observed the following phenomenon in the performance of both groups (UML and UML+): the more changes are required to realize a programming task (Section 3.3), the more prone are the participants to introduce mistakes in their answers. Figure 8 illustrates the existing correlation between the performance of the participants in terms of number of changes (change quality) and the percentage of changes required to realize the programming tasks. It is important to highlight that there is not only one way in which each task can be realized to obtain a solution. For this reason, the number of changed program elements can vary in each task.

However, taking Figure 7 into consideration, we can observe that the percentage of changes was lower in the UML+ group when compared to the UML group. For instance, when UML+ participants were required to add the method `int saveScore()` to the class `PersistentData` using AspectJ's mechanisms, some of them decided to add as well a new pointcut in the aspect `BoardTracing` instead of modifying both the existing pointcut

(line 08 – Figure 9) and advice (lines 10 to 12 – Figure 9). On the other hand, UML participants realized the same tasks changing the pointcut expression and also the advice of the aspect `Boardtracing`. In particular, this means that the UML participants changed more frequently the existing source code when compared to the UML+ group.

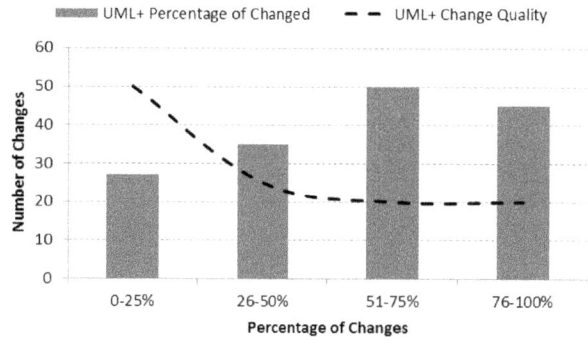

**Figure 8. Percentage of changes realized by UML+ group**

```
01 class PersistentData {
02     …
03     int saveJPG( ) { … }
04     int saveXML( ) { … }
05     …
06 }
07 aspect BoardTracing{
08     pointcut traceSaveBoard():call(int
               PersistentData.save*( ));
09     ...
10     after( ): traceSaveBoard (){
11         System.out.println("Board Saved");
12     }
13     …
14 }
```

**Figure 9. Slice of code from Task T4**

## 5. RELATED WORK

Related work can be divided into two categories: the impact of UML models on software maintenance (Section 5.1), and quality analysis of programs built with post object-oriented techniques (Section 5.2). None of related work discussed below analysed the impact of UML models designed with explicit composition properties. We provide an overview of these works and highlight the main differences when compared to ours.

**Table 8. Summary of Quality Questions**

| Question (Q) | UML | UML+ |
|---|---|---|
| Q1: Did the scope of the pointcuts affect the realization of tasks? | 2/14 | 3/14 |
| Q2: Did the occurrence of multiples aspects sharing the same joinpoint make the realization of tasks harder? | 3/14 | 2/14 |
| Q3: Did the existence of different types of modules, i.e., classes and aspects affect the realization of tasks? | 0/14 | 0/14 |
| Q4: Did the dependency among aspects and classes, which are highly based on the program language syntax, affect the realization of tasks? | 10/14 | 12/14 |
| Q5: Did the occurrence of methods with similar signature pattern (e.g., names starting with the prefixes "set" and "get") make the realization of tasks harder? | 9/14 | 7/14 |
| Q6: Did you take advantage of Class diagrams? | 12/14 | 12/14 |
| Q7: Did you take advantage of Sequence Diagrams? | 10/14 | 12/14 |
| Q8: Did you take advantage of composition properties specifications? | - | 11/14 |

## 5.1 The Impact of UML Models

Previous research work explored the impact of UML models on software maintenance [8][9][10][14][15]. All of them claim that the availability of UML design models provides developers with a more effective representation of several system properties. In their experimental investigation they evaluate whether using UML is profitable in a realistic context for a large project system. In particular, according to them, all of the participants found the UML diagrams useful in terms of maintenance. In addition, they also concluded that the use of UML models establishes a better communication among software developers. Nevertheless, all these approaches are different from ours because they have only focused on evaluating the general benefits of UML models in object-oriented systems. There is no evidence on what extent the use of UML models with proper composition specification improves the maintenance of software systems when post object-oriented techniques are used. For this reason, our work is a first attempt to empirically investigate the impact on software maintenance when developers use UML models enriched with composition properties.

## 5.2 Analysis of Advanced Mechanisms

Several works have analysed the use of AOP on software maintenance and development [7][23-27]. For instance, Figueiredo *et al.* [7] investigated the impact of AspectJ mechanisms on software evolution. The authors focused on the source code analysis based on a suite of modularity metrics. However, they do not discuss at all in terms of the use of modelling techniques for making the composition code properties explicit. Other works have evaluated the use of AspectJ on refactoring software systems (e.g., [23][24]). There are other works that have investigated specific attributes; e.g., how faults are introduced during maintenance tasks involving aspects [25] and the time spent by developers to evolve and manage the complexity of composition code [26][45]. However, similar to Figueiredo *et al.* [7], all these works only investigated the impact of AspectJ mechanisms on the quality of the source code without taking into consideration the impact of composition code properties on the software maintenance. Also, they do not provide any evaluation in terms of the impact of using modelling techniques on software maintenance. Thus, our work is different from the aforementioned ones as we are interested in analysing how the UML models with composition properties are effective to assist developers on software maintenance tasks.

## 6. THREATS TO VALIDITY

Some relevant threats to validity and the manners in which we have addressed them are discussed as follows [29].

**Internal Validity.** Threats to internal validity reside on the subject's expertise, subject's assignment bias, task bias and time restriction. Regardless of the fact that the participants may not be sufficiently expert, we have reduced this threat through *a priori* application of a technical skills questionnaire. It intends to assess the participants' expertise degree on dealing with the technologies involved in the study. According to the results of this questionnaire, we selected the participants with score equal or superior to 2 for all fields, which means a similar and high knowledge level. These procedures were important given the between-subject nature of the experiment design. Another question concerns the fair distribution of the subject's knowledge. In fact, they were randomly distributed to each group. However,

this threat was minimized since all of them have a high level of knowledge (score >=2) about the involved technologies (see Figure 4 and Table 4). Another prominent criticism is about the motivation of the participants and also their knowledge about the experimental goal. These criticisms are mitigated by the fact that all participants participated on a voluntary basis and although they could guess, they were not familiar with the research questions and hypothesis (Section 3.1). Another threat is associated with the assignment bias as we have a Between-Subject experiment. This threat was reduced as we selected participants with good and similar expertise. A final threat relies on the time to perform the tasks and their degree of difficulty. We attempted to mitigate this threat performing one pilot study with four extra participants who had the same level of knowledge required to attend the experiment. Based on their performance it was possible to adjust the level of difficulty and the required time to perform each task. With respect to the composition specification, it is true that the use of a detailed composition specification (UML+) could make the program maintenance hard to perform. In fact, we tried to select design slices that had many non-trivial characteristics of aspect-oriented designs (aspect inheritance, multiple aspects with global quantification - i.e., resulting in many crosscut relationships). On the other hand, we also considered that some good modelling practices were applied. In addition, we are considering the context here where the programmer is supported by a tool to navigate through the models and code, and it is, therefore, focusing only on the modules involved in maintenance task.

**External Validity.** External validity involves the extent to which the results of a study can be applied. . In order to minimize this threat, we chose developers with heterogeneous backgrounds who are, at most, professional developers and postgraduate students as well [42][43][44]. In addition, the evolving tasks by themselves can be considered a threat as they can be considered far from real evolving software scenarios. We tried to neutralize this threat with the choice of the application. When we decided to work with applications from well-known domains such as mobile devices and games, we took into consideration the facility of simulating real scenarios. This means that the functionalities discussed in our experiment mitigate the real functionality of cell phones and card games. In addition, as we are using programming scenarios from different subject applications (Section 3.2), we tend to generate more reliable results.

Generalization of results also depends on the scale of the tasks considered in the experiment. For context of this experiment design, the changes are, in fact, not expected to affect too many modules. On the contrary, the context of the tasks was assumed to one where developers were concerned with modularity principles; this is also why post object-oriented programming techniques were applied in the first place. A narrow scope of changes was also adopted for the experiment tasks as the target program was representative of those where good programming/design practices are adopted (e.g., the MVC architecture style and other design patterns in the case of the application used in the experiment). Hence, the maintenance tasks indeed did not require a high number of changes. We also tried to reduce the threat associated with this point mimicking maintenance scenarios of real applications (Section 3.2).

**Construct Validity.** The threat to construct validity includes the methodology used both for quantifying changes and reusability degree. In order to reduce this threat, we used a set of composition metrics that allowed us to predict changes and reusability from each task point of view. We adopted these metrics because all of them were empirically found to have correlation with design change [6]. In addition, they enabled us to make a more objective comparison with outcomes of relevant previous studies [6][7][18]. Another factor that contributes to neutralize is the data collection process. In this process, we consider partially correct answers following the criteria presented in Section 3.5.

# 7. FINAL CONSIDERATIONS AND FUTURE WORK

An Internet-based experiment was carried out in order to investigate the contributions associated with the availability of composition-enriched (UML+) design models during maintenance programming tasks. The UML models were used to support the program change tasks and they had a different degree of module composition information. Half of the participants realized the programming tasks using plain UML models. The second half realized the same programming tasks using UML+ diagrams.

Our results showed that, regardless of participants' expertise, developers who had an UML+ model outperformed the other case in terms of leading to better change quality. In addition, a complementary analysis highlighted the correlation between the change quality and the number of required changes to perform the tasks. These initial results demonstrated the benefits of explicit composition modelling and its usefulness to help participants in code maintenance tasks mainly with respect to change quality.

Further studies are still required to validate our results and thus the impact of explicit composition modelling on the realization of maintenance programming tasks. The main reason for this is that we need to better understand whether UML+ models also play a role when using other advanced programming techniques, such as feature-oriented programming. We hope that the issues outlined throughout the paper encourage researchers to replicate our study in the future under different circumstances.

# 8. ACKNOWLEDGMENT

This work has received full or partial funding from the following agencies and projects: DANSis/Faperj project (grant E-26/111.152/2011), Faperj (JCNE grant), CNPq (productivity grant 305526/2009- 0), PUC-Rio (productivity grant), Universal Project (grants 483882/2009-7, 483699/2009-8, 485348/2011-0), CAPES: international collaboration scheme (grant 5688-09) and CITI – PEst-OE/EEI/UI0527/2011.

# 9. REFERENCES

[1] Kiczales, G. *et al*. 1997. Aspect-oriented programming. In *Proceedings of the ECOOP*, 220-242.

[2] Aracic, I. *et al*. 2006. Overview of CaesarJ. *Transactions on AOSD I*, LNCS, 3880, 135-173.

[3] Rajan, H. and Sullivan, K. 2005. Classpects: unifying aspect- and object-oriented language design. In *Proceedings of the ICSE*, 59- 68.

[4] Bergmans, M. *et al*. 1992. Composition filters: extended expressiveness for OOPLs. Position paper for the OOPSLA: The Next Generation, Vancouver.

[5] Kuzniarz, L. et al. 2004. An empirical study on using stereotypes to improve understanding of UML models. In *Proceedings. of the IWPC*, IEEE Computer Society, 14–23.

[6] Dantas, F. and Garcia, A. 2010. Software Reuse versus Stability: Evaluating Advanced Programming Techniques. In *Proceedings. of the SBES* IEEE Computer Society, Washington, DC, USA, 40-49.

[7] Figueiredo, E. *et al*. 2008. Evolving software product lines with aspects: an empirical study on design stability. In *Proceedngs of the ICSE*, ACM, New York, NY, USA, 261-270.

[8] Dzidek, W. *et al*. 2008. A realistic empirical evaluation of the costs and benefits of UML in software maintenance. *IEEE Trans. Softw. Eng.*, vol. 3, 407-432.

[9] Bente Anda, Kai Hansen, Ingolf Gullesen, and Hanne Kristin Thorsen. 2006. Experiences from introducing UML-based development in a large safety-critical project. *Empirical Softw. Engg.* 11, 4 (December 2006), 555-581.

[10] Arisholm, E. *et al*. 2006. The impact of UML documentation on software maintenance: an experimental evaluation. *IEEE Trans. Software Eng.*, vol. 32, 365-381.

[11] Kuzniarz, L., Staron, M. and Wohlin, C. 2004. An empirical study on using stereotypes to improve understanding of UML models. In *Proceedings. of the IWPC*, IEEE Computer Society, 14–23.

[12] Tryggeseth. E. 1997. Report from an experiment: impact of documentation on maintenance. *Empirical Software Engineering: An International Journal*, vol. 2 (2), 201-207.

[13] Briand, L. C. 2003. Software documentation: how much is enough? *In Proceedings of the CSMR*, 13-15.

[14] Briand, L. C. *et al*. 2005. An experimental investigation of formality in UML-based development. *IEEE Transactions on Software Engineering*, vol. 31 (10), 833-849.

[15] Agarwal, R. and Sinha, A. 2003. Object-oriented modeling with UML: a study of developers' perceptions. *Commun ACM* 46(9): 248–256, September.

[16] Clarke, S. Composition of Object-Oriented Software Design Models, Ph.D. Thesis, Dublin City University, January 2001.

[17] Farias, K., Garcia, A. and Lucena, C.. Evaluating the Impact of Aspects on Inconsistency Detection Effort: A Controlled Experiment. In *Proc. of the 15th MODELS 2012*, Foundations Track, Innsbruck, Austria, September 30 - October 5, 2012.

[18] Dantas, F. *et al*. 2012. On the role of composition code properties on evolving programs. *In Proceedings of the ACM-IEEE international symposium on Empirical software engineering and measurement* (ESEM '12). ACM, New York, NY, USA, 291-300

[19] Chavez, C. *et al*. 2000. Taming heterogeneous aspects with crosscutting interfaces. In *Proceedings. of the SBES*, Uberlândia, Brazil, 216-231.

[20] Chavez, C and Lucena, C. 2002. A metamodel for aspect-oriented modeling. In *Proceedings of the AOM at AOSD*, Enschede, the Netherlands, 10-18.

[21] Buschmann, F. *et al*. 1996. Pattern-Oriented Software Architecture: a System of Patterns. John Wiley & Sons, Inc.

[22] Di Penta, M. *et al.* 2007. Designing your next empirical study on program comprehension. In Proceedings. of the ICPC IEEE Computer Society, Washington, DC, USA, 281-285.

[23] Kastner, C. *et al.* 2007. A case study implementing features using aspectj. In *Proceedings of the SPLC*, IEEE, USA, 223-232.

[24] Alves, V. *et al.* 2006. Refactoring product lines. In *Proceedings of the 5th international conference on Generative programming and component engineering* (GPCE '06). ACM, New York, NY, USA, 201-210.

[25] Burrows, R. *et al.* 2011. Reasoning about faults in aspect-oriented programs: a metrics-based evaluation. In *Proceedings of the ICPC*, 131-140.

[26] Endrikat, S. and Hanenberg. S. 2011. Is aspect-oriented programming a rewarding investment into future code changes? a socio-technical study on development and maintenance time. In *Proceedings of the ICPC*, Kingston, Canada, 51-60.

[27] Greenwood, P. *et al.* 2007. On the impact of aspectual decompositions on design stability: an empirical study. In *Proceedings of the ECOOP*, v. 4609, LNCS, 176-200, Springer.

[28] Enhancing Design Models with Composition Properties: A Software Maintenance Study, Website Study. http://www.inf.puc-rio.br/~fneto/aosd13/.

[29] Wohlin, C. *et al.* 2000. Experimentation in software engineering: an introduction. Kluwer Academic Publishers: Norwell, USA.

[30] Gurgel, A. *et al.* 2012. Integrating Software Product Lines: A Study of Reuse versus Stability. In *Proceedings of COMPSAC* 2012, 89-98.

[31] Camilleri, A., Coulson, G. and Blair, L. 2009. CIF: A Framework for Managing Integrity in Aspect-Oriented Composition. In *Proc. of the TOOLS* (47), 18-36

[32] Prehofer, C. 1997. Feature-Oriented Programming: A Fresh Look at Objects. In *Proc. of ECOOP*, 419-443

[33] Katz, E. and Katz, S. 2009. Modular verification of strongly invasive aspects: summary. In *Proc. of the FOAL*, 7-12

[34] Ferrari, F. *et al.* 2010. An Exploratory Study of Fault-Proneness in Evolving Aspect-Oriented Programs. In *Proc. of ICSE 2010*, 65 – 74.

[35] Schaefer, I. *et al.* 2010. Delta-Oriented Programming of Software Product Lines. In *Proc. of SPLC*, 77-91

[36] Macia, I. *et al.* 2012. On the Relevance of Code Anomalies for Identifying Architecture Degradation Symptoms. In *Proc. of CSMR*, 277-286

[37] Goldman, M. *et al.* 2010. MAVEN: modular aspect verification and interference analysis. Formal Methods in System Design 37(1): 61-92.

[38] Burrows, R. *et al.* 2010. The Impact of Coupling on the Fault-Proneness of Aspect-Oriented Programs: An Empirical Study. In *Proc. of ISSRE*, 329-338.

[39] Tourw, T. et al.. 2003. On the existence of the AOSD-evolution paradox. In *Proceedings of SPLAT: Software engineering Properties of Languages for Aspect Technologies*, 1-5.

[40] JBoss. Available at: http://www.jboss.org/

[41] Spring. Available at: http://www.springsource.org/

[42] M. Höst, B. Regnell, and C. Wohlin. 2000. Using Students as Subjects—A Comparative Study of Students and Professionals in Lead-Time Impact Assessment. *Empirical Software Engineering*, vol. 5, 201-214.

[43] E. Arisholm and D. I. K. Sjoberg. 2004. Evaluating the Effect of a Delegated versus Centralized Control Style on the Maintainability of Object-Oriented Software. *IEEE Transactions on Software Engineering*, vol. 30, 521-534.

[44] R. W. Holt, D. A. Boehm-Davis, and A. C. Shultz. 1987. Mental Representations of Programs for Student and Professional Programmers. In Proceedings of Empirical Studies of Programmers: Second Workshop33-46.

[45] V. R. Basili, F. Shull, and F. Lanubile, "Building Knowledge through Families of Experiments. 1999. *IEEE Transactions on Software Engineering*, vol. 25, 456-473.

[46] Nelio Cacho *et al.* 2006. Composing design patterns: a scalability study of aspect-oriented programming. In *Proceedings of the 5th international conference on AOSD*. ACM, New York, NY, USA, 109-121.

[47] Alessandro Garcia *et al.* 2005. Modularizing design patterns with aspects: a quantitative study. In *Proceedings of the 4th international conference on AOSD*. ACM, New York, NY, USA, 3-14.

[48] C. Lobato et al. Evolving and Composing Frameworks with Aspects: The MobiGrid Case. In *Proc. of the 7th ICCBSS*, Madrid, Spain, 25-29.

# Model-Driven Adaptive Delegation*

Phu H. Nguyen, Gregory Nain, Jacques Klein, Tejeddine Mouelhi, and Yves Le Traon
Interdisciplinary Centre for Security, Reliability and Trust (SnT), University of Luxembourg
4, rue Alphonse Weicker, L-2721 Luxembourg
{phuhong.nguyen, gregory.nain, jacques.klein, tejeddine.mouelhi,
yves.letraon}@uni.lu

## ABSTRACT

Model-Driven Security is a specialization of Model-Driven Engineering (MDE) that focuses on making security models productive, i.e., enforceable in the final deployment. Among the variety of models that have been studied in a MDE perspective, one can mention access control models that specify the access rights. So far, these models mainly focus on static definitions of access control policies, without taking into account the more complex, but essential, delegation of rights mechanism. User delegation is a meta-level mechanism for administrating access rights, which allows a user without any specific administrative privileges to delegate his/her access rights to another user. This paper analyses the main hard-points for introducing various delegation semantics in model-driven security and proposes a model-driven framework for 1) specifying access control, delegation and the business logic as separate concerns; 2) dynamically enforcing/weaving access control policies with various delegation features into security-critical systems; and 3) providing a flexibly dynamic adaptation strategy. We demonstrate the feasibility and effectiveness of our proposed solution through the proof-of-concept implementations of different systems.

## Categories and Subject Descriptors

D.2.11 [**Software Engineering**]: Software Architectures; K.6.m [**Management of Computing and Information Systems**]: Miscellaneous—*Security*

## General Terms

Design, Security

## Keywords

Model-driven security, model-driven engineering, model composition, delegation, access control, dynamic adaptation

*This work is supported by the Fonds National de la Recherche (FNR), Luxembourg, under the MITER project C10/IS/783852.

## 1. INTRODUCTION

Software security is a polymorphic concept that encompasses different viewpoints (hacker, security officer, end-user) and raises complex management issues when considering the ever increasing complexity and dynamism of modern software. In this perspective, designing, implementing and testing software for security is a hard task, especially because security is dynamic, meaning that a security policy can be updated at any time and that it must be kept aligned with the software evolution.

Managing access control to critical resources requires the dynamic enforcement of access control policies. Access control policies stipulate actors access rights to internal resources and ensure that users can only access the resources they are allowed to in a given context. A sound methodology supporting such security-critical systems development is extremely necessary because access control mechanisms cannot be "blindly" inserted into a system, but the overall system development must take access control aspects into account. Critical resources could be accessible to wrong (or even malicious) users just because of a small error in the specification or in the implementation of the access control policy.

Several design approaches like [20] [4] have been proposed to enable the enforcement of classical security models, such as Role-Based Access Control (RBAC) [25]. These approaches bridge the gap from the high-level definition of an access control policy to its enforcement in the running software, automating the dynamic deployment of a given access control policy. Although such a bridge is a prerequisite for the dynamic administration of a given access control policy, it is not sufficient to offer the advanced administration instruments that are necessary to efficiently manage access control. In particular, delegation of rights is a complex dimension of access control that has not yet been addressed by the adaptive access control mechanisms. User delegation is necessary for assigning permissions from one user to another user. An expressive design of access control must take into account all delegation requirements.

Delegation models based on RBAC management have been known as *secure*, *flexible* and *efficient* access management for resources sharing, especially on distributed environment. Flexible means that different subjects for delegation should be supported, i.e. delegation of roles, specific permissions or obligations. Also, different features of delegation should be supported, like temporary and recurrent delegations, transfer of role or permissions, delegation to multiple users, multi-step delegation, revocation, etc. However, the addition of flexibility for delegation must come with mechanisms to make sure that the security policy of the system is securely consistent. And last but not least, the administration of delegations must remain simple to be efficient.

Delegation is a complex problem to solve and to our best knowledge, there has been no complete approach for both specifying and dynamically enforcing access control policies by taking into account various characteristics of delegation. Having such an expressive security model is crucial in order to simplify the administrator task and to manage collaborative work securely, especially with the increase in shared information and distributed systems.

Based on previous work [20], in this paper we propose a new Modular Model-Driven Security solution to easily and separately specify 1) the business logic of the system without any security concern using a Domain Specific Modeling Language (DSML) for describing the architecture of a system in terms of components and bindings; 2) the "traditional" access control policy using a DSML based on a RBAC-based metamodel; 3) an advanced delegation policy based on a DSML dedicated to the delegation management. In this third DSML, delegation can be seen as a "meta-level" mechanism which impacts the existing access control policies similarly as an aspect can impact a base program. The security enforcement is enabled by leveraging automated model transformation/composition (from security model to architecture model). Consequently, in addition to [20], an advanced model composition is required to correctly handle the new delegation features.

To be more specific, only basic delegation features have been considered in [20]. Moreover, these delegation features have been handled as traditional access control rules. In this paper, we claim that delegation needs to be clearly separated from access control since a delegation impacts access control rules. Therefore, delegation and access control are not at the same level and should be separated. This separation involves an advanced model composition approach to dynamically know, at any time, which are the set of new access controls that have to be considered, i.e., the "normal" access control rules as well as the access control rules modified by the delegations. From a more technical point of view, the security enforcement is dynamically done by leveraging automated model transformation/composition (from security model to architecture model) and dynamic reconfiguration ability of modern adaptive execution platforms.

The contributions of this paper are the followings: 1) A metamodel/DSML dedicated to the delegation management, for specifying RBAC and RBAC-based delegation features. OCL constraints are used to check the consistency of the security policy (access control + delegation); 2) A model-driven framework for dynamically enforcing access control and delegation mechanisms specified with our DSMLs. In this framework, newly defined model transformation rules play an important role in the dynamic enforcement of security policies. We claim that to handle advanced delegation features, an ideal solution is to separate the delegation rules from the access control policy, each being specified in isolation, and then compose/weave them together to obtain a new access control policy reflecting the delegation-driven policy; 3) A flexibly dynamic adaptation strategy with better support for delegation, and the co-evolution of the security policy and the security-critical system.

The rest of this paper is organized as follows. Section 2 briefly presents the background on RBAC, delegation, and the security-driven model-based dynamic adaptation. Next, a running example is given in Section 3. It will be used throughout the paper to describe the diverse characteristics of delegation and illustrate the various aspects of our approach. In Section 4, we first give an overview of our approach. Then, we formalize our delegation mechanism based on RBAC and show how our delegation metamodel can be used to specify expressive access control policies that take into account various features of delegation. Based on the delegation metamodel, we describe our model transformation/composition rules used for transforming and weaving security policy into the architecture model. This section ends with a discussion of several security policy dynamic adaptation and evolution strategies. Section 5 describes three case studies that have been used for evaluating our approach. It is followed by Section 6 which presents related work. Section 7 concludes the paper and discusses the future work.

## 2. BACKGROUND

### 2.1 Access Control

Access Control [11] is known as one of the most important security mechanisms. It enables the regulation of user access to system resources by enforcing access control policies. A policy defines a set of access control rules which expresses: who has the right to access a given resource or not, and the way to access it, i.e. which actions a user can access under which conditions or contexts.

### 2.2 Delegation

In the field of access control, delegation is a very complex but important aspect that plays a key role in the administration mechanism [5]. A software system, which supports delegation, should allow its users without any specific administrative privileges to grant some authorizations. Delegation of rights allows a user, called the delegator, to delegate his/her access rights to another user, called the delegatee. By this delegation, the delegatee is allowed to perform the delegated roles/permissions on behalf of the delegator [9]. The delegator has full responsibility and accountability of the delegated accesses since he/she provides the accesses to the resources to other users, who are not initially authorized by the access control rules to access these resources.

Delegation is a powerful and very useful way to perform policy administration. On one hand, it allows users to temporarily modify the access control policy by delegating access rights. By delegation, a delegatee can perform the delegated job, without requiring the intervention of the security officer. On the other hand, the delegator and/or some specific authorized users should be supported to revoke the delegation either manually or automatically. In this way, the administrative task can be simplified and collaborative work can be managed securely, especially with the increase in shared information and distributed systems [1]. However, the simpler the administrative task is, the more complex features of delegation have to be properly specified and enforced in the software system. To the best of our knowledge, there is no approach for both specifying and dynamically enforcing access control policies taking into account all delegation features like temporary delegation, transfer delegation, multiple delegation, multi-step delegation, etc.

## 2.3 Security-Driven Model-Based Dynamic Adaptation

In [20], the authors propose to leverage MDE techniques to provide a very flexible approach for managing access control. On one hand, access control policies are defined by security experts, using a DSML, which describes the concepts of access control, as well as their relationships. On the other hand, the application is designed using another DSML for describing the architecture of a system in terms of components and bindings. This component-based software architecture only contains the business components of the application, which encapsulate the functionalities of the system, without any security concern. Then, they define mappings between both DSMLs describing how security concepts are mapped to architectural concepts. They use these mappings to fully generate an architecture that enforces the security rules. When the security policy is updated, the architecture is also updated. Finally, the proposed technique leverages the notion of models@runtime [17] in order to keep the architectural model (itself synchronized with the access control model) synchronized with the running system. This way, the running system can be dynamically updated in order to reflect changes in the security policy. Only users who have the right to access a resource can actually access this resource. The different steps of this approach are summed up in Fig. 1.

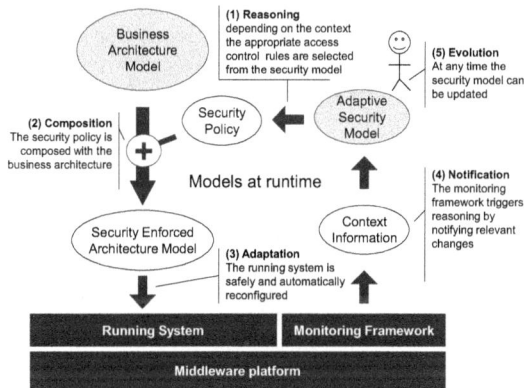

**Figure 1: Overview of the Model-Driven Security Approach of [20]**

## 3. A RUNNING EXAMPLE

In this section, we give a motivating example which will be used throughout the paper for describing the diverse characteristics of delegation and illustrating the various aspects of our approach.

Let us consider a library management system (LMS) providing library services with security concerns like access control and delegation management. There are two types of user account: personnel accounts (director, secretary, administrator and librarian) are managed by administrator; and borrower accounts (lecturer and student) are managed by secretary. The director of the library has the same accesses as the secretary, but additionally, he can also consult the personnel accounts. The librarian can consult the borrower accounts. A secretary can add new books in the LMS when they are delivered. Lecturers and students can borrow, reserve and return books, etc. In general, the library is organized with the following entities and security rules.

*Roles (users): access rights (e.g. working days)*
**Director** (*Bill*): consult personnel account, consult, create, update, and delete borrower account.
**Secretary** (*Bob* and *Alice*): consult, create, update, and delete borrower account, deliver book.
**Administrator** (*Sam* and *Tom*): consult, create, update, and delete personnel account.
**Librarian** (*Jane* and *John*): consult borrower account, find book by state, find book by keyword, report a book damaged, report a book repaired, fix a book.
**Lecturer** (*Paul*) and **Student** (*Mary*): find book by keyword, reserve, borrow and return book.

*Resources and actions to be protected*
**Personnel Account**: consult, create, update, and delete personnel account.
**Borrower Account**: consult, create, update, and delete borrower account.
**Book**: report a book damaged, report a book repaired, borrow a book, deliver a book, find book by keyword, find book by state, fix a book, reserve a book, return a book

In this organization, users may need to delegate some of their authorities to other users. For instance, the director may need the help of a secretary to replace him during his absence. A librarian may delegate his/her authorities to an administrator during a maintenance day.

It is possible to only specify role or action delegations by using the DSML described in [20]. For instance, a role delegation rule can be created to specify that *Bill*, the director (prior to his vacation) delegates his role to *Bob*, one of his secretaries. But it is impossible for *Bill* to define whether or not *Bob* may re-delegate the *director* role to someone else (in case *Bob* is also absent for some reason). The role delegation of *Bill* to *Bob* is also handled manually: it is enforced when *Bill* creates the delegation rule and only revoked when *Bill* deletes this rule. There is no way for *Bill* to define a temporary delegation, i.e. its active duration is automatically handled. Obviously the DSML described in [20] is not expressive enough to specify complex characteristics of delegation.

There are many delegation situations that motivate our work. We consider in the following some delegation situations:
**1.** The director (*Bill*) delegates his role to a secretary (*Bob*) during his vacation (the delegation is automatically activated at the start of his vacation and revoked at the end of his vacation).
**2.** A secretary (*Alice*) delegates her task/action of create borrower account to a librarian (Jane).
**3.** A secretary (*Bob*) transfers his role to an administrator (*Sam*) during maintenance day. In case of a transfer delegation, the delegator temporarily looses his/her rights during the time of delegation.
**4.** The role administrator is not delegable.
**5.** The permission of deleting borrower account is not delegable.
**6.** The director can delegate, on behalf of a secretary, the secretary's role (or some his/her permitted actions) to a librarian (e.g. during the secretary's absence).
**7.** If a librarian empowered in role *secretary* by delegation is no longer able to perform this task, then he/she can delegate, again, this role to another librarian.

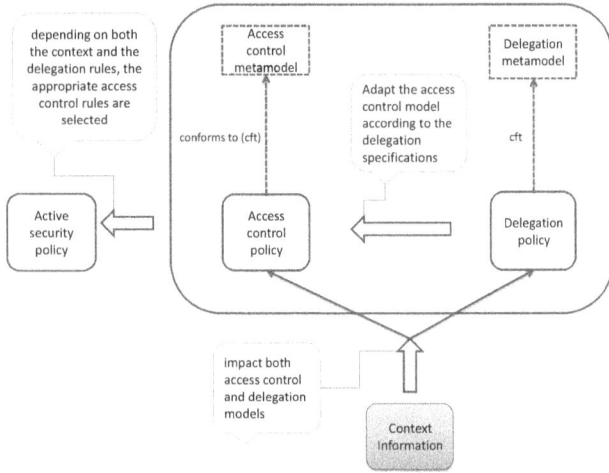

Figure 2: Delegation impacting Access Control

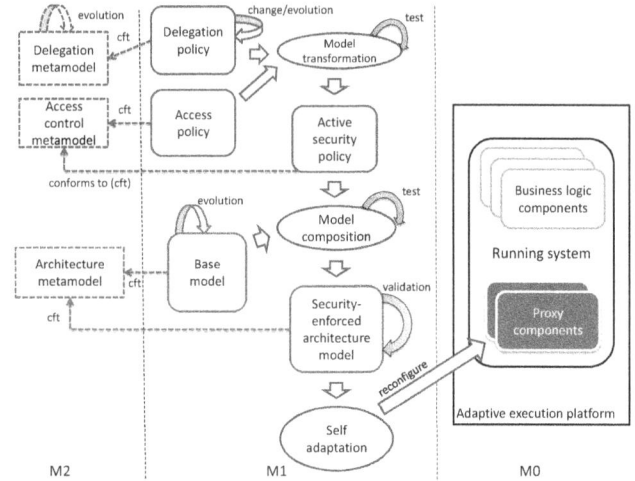

Figure 3: Overview of our approach

**8.** The secretary empowered in role *director* by delegation is not allowed to delegate/transfer, again, this role to another secretary.

**9.** A secretary is allowed to delegate his/her role to a librarian only and to one librarian at a given time.

**10.** A secretary is allowed to delegate his/her task of book delivery to a librarian only and scheduled on every Monday.

**11.** *Bill* can delegate his role and permitted actions only to *Bob*

**12.** *Bob* is not allowed to delegate his role.

**13.** *Alice* is not allowed to delegate her permitted action of book delivery.

**14.** Users can always revoke their own delegations.

**15.** The director can revoke users from their delegated roles.

**16.** A secretary can revoke librarians empowered in *secretary* role by delegation, even if he/she is not the grantor of this delegation (e.g. the grantor is the director or another librarian).

It can be seen that there are two levels of delegation rules: user-level (rules defined by a user: e.g. situations 1, 2, 3) and master-level (rules defined by a security officer: e.g. 4, 5, 6). Normally, delegations at user-level have to conform to rules at master-level. For example, the security officer can define that users of role *director* are able to delegate on behalf of users of role *secretary*. Then at user-level, *Bill* (director) can create a delegation rule to delegate, on behalf of *Alice*, her role (secretary) to *Jane* (librarian).

# 4. A MODEL-DRIVEN APPROACH FOR ADAPTIVE DELEGATION

## 4.1 Overview of our approach

In our approach, delegation is considered as a "meta-level" mechanism which impacts the existing access control policies, like an aspect can impact a base program. We claim that to handle advanced delegation rules, an ideal solution is to separate the delegation rules from the access control policy, each being specified in isolation, and then compose/weave them together to obtain a new access control policy (called active security policy) reflecting the delegation-driven policy (Fig. 2). We present our metamodel (DSML) for specifying delegation based on RBAC in Section 4.2.

The separation of concerns is not only between delegation and access control, but also with the business logic of the system. Fig. 3 presents a wider view of the overall approach. In order to enforce security policy to the system, the core business architecture model of the system is composed with the active security policy previously obtained. The architecture model is expressed in another DSML, called architecture metamodel (an architecture modeling language described in [20]). The idea is to reflect security policy into the system at the architecture level. Section 4.3 defines transformation rules to show how security concepts are mapped into architectural concepts.

The security-enforced architecture model obtained above is a pure architecture model which by itself reflects how the security policy is enforced in the system. It is important to note that the security-enforced architecture model is not used for generating the whole system but only the proxy components. These proxy components can be adapted and integrated with the running system at runtime to physically enforce the security policy. The adaptation and integration can be done by leveraging the runtime adaptation mechanisms provided by modern adaptive execution middleware platforms. The approach of generating proxy components overcomes some main limitations of [20]. Section 4.4 is dedicated to discuss our strategy for adaptation and evolution.

## 4.2 Delegation metamodel

Our metamodel displayed in Fig. 4 defines the conceptual elements and their relationships that can be used to specify access control and delegation policies. Because delegation mechanism is based on RBAC, we first explain the main conceptual elements of role-based access control. Then, we show how our conceptual elements of delegation, based on the RBAC conceptual elements, can be used to specify various delegation features.

As shown in Fig. 4, the root element of our metamodel is the *Policy*. It contains *Users*, *Roles*, *Resources*, *Rules*, and *Contexts*. Each user has one role. A security officer can specify all the roles in the system, e.g. *admin*, *director*, etc., via the *Role* element. In order to specify an access control policy, the security officer should have defined in advance the *resources* that should be protected from unauthorized access. Each resource contains some *actions* which are only

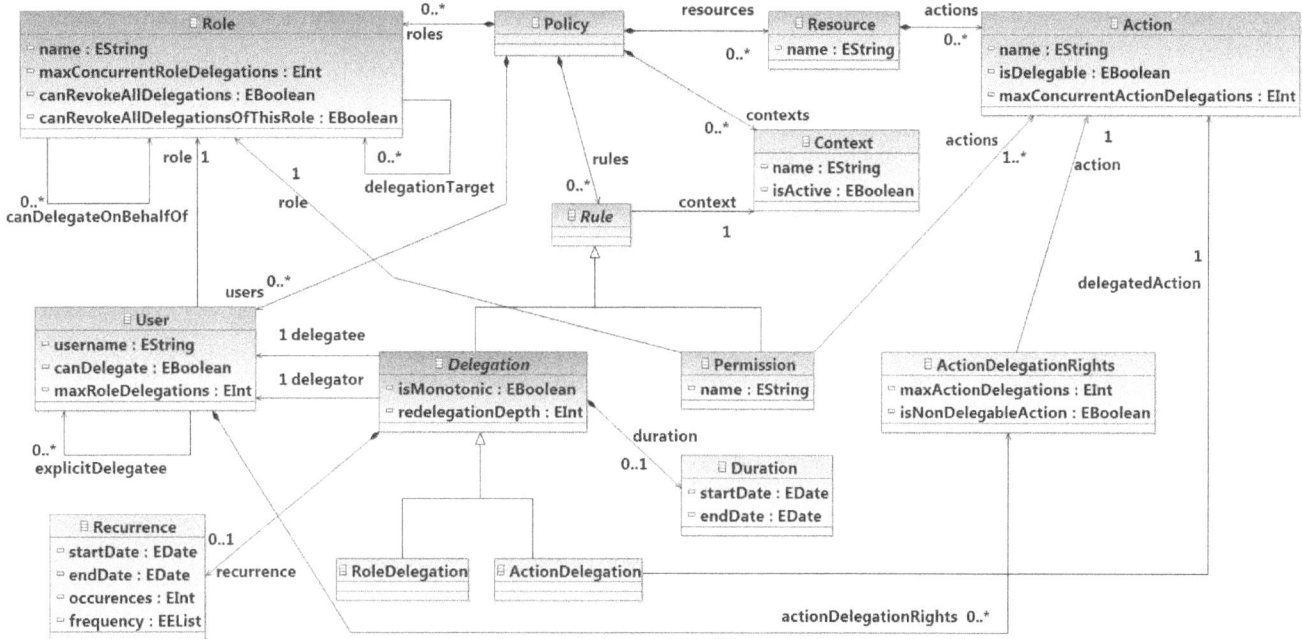

Figure 4: The Delegation Metamodel

accessible to authorized users. These protections are defined in rules: permission rules and delegation rules. Permission rules are used to specify which actions are accessible to users based on their *roles*. That means, without delegation rules or user-specific access control rules, every user is able to access the actions associated with his/her role only. Delegation rules are used to specify which actions are accessible to users by delegation. There are two basic types of delegation:

- **Role delegation**: When users empowered in role(s) delegated by other user(s), they are allowed to access not only actions associated with their roles but also actions associated with the delegated role(s).

- **Action delegation**: Instead of delegating their roles, users may want to delegate only some specific actions associated with their roles.

Another important aspect of our access control and delegation framework is the notion of *context*. It can be seen from our metamodel that every permission/delegation rule is associated with a context. A rule is only active within its context. The concept of context actually provides our model with high flexibility. Security policies can be easily adapted according to different contexts.

The full metamodel for specifying delegation is displayed in Fig. 4. It depicts the features that are supported by our delegation framework. All delegation management features are developed based on two basic types of delegation mentioned above. In the following, we present the delegation features and show how they can be specified, w.r.t. our metamodel.

- **Temporary delegation**: This is one of the most common types of delegation used by users. It describes when the delegation *starts* to be active and when it *ends*. The grantor can specify that the delegated role/action is authorized only during a given time interval, e.g. situation 1 of the running example in Section 3. Actually, this can be specified using the recurrence of delegation described below, but we want to define it separately because of its common use.

- **Monotonicity (Transfer of role or permissions)**: A property *isMonotonic* can be used to specify if a delegation is monotonic or non-monotonic. The former (*isMonotonic = true*) specifies that the delegated access right is available to both the delegator and delegatee after enforcing this delegation. The latter (*isMonotonic = false*) means the delegated role/action is transferred to the delegatee, and the delegator temporarily loses his rights while delegating, e.g. situation 3. In this case, the delegation is called a transfer.

- **Recurrence**: It refers to the repetition of the delegation. A user may want to delegate his role to someone else for instance every week on Monday. Recurrence defines how the delegation is repeated over time. It is similar to what is implemented in calendar system and more precisely the icalendar standard (RFC2445[1]). It has several properties; the *startDate* and *endDate* are the starting and ending dates of the recurrence. In addition, the *startDate* defines the first occurrence of the delegation. The *frequency* indicates one of the three predefined types of frequency, daily, weekly or monthly. The *occurrences* is the number of times to repeat the delegation. If the *occurrences* is for instance equals to 2 it means that it should only be repeated twice even when the *endDate* is not reached. An example of this delegation is situation 10.

- **Delegable roles and delegable actions**: These kinds of delegation define which roles and actions can be delegated and how. A policy officer can specify that a role can only be delegated/transferred to specific role(s), e.g. situation 9. If no *delegationTarget* is defined for a role, this role cannot be delegated/transferred, e.g. situation 4. If a role or action (*isDelegable = false*) is not delegable, it should never be included in a delegation rule. Moreover, a role can also be delegated by a user not having this role but his/her own role is specified as can delegate on behalf of a user in this role (*canDelegateOnBehalfOf = true*), e.g. situation 6.

- **Multiple delegations**: It should be possible to define the

[1]http://www.rfc-editor.org/info/rfc2445

max number of concurrent delegations in which the same role or action can be delegated at a given time. The properties *maxConcurrentRoleDelegations* and *maxConcurrentActionDelegations* define how many concurrent delegations of the same role/action can be granted, e.g. situation 9. Moreover, it is possible to define for each specific user a specific maximum number of concurrent delegations of the same role/action: *maxRoleDelegations* and *maxActionDelegations*.
- **User specific delegation rights**: All user-specific elements are used to define more strict rules for a specific user rather for his/her role. There are other user-specific delegations than *maxRoleDelegations* and *maxActionDelegations*. It is possible to define that a specific user is allowed to delegate his role/permitted action(s) or not (*canDelegate = true* or *false*), e.g. situation 12. The property *isNonDelegableAction* specifies an action that a specific user cannot delegate, e.g. situation 13. Moreover, the security officer can define to which explicit user(s) only (*explicitDelegatee*) a user can delegate/transfer his role to, e.g. situation 11.
- **Multi-step delegation**: It provides flexibility in authority management, e.g. situations 7, 8. The property *redelegationDepth* is used to define whether or not the role/action of a delegation can be delegated again. When a grantor creates a new delegation, he/she can specify how many times the delegated role/action can be re-delegated. If the *redelegationDepth = 0*, it means that the role/action cannot be delegated anymore, e.g. situation 8. If the *redelegationDepth > 0*, that means the role/action can be delegated again and each time it is re-delegated, the *redelegationDepth* is decreased by 1.
- **Revocations**: All users can revoke their own delegations, e.g. situation 14. Security officer may set *canRevokeAllDelegations = true* for a role with a super revocation power in such a way that a user empowered in this role can revoke all delegations, e.g. situation 15. Moreover, a role can also be defined such that every user empowered in this role can revoke any delegation from this role (*canRevokeAllDelegationsOfThisRole = true*), even he/she is not the grantor of the delegation, e.g. situation 16.

Moreover, each possible instance of the security policy has to satisfy all necessary validation condition expressed as OCL invariants. For example[2], we can make sure that no delegation is out of target, meaning that delegatee's role has to be a delegation target of delegator's role:
context Delegation **inv** NoDelegationOutOfTarget: self.delegator.role.delegationTarget −>exists (t | t = self.delegatee.role)
Or to check that for every user, the number of concurrent role delegations cannot be over its thresholds:
**context** User **inv** NoRoleDelegationOverMax: RoleDelegation.allInstances −>select (d | d.delegator = self) −>size() ≤ self.role.maxConcurrentRoleDelegations and RoleDelegation.allInstances −>select (d | d.delegator = self) −>size() ≤ self.maxRoleDelegations
Other examples are to restrict the value of the *redelegationDepth* must not be negative, or *startDate* cannot be later than *endDate*:
**context** Delegation **inv** NonNegativeDeleDepth: self.redelegationDepth ≥ 0
**context** Duration **inv** ValidDates: self.startDate ≤ self.endDate

---

[2]Due to space restrictions, the OCL expressions presented here are not exhaustive.

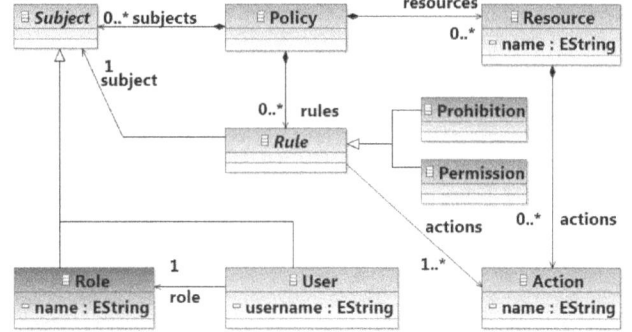

**Figure 5: A pure RBAC metamodel**

## 4.3 Transformations/Compositions

After specifying security policies by the DSML described in Section 4.2, it is crucial to dynamically enforce these policies into the running system. Transformations play an important role in the dynamic enforcement process. Via model transformations, security models containing delegation rules and access control rules are automatically transformed into component-based architecture models. Note that instances of security models and architecture models are checked before and after model transformations, using predefined OCL constraints.

The model transformation is executed according to a set of transformation rules. The purpose of defining transformation rules is to correctly reflect security policy at the architectural level. Based on transformation rules, security policy is automatically transformed to proxy components, which are then integrated to the business logic components of the system in order to enforce the security rules. The metamodel of component-based architecture can be found in [20] and an instance of it can be seen in Fig. 7. We first describe the transformation that derives an access control model according to delegation rules (Section 4.3.1), and then describe another transformation to show how security policy can be reflected at the architecture level (Section 4.3.2). Moreover, we also show an alternative way of transformation that combines two steps into one step.

### 4.3.1 Adapting RBAC policy model to reflect delegation

Within the security model shown in Fig. 2, delegation rules are considered as "meta-level" mechanisms that impact the access control rules. The appropriate access control rules and delegation rules are selected depending on the context information and/or the request of changing security rules coming from the system at runtime. According to the currently active context (e.g. WorkingDays), only **in-context** delegation rules and **in-context** access control rules of the security model (e.g. rules that are defined with context = WorkingDays) are taken into account to derive the active security policy model (Fig. 2). Theoretically, we could say that delegation rules impact the core RBAC elements in the security model in order to derive a pure RBAC model (without any delegation and context elements) which conforms to a "pure" metamodel of RBAC (Fig. 5). Delegation elements of a security policy model are transformed as follows:
**A.1**: Each **action delegation** is transformed into a new permission rule. The *subject* of the permission is *user* (dele-

gatee) object. The set of *actions* of the permission contains the delegated action.

**A.2**: Each **role delegation** is transformed as follows. First, a set of actions associated to a role is identified from the permissions of this role. Then, each action is transformed into a permission like transforming an **action delegation** described above.

**A.3**: A **temporary delegation** is only taken into account in the transformation if it is in active duration defined by the **start** and **end** properties. In fact, when its active duration starts the (temporary) action/role delegation is transformed into permission rule(s) as described above. When its active duration ends the temporary delegation is removed from the policy model.

**A.4**: If an action delegation is of type transfer delegation (**monotonic**), then it is transformed into a permission rule and a prohibition rule. The *subject* of the permission is the *user*-delegatee object. The set of *actions* of the permission contains the delegated action. The *subject* of the prohibition is the *user*-delegator object. The set of *actions* of the prohibition contains the delegated action.

**A.5**: If a role delegation is of type transfer delegation, then it is also transformed into a permission rule and a prohibition rule. The *subject* of the permission is the *user*-delegatee object. The set of *actions* of the permission contains the delegated actions. The delegated actions here are the actions associated with this role. The *subject* of the prohibition is the *user*-delegator object. The set of *actions* of the prohibition also contains the delegated actions.

**A.6**: If a delegation rule is defined with a **recurrence**, based on the values set to the recurrence, the delegation rule is only taken into account in the transformation within its *fromDate* and *untilDate*, repeated by *frequency* and limited by *occurrences*. In other words, only active (during recurrence) delegation rules are transformed.

**A.7**: (User-specific) If a user is associated with any **nondelegable action**, the action delegation containing this action and this user (as delegator) is not transformed into a permission rule. Similarly, if a user is specified as he/she **cannot delegate** his/her role/action, no role/action delegation involving this user is transformed.

**A.8**: (Role/action-specific) Any delegation rule with a **nondelegable** role/action will not be transformed. In fact, a delegation rule is only transformed if it satisfies (at least) both user-specific and role/action-specific requirements.

**A.9**: Only a role delegation to a user (delegatee) whose role is in the set of **delegationTarget** will be considered in the transformation.

**A.10**: Before any delegation is taken into account in the transformation, it has to satisfy the requirements of **max concurrent action/role delegations**. Note that the user-specific values have higher priorities than the role-specific values.

**A.11**: A delegation is only transformed if its **redelegationDepth** > 0. Whenever a user empowered in a role/an action by delegation re-delegates this role/action, the newly created delegation is assigned a **redelegationDepth** = the previous **redelegationDepth** - 1.

After transforming all delegation rules, we obtain a pure RBAC model which reflects both the delegation model and access control model. This pure RBAC model is then transformed into a security-enforced architecture model as described next.

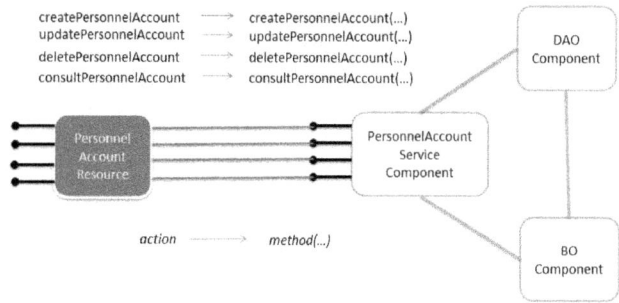

**Figure 6: Mapping Resources to Business Logic Components**

### 4.3.2 Transformation of Security Policy to Component-based Architecture

The transformation rules are defined below. The goal is to transform every security policy model (pure RBAC model obtained in step 1) which conforms to the metamodel shown in Fig. 5 to a component-based architecture model which conforms to the metamodel described in [20]. However, both the security policy model and the *base model* provided by a system designer are used as inputs for the model transformation/composition. Via a graphical editor, the security designer must define in advance how the resource elements in the policy model are related to the business components in the *base model*. Fig. 6 shows how each action in the policy can be mapped to the Java method in the business logic.

Because the *base model* already conforms to the architecture metamodel, we now only focus on transforming the security policy model into the security-reflected architecture model. As we know, this transformation/composition process will also weave the security-reflected elements into the base model in order to obtain the security-enforced architecture model.

The core elements of RBAC like *resource*, *role*, and *user* are transformed following these transformation rules. All the transformation rules make sure that the security policy is reflected at the architectural level.

**R-A.1**: Each *resource* is transformed into a component *instance*, called a *resource* proxy component. According to the relationship between the resource elements in the policy model and the business components in the *base model*, each *resource* proxy component is connected to a set of business components via bindings. To be more specific, each action of a resource element is linked to an operation of a business component (Fig. 6). By connecting to business components, a *resource* proxy component provides and requires all the services (actions) offered by the resource.

**R-A.2**: Each *role* is also transformed into a *role* proxy component. According to the granted accesses (permission rules associated with this role) to the services provided by the resources, the corresponding *role* proxy component is connected to some *resource* proxy component(s) (Fig. 7). A *role* proxy component is connected to a *resource* proxy component by transforming granted accesses into ports and bindings. Each (active) access granted to a role is transformed into a pair of ports: a client port associated with the *role* proxy component, a server port associated with the *resource* proxy component, and a binding linking these ports.

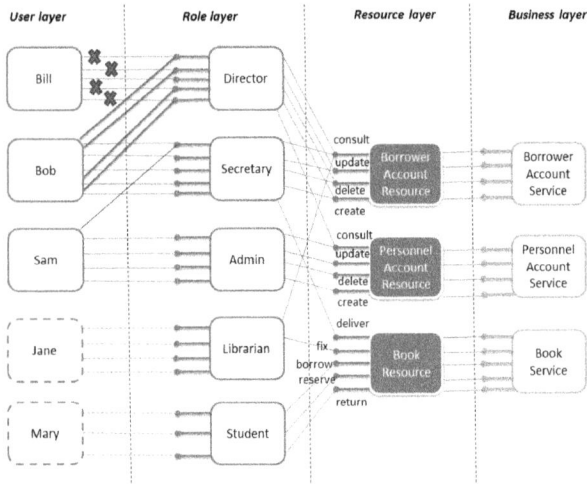

**Figure 7: Architecture reflecting security policy before and after adding a delegation rule (bold lines)**

**R-A.3**: Each *user* element defined in the policy model is also transformed into a *user* proxy component. Because each user must have one role, each *user* proxy component is connected to the corresponding *role* proxy component. However, each user may have access to actions associated to not only his/her role but also to actions associated to other roles by delegation. Thus, each *user* proxy component may connect to several *role* proxy components. The connection is established by transforming each access granted to a user into a pair of ports: a client port associated with the *user* proxy component, a server port associated with the corresponding *role* proxy component (providing the access/port), and a binding linking these ports (Fig. 7). Actually, the granted accesses are calculated not only from *permission* rules but also from *prohibition* rules. Simply, the granted accesses that equal permissions exclude prohibitions.

In our approach, revocation of a delegation simply consists in deleting the corresponding delegation rule. In this way, the revocation is reflected at the architectural level and physically enforced in the running system. Moreover, both the delegator and delegatee elements will be removed if these users are not involved in any delegation rules. As described above, user elements are transformed into proxy components. However, it is important to stress that only users involved in delegation rules (e.g. Bill, Bob and Sam in Fig. 7) are created in the security policy model and transformed into proxy components. Users who are not involved in any delegation rules (e.g. Jane and Mary in Fig. 7), are manipulated as session objects which directly access the services offered by the corresponding role proxy components.

Two steps described above are two separate model transformations that mainly used to explain how delegation can be considered as a "meta-level" mechanism for administrating access rights. The first model transformation is to transform a delegation-driven security model into a pure RBAC model. The second model transformation is to transform the RBAC model into an architecture model. In fact, these two steps could be done in only one model transformation that directly transforms the delegations, the access control policy and the business logic model into an architecture model reflecting the security policy. However, this alternative way

(described in the following) has the disadvantage of losing the intermediate security model (the active security policy) that could be useful for traceability purpose.

### 4.3.3 An alternative way: using only one transformation

In this way, we have to define other transformation rules to transform directly every security policy model which conforms to the metamodel, shown in Fig. 4, to a component-based architecture model which conforms to the architecture metamodel described in [20].

Core elements of RBAC like *resources*, *roles*, and *users* are transformed following these transformation rules:

**R-B.1**: Each *resource* is transformed into a component *instance*, called a *resource* proxy component (already presented).

**R-B.2**: Each *role* also is transformed into a *role* proxy component (already presented). The only difference is the *context* has to be taken into account (in Section 4.3.2, no *context* existed because *context* already dealt with in Section 4.3.1). Because every permission is associated with a *context*, we only transform permissions with the *context* that is active at the moment.

**R-B.3**: Each *user* element defined in the policy model is also transformed into a *user* proxy component. However, the connection (via bindings) from a *user* proxy component to the *role* proxy component(s) is not only depended on the user's role but also delegation rules that the corresponding user involved in. The transformation of delegation rules is presented below.

All the transformation rules above make sure that access control rules are reflected at the architecture level. However, the delegation rules will impact this transformation process in order to derive the security-enforced architecture model reflecting both access control and delegation policy. Delegation elements of a policy model are transformed as follows:

**R-B.4**: Each action involved in an **action delegation** is transformed into a pair of ports and a binding. A client port (representing the required action) is associated with the user (*delegatee*) proxy component. The binding links the client port to the corresponding server port (representing the same action provided) that associated with the *role* proxy component reflecting the role of the *delegator*.

**R-B.5**: Each **role delegation** is transformed in a similar way as **action delegation**. First, a set of actions associated to a role can be identified from the permissions of this role. Then, each action in the set is transformed into a pair of ports and a binding as transforming an action delegation.

**R-B.6**: A **temporary delegation** is only transformed into bindings if it is still in active duration defined by **start** and **end** properties.

**R-B.7**: If a delegation is of type **transfer delegation**, then both user elements (delegator and delegatee) are transformed into delegator and delegatee proxy components as described above. The delegator proxy component is not connected to the corresponding role proxy component because he/she already transfered his/her access rights to the delegatee. Fig. 7 shows a change in the architecture when *Bill* transfers his role to *Bob*.

**R-B.8**: If a delegation is defined with a **recurrence**, based on the values set to recurrence, the delegation rule is only active during the recurrence (similar to A.6).

**R-B.9**: If a user is associated with any **non-delegable ac-**

**tion**, the delegation of this action is not taken into account while doing the transformation. Similarly, if a user is specified as he/she **can not delegate** his/her role/action, no delegation requested by this user will be transformed.

**R-B.10**: Only a role delegation to a user (delegatee) whose role is in the set of **delegationTarget** will be consider in the transformation.

**R-B.11**: Before any delegation is taken into account in the transformation, it has to satisfy the requirements of **max concurrent action/role delegations**. Note that the user-specific values have higher priorities than the role-specific values.

**R-B.12**: A delegation is only transformed if its **redelegationDepth** $> 0$. Whenever a user empowered in a role/an action by delegation re-delegates this role/action, the newly created delegation is assigned a **redelegationDepth** = the previous **redelegationDepth** - 1.

By taking into account delegation rules while transforming access control rules of policy model into security-enforced architecture model, both delegation and access control rules are reflected at the architecture level.

## 4.4 Adaptation and Evolution strategies

The model transformation/composition presented in Section 4.3 ensures that the security policies are correctly and automatically reflected in an architectural model of the system. The key steps to support delegation (i.e. specifications and transformations) are already presented in Sections 4.2 and 4.3. The last step consists in a physical enforcement of the security policy by means of a dynamic adaptation of the running system. In this section, our adaptation and evolution strategy is discussed.

### 4.4.1 Adaptation

The input for the adaptation process is a newly created security-enforced architecture model (Fig. 8). First, this new architecture model is validated using simulation or invariant checking [18]. This valid architectural model actually represents the new system state the runtime must reach to enforce the new security policy of the system. According to the classical MAPE control loop of self-adaptive applications, our reasoning process performs a comparison (using EMFCommpare) between the new architecture model (target configuration) and the current architecture model (kept synchronized with the running system) [19]. This process triggers a code generation/compilation process, and also generates a safe sequence of reconfiguration commands [18]. Actually, the code generation/compilation process is only triggered if there are new proxy components, e.g. new user proxy components involved in delegation, that need to be introduced into the running system. The dynamic adaptation of the running system is possible thanks to modern adaptive execution platforms like OSGi [26] or Fractal [8] which provide low-level APIs to reconfigure a system at runtime. The running system is then reconfigured by executing the safe sequence of commands, compliant to the platform API, issued by the reasoning process. In this process, the generation/compilation phase is time consuming. However, this phase has no impact on the running system, which remains stable until being adapted by executing the reconfiguration script. Thus, the actual adaptation phase only lasts for several milliseconds, during which the system is not available.

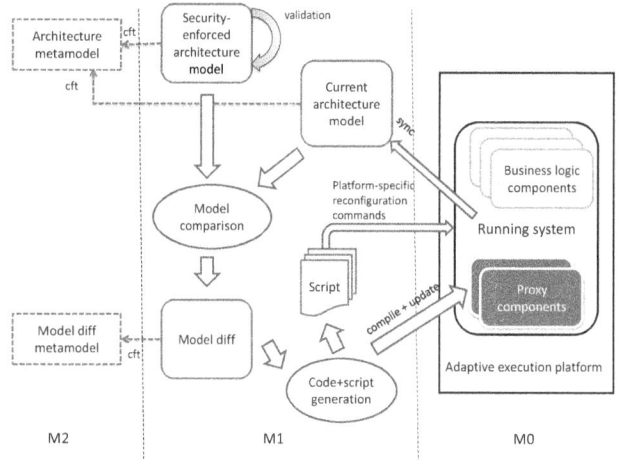

**Figure 8: Overview of our adaptation strategy**

In [20], the adaptation is entirely based on executing platform - specific reconfiguration scripts specifying which components have to be stopped, which components and/or bindings should be added and/or removed. This results in several limitations regarding delegation mechanisms:

**L.1**: Using reconfiguration scripts only implies to create all the potentially needed ports (used for bindings between *user* proxy components) beforehand. But all the combinations of users, roles, resources, actions could lead to a combinatorial explosion and make it infeasible for implementation.

**L.2**: In [20], the delegation between users are reflected using bindings connecting one *user* proxy component to another. But this approach is not suitable for supporting complex delegation features. For example, a transfer delegation will be reflected such as adding bindings between the *delegator* and *delegatee* but removing bindings between *delegator* and the corresponding *role* proxy component. Consequently both *delegator* and *delegatee* cannot access the resource, that does not correctly reflect a transfer delegation.

**L.1** can be solved by the automatic re-generation of proxy components and bindings between them according to changes in the architectural model. Moreover, as mentioned in Section 4.3.2, only users involved in a delegation are transformed into *user* proxy components with necessary ports and bindings. In this way, only required ports and bindings are created dynamically. **L.2** is solved by our model transformation approach. All complex delegation features are considered as "meta-level" mechanisms that impact access control rules. In this way, a transfer delegation will be reflected such as adding bindings between the *delegatee* and the corresponding delegated *role* proxy component, but removing bindings between *delegator* and the corresponding *role* proxy component.

Our adaptation strategy could take more time than simply running a reconfiguration script because of the generation and compilation time of newly generated proxy components. But the process of generating and compiling new proxy components could not harm the performance because each proxy component is very light-weight and only necessary proxy components are generated (see Section 5).

### 4.4.2 Evolution

In [20], the evolution of the security policy is not totally dealt with. It is possible to run a reconfiguration script to reflect the changes like adding, removing and updating rules. But adding a new user, role or resource requires the generation and compilation of new proxy components, which is impossible using only reconfiguration scripts. Thus, our strategy of automatically generating and compiling proxy components (see Section 5) is more practical w.r.t evolution.

Another important aspect of evolution relates to the addition, removal or update of resources and actions in the business logic. The base architecture model can be updated with the changes in the business logic, e.g. a new resource is added. On the other side, security officers can manually update the mappings (Fig. 6) following the changes of resources/actions in the base architecture model. By composing the security model with the base architecture model as described earlier, the security policy is evolved together with the business logic of the system.

# 5. EVALUATION

To evaluate the feasibility of our approach, we have applied it on three different Java-based case studies, which have also been used in our previous research work on access control testing [21]:

**1) LMS**: already described in our running example.

**2) VMS[3]**: The Virtual Meeting System offers simplified web conference services. The virtual meeting server allows the organization of work meetings on a distributed platform. When connected to the server, a user can enter (exit) a meeting, ask to speak, eventually speak, or plan new meetings. There are three resources (Meeting, Personnel Account, User Account) and six roles (Administrator, Webmaster, Owner, Moderator, Attendee, and Non-attendee) defined for this system with many access control rules, and delegation situations between the users of each role.

**3) ASMS**: The Auction Sale Management System allows users to buy and sell products online. Each user in the system has a profile including some personal information. Users wanting to sell a product (sellers) are able to start a new auction by submitting a description of the product, the starting and ending date of the auction. There are five resources (Sale, Bid, Comment, Personnel Account, User Account) and five roles (Administrator, Moderator, Seller, Senior Buyer, and Junior Buyer) defined for this system, also with many access control rules, and delegation situations between users of each role.

Table 1: Size of each system in terms of source code.

|  | # Classes | # Methods | # LOC |
|---|---|---|---|
| LMS | 62 | 335 | 3204 |
| VMS | 134 | 581 | 6077 |
| ASMS | 122 | 797 | 10703 |

We applied our approach to enable dynamic security enforcement for these systems, and examined how effective our approach is. Table 1 provides some information about the size of these three applications (the number of classes, methods and lines of code). In terms of security policies, Table 2 shows the number of access control (AC) rules and delegation rules defined for each system, used in our experiments.

Table 2: Security rules defined for each system.

|  | # AC rules | # Delegations | Total |
|---|---|---|---|
| LMS | 23 | 4 | 27 |
| VMS | 36 | 8 | 44 |
| ASMS | 89 | 8 | 97 |

All these systems are component-based systems. The business components of each system contain the business logic, e.g. Book Service component, Personnel Account component, Meeting, Sale, Authenticate component, Data Access Object components, etc. To enable dynamic security enforcement for a system, the resources (components that have to be controlled) are specified in the base model, and mapped to the resources of the security policies. Our metamodels are applicable for different systems without any modification or adaptation. The structure of delegation and access control policies for all case studies is the same, only roles, users, resources, actions are specific to each case study. The proxy components are automatically generated and synchronized with the security policy model via model transformations and reconfiguration at runtime. The model-to-model transformation and model-to-text transformation (code generation) can be implemented using any transformation engines like Kermeta [22] (or ATL [4]) and Xpand [14]. The security policy models that are stored in eXistDB [16], a native XML database, can be easily modified by using XPath, XQuery, and XUpdate. For experimenting with performance of adapting the running system, we have implemented the model transformation/composition rules using not only Kermeta but also ATL.

There are two kinds of response time we would like to measure in our case studies: the authorization mechanism and the dynamic adaptation according to changing security policies. The experiments were performed on Intel Core i7 CPU 2.20 GHz with 2.91 GB usable RAM running on Windows 7 and Equinox [5]. Because all our access control and delegation rules are transformed to proxy components reflecting our security policy, response times to an access request only depends on method calls between these proxy components and business components (Fig. 7). Unsurprisingly, response time to every resource access is a constant, only about 1 millisecond, because the access is already possible or not by construction. In other words, our 3-layered architecture reflecting security policy enables very quick response, independently from the number of access control and delegation rules.

Table 3: Performance of weaving Security Policies using Kermeta and ATL.

|  | # Rules | Kermeta 1.4.1 | ATL 3.2.1 |
|---|---|---|---|
| LMS | 27 | 4s | 0.048s |
| VMS | 44 | 7s | 0.055s |
| ASMS | 97 | 18s | 0.140s |

Regarding the adaptation process, Table 3 shows results of each case study for performing the model transformations of security policies mentioned in Table 2, using Kermeta 1.4.1 and ATL 3.2.1 correspondingly. At first, we used Kermeta 1.4.1 to implement our model transformations. However, the results shown in Table 3 have been disappointing. We have

---

[3]For more information about VMS (server side), please refer to http://franck.fleurey.free.fr/VirtualMeeting.

[4]http://www.eclipse.org/atl/

[5]http://www.eclipse.org/equinox/

tried to use Kermeta 2.0.4 (the latest version of Kermeta at this moment, compiled to byte code, which means much better performances), but the tool is not mature yet. To know if this performance problem is inherently linked to our approach or simply linked to the use of Kermeta 1.4.1, we decided to also implement our model transformations using ATL 3.2.1. Our experiments show that the implementation using ATL 3.2.1 is much more efficient. We can conclude that the initial performance issue was due to Kermeta 1.4.1.

Note that the transformation, code generation and compilation are performed "offline" meaning that the running system is not yet adapted. The actual adaptation happens when the newly compiled proxy components are integrated into the running system to replace the current proxy components. This actual adaptation process takes only some milliseconds by using the low-level APIs to reconfigure a system at runtime provided by the modern adaptive execution platforms, e.g. OSGi [26]. Right after the new proxy components are up and running, the new security policy is really enforced in the running system.

## 6. RELATED WORK

There is a substantial work related to delegation as extension of existing access control models. Most researchers focused on proposing models solely relying on the RBAC formalism [25], which is not expressive enough to deal with all delegation requirements. Therefore, some other researchers extended the RBAC model by adding new components, such as new types of roles, permissions and relationships [2, 27, 1, 9, 23]. In [5], the authors proposed yet another delegation approach for role-based access control (more precisely for OrBAC model) which is more flexible and comprehensive. However, no related work has provided a model-driven approach for both specifying and dynamically enforcing access control policies with various delegation requirements. Compared to [20], we extend the model-based dynamic adaptation approach of [20] with some key improvements. More specifically, we propose a new DSML for delegation management, but also new composition rules to weave delegation in a RBAC-based access control policy. In addition, we present a new way (by generating proxy) to implement the adaptation of the security-enforced architecture of the system. Indeed, we provide an extensive support for delegation as well as co-evolution of security policy and security-critical system. That means our approach makes it possible to deeply modify the security policy (e.g. according to evolution of the security-critical system) and dynamically adapt the running system, which is often infeasible using the other approaches mentioned above.

In addition, several researchers proposed new flexible access control models that may not include delegation, but allow to have a flexible and easy to update policy. For instance, Bertino et al. [6] proposed a new access control model that allows expressing flexible policies that can be easily modified and updated by users to be adapted to specific contexts. The advantage of their model resides in the ability to change the access control rules by granting or revoking the access based on specific exceptions. Their model provides a wide range of interesting features that increase the flexibility of the access control policy. It allows advanced administrative functions for regulating the specification of access controls rules. More importantly, their model supports delegation, enabling users to temporarily grant other users some

of their permissions. Furthermore, Bertolissi et al. proposed DEBAC [7] a new access control model based on the notion of event that allows the policy to be adapted to distributed and changing environments. Their model is represented as a term rewriting system [3], which allows specifying changing and dynamic access control policies. This allows having a dynamic policy that is easy to change and update.

As far as we know, no previous work tackled the issue of enforcing adaptive delegation. Some previous approaches were proposed to help modeling more general access control formalisms using UML diagrams (focusing on models like RBAC or MAC). RBAC was modeled using a dedicated UML diagram template [13], while Doan et al. proposed a methodology [10] to incorporate MAC in UML diagrams during the design process. All these approaches allow access control formalisms to be expressed during the design. They do not provide a specific framework to enable adaptive delegation at runtime. Concerning the approaches related to applying MDE for security, we can cite UMLsec [12], which is an extension of UML that allows security properties to be expressed in UML diagrams. In addition, Lodderstedt et al. [15] propose SecureUML which provides a methodology for generating security components from specific models. The approach proposes a security modeling language to define the access control model. The resulting security model is combined with the UML business model in order to automatically produce the access control infrastructure. More precisely, they use the Meta-Object facility to create a new modeling language to define RBAC policies (extended to include constraints on rules). They apply their technique in different examples of distributed system architectures including Enterprise Java Beans and Microsoft Enterprise Services for .NET. Their approach provides a tool for specifying the access control rules along with the model-driven development process and then automatically exporting these rules to generate the access control infrastructure. However, they do not directly support delegation. Delegation rules should be taken into account early and the whole system should be generated again to enforce the new rules. Our approach enables supporting directly the delegation rules and dynamically enforcing them by reconfiguring the system at runtime.

## 7. CONCLUSION AND FUTURE WORK

In this paper, we have proposed an extensive model-driven approach for RBAC-based delegation. It has been shown that various delegation requirements can be specified using our delegation DSML. Our DSML supports complex delegation characteristics like temporary, recurrence delegation, transfer delegation, multiple and multi-step delegation, etc. We have shown that we can deal with revocation in a simple manner. Another main contribution of this paper is to provide adaptive delegation enforcement in which delegation is considered as a "meta-level" mechanism that impacts the access control rules. A complete model-driven framework has been proposed to enable dynamic enforcement of delegation and access control policies that allows the automatic configuration of the system according to the changes in delegation/access control rules. Moreover, our framework also enables an adaptation strategy that better supports co-evolution of security policy and the security-critical system. Our approach has been validated via three different case studies with consideration of performance and extensibility issues.

Our approach could be better supported using an optimized models@runtime framework such as Kevoree[6] instead of Equinox. We have not dealt with this idea yet in this paper, but keep it for our future work. Moreover, revocation mechanism in our current approach has not been completely taken into account, i.e. without options of strong/weak revocation. So far, we only focused on the delegation of rights, further work will also be dedicated to the delegation of obligations and the support for usage control [24].

# 8. REFERENCES

[1] G.-J. Ahn, B. Mohan, and S.-P. Hong. Towards secure information sharing using role-based delegation. *J. Netw. Comput. Appl.*, 30(1):42–59, Jan. 2007.

[2] E. Barka and R. Sandhu. Role-based delegation model/ hierarchical roles (rbdm1). In *Proceedings of the 20th Annual Computer Security Applications Conference*, ACSAC '04, pages 396–404, Washington, DC, USA, 2004. IEEE Computer Society.

[3] S. Barker and M. Fernández. Term rewriting for access control. In *DBSec*, pages 179–193, 2006.

[4] D. Basin and J. Doser. Model driven security: From UML models to access control infrastructures. *ACM Transactions on Software*, (1945):353–398, 2006.

[5] M. Ben-Ghorbel-Talbi, F. Cuppens, N. Cuppens-Boulahia, and A. Bouhoula. A delegation model for extended rbac. *Int. J. Inf. Secur.*, 9(3):209–236, June 2010.

[6] E. Bertino, S. Jajodia, and P. Samarati. A flexible authorization mechanism for relational data management systems. *ACM Trans. Inf. Syst.*, 17(2):101–140, 1999.

[7] C. Bertolissi, M. Fernández, and S. Barker. Dynamic event-based access control as term rewriting. In *DBSec*, pages 195–210, 2007.

[8] E. Bruneton, T. Coupaye, M. Leclercq, V. Quéma, and J. Stefani. The FRACTAL Component Model and its Support in Java. *Software Practice and Experience, Special Issue on Experiences with Auto-adaptive and Reconfigurable Systems*, 36(11-12):1257–1284, 2006.

[9] J. Crampton and H. Khambhammettu. Delegation in role-based access control. *Int. J. Inf. Sec.*, 7(2):123–136, 2008.

[10] T. Doan, S. Demurjian, T. C. Ting, and A. Ketterl. Mac and uml for secure software design. In *FMSE '04: Proceedings of the 2004 ACM workshop on Formal methods in security engineering*, pages 75–85, New York, NY, USA, 2004. ACM.

[11] S. Jajodia, P. Samarati, M. L. Sapino, and V. S. Subrahmanian. Flexible support for multiple access control policies. *ACM Trans. Database Syst.*, 26(2):214–260, 2001.

[12] J. Jürjens. UMLsec: Extending UML for Secure Systems Development. In *UML '02: 5th International Conference on The UML*, pages 412–425, Dresden, Germany, 2002. Springer-Verlag.

[13] D.-K. Kim, I. Ray, R. B. France, and N. Li. Modeling role-based access control using parameterized uml models. In *FASE*, pages 180–193, 2004.

[14] B. Klatt. Xpand: A closer look at the model2text transformation language. *Language*, (10/16/2008), 2007.

[15] T. Lodderstedt, D. Basin, and J. Doser. SecureUML: A UML-Based Modeling Language for Model-Driven Security. In *UML '02: 5th International Conference on The UML*, pages 426–441, Dresden, Germany, 2002. Springer-Verlag.

[16] W. Meier. exist: An open source native xml database. In *Web-Services, and Database Systems, NODe 2002 Web and Database-Related Workshops*, pages 169–183. Springer, 2002.

[17] B. Morin, O. Barais, J.-M. Jezequel, F. Fleurey, and A. Solberg. Models@ run.time to support dynamic adaptation. *Computer*, 42(10):44–51, Oct. 2009.

[18] B. Morin, O. Barais, G. Nain, and J. Jézéquel. Taming Dynamically Adaptive Systems with Models and Aspects. In *ICSE'09: 31st International Conference on Software Engineering*, Vancouver, Canada, May 2009.

[19] B. Morin, F. Fleurey, N. Bencomo, J.-M. Jézéquel, A. Solberg, V. Dehlen, and G. Blair. An aspect-oriented and model-driven approach for managing dynamic variability. In *Proceedings of the 11th international conference on Model Driven Engineering Languages and Systems*, MoDELS '08, pages 782–796, Berlin, Heidelberg, 2008. Springer-Verlag.

[20] B. Morin, T. Mouelhi, F. Fleurey, Y. Le Traon, O. Barais, and J.-M. Jézéquel. Security-driven model-based dynamic adaptation. In *Proceedings of the IEEE/ACM international conference on Automated software engineering*, ASE '10, pages 205–214, New York, NY, USA, 2010. ACM.

[21] T. Mouelhi, Y. L. Traon, and B. Baudry. Transforming and selecting functional test cases for security policy testing. In *Proceedings of the 2009 International Conference on Software Testing Verification and Validation*, ICST '09, pages 171–180, Washington, DC, USA, 2009. IEEE Computer Society.

[22] P.-A. Muller, F. Fleurey, and J.-M. Jézéquel. Weaving executability into object-oriented meta-languages. In *International Conference on Model Driven Engineering Languages and Systems (MoDELS)*, LNCS 3713, pages 264–278. Springer, 2005.

[23] S. Na and S. Cheon. Role delegation in role-based access control. In *Proceedings of the fifth ACM workshop on Role-based access control*, RBAC '00, pages 39–44, New York, NY, USA, 2000. ACM.

[24] R. Sandhu and J. Park. Usage control: A vision for next generation access control. *V. Gorodetsky et al. (Eds.): MMM-ACNS 2003, LNCS 2776*, 1:17–31, 2003.

[25] R. S. Sandhu, E. J. Coyne, H. L. Feinstein, and C. E. Youman. Role-based access control models. *Computer*, 29(2):38–47, Feb. 1996.

[26] O. The OSGi Alliance. Osgi service platform core specification, release 4.1. 2007.

[27] X. Zhang, S. Oh, and R. Sandhu. Pbdm: a flexible delegation model in rbac. In *Proceedings of the eighth ACM symposium on Access control models and technologies*, SACMAT '03, pages 149–157, New York, NY, USA, 2003. ACM.

---

[6]http://www.kevoree.org/, last access March 2012

# Refactoring Delta-Oriented Software Product Lines

Sandro Schulze
TU Braunschweig
Braunschweig, Germany
sanschul@tu-bs.de

Oliver Richers
TU Braunschweig
Braunschweig, Germany
oliver.richers@gmail.com

Ina Schaefer
TU Braunschweig
Braunschweig, Germany
i.schaefer@tu-bs.de

## ABSTRACT

*Delta-oriented programming (DOP)* is an implementation approach to develop *software product lines (SPL)*. Delta-oriented SPLs evolve over time due to new or changed requirements and need to be maintained to retain their value. *Refactorings* have been proposed as behavior-preserving program transformations that improve the design and structure of (object-oriented) software systems. However, there is a lack of refactoring support for software product lines since refactoring of SPLs is more complex than of single systems. For refactoring SPLs, we have to preserve the behavior of probably thousands of programs instead of only one. In this paper, we address the refactoring of software product lines by presenting a catalogue of refactorings for delta-oriented SPLs. Additionally, we propose *code smells* to guide developers to potential refactoring opportunities. We show how code smells can aid the identification of SPL refactorings and how these refactorings improve the evolvability and maintainability of delta-oriented SPLs.

## Categories and Subject Descriptors

D.2.7 [**Software Engineering**]: Distribution, Maintenance, and Enhancement; D.2.13 [**Software Engineering**]: Reusable Software; D.3.3 [**Programming Languages**]: Language Constructs and Features

## General Terms

Design, Languages

## Keywords

Software Product Lines, Evolution, Delta-oriented Programming, Refactoring

## 1. INTRODUCTION

*Software Product Lines (SPL)* [4, 19] gained momentum in recent years, in both, academia as well as industry. An SPL facilitates the development of a whole product portfolio instead of developing each product in isolation from scratch.

To this end, a developer can manage a common set of features by means of a product line. In this context, a *feature* is an increment in functionality that is visible to any stakeholder. As a result, a product line increases the reusability and decreases maintenance effort and time-to-market. Recently, *delta-oriented programming (DOP)* [21, 23] has been proposed as a modular, yet flexible implementation technique for software product lines. DOP modularizes the features of an SPL into *delta modules*, which can be applied to create particular variants of the SPL.

However, similar to standalone software systems, SPLs undergo changes during their lifecycle. This process is called *software evolution*, and it is inherent to every software system (and every SPL) due to new or changing requirements [12, 13]. However, this evolution may lead to *software aging* that can potentially increase the complexity of anSPL. As a result, maintainability or reusability may decrease, which increases the effort for further SPL development. In particular, we focus on changes regarding variability modeling and implementation within DOP and do not consider other parts of the SPL engineering process [19].

An established technique to reduce complexity and to counter the process of software aging of single software systems on implementation level are *refactorings* [17, 7]. Refactorings are program transformations that change the internal structure of a program without altering its external (visible) behavior. Generally, such refactorings are applied in preparation for an evolutionary modification or during the process of maintenance (e.g., to clean-up the source code). When used as preparation, the aim of the refactoring is to change the internal structure of the program in a way, that the subsequent modification of its external behavior can be executed with minimum effort. When used as a clean-up, its aim is to reduce the complexity of the program and to improve the source code quality with respect to certain design guidelines. Furthermore, Fowler introduced the concept of *code smells*, which indicate refactoring opportunities [7].

Unfortunately, only little is known about both, refactoring and code smells, in software product lines. Moreover, tool support, which is crucial for refactoring, is non-existent. In this paper, we address this lack of refactoring support by providing refactorings for delta-oriented software product lines. In particular, we make the following contributions:

- We provide a catalogue of 23 refactorings for DOP that aim at cleaning up the SPL implementation, but also at supporting the efficient evolution of SPLs.

- To support the decision of applying refactorings, we propose code smells that are specifically tailored to delta-oriented SPL implementations.
- We show examples of evolutionary changes to delta-oriented SPLs as use cases and apply exemplary refactorings to retain a clear and maintainable structure of the SPL.
- We provide tool support for most of our refactorings as an Eclipse Plugin, which enables other researchers to reproduce or even extend our use cases.

The paper is structured as follows: In Section 2, we provide background on product lines and delta-oriented programming. In section 3, we motivate the necessity for refactorings in SPLs. Afterwards, we present our catalogue of refactorings and provide an example of how and when to use such refactorings in Section 4. We present the implementation of our refactorings in Section 5. In Section 6 we acknowledge related work before we conclude our results in Section 7.

## 2. SOFTWARE PRODUCT LINES

A *software product line* is a set of related software products that share a common set of features [4, 19]. A *feature* is any system property relevant (and visible) to some stakeholder, including not only the product customers, but also the software developers, analysts, system architects, administrators, etc. To create a concrete program of an SPL, the stakeholder selects the respective features (according to some requirements), and the desired *product variant* is generated. Through the systematic management of the commonalities and variabilities in the process of *software product line engineering*, a high degree of software reuse can be achieved, which increases the number of products while time to market of maintenance costs decreases [19].

To describe commonalities and variabilities in SPLs, *feature models* are commonly used [5, 2, 6]. Feature models can be described by *feature diagrams* [8] or *propositional formulas* [2]. In Figure 1, we show an example for a feature diagram of an ATM software product line. Within such feature diagrams, we distinguish between two kinds of features: *compound features* such as the feature *Languages* and *primitive features* such as the feature *English*. Compound features are used to group other features. Within our example, feature *Cash* corresponds to the functionality to withdraw money while the features *English* and *German* refer to the language used by the ATM. To express dependencies between features, we can define different constraints on both, compound and primitive features. For instance, in our example, the features *Cash* and *Language* are *mandatory*, which indicates that these features have to be selected in each variant. In contrast, a feature can also be *optional* for particular variants. Beside these constraints, we can also define constraints on feature groups, that are, sub-features of a compound feature. For example, the features *English* and *German* form an *Alternative-Group*, indicating that these features are mutually exclusive. Moreover, additional *cross-tree constraints* can be specified by means of propositional formulas capturing *requires* and *excludes* dependencies. For example, the formula $A \implies B \wedge C$ requires that whenever feature $A$ is selected, features $B$ and $C$ have to be selected as well.

The implementation of SPLs can can be distinguished into three categories: *compositional, transformational* and *annotative* approaches. Examples for compositional ap-

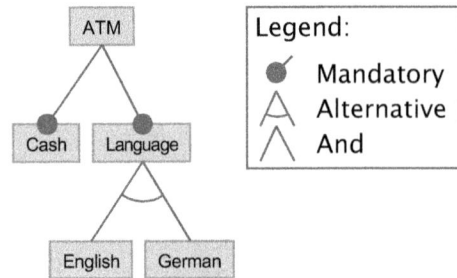

Figure 1: Feature diagram of an ATM SPL

proaches are *feature-oriented programming (FOP)* [14] or *aspect-oriented programming (AOP)* [9, 25], whereas the *C preprocessor* is an example for annotative approaches. In this paper, we focus on *Delta-oriented programming (DOP)* [23], a transformational approach, that subsumes feature-oriented programming. With feature-oriented programming, the general idea is to implement a specific feature of the product line in a distinct, cohesive unit, called *feature module*. A feature module can only extend or replace parts of other feature modules. In contrast, the delta-oriented programming approach uses the concept of *delta modules* which can be seen as generalized feature modules. In Table 1, we show the differences between feature and delta modules. A delta module may implement more than one feature, but may also implement only fragments of one or more features. Furthermore, a delta module can not only extend or replace existing functionality, but also remove parts of other implementations and thus the corresponding functionality. A particular product variant is generated by applying the modification of a selected set of deltas modules in a specified ordering.

| Approach | Features | | | Actions | | |
|---|---|---|---|---|---|---|
| | One | Multiple | Fragments | Extend | Replace | Remove |
| Feature Module | X | - | - | X | X | - |
| Delta Module | X | X | X | X | X | X |

Table 1: Feature Modules versus Delta Modules

Delta-oriented programming supports *proactive, extractive,* and *reactive* development of software product lines [10]. In proactive development, the complete set of product variants has to be known in advance and is developed from scratch. The reactive approach develops only a basic set of products initially, and evolves the product line when new requirements or products are requested. The extractive approach starts with a set of existing, independently developed products. This set of legacy products is gradually transformed into a product line. New products can then be derived by modifying the existing products.

In Figure 2, we show a small example to illustrate the syntax of DELTAJ, an extension of JAVA for implementing delta-oriented SPLs. A delta module (here:*DCash*) is defined by the keyword **delta**. To add a class, method, or field to a delta module, we use the keyword **adds**. In our example, we add four classes in Line 3–8 and their initial implementation, indicated by {...} which together implement the functionality to withdraw cash from the ATM. We can also alter an existing class within a delta module, for which we use the `modifies` keyword as in the delta modules *DEnglish* and *DGerman* which modify the *Screen* class to produce output in the selected language. We can also remove classes, fields, and methods. While removal of classes is indicated by the `removes` keyword, removal of fields and methods is indicated by their corresponding keywords `removesMethod` and `removesField`, respectively. Adding or removing fields and methods is only possible within classes that are referenced by `adds` or `modifies`.

The SPL declaration starts with the keyword `spl`, specifying which delta modules are applied for which product variant. First, using the keyword `features`, we can list all features of the feature model. Then, in a propositional formula following the keyword `configurations`, the set of valid feature selections are provided along the lines of [2]. After the keyword `deltas`, the set of delta modules comprising the code base of the SPL is provided. An *application condition* is attached to every delta module in the `when` clause, in the form of a *propositional formula* over the features. If the complete formula evaluates to true for a given feature selection, the delta module is *applied* during product generation, otherwise it is ignored. In the example in Figure 2, each delta module is applied when its corresponding feature is selected. Additionally, the ordering in which the selected delta modules are applied, is specified by an ordered partition over the set of delta modules where a partition is enclosed by [...]. In order to guarantee that for each feature selection, a unique product variant is generated, the delta modules within one partition must be consistent, meaning that they may not change the same program entity, i.e., class, method or field. This restriction simplifies checking that the SPL is unambiguous. For generation of a particular program variant, the delta modules within one partition may be swapped, while the partitions must be applied in the specified ordering.

# 3. MOTIVATION

Software product lines evolve in the same way as ordinary single software systems to address new or (constantly) changing requirements [12]. For instance, a SPL for database management systems (DBMS SPL) may initially only contain some basic features, concerning the storage and retrieval of information. Later on, during its utilization and probably after some distressing experience, additional requirements such as an exhaustive logging of user activity or transaction management may be requested. The implementation of such supplementary and especially unanticipated requirements can require a complete redesign of some program parts, causing a tremendous amount of development time and cost. Even for its initial implementation, a proactive development may not be feasible, because of unclear requirements, especially regarding the future evolution of the SPL. Hence, a reactive development approach, based on an existing system, may become necessary. This, in turn, requires a flexible and modifiable code base, that can be seamlessly adapted to integrate new features or to modify existing ones.

Delta-oriented programming [23] is a viable approach to implement software product lines with mechanisms that provide both, flexibility (regarding evolutionary changes) and reusability of the code base. To implement new or to change existing functionality, a new delta module can be added that

```
 1  delta DCash {
 2      adds class Bank {...}
 3      adds class Screen {...}
 4      adds class CashDispenser {...}
 5      adds class Controller {
 6          Bank bank;
 7          Screen screen;
 8          CashDispenser cashDispenser;
 9          void startExecution() {
10              this.executeWithdrawal();
11          }
12          void executeWithdrawal() {
13              int amount = this.screen.askForAmount();
14              this.bank.withdrawCash(amount);
15              this.cashDispenser.dispense(amount);
16          }
17      }
18  }
19
20  delta DEnglish {
21      modifies Screen {...}
22      }
23  }
24  delta DGerman {
25      modifies Screen {...}
26      }
27  }
28
29  spl Atm {
30      features Cash, English, German
31      configurations Cash && (English && !German || German &&
              !English)
32      deltas
33          [DCash when Cash]
34          [DEnglish when English]
35          [DGerman when German]
36  }
```

**Figure 2: Implementation of the ATM SPL**

contains the required modifications to the existing code base. However, a series of such modifications, as usual during software evolution, almost inevitably increases the complexity of the internal SPL structure, especially if the modifications do not refer to an isolated new feature.

We illustrate the problems of such evolutionary changes to delta-oriented SPLs with the help of the ATM product line considered in Section 2. We suppose that the bank demands for a more advanced solution with the additional requirement that a customer can ask for his current balance.

In Figure 3, we show the implementation of the corresponding delta module *DBalance*, which extends the original ATM SPL by the feature *Balance*. This delta module removes some classes, methods and fields introduced by the delta module *DCash* and modifies the *startExecution()* method to call *printBalance()*. The *printBalance()* method implements the balance checking process: The bank is asked for the current balance of the banking account and the returned value is shown to the customer.

In the feature model part of the product line declaration in Figure 3 and in the feature digram in Figure 4, it is specified that the features *Balance* and *Case* form an alternative-group. Unfortunately, this contradicts with application conditions in the SPL declaration of the delta module *DBalance*, which states that the delta module *DCash* is always applied while *DBalance* is only applied if the corresponding feature *Balance* is selected. The reason for that incompatibility is that delta module *DCash* not only implements the feature *Cash*, but also the common base of both features *Balance* and *Cash*.

```
1   delta DCash {...}
2   delta DBalance {
3       removes CashDispenser;
4       modifies Controller {
5           removesField cashDispenser;
6           removesMethod executeWithdrawal;
7           modifies void startExecution() {
8               this.printBalance();
9           }
10          adds void printBalance() {
11              int balance = this.bank.getBalance();
12              this.screen.printBalance(balance);
13          }
14      }
15  }
16  spl Atm {
17      features Cash, Balance, English, German
18      configurations (Cash && !Balance || !Cash && Balance) &&
                (English && !German || German && !English)
19      deltas
20          [DCash when Cash || Balance]
21          [DBalance when Balance]
22          [DEnglish when English]
23          [DGerman when German]
24  }
```

**Figure 3: Implementation of Evolved ATM SPL**

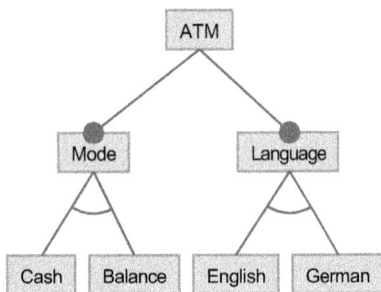

**Figure 4: Feature Diagram of Evolved ATM SPL**

Both, the aforementioned implementation of delta module *DCash* (encompassing to much functionality) as well as the described incompatibility between the feature model and the SPL declaration is the result of an improper (or even bad) design, which hinders evolution and maintenance of software product lines and thus contributes to its decay, referred to as *software aging* [18]. As a result, the features *Cash* and *Balance* cannot be selected together, which makes the possible products of the SPL somewhat useless. The reason is, that only variants can be generated that either enable customers to withdraw money (without checking the balance) or to check their account balance (without withdrawing money). However, the desired solution is to allow variants with *both* features, as reflected by the feature diagram in Figure 5.

One possibility to achieve this result on implementation level and to solve the aforementioned incompatibility, is to extract the common source code from *DCash* into a separate delta module. However, performing such a restructuring task manually, known as *refactoring*, is very tedious and error-prone. Hence, it is inevitable to support the developer with predefined refactorings that are automated as far as possible. To counter this problem, we propose a catalogue of refactorings for delta-oriented SPLs, which we introduce in the following section. With these refactorings, we address both issues caused by SPL evolution. First, we propose

refactorings that aim at restructuring an SPL to support product line evolution *in advance* such as extracting common parts *before* introducing a new feature that relies on this parts. Second, we describe refactorings that aim at improving the design of a product line after one or several consecutive changes to increase the maintainability of an SPL.

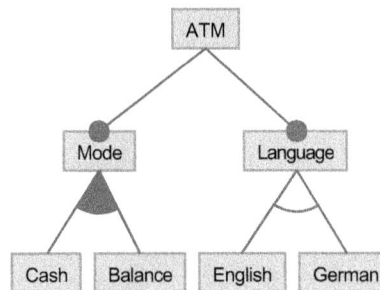

**Figure 5: Feature diagram that allows the selection of both features *Cash* and *Balance*.**

## 4. DELTA-ORIENTED REFACTORINGS

*Refactorings* [17, 7] for single programs improve the internal program structure without changing its external behavior, which is an important technique to support the evolution of software (and software product lines). Refactorings of software product line improve the structure of the product line without changing the behavior of the single products that can be derived. In this section, we present a catalogue of 23 refactorings of delta-oriented SPLs. First, we give an overview of these refactorings, which are summarized in Table 2 and 3. Second, we present use cases and examples for the application of particular refactorings by means of the introduced ATM SPL. Finally, we discuss important aspects of our proposed refactorings such as generalizability.

### 4.1 Overview

The proposed refactorings address *all* entities of delta-oriented SPLs, that is the delta modules and the SPL declaration, i.e., the feature model constraint, the application conditions and the delta module ordering. Hence, we first introduce the terminology for the proposed refactorings which refers to the respective target of each refactoring. Whenever the name of a refactoring refers to *Feature*, the refactoring addresses the feature model. In the context of DELTAJ, such a refactoring affects the part which starts with the **features** keyword (e.g., Figure 3, Line 17). In the same way, a refactoring that refers to *Product Line* targets at the product line declaration itself, indicated by the **spl** keyword (e.g., Figure 3, Line 16). A refactoring that refers to *Delta Module* targets a delta module (indicated by the keyword **delta**). A couple of refactorings refer to an *Action*. These refactorings refer to a concrete action *within* a delta module which is indicated by the keywords **add, modifies,** or **removes**. In a more general way, refactorings that refer to *Delta Actions* also address *logical* actions in delta modules. A logical delta action consists of statements that do not contain any of the aforementioned keywords, but are implicitly part of a modification specified in a delta module. For instance, in Figure 3, the statement in Line 11 is a usual variable assignment, but because it is part of an action (indicated by **add void printBalance()**), it is considered a delta action as well.

| Name of Refactoring | Typical Situation | Recommended Action |
| --- | --- | --- |
| *Rename Feature* | The meaning of a feature has changed or its name is badly chosen. | Assign a new and appropriate name to the feature. |
| *Rename Delta Module* | The content of a delta module has changed or its name is badly chosen. | Assign a new and appropriate name to the delta module. |
| *Rename Product Line* | The name of a software product line does not reflect its purpose. | Assign a new and appropriate name to the software product line. |
| *Extract Delta Action* | The application condition of a single delta action is subject to be changed. | Move the delta action into a new delta module, retaining its application condition in the first step. Afterwards, the application condition can be changed as necessary. |
| *Extract Conflicting Actions* | Two delta modules with conflicting delta actions shall be made consistent, so that they can be used in the same product. Two delta actions are in conflicting if they modify the same program entity in a different way. | Extract the conflicting delta actions into separate modules. |
| *Resolve Modification Action* | A modification action shall be eliminated without changing the behavior of the software product line. | Change the application conditions of preceding delta actions so that the modification action can be converted into an addition action. |
| *Resolve Removal Action* | A removal action shall be eliminated without changing the behavior of the software product line. | Change the application conditions of preceding delta actions so that the removal action can be removed. |
| *Merge Delta Modules with Equivalent Conditions* | Two delta modules of the same or consecutive partition parts have equivalent application conditions. | Move the delta actions from one delta module into the other delta module and remove the remaining empty delta module. |
| *Merge Delta Modules with Equivalent Content* | Two delta modules have the same content, and thus one delta module is redundant. | Combine the application conditions of the delta modules using the boolean *or*-operator and remove one of the modules. |
| *Merge Delta Modules with Inverse* | The delta actions of one delta module shall be incorporated into another delta module. | The second delta module is merged into the first delta module by applying its delta actions to the first delta module. An inverse delta module is created to revert this merge. |
| *Merge Configurations into Conditions* | The feature model of the product line shall be changed to support more feature configurations. | Merge the propositional formula describing all valid feature configurations into all application conditions, using the boolean and-operator. |
| *Extract Configurations from Conditions* | The application conditions of the delta modules are redundant with respect to the feature model constraint, e.g. some unnecessary checks for invalid feature conditions are made. | Simplify the application conditions with respect to the valid feature configurations defined by the feature model. |

**Table 2: Refactorings of DeltaJ Software Product Lines (1/2)**

We present our refactorings in the spirit of the well-known and commonly accepted object-oriented refactorings of Fowler et al. [7]. Hence, in Table 2 and 3, we group our refactoring by their purpose and provide a short explanation regarding typical situations and recommended actions for each of them.

In the following, we roughly describe each group of refactorings. The first group[1] in Table 2 focuses on renaming different entities of delta-oriented SPLs such as features or delta modules. Renaming is a common task during evolution, for example, to better reflect the purpose of a particular entity or to avoid duplicated names. The second group extracts delta actions into a new delta module. This may be necessary due to conflicts between particular actions or for changing application conditions. In the case, that a certain action is not needed anymore, for example, due to changed requirements that affect a certain delta module, it is desirable to remove this action to clean up the code. To this end, we propose the refactorings of group three. During evolution, we may reach the point where particular delta modules are redundant or interchangeable. Hence, it is desirable to merge these modules in a specific way, which is addressed by the fourth group of Table 2. Another issue, specific to software product lines, are the valid features configurations and, thus, the possible products that can be derived from an SPL. With the refactorings of the fifth group, we support the addition of possible products by refactoring the valid features configurations or application conditions of the delta modules.

---

[1]The different groups of refactorings are separated by horizontal lines in the respective tables.

| Name of Refactoring | Typical Situation | Recommended Action |
|---|---|---|
| *Resolve Duplicated Actions* | Two delta actions execute the same modification, e.g. add equivalent methods or remove the same class. | Extract the delta actions into a single new delta module, combining the application conditions using the boolean *or*-operator. |
| *Remove Dead Delta Action* | A delta action has no effect on any product, e.g. a field added by a delta action is always removed by another delta action. | If the delta action is a removal or modification action, resolve it by the refactorings *Resolve Removal Action* or *Resolve Modification Action*. If the delta action is an addition action, resolve all succeeding removal or modification actions. Finally, remove all dead delta modules by executing the refactoring *Remove Dead Delta Module*. |
| *Remove Dead Delta Module* | A delta module is never used because its application condition is a contradiction, either by itself or with respect to the valid feature configurations. | Remove the dead delta module and all references to it. |
| *Remove Empty Delta Module* | All delta actions of a delta module have been extracted or moved into other delta modules. | Remove the empty delta module and all references to it. |
| *Merge Compatible Partition Parts* | Two partition parts contain only pairwise commuting delta modules. | Merge the two partition parts into a single partition part. |
| *Remove Empty Feature* | A feature is defined in the feature model, but has no effect on the application conditions of the delta modules. | Remove the feature from the feature list and replace all references to the feature in the application conditions and feature model constraint with the boolean constant *false*. |
| *Remove Unused Feature* | A feature is defined in the feature model, but not used by any product of the product line. | Remove the feature from the feature list and replace all references to the feature in the application conditions and feature model constraints with the boolean constant *false*. |
| *Merge Duplicated Features* | Two features are defined in the feature mode, but each feature has the same effect in the application conditions of the delta modules. | Merge both features into a single feature by replacing any reference to one of the original features with a reference to the new feature. |
| *Merge Joined Features* | Two features are defined in the feature model, but each feature has only any effect on the application conditions of the delta modules if both of them are selected. | Remove references to the features from products, referencing only one of the features. Thereafter, merge both features into a single feature by replacing any reference to one of the original features with a reference to the new feature. |
| *Simplify Application Conditions* | After executing some refactorings, e.g. *Resolve Removal Action* or *Remove Unused Feature*, the application condition of a delta module got unnecessarily complex. | Replace the application condition of the delta module with an equivalent, but simplified version. |
| *Simplify Feature Configurations* | After refactoring the feature model, the propositional formula of the valid feature configurations became unnecessarily complex. | Replace the propositional formula with an equivalent, but simplified version. |

**Table 3: Refactorings of DeltaJ Software Product Lines (2/2)**

Duplicated or dead code may occur in product lines as well. With the first two groups of refactorings in Table 3, we address this problem by proposing refactorings that remove dead or duplicated delta actions as well as complete delta modules. Similarly to code artifacts, a feature model of an SPL also evolves. This can yield that the dependencies amongst features are hard to understand. With the refactorings of the third group of Table 3, we provide means to restructure feature models such as removing unused or duplicated features. Such refactorings do not affect the number of possible products, because all existing products can still be generated. Finally, we propose two refactorings that deal with simplifying application conditions and feature configuration constraints in case that they become too complex, for instance, as a consequence of other refactorings.

## 4.2 Selected Refactorings

In the following, we describe the mechanisms for two selected refactorings used again in the next subsection. For a comprehensive overview and description of *all* refactorings, we refer to [20].

### Simplify Application Conditions

Many automated refactorings modify the application conditions of delta modules mechanically, which can produce very complex propositional formulas. Because of that, the

78

*Simplify Application Conditions* refactoring can be executed as a final clean-up after the execution of other refactorings.

*Strategy.* This refactoring can be triggered automatically by other refactorings, or the developer can trigger it manually for a single application condition or the complete software product line.

*Input.* Either an application condition must be selected, or if all application conditions of a product line shall be simplified the software product line must be selected.

*Preconditions.* No specific preconditions exist for this refactoring. The DELTAJ program must be well-formed, though.

### Mechanics

- For each application condition, the propositional formula following the *when*-statement is analyzed, and a simplified version is computed, using a suitable minimization algorithm.
- The original propositional formula of the *when*-statement is replaced with the new simplified version.

### Resolve Removal Action

Every removal action can be eliminated by modifying the application conditions of preceding delta actions, affecting the same target entity. This is true for all class, method, or field removal actions.

*Execution Strategy.* The developer can manually trigger this refactoring for each class, method or field removal action. Alternatively, the transformation can be automatically triggered for all removal actions in a given delta module, or even the complete software product line.

*Input.* The developer has to select a removal action, a delta module or the software product line.

*Preconditions.* No specific preconditions exist for this refactoring. The DELTAJ program must be well-formed, though.

*Mechanics.* Suppose that a delta action contained in the delta module *DRemove* shall be resolved.

- First, all *preceding* delta actions (i.e., actions executed *before* the removal action) that affect the same target (class, method or field), have to be extracted. This can be done by using the refactoring *Extract Delta Action*, internally. Here, *preceding delta actions* refers to those delta actions, which are contained in delta modules preceding *DRemove* with respect to the delta module ordering.
- The extracted delta modules with the preceding delta actions can now be disabled for those feature configurations, in which the application conditions of the delta module *DRemove* evaluate to true. That is, given the declaration `DRemove when R`, the module references of all preceding extracted delta actions need to be converted from `DPreceding when C` to `DPreceding when C && !R`.
- Thereafter, the removal action can safely be deleted from the delta module *DRemove*.

- As a final clean-up step, the refactorings *Merge Delta Modules with Equivalent Conditions*, *Remove Empty Delta Module* and *Simplify Application Conditions* can be executed, if required.

## 4.3 Delta-Oriented Refactorings in Action

We now illustrate refactorings of delta-oriented SPLs in more detail to prove their applicability and usefulness using the ATM product line, introduced in Section 2. Our starting point is the incompatibility of the features *Cash* and *Balance* mentioned in Section 3. Our goal is to make them compatible so that both features can be selected for certain product variants, according to the feature diagram in Figure 5. To achieve this goal, several refactoring steps are necessary, which we explain in the following. We explicitly point out that the actual application of the refactorings is performed (semi-)automatically, while determining the code, subject to refactoring, is primarily a manual task.

### Preparing the Evolution

Within the initial implementation of the delta module *DBalance* (and thus of the feature *Balance*), different `remove` and `modifies` actions disable some functionality of delta module *DCash* in case the feature *Balance* is selected (cf. Figure 3). Consequently, the first step to make both features compatible is to undo these actions. To this end, we apply the refactorings *Resolve Removal Action* and *Resolve Modification Action* to the delta module *DBalance*.

In Figure 6, we show the implementation of the ATM SPL after applying these the refactorings to the implementation shown in Figure 3. The applied refactorings automatically extract the delta actions for the class `CashDispenser`, for the field `Controller.cashDispenser` and for the two methods `startExecution()` and `executeWithdrawal()` from the delta module *DCash* into separate delta modules. The application conditions of the (newly created) delta modules are defined such that the delta modules are not applied if the feature *Balance* is selected which is achieved by adding the negation of this feature to the existing application conditions. This makes the removal and modification actions in *DBalance* obsolete such that they are deleted.

### Removing Redundancy

As a result of the previous refactoring, we obtain several micro delta modules with complex application conditions. Furthermore, different delta modules have identical application conditions, which can lead to inconsistent changes in future evolution steps. To reduce the complexity and to remove the redundancy, we can merge delta modules with the same application conditions by applying the refactoring *Merge Delta Modules with Equivalent Conditions*. As a result, all delta modules of Figure 6 that start with the prefix `DCash_` are merged into the delta module *DCashExtracted*, which we show in Figure 7.

### Improving Readability

To improve readability and to support developers with understanding the source code, the names of different program entities, such as the delta modules, should correctly reflect their content [15]. To this end, we apply the refactoring *Rename Delta Module* to rename the delta module *DCash* into *DCommon* and *DCashExtracted* into *DCash*. For the same purpose, we apply the refactoring *Simplify*

```
1
2   delta DCash {
3       adds class Bank {...}
4       adds class Screen {...}
5       adds class Controller {
6           Bank bank;
7           Screen screen;
8       }
9   }
10  delta DBalance {
11      modifies Controller {
12          adds void startExecution() {
13              this.printBalance();
14          }
15          adds void printBalance() {
16              int balance = this.bank.getBalance();
17              this.screen.printBalance(balance);
18          }
19      }
20  }
21  delta DCash_CashDispenser {
22      adds class CashDispenser {...}
23  }
24  delta DCash_Controller_cashDispenser {
25      modifies Controller {
26          adds CashDispenser cashDispenser;
27      }
28  }
29  delta DCash_Controller_executeWithdrawal {
30      modifies Controller {
31          void executeWithdrawal() {
32              int amount = this.screen.askForAmount();
33              this.bank.withdrawCash(amount);
34              this.cashDispenser.dispense(amount);
35          }
36      }
37  }
38  delta DCash_Controller_startExecution {
39      modifies Controller {
40          void startExecution() {
41              this.executeWithdrawal();
42          }
43      }
44  }
45  spl Atm {
46      features Cash, Balance, English, German
47      configurations (Cash && !Balance || !Cash && Balance) &&
                       (English && !German || German && !English)
48      deltas
49          [DCash when Cash || Balance,
50              DCash_CashDispenser
51                  when (Cash || Balance) && !Balance,
52              DCash_Controller_cashDispenser
53                  when (Cash || Balance) && !Balance,
54              DCash_Controller_executeWithdrawal
55                  when (Cash || Balance) && !Balance,
56              DCash_Controller_startExecution
57                  when (Cash || Balance) && !Balance]
58          [DBalance when Balance]
59          [DEnglish when English]
60          [DGerman when German]
61  }
```

**Figure 6: ATM SPL after resolving removal and modification actions**

```
1   delta DCash {
2       adds class Bank {...}
3       adds class Screen {...}
4       adds class Controller {
5           Bank bank;
6           Screen screen;
7       }
8   }
9   delta DCashExtracted {
10      adds class CashDispenser {...}
11      modifies Controller {
12          adds CashDispenser cashDispenser;
13          adds void startExecution() {
14              this.executeWithdrawal();
15          }
16          adds void executeWithdrawal() {
17              int amount = this.screen.askForAmount();
18              this.bank.withdrawCash(amount);
19              this.cashDispenser.dispense(amount);
20          }
21      }
22  }
23  delta DBalance {
24      modifies Controller {
25          adds void startExecution() {
26              this.printBalance();
27          }
28          adds void printBalance() {
29              int balance = this.bank.getBalance();
30              this.screen.printBalance(balance);
31          }
32      }
33  }
34  spl Atm {
35      features Cash, Balance, English, German
36      configurations (Cash && !Balance || !Cash && Balance) &&
                       (English && !German || German && !English)
37      deltas
38          [DCash when Cash || Balance,
39              DCashExtracted when (Cash || Balance) && !Balance]
40          [DBalance when Balance]
41          [DEnglish when English]
42          [DGerman when German]
43  }
```

**Figure 7: ATM SPL after merging delta modules**

method startExecution() in Line 13 and Line 25 of Figure 7 such that the resulting program is not well-formed. To eliminate this conflict, we apply the refactoring *Extract Conflicting Actions*, which extracts the two conflicting methods into separate delta modules *DCash_startExecution* and *DBalance_startExecution*, respectively (cf. Figure 9).

### Adapting the Feature Model

In order to add the configuration where the features *Balance* and *Cash* are simultaneously selected, we have to adapt the feature model of the ATM SPL. We apply the refactoring *Merge Configurations into Conditions* to add the feature model constraint to the application conditions of the delta modules in order to maintain the applicability of the delta modules if the feature model is changed. This refactoring merges the propositional formula describing the feature model into all application conditions, using the logical AND operation which is illustrated in Figure 10.

### Evolving the ATM SPL

The applied refactorings have prepared the existing implementation of the ATM SPL in a way that allows easily extending the product line to comprise also a product variant containing the *Cash* **and** the *Balance* feature. First,

*Application Conditions* to reduce the propositional formula $(Cash \vee Balance) \wedge \neg Balance$ to $Cash$. We show the resulting implementation in Figure 8.

### Eliminating Conflicts

After the applied refactorings, it is possible to select the feature *Cash* and *Balance* for the same program variant by the application conditions of the delta modules. However, the content of the corresponding delta modules *DCash* and *DBalance* contains conflicting delta actions, regarding the

```
 1  delta DCommon {...}
 2  delta DCash {...}
 3  delta DBalance {...}
 4  spl Atm {
 5      features Cash, Balance, English, German
 6      configurations (Cash && !Balance || !Cash && Balance) &&
 7          (English && !German || German && !English)
 8      deltas
 9          [DCommon when Cash || Balance,
10              DCash when Cash]
11          [DBalance when Balance]
12          [DEnglish when English]
13          [DGerman when German]
    }
```

**Figure 8: ATM SPL after renaming delta modules and simplifying application conditions**

we create a delta module *DCashAndBalance* for the part that is common for both, *DCash* and *DBalance*, in method `startExecution`. Second, the feature model is changed to contain the combination of the *Cash* and *Balance* features. As a clean-up step, the refactoring *Extract Configurations from Conditions* can be used to reduce again the verbosity of the application conditions. Listing 11 shows the evolved software product line, realizing the feature model in Figure 5.

With this example, we have shown refactorings that support tasks required regularly in real-world SPLs. We applied refactorings for both purposes, i.e., for supporting SPL evolution (e.g., in *Preparing the Evolution*) as well as for increasing maintainability (and readability) of code (e.g., in *Remove Redundancy*). However, even within our small example, it is difficult to decide *when* to apply a certain refactoring. To support developers in making this decision, we briefly introduce code smells for delta-oriented SPLs in Appendix A.

## 4.4 Discussion

In this section, we discuss issues that play a pivotal role for the applicability or the proposed refactorings.

*Scope of refactorings.*

Most of the presented refactorings are based on the well-known object-oriented refactorings proposed by Fowler et al [7]. We argue that this is reasonable because of the similarities in evolution and maintenance of product lines in relation to single object-oriented systems. For instance, renaming is a common task for different reasons, such as better reflection of purposes or avoidance of name conflicts. Moreover, we argue that extracting code fragments is a common task in SPLs, *independent* of the actual approach. For example, in AOP it may be necessary to extract code from an advice to split this advice to provide a more fine-grained composition of an SPL. Similarly, extracting methods or classes from features is particularly useful, if it is necessary to split this feature either for supporting more variants or for reducing the size of a (too dominant) feature. However, we also provide refactorings that are specific to SPLs such as those concerning features or application conditions of delta modules. Nevertheless, there may be refactorings which are currently not considered, but turn out to be important during evolution of product lines. To find such refactorings, analyzing the evolution of SPLs (and recurring patterns along this process) is a promising approach.

```
 1  delta DCommon {...}
 2  delta DCash {...}
 3  delta DBalance {...}
 4  delta DCash_startExecution {
 5      modifies Controller {
 6          adds void startExecution() {
 7              this.executeWithdrawal();
 8          }
 9      }
10  }
11  delta DBalance_startExecution {
12      modifies Controller {
13          adds void startExecution() {
14              this.printBalance();
15          }
16      }
17  }
18
19
20  spl Atm {
21      features Cash, Balance, English, German
22      configurations (Cash && !Balance || !Cash && Balance) &&
                (English && !German || German && !English)
23      deltas
24          [DCommon when Cash || Balance,
25              DCash when Cash,
26              DCash_startExecution when Cash]
27          [DBalance when Balance,
28              DBalance_startExecution when Balance]
29          [DEnglish when English]
30          [DGerman when German]
31  }
```

**Figure 9: ATM SPL after extracting conflicting actions**

```
 1  delta DCommon {...}
 2  delta DCash {...}
 3  delta DBalance {...}
 4  delta DCash_startExecution {...}
 5  delta DBalance_startExecution {...}
 6  delta DCashAndBalance {...}
 7  spl Atm {
 8      features Cash, Balance, English, German
 9      configurations (Cash && !Balance || !Cash && Balance) &&
                (English && !German || German && !English)
10      deltas
11          [DCommon when Cash && !Balance || !Cash && Balance,
12              DCash when Cash && !Balance,
13              DCash_startExecution when Cash && !Balance]
14          [DBalance when Balance && !Cash,
15              DBalance_startExecution when Balance && !Cash]
16          [DEnglish when English && (Cash && !Balance || !Cash &&
                Balance)]
17          [DGerman when German && (Cash && !Balance || !Cash &&
                Balance)]
18  }
```

**Figure 10: ATM SPL after merging feature configurations into application conditions**

*Scalability.*

To illustrate the applicability of our refactorings, we use the example of an ATM product line. While this is a small and intuitive SPL, real-world product lines exhibit more features and code and, thus, are more complex. This raises the question to what extent our refactorings are scalable to such SPLs. The answer is twofold. Generally, applying the refactorings is possible even for large-scale SPLs, because the refactorings are executed automatically. Nevertheless, there may be limitations in understanding what has been changed by a refactoring in such SPLs, because the changes

```
1   delta DCommon {...}
2   delta DCash {...}
3   delta DBalance {...}
4   delta DCash_startExecution {...}
5   delta DBalance_startExecution {...}
6
7   delta DCashAndBalance {
8       modifies Controller {
9           adds void startExecution() {
10              if(this.screen.askForTypeOfRequest())
11                  this.executeWithdrawal();
12              else
13                  this.printBalance();
14          }
15      }
16  }
17
18  spl Atm {
19      features Cash, Balance, English, German
20      configurations (Cash || Balance) && (English && !German ||
                German && !English)
21      deltas
22          [DCommon when Cash || Balance,
23              DCash when Cash,
24              DCash_startExecution when Cash && !Balance]
25          [DBalance when Balance,
26              DBalance_startExecution when Balance && !Cash]
27          [DCashAndBalance when Cash && Balance]
28          [DEnglish when English]
29          [DGerman when German]
30  }
```

**Figure 11: Evolved ATM SPL**

may affect different parts of the SPL, e.g., the feature model and different delta modules. However, these problems also occur with existing object-oriented refactorings and are not specific to DOP refactorings. A solution could be to guide the developer through the refactoring process in a step-wise manner. We argue that this is a tooling problem rather than a technical one. In the same way, identifying refactoring opportunities may become difficult in large, complex SPLs, which could be addressed by automated code smell detection.

*Generalizability.*

Although the proposed refactorings are tailored to DOP, we argue that they are applicable for other compositional SPL implementation approaches, such as FOP or AOP, as well. To adopt our refactorings for these alternative approaches, we have to distinguish two kinds of refactoring targets: features and source code. For the former, our proposed refactorings can be used "as is", because features are independent of the underlying language. We only have to take the mapping between the feature model and the corresponding source code (feature modules or aspects) into account. For instance, in case of the *Rename Feature*, we have to ensure that the renaming is performed at all places (which includes possible mappings). Moreover, for FOP we can even apply the refactorings that target on delta modules by only replacing delta by feature modules, since DOP encompasses FOP [23]. While the effort for adapting the source code refactorings to the other approaches may be higher, the underlying concepts can be reused anyway. For instance, extracting delta actions corresponds to extracting parts of an aspect and changing the corresponding advice (or any expression that maps the aspect to a certain feature) in AOP. Another example is the merging of delta modules in case of equivalent application conditions. For AOP, we would have to search for aspects

that contribute to the same feature or address identical join points. Hence, even on the source code level, our refactorings can be generalized.

## 5. IMPLEMENTATION

We provide tool support for applying the refactorings presented in this paper. The primary goal of this implementation is to prove the feasibility of the proposed refactorings and to evaluate their benefits in practice. The implementation of the delta-oriented refactorings is based on the prototype implementation of the DELTAJ language, presented in [22]. The prototype is implemented as a set of *Eclipse* plug-ins, using the *Xtext* framework with Eclipse 3.7.1 and Xtext 2.2.1. It can be obtained from our project page [2].

The *Xtext* framework [26][3] is an open-source framework for the development of *domain-specific languages* within the Eclipse framework. The goal of Xtext is to make the development of programming language as easy and efficient as possible. This is achieved by generating big parts of the language infrastructure from the language grammar file, which is a domain-specific language itself.

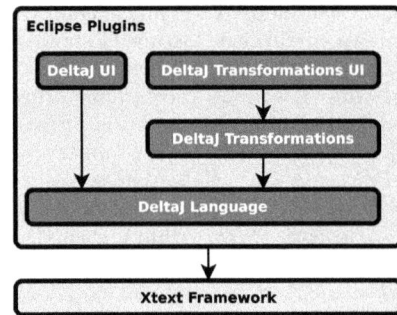

**Figure 12: Architecture of the implementation**

In Figure 12, we illustrate the architecture and the dependencies between the developed Eclipse plug-ins. The prototype implementation for the DELTAJ language is realized by the plug-ins *DeltaJ Language* and *DeltaJ UI*. The *DeltaJ Language* plug-in defines the grammar of the DELTAJ language and provides corresponding source code parsers and validators, while the *DeltaJ UI* plug-in extends the Eclipse UI with a source file editor and a project creation wizard.

Based on the DELTAJ language plug-ins, the delta-oriented SPL refactorings are realized by the *DeltaJ Transformations* plug-in. The refactorings are implemented by altering the *abstract syntax tree* of the parsed DELTAJ program. New tree nodes can be created and added to the abstract syntax tree, while existing nodes can be moved within or removed from the abstract syntax tree. The Xtext framework automatically applies the transformations of the abstract syntax tree to the source code shown in the editor. For example, if a tree node of type *DeltaModule* is inserted into the abstract syntax tree, the corresponding code for the delta module is automatically inserted into the source code of the DELTAJ program.

To enable the user to trigger the implemented refactorings, the *DeltaJ Transformations UI* plug-in extends the Eclipse outline view with a dynamic context menu. In the outline

[2] https://svn.isf.cs.tu-bs.de/redmine/projects/
deltaj_ext-refactoring/files
[3] http://xtext.itemis.com/

82

view, for each language entity, a context menu with a set of possible refactoring transformations is offered. For example, in Figure 13, a selected modification action can be resolved, removed or extracted into a separate delta module.

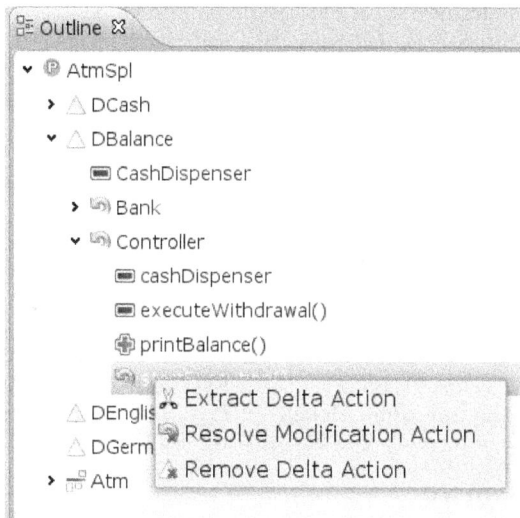

**Figure 13: Outline view of the DeltaJ user interface**

## 6. RELATED WORK

While Opdyke and Fowler initially proposed refactorings for object-oriented programming [17, 7], approaches for refactoring of software product lines have been proposed as well. In the following, we mention important work and compare it to ours. Liu et al. proposed feature-oriented refactoring for FOP [14]. However, they rather aim at decomposing existing programs into features and, thus, support the extractive product line development approach [10]. Beside this, they rather propose formal foundations for this process than practical refactorings or even tool support. In contrast, we aim at improving the structure of *existing* delta-oriented SPLs and propose concrete refactorings.

Similarly, Alves et al. propose refactorings for product lines with a special focus on extractive SPL development [1]. Additionally, they consider refactorings for feature models that aim at improving the model structure. In the same way, Thüm et al. provide a sound reasoning on refactoring feature models [27]. While we support the latter as well (e.g., by simplifying feature configurations), we additionally provide *real* source code transformations. Furthermore, we aim at restructuring instead of creating software product lines, which is also different from Alves et al. [1].

Kuhlemann et al. propose the concept of *refactoring feature modules (RFM)* for refactoring software product lines [11]. RFM's are specific feature modules that contain all code necessary for applying an object-oriented refactoring. However, while this approach can only apply refactoring for certain variants (those, that have an RFM in their configuration), our refactorings affect the whole SPL.

Recently, we proposed variant-preserving refactorings in a catalogue-like manner for FOP [24]. While this is similar to this paper, we only provided four refactorings, tailored to FOP, which is less comprehensive than the refactorings

in this paper. Moreover, we provide no tool support nor an evaluation for the refactorings in FOP.

Finally, two other papers represent important work, because they are complementary to ours. First, Borba proposed a theory of product line refinement that enables stepwise SPL evolution [3]. To show the usability, he explicitly describes how this could work for exemplary artifacts of a product line. However, the focus of this work is rather theoretical and, thus, no implementation or case study exist. Nevertheless, it has the same focus as our work, that is, supporting evolution and refactorings of product lines. Second, Monteiro presented a catalogue of refactorings for aspect-oriented programming [16]. Also code smells specific to AOP are considered, and it is shown how existing code smells indicate crosscutting concerns. Although this work is very similar to ours regarding refactoring and code smells (except for the paradigm), two differences remain: First, we explicitly focus on software product lines and, thus, consider *all* entities of SPLs. In contrast, Monteiro solely supports aspect-oriented programs without taking variability into account. Second, we provide tool support, integrated in an IDE, for applying most of the proposed refactorings (semi-)automatically.

## 7. CONCLUSION

Software, and in particular SPLs, evolves over time in order to meet new or changing requirements, which increases its complexity and impedes maintainability. As a result, it is crucial for the success of SPLs to counter the implications of evolutionary changes to ensure the reusability and maintainability of the artifacts encompassed by the SPL. In this paper, we tackle this problem by proposing a catalog of refactorings for delta-oriented software product lines to address the preparation of evolution or the improvement of maintainability. Additionally, we proposed code smells for delta-oriented SPLs for identifying possible refactoring opportunities. Since refactoring SPLs is manually tedious or even infeasible, we provide an implementation to apply the refactorings semi-automatically.

Beyond that, we discuss several aspects of the proposed refactorings that are crucial for a broader acceptance. Amongst others, we provided arguments why and how our refactorings are generalizable regarding other compositional approaches and how they make up with large-scale SPLs.

## 8. REFERENCES

[1] V. Alves, R. Gheyi, T. Massoni, U. Kulesza, P. Borba, and C. Lucena. Refactoring Product Lines. In *Proc. Int'l Conf. Generative Programming and Component Engineering*, pages 201–210. ACM, 2006.

[2] D. Batory. Feature Models, Grammars, and Propositional Formulas. In H. Obbink and K. Pohl, editors, *Software Product Lines*, volume 3714 of *Lecture Notes in Computer Science*, pages 7–20. Springer, 2005.

[3] P. Borba, L. Teixeira, and R. Gheyi. A Theory of Software Product Line Refinement. In *Proc. Int'l Colloq. on Theoretical Aspects of Computing*, pages 15–43. Springer, 2010.

[4] P. Clements and L. Northrop. *Software Product Lines: Practices and Patterns.* Addison-Wesley Professional, 3rd edition, Aug. 2001.

[5] K. Czarnecki and U. Eisenecker. *Generative Programming: Methods, Tools, and Applications.* Addison-Wesley Professional, June 2000.

[6] K. Czarnecki, S. Helsen, and U. W. Eisenecker. Formalizing Cardinality-Based Feature Models and their Specialization. *Soft. Proc. Improvement and Practice*, 10(1):7–29, 2005.

[7] M. Fowler. *Refactoring: Improving the Design of Existing Code.* Addison-Wesley, Boston, MA, USA, 1999.

[8] K. C. Kang, S. G. Cohen, J. A. Hess, W. E. Novak, and A. S. Peterson. Feature-Oriented Domain Analysis (FODA) Feasibility Study. Technical report, Carnegie-Mellon University Software Engineering Institute, November 1990.

[9] G. Kiczales, J. Lamping, A. Mendhekar, C. Maeda, C. Lopes, J.-M. Loingtier, and J. Irwin. Aspect-Oriented Programming. In M. Aksit and S. Matsuoka, editors, *Proc. Europ. Conf. Object-Oriented Programming*, volume 1241 of *Lecture Notes in Computer Science*, pages 220–242. Springer, 1997.

[10] C. Krueger. Eliminating the Adoption Barrier. *IEEE Software*, 19(4):29–31, 2002.

[11] M. Kuhlemann, D. Batory, and S. Apel. Refactoring feature modules. In *Proc. Int'l Conf. Software Reuse*, pages 106–115. Springer, 2009.

[12] M. Lehman. Program Evolution. *Information Processing & Management*, 20(1–2):19–36, 1984. Special Issue Empirical Foundations of Information and Software Science.

[13] M. M. Lehman and J. F. Ramil. Evolution in Software and Related Areas. In *Proc. Int'l Workshop on Principles of Software Evolution*, pages 1–16. ACM, 2001.

[14] J. Liu, D. Batory, and C. Lengauer. Feature Oriented Refactoring of Legacy Applications. In *Proc. Int'l Conf. Software Engineering*, pages 112–121. ACM, 2006.

[15] R. C. Martin. *Clean Code: A Handbook of Agile Software Craftsmanship.* Prentice Hall, 1 edition, 8 2008.

[16] M. P. Monteiro and J. a. M. Fernandes. Towards a Catalog of Aspect-Oriented Refactorings. In *Proc. Int'l Conf. Aspect-Oriented Software Development*, pages 111–122. ACM, 2005.

[17] W. F. Opdyke. *Refactoring Object-Oriented Frameworks.* PhD thesis, Champaign, IL, USA, 1992. UMI Order No. GAX93-05645.

[18] D. L. Parnas. Software Aging. In *Proc. Int'l Conf. Software Engineering*, pages 279–287. IEEE, 1994.

[19] K. Pohl, G. Böckle, and F. van der Linden. *Software Product Line Engineering: Foundations, Principles and Techniques.* Springer, 2005.

[20] O. Richers. Transformation and Evolution of DeltaJSoftware Product Lines. Master thesis (Diplomarbeit), TU Braunschweig, 2012. (`https://svn.isf.cs.tu-bs.de/redmine/documents/1`).

[21] I. Schaefer, L. Bettini, V. Bono, F. Damiani, and N. Tanzarella. Delta-oriented Programming of Software Product Lines. In *Proc. Int'l Software Product Line Conference*, pages 77–91. Springer, 2010.

[22] I. Schaefer, L. Bettini, and F. Damiani. Compositional Type-Checking for Delta-Oriented Programming. In *Proc. Int'l Conf. Aspect-Oriented Software Development*, pages 43–56. ACM, 2011.

[23] I. Schaefer and F. Damiani. Pure Delta-Oriented Programming. In *Proc. Int'l Workshop on Feature-Oriented Software Development*, pages 49–56. ACM, 2010.

[24] S. Schulze, T. Thüm, M. Kuhlemann, and G. Saake. Variant-Preserving Refactoring in Feature-Oriented Software Product Lines. In *Proc. Int'l Workshop on Variability Modeling in Software-intensive Systems*, pages 73–81. ACM, 2012.

[25] Steffen Zschaler et al. Variability management. In *Aspect-Oriented, Model-Driven Software Product Lines.* CUP, 2011.

[26] The Eclipse Foundation. Xtext homepage. http://www.eclipse.org/Xtext/documentation/, 2012.

[27] T. Thüm, D. Batory, and C. Kastner. Reasoning About Edits to Feature Models. In *Proc. Int'l Conf. Software Engineering*, pages 254–264. IEEE, 2009.

# APPENDIX

## A. CODE SMELLS

In the following, we shortly explain code smells in delta-oriented SPLs and their countermeasures in terms of refactorings. A more comprehensive overview of these code smells can be found in [20].

***Duplicated Delta Action.*** Two delta actions with identical content exist in the SPL. Recommended refactoring: *Resolve Duplicated Actions.*

***Dead Delta Action.*** The effect of a delta action is always overwritten by another delta action. Recommended refactoring: *Remove Dead Delta Action*

***Dead Delta Module.*** The application condition of a delta module is a contradiction, either by itself or with respect to the valid feature configurations. Recommended refactoring: *Remove Delta Module*

***Empty Delta Module.*** A delta module contains no delta actions. Recommended refactoring: *Remove Empty Delta Module*

***Unused Feature.*** A feature is defined, but not used by any product. Recommended refactoring: *Remove Unused Feature*

***Empty Feature.*** Selection of a feature has no effect on generated product. Recommended refactoring: *Remove Empty Feature*

***Duplicated Features.*** Two distinct features have the same effect. Recommended refactoring: *Merge Duplicated Features*

***Joined Features.*** Two distinct features have only any effects if both are selected. Recommended refactoring: *Merge Joined Features*

***Complex Feature Configurations.*** The propositional formula, specifying the valid feature configurations, is more complex than necessary. Recommended refactoring: *Simplify Feature Configurations*

***Complex Application Conditions.*** An application condition is more complex than necessary. Recommended refactoring: *Simplify Application Conditions*

# Modular Specification and Checking of Structural Dependencies

Ralf Mitschke
Technische Universität
Darmstadt
Darmstadt, Germany
mitschke@st.informatik.tu-
darmstadt.de

Michael Eichberg
Technische Universität
Darmstadt
Darmstadt, Germany
eichberg@informatik.tu-
darmstadt.de

Mira Mezini
Technische Universität
Darmstadt
Darmstadt, Germany
mezini@informatik.tu-
darmstadt.de

Alessandro Garcia
Pontifical Catholic University
of Rio de Janeiro
Rio de Janeiro, Brazil
afgarcia@inf.puc-rio.br

Isela Macia
Pontifical Catholic University
of Rio de Janeiro
Rio de Janeiro, Brazil
ibertran@inf.puc-rio.br

## ABSTRACT

Checking a software's structural dependencies is a line of research on methods and tools for analyzing, modeling and checking the conformance of source code w.r.t. specifications of its intended static structure. Existing approaches have focused on the correctness of the specification, the impact of the approaches on software quality and the expressiveness of the modeling languages. However, large specifications become unmaintainable in the event of evolution without the means to modularize such specifications. We present Vespucci, a novel approach and tool that partitions a specification of the expected and allowed dependencies into a set of cohesive slices. This facilitates modular reasoning and helps individual maintenance of each slice. Our approach is suited for modeling high-level as well as detailed low-level decisions related to the static structure and combines both in a single modeling formalism. To evaluate our approach we conducted an extensive study spanning nine years of the evolution of the architecture of the object-relational mapping framework Hibernate.

## Categories and Subject Descriptors

D.2.2 [**Software Engineering**]: Design Tools and Techniques; D.2.4 [**Software Engineering**]: Software/Program Verification—*Validation*; D.2.11 [**Software Architectures**]: Information Hiding

## Keywords

Software Architectures, Modularity, Scalability, Structural Dependency Constraints, Static Analysis

## 1. INTRODUCTION

A documented software architecture is an acknowledged success factor for the development of large, complex systems [29]. Traditionally, architecture description languages (ADLs) have been used to specify the architecture and verify its properties. Generally, this process has been detached from coding and the architecture specification has been considered as a means to prescribe the structure of the code resulting from programming or eventually to generate a first skeleton of that code. However, as systems evolve over time, due to new requirements or corrections, the implemented architecture starts to diverge from the intended architecture [11, 15, 22] — resulting in *architecture erosion* [26].

To combat architecture erosion, several approaches have emerged that focus on structural dependencies [10, 25, 28, 32] and whose proponents argue for automated checking of architecture specifications w.r.t. the static structure of the source code. These approaches generally allow to group[1] source code elements into *building blocks* — cohesive units of functionality in the software system — and to specify in which way a building block is allowed to statically depend on which other building block. The specification formalisms in these approaches vary and can be summarized as: (i) a flat graph with building blocks as nodes and allowed dependencies as edges [25]; (ii) a matrix notation with building blocks in rows/columns and their dependencies in the cells [28]; (iii) a graph with hierarchical nodes and component-connector style ports to manage internal/external dependencies [10]; (iv) a textual specification of access restrictions on target building blocks [32].

Such specifications are used either analytically [25] — to analyze already written code for conformance with an intended static structure — or constructively [10, 32] to enforce the code's compliance with the specification of the static structure continuously during development. Constructive approaches were proven to help developers in realizing the intended architecture. Several case studies [16, 19, 20] show that constructive approaches can prevent structural erosion [27, 33].

Though current approaches have proven to be valuable,

---

[1] using, e.g., regular expressions over classes or source files

they all share the property that a single monolithic specification is used and – as in case of a monolithic software system — a monolithic description of the structure does not scale and becomes unmaintainable once the software reaches a certain complexity. Sangal et al. [28] explicitly try to solve the maintainability and scalability issues using a special notation called dependency structure matrices (DSMs). However, we believe that the problem is not so much the notation. The root of the problem is the monolithic nature of the specifications. Based on some preliminary experience with modeling the architecture of real systems, such as Hibernate [5], we doubt that any such approach can scale, even with compact notations such as dependency structure matrices. As a result, typically only the highest level of components and/or libraries is considered [32, 27]; requiring different notations and tools for different levels of the design. This precludes a seamless design at various granularity levels.

In this paper, we argue that modeling a software's static structure should consist of multiple views, that focus on different parts and on different levels of detail. We take the position that like programming languages, architecture modeling languages in general should support modularity and scoping mechanisms to support modular reasoning about different architectural concerns and information hiding to facilitate evolution.

Accordingly, we propose a novel modeling approach and tool, called Vespucci, that allows to separate the specification of a software's static structure into multiple complementary views, called *slices* throughout this paper. Each slice can be reasoned over in separation. Multiple slices can express different views on the same part of the software and each slice can be evolved individually. Hence, evolution of large scale specifications consisting of several slices is facilitated by distributing work to systematically update the architecture in a modular fashion. Contrary to a monolithic specification, our approach also has the benefit that individual concerns can remain stable. Stable parts can be modularized into different slices to be separated from architectural "hot-spots", i.e., slices that require frequent changes during the evolution.

The contributions of this paper are:

- A first approach towards the specification of a software's structural dependencies that supports a modularized specification by means of individual slices.

- A new approach for modeling a software's structural dependencies that combines the advantages of hierarchical and graph-based modeling approaches to enable reasoning over a software's static structure at different abstraction levels.

- Discussion of an implementation of the proposed approach that enables the specification and checking of a software's structural dependencies.

The remainder of the paper is organized as follows. In Section 2, we briefly introduce the Hibernate framework [5], which we use to illustrate concepts of the proposed approach and to evaluate its effectiveness. Section 3 introduces Vespucci's specification language. In Section 4 we present an in-depth evaluation of Vespucci. After that, we discuss related work in Section 5. Finally, we give a summary and discuss future work.

## 2. ARCHITECTURE OF HIBERNATE

As part of the development of Vespucci, we did a comprehensive analysis of the architecture of the object-relational mapping framework Hibernate [5]. We provide a short overview of Hibernate and its architecture in this section since we will refer to it to discuss and motivate various features of Vespucci.

We chose Hibernate as it is a large, mature, widely-adopted software system, which has been continually updated and enhanced. We reengineered the architecture of the core of Hibernate in version 1.0.1 (July 2002) and played back its evolution until version 3.6.6 (July 2011)[2]. During this time the core grew from 2 000 methods in over 255 classes organized in 18 packages to 17 700 methods in over 1 954 classes in 100 packages.

In the following, the major building blocks of Hibernate's architecture are presented. A *building block* is a logical grouping of source code elements that provide a cohesive functionality, independent of the program's structuring, e.g., in packages or classes. The scope of a building block depends on the considered abstraction level and ranges from a few source code elements up to several hundreds. For example, Hibernate's support for different SQL dialects is represented by one top-level building block with many source code elements, but further structured into smaller building blocks for elements that abstract over the support for concrete dialects and those that actually implement the support.

Table 1: Overview of Hibernate 1.0

| Top-level Building Block | 2L Building Blocks | Classes contained | Elements contained | Relation to Packages |
|---|---|---|---|---|
| Cache | 4 | 6 | 60 | ≡ |
| CodeGeneratorTool | 0 | 9 | 68 | ⊂ |
| ConnectionProvider | 3 | 5 | 51 | ≡ |
| DatabaseActions | 3 | 9 | 59 | ⊂ |
| DataTypes | 10 | 37 | 410 | ≡ |
| DeprecatedLegacy | 2 | 2 | 6 | ⊂ |
| EJBSupport | 0 | 1 | 22 | ≡ |
| HibernateORConfiguration | 2 | 2 | 39 | |
| HibernateORMapping | 12 | 33 | 389 | ≡ |
| HQL (Hibernate Query Lang.) | 3 | 9 | 130 | ≡ |
| IdentifierGenerators | 4 | 12 | 92 | ≡ |
| MappingGeneratorTool | 0 | 19 | 233 | ⊂ |
| PersistenceManagement | 6 | 35 | 674 | |
| PropertySettings | 0 | 1 | 43 | |
| Proxies | 0 | 3 | 23 | ⊂ |
| SchemaTool | 2 | 5 | 34 | ⊂ |
| SessionManagement | 6 | 10 | 312 | |
| SQLDialects | 3 | 12 | 119 | ≡ |
| Transactions | 2 | 4 | 37 | ≡ |
| UserAPI | 9 | 9 | 63 | |
| UtilitiesAndExceptions | 2 | 33 | 235 | |
| XMLDatabinder | 0 | 2 | 21 | |

The architectural model of Hibernate 1.0 consists of the 22 top-level building blocks shown in Table 1. Of these 22 top-level building blocks, 16 were further structured. In total, we identified 73 second-level building blocks. Given the size of Hibernate 1.0, we did not analyze lower levels. On

---

[2]Hibernate 4.0 was released after the case study.

average each top-level building block already only contains 11 classes and the 2nd level building blocks consist of even fewer classes. The key figures of the architecture are given in Table 1. In the following, we discuss those elements of the architecture that are most relevant when considering the modeling of architectures. The complete architecture can be downloaded from the project's website [2].

For Hibernate 1.0 nine of the building blocks have a one-to-one mapping to a package (cf. Table 1 – Relation to Packages ≡). Six building blocks map to a subset (⊂) of the code of some non-cohesive package. For example, the package `cirrus.hibernate.impl` contains classes for creating proxies as well as classes related to database actions. These sets of classes have no interdependencies and belong to different building blocks. The source code elements of the remaining building blocks are spread across several packages. For example, the code related to session handling is spread across two packages in version 1.0.

Overall, the architecture features several well modularized building blocks, such as the Cache, HQL or Transactions building blocks, which are only coupled with at most three other building blocks. The number of well modularized building blocks with few dependencies is, however, small. The majority of Hibernate's functionality belongs to building blocks that exhibit high coupling, such as PersistenceManagement, SessionManagement and DataTypes.

## 3. THE VESPUCCI APPROACH

In this section, we first describe the three major parts Vespucci [2] consists of: (1) a declarative source code query language to overlay high-level abstractions over the source code, (2) an approach that enables the modular, evolvable, and scalable modeling of an application's structural dependencies, and (3) a runtime for checking the consistency between the modeled and the implemented dependencies. After that, we present in Section 3.4 the different modeling approaches supported by Vespucci. Finally, we discuss in Section 3.5 how the proposed approach facilitates the evolution of the specification and the underlying software and how it supports large(r) scale software systems.

### 3.1 High-level abstractions over source code

Vespucci is concerned with modeling and controlling structural dependencies at the code level. But, it does so at a high-level of abstraction.

**Ensembles** are Vespucci's representation of high-level building blocks of an application, whose structural dependencies are modeled and checked. Specifically, Vespucci's ensembles are groups of source code elements, namely type, method, and field declarations. The definition of an ensemble involves the specification of source code elements that belong to it by means of source code queries. We refer to the set of source code elements that belong to an ensemble as the *ensemble's extension*.

The visual notation of an ensemble is a box with a name label. For example, Figure 1 shows two ensembles, one called SessionManagement and one called HQL. Vespucci explicitly predefines the so-called *empty ensemble* that never matches any source elements and is depicted using a simple gray box (▦). The empty ensemble supports some common modeling tasks, e.g., to express that a utility package should not have any dependencies on the rest of the application's code.

**The source code query language** is introduced – mostly example-driven – in the following paragraphs. The language is not the primary focus of this paper, which is rather on modularity mechanisms for modeling structural dependencies. In fact, the approach as a whole is parameterized by the query language, in the sense that the modularization mechanisms can be reused with other more expressive query languages and more sophisticated query engines. For a more systematic definition of the current query language, the interested reader is referred to the website of the project [2].

The query language provides a set of basic predicates that can select individual fields or methods, entire classes, packages, or source files. Predicates take quoted parameters, which filter respective code elements by their signature, e.g., the predicate `package('cirrus.hibernate.helpers')` selects code elements in Hibernate's `helpers` package, using the package name as the filter. The query defines the Utilities ensemble, which we have used in modeling Hibernate's structural dependencies.

In the above example, source code elements are precisely specified by their fully qualified signature. Furthermore, wildcards ("*") can be used to abstract over individual predicate parameters. For example, the `field` predicate below selects field declarations in class `Hibernate` with any name (the second parameter is "*"), of a type that ends with the suffix `Type`. We have used the query to define an ensemble called TypeFactory which serves as a factory for Hibernate's built-in types.

```
field('*.Hibernate','*','*Type')
```

Queries can be composed using the standard set theoretic operations (union, intersection, difference), or by passing a query as an argument to a type parameter of another query. This form of composition is useful to reason over inheritance for selecting all sub-/supertypes of a given type. For example, consider the query:

```
class_with_members(subtype+('Dialect'))
```

It uses the basic predicate `class_with_members`, which selects a class and all it's members. Since the predicate expects a type to be selected, we can instead pass a subquery. The `subtype+` query returns the transitive closure of all subtypes of the class `Dialect`. Hence, the example query selects all classes (and their members) that are a subtype of the class `Dialect`. In Hibernate these represent all supported SQL dialects – the shown query actually defines the ensemble ConcreteDialects.

As already mentioned, the query language is interchangeable. What is interesting about the use of the source query language as an ingredient of our approach is that it enables modeling structural dependencies at a high-level of abstraction. Furthermore, it supports the definition of ensembles that cut across the modular structure of the code, e.g., TypeFactory cuts across the class-based decomposition of code. This enables feature-based control of structural dependencies.

Vespucci provides an **ensemble repository** that stores the definitions of all ensembles. It serves as a project-wide repository and provides the starting point for modeling an application's intended structural dependencies. Capturing all ensemble definitions in a single repository serves two purposes. First, it enables a model of intended structural dependencies to be modularized with the guarantee that all modules refer to the same extension for a particular ensemble. Second, it allows modules to pose global constraints quantifying over all defined ensembles (see the discussion about global constraints in the following section).

## 3.2 Modeling Structural Dependencies

**Dependency slices** are Vespucci's mechanism to support the modularized specification of an application's structural dependencies. A slice captures one or more specific design decisions, by expressing one or more constraints over ensemble inter-dependencies, e.g., which ensemble(s) is/are allowed to use a certain other ensemble.

For illustration, Figure 1 shows an an exemplary slice, which governs dependencies to source code elements that implement the Hibernate query language, represented by the HQL ensemble. Specifically, it states that elements pertaining to HQL may be used ONLY by those pertaining to the SessionManagement ensemble. The cycle attached to the arrow pointing to HQL states that globally, i.e., for all ensembles in the ensemble repository, this is the only dependency on HQL's elements that is allowed.

Figure 1: Dependency rule for Hibernate Query Language

Figure 2 shows another example slice, which states that source code elements pertaining to SQLDialects are only allowed to be used by PersistenceManagement's or SessionsManagement's source code elements.

Figure 2: Users of SQL Dialects

There can be an arbitrary number of slices in a model of structural dependencies; the set of ensembles referred to in different slices may overlap. Deciding about the number/kind of slices, in which one may want to break down the specification of an application's structural dependencies is a matter of modeling methodology, as we elaborate on in section 3.4. Yet, we envision the default strategy to be one in which each slice is used to express allowed and expected dependencies from the perspective of a single ensemble; this strategy was used in the case study and in the examples shown in the paper. For this purpose, the visual notation features arrow symbols that are shown next to the ensemble that is constrained[3]. For example, by looking at Figure 1 we can reason about *all* dependencies that are allowed for HQL and looking at Figure 2 we can reason about *all* dependencies that are allowed for SQLDialects.

Ensembles that participate in a slice but which have no arrow symbols next to their box are not constrained. For example, both slices refer to SessionManagement, but make no statement w.r.t. the total of its allowed dependencies. From these two slices we can see that SessionManagement's source code elements are allowed to depend on both SQLDialect's

---

[3]For this paper the visual models were compressed to save space. Hence the distinction may not be as obvious as it is when you use the Vespucci tool.

and HQL's source code elements. However, SessionManagement and PersistenceManagement are not constrained.

**Constraint types** are classified into two basic categories: constraints that are defined w.r.t. the allowed and those w.r.t. the not-allowed dependencies. Constraints on allowed dependencies are further classified as *Outgoing and Incoming Constraints* and *Local and Global Constraints*. The rationale for distinguishing between the above types of constraints relates to enabling modular reasoning about individual architectural concerns. Modular reasoning fosters scalability by allowing each slice to be understood as a single unit of comprehension, and also fosters evolvability as each slice can be adapted without the need to refer to other slices. We elaborate on the role that different constraint types play with these respects in the following section. Here, we exclusively focus on explaining the meaning of these different constraints.

An *incoming constraint* restricts the set of source code elements that may use the elements of a particular ensemble (target ensemble). Incoming constraints are denoted by the symbol "->" shown next to the target ensemble (cf. Figure 1, Figure 2). For example, the constraint in Figure 1 restricts source code dependencies, of which the target element belongs to HQL: the source of the dependency must belong to SessionManagement; source code dependencies from and to the source code elements belonging to SessionManagement are — w.r.t. that slice — unrestricted.

An *outgoing constraint* restricts the set of source code elements on which code elements of a specific ensemble (source ensemble) may depend. Outgoing constraints are visually denoted by the symbol ">" shown next to the source ensemble. For example, the slice in Figure 3 features two outgoing

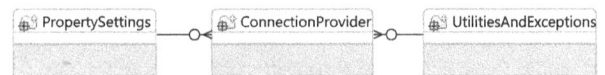

Figure 3: Constraints on the Connection Provider

constraints; from ConnectionProvider to PropertySettings, respectively to UtilitiesAndExceptions. Outgoing constraints only affect code elements of their source ensemble. Hence, the slice in Figure 3 governs the dependencies of code elements involved in providing connections (captured by the ConnectionProvider ensemble). They may only use generic functionality (captured by the UtilitiesAndExceptions ensemble), or functionality for getting and setting properties (captured by PropertySettings). The targets of the constraint (PropertySettings and UtilititiesAndExceptions) can — w.r.t. the slice in Figure 3 — depend on any other ensemble.

*Global constraints* quantify over all defined ensembles. Visually they are denoted by a "o" attached to a constraint. All constraints considered in the examples so far were global. For example, the constraint shown in Figure 1 affects code elements that belong to any ensemble defined in the repository of Hibernate, even if not referred to by the slice, e.g., ConnectionProvider or PropertySettings in Figure 3. Code elements of the latter ensembles are not allowed to depend on elements in HQL.

Global constraints are hard constraints w.r.t. the addition of new ensembles into the architecture. Whenever new ensembles are defined in the ensemble repository, they are included when checking a global constraint. The purpose is to provide tight control over the evolution of the architecture. If a new ensemble has dependencies that violate a global constraint, then architects can asses whether the violation needs to be removed from the code or, whether the currently defined architectural rules are too narrow. The essential point is that an architect has assessed the situation and no uncontrolled erosion of the software's structure has occurred.

*Local constraints* quantify only over ensembles that are referenced in one particular slice. Visually, they are characterized by the lack of the "○" symbol. Figure 4 depicts local constraints on the implementation of Hibernate's support for different SQL dialects (e.g., "Oracle SQL","DB2 SQL"). Each dialect is realized by implementing a common interface. Elements of this interface are captured by the AbstractDialect ensemble. Support for specific dialects is captured by ConcreteDialects. The TypeNameMap ensemble captures code elements involved in implementing a specialized dictionary for mapping database type names to a common set of names. The defined constraints specify that only code pertaining to ConcreteDialects is allowed to depend on code pertaining to AbstractDialect and code in the latter is only allowed to depend on TypeNameMap's code. Furthermore, neither source code elements of AbstractDialect nor TypeNameMap are allowed to depend on elements of ConcreteDialects due to the incoming constraint between the empty ensemble and ConcreteDialects. However, the constraints of the slice in Figure 4 do not restrict in any way code elements belonging to ensembles that are not referenced by this slice, e.g., code pertaining to HQL (slice in Figure 1) could use code pertaining to ConcreteDialects.

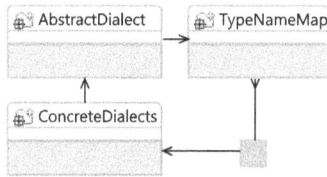

Figure 4: Supporting Multiple SQL Dialects

Local constraints provide tight control over the evolution of source code w.r.t. the scope of the ensembles referenced in a slice. Their purpose is to capture localized rules that reason only over a part of all the dependencies in the architecture, e.g., as in Figure 4 where only dependencies pertaining to the implementation of multiple SQL dialects are considered. The implementation details of involved ensembles can change (and respectively their extensions), but the changes are guaranteed to adhere to the specified allowed/expected dependencies. The rest of the architecture can evolve independently, i.e., new ensembles and dependencies can be introduced as long as they do not violate the localized rules. **Different kinds of dependencies** can be constrained individually by annotating constraints. The kinds of dependencies are those that can be found in Java code (e.g., Field Read Access, Field Write Access, Inherits, Calls, Creates,...; a complete reference is available online [2]); by default, all kinds of dependencies are constrained and no further annotation of a constraint is necessary. Dependency kinds are important when documenting detailed design choices.

For example, Figure 5 restricts only dependencies of the create kind (i.e., object creations) to the ConcreteConnectionProviders; only the ConnectionProviderFactory is allowed

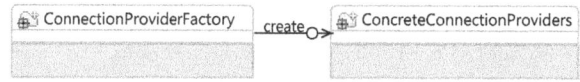

Figure 5: Restricting connection provider creation to a factory

to create new connection providers. All other dependencies are allowed for all ensembles, hence clients may use the created provider, e.g., by calling its methods. The range of possible applications is broad, e.g., one can also disallow classes from throwing particular exceptions, while allowing their methods to catch them.

**Nesting of ensembles** is also enabled in Vespucci to reflect part-whole relationships. The information about child/parent relationships between ensembles is stored in the global repository. For illustration consider that the slice shown in Figure 4 actually models the internal architecture of Hibernate's support for SQL dialects. One can express this relation by making the ensembles referred to in Figure 4 children of the SQLDialects ensemble, as shown in Figure 6.

Figure 6: (Sub-)Ensembles of SQL Dialects

The extension of an ensemble that has inner ensembles is the union of the extension of its inner ensembles; i.e., an ensemble with inner ensembles does not define its own query to match source elements, but instead reuses the queries of its inner ensembles. Hence, the semantics of nesting is that constraints defined for parent ensembles implicitly apply to source code elements of all their children, e.g., constraints defined for SQLDialects in the slice in Figure 2 apply to all its children ensembles.

Constraints that cross an ensemble's border are disabled in Vespucci for keeping the semantics simple. Due to slices, this can be done without loss of expressivity. If an architect needs to define a constraint between two ensembles that do not have the same ancestor ensemble, it is always possible to specify the constraint in a new slice that just refers to the directly relevant ensembles.

With hierarchical modeling, architects can distinguish between ensembles that are involved in the architectural-level modeling of dependencies (SQLDialects) and those involved in modeling decisions at lower design levels (ensembles in Figure 4). In the following sections, we discuss how the combination of slices and hierarchies facilitates the incremental refinement of a software's architecture and is advantageous in case of software evolution.

### 3.3 Constraint Enforcement and Tooling

Conceptually, checking the implementation against the modeled dependencies is done as described next.

First, for each ensemble its extension along with the set of source code dependencies related to it (those that have the ensemble as source or target) is calculated; self-dependencies, i.e., source code dependencies, where the source and target elements belong to the same ensemble are filtered out. Furthermore, dependencies from and to source code elements that do not belong to any ensemble are ignored.

Second, each slice is checked on its own. To do so, Vespucci iterates over all ensembles of each slice and checks that none of the dependencies between respective source code elements violates a defined constraint. For example, to check the compliance of an application's source code with the slice in Figure 3, Vespucci effectively checks that the target of all code dependencies, starting at a code element in ConnectionProvider, either belongs to PropertySettings or to UtilitiesAndExceptions.

The implementation of Vespucci's dependency checker is integrated into the Eclipse IDE. Checking is done as part of the incremental build process and incremental checking ([9]) is efficient enough for (at least) mid-sized projects such as Hibernate. Likewise the modeling side is incrementalized, hence, changes to queries are immediately reflected in the IDE.

The rationale behind the decision to ignore dependencies to source code elements that do not belong to any ensemble is that dependencies to an application's essential libraries and frameworks are most often not of architectural relevance and should not clutter the overall specification. Nevertheless, it is always possible to create an ensemble that covers some fragment of a fundamental library to restrict its usage. E.g., while it generally does not make sense to restrict the usage of the JDK, it may still be useful to restrict the usage of the `java.util.logging` API, because the project as a whole uses a different API for logging and it has to be made sure that no one accidentally uses the default logging API. One possibility to model such a decision is to create a global incoming constraint from an empty ensemble to the ensemble representing the `java.util.logging` API. However, Vespucci provides a specialized view that lists source code elements that do not belong to any ensemble to make it easily possible to find unintended holes in the specification.

### 3.4  On Modeling Methodology

Figure 7 schematically shows four principal ways to model the architecture of a hypothetical system consisting of four ensembles (boxes labeled 1 to 4) with Vespucci. In (A), all constraints are modeled in a single model. In (B), the model makes use of hierarchical structuring – specifically, ensembles 1 and 2 are nested into an ensemble 1&2. In (C), the model makes use of slicing; specifically, per ensemble one slice is defined, modeling only decisions related to that ensemble, but slicing at other granularity levels is conceivable (see below). In (D), the model makes use of both slices and hierarchies, which is the expected typical usage of Vespucci.

In general, the structural dependency model of a system in Vespucci consists of an arbitrary number of slices. It is a matter of modeling decisions – taken by the architect – in how many slices she breaks down the overall architectural specification. As part of this process, a trade-off is to be made between (i) creating (large(r)) slices that capture several architectural rules related to multiple ensembles that conceptually belong together and (ii) creating one slice per ensemble that just captures the architectural rules related to that ensemble. In the former case cohesiveness is fostered while in the latter case (local) comprehensibility of the architecture and evolvability of the specification is fostered.

In the Hibernate case study, as a rule of thumb, each high-level slice focused on design decisions concerning one

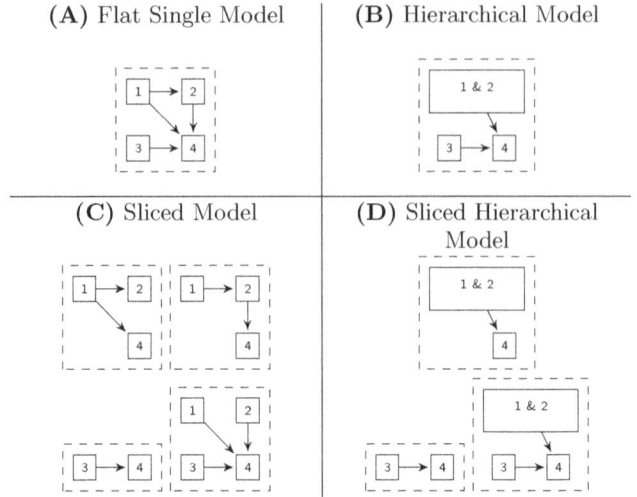

Figure 7: Alternative Architectural Models of Dependencies

ensemble. For instance, the slices in Figure 1 and Figure 3 focus on specific design decisions related exclusively to allowed *incoming dependencies* to HQL, respectively allowed *outgoing dependencies* of ConnectionProvider. Internal dependencies for ensembles with nested sub-ensembles were in general related to a small set of ensembles and hence captured in a single slice, as e.g., in Figure 4, where the internals of SQLDialects were captured.

The one-slice-per-high-level-ensemble strategy for breaking down specifications is just a first approximation. For reasons of better managing complexity and evolvability as well as understandability, it may make sense to chose more fine-grained or coarse-grained strategies. One such strategy is to split the specification of incoming and outgoing dependencies of an ensemble, if those are too complex or evolve in different ways. On the other hand, slices of related ensembles may be merged, when their separated specifications are too simple to justify separate slices or hard to understand in isolation.

One may criticize that a specification becomes complex with an increasing number of slices. However, a single specification that controls the dependencies to the same degree is no less complex and includes all information that are captured in the slices. For example, if internal dependencies are controlled, they need to be specified and maintained in a single specification as well. The focus here is to make a case for enabling the architects to break down specifications of structural dependencies in several modules that are more manageable w.r.t. scalability and evolvability and can be reasoned over in isolation. Hence, slices also facilitate distribution of work, such that large architectures can be maintained by a team rather than a single architect.

Per ensemble slicing of the dependency model may also impair understandability of dependencies pertaining to several modules. A view of the dependencies for multiple ensembles (in contrast to their individual constraints) can be advantageous for the exploration of the architecture, e.g., if one wants to follow transitive dependencies such as the path of communication from ensemble A to B. Note that if such a path is relevant to the architect, it can also be encoded as a slice. A second scenario for global comprehension is to find all slices in which an ensemble participates. This can be supported by a simple analysis over the defined slices.

All the above said, systematically deriving guidelines for structuring architectural decisions into slices and distributing the work is a matter of performing comprehensive studies and is out of the scope of this paper.

## 3.5 Scalability and Evolvability

Vespucci enables architects to reason about architectural decisions concerning structural dependencies of a set of ensembles in isolation, while treating the rest of the system as a black-box, and to do so in a top-down manner. This is due to (1) Vespucci's support for breaking down the specification into slices, (2) mechanisms for expressing structural rules via a constraint system, (3) a scoping mechanism that enables to quantify locally or globally over the set of affected ensembles, and (4) Vespucci's support for enabling the hierarchical organization of specifications. The latter is a traditional mechanism to govern complexity [31] and will for this reason not be further considered in the following discussion.

*Support for modular reasoning.*

Slices enable the architect to focus on constraints that concern individual ensembles or a set of strongly related ensembles. This makes it possible to isolate a small set of related architectural decisions from the rest for the purpose of modularly reasoning about them, while treating the rest as a black-box.

This fosters scalability by reducing the number of ensembles and constraints that need to be considered at once: Each slice in Figure 7 (C) contains less ensembles and constraints than the model in Figure 7 (A). One may argue that slicing actually increases the overall number of elements (ensembles/constraints) — since some of them are mentioned in multiple slices. However, as they represent the same abstractions in all slices, the overall number of elements that need to be understood remains the same as in the model A.

Consider for illustration the slice depicted in Figure 2. It expresses that only SessionManagement and PersistenceManagement may use SQLDialects with the minimum amount of explicitly mentioned ensembles and constraints. No rules governing dependencies between SessionManagement and PersistenceManagement, respectively between those and other ensembles, are specified. The slice in Figure 2 models architectural constraints from the perspective of SQLDialects. Dependencies between SessionManagement and PersistenceManagement or between those and other ensembles are irrelevant from this perspective and are, thus, left unspecified. Further, we do not explicitly enumerate all ensembles that are not allowed to depend on SQLDialects.

Vespucci's constraint system for modeling dependencies and the way checking for architecture compliance operates (see previous section) is key to the conciseness of specifications. Slices are checked in isolation. The constraint system interprets the lack of a constraint in a slice as "don't care" in the sense that the presence or absence of code dependencies is ignored. E.g., potential dependencies between SessionManagement and PersistenceManagement are ignored when checking compliance with rules in slice in Figure 2. They may well be the subject of specification in other slices to be reasoned on separately.

The role played in this respect by our distinction of incoming and outgoing constraints needs to be highlighted here. It is the use of the incoming constraints in Figure 2 that enables us to talk about constraints from the perspective of SQLDialects – excluding from consideration any further dependencies in which, e.g., SessionManagement may engage. Incoming/outgoing constraints are "unilateral" – they belong to one ensemble. Without this distinction, we would be left with "bilateral" constraints; mentioning one such constraint that affects SessionManagement would require to mention all other constraints affecting SessionManagement; hence, making it impossible to slice specifications.

The ability to abstract over any dependencies that are not explicitly constrained comes in also very handy when handling ensembles that are expected to be ubiquitously used, e.g., Hibernate's Utilities ensemble. Such ensembles would typically contribute a significant amount of complexity to architectural specifications, if the specification approach requires to explicitly mention allowed dependencies. By using a constraint system this complexity can be avoided. The specification would make no mention of dependencies to Utilities, in order to leave it unconstrained.

The ability to state a constraint that affects arbitrary many ensembles without having to enumerate those explicitly is due to the ability to make global statements. Ensembles that are not explicitly mentioned in a slice are reasoned over by global constraints, e.g., the slice shown in Figure 2 implicitly states that all other ensembles mentioned in Figure 1, 2, 3, and many more, are not allowed to use SQL Dialects. This specification is much smaller compared to enumerating this fact for all other ensembles constituting the rest of Hibernate. The latter would be necessary, if Vespucci only had allowed and not-allowed constraints and no distinction between local and global scopes.

*Support for evolution.*

Due to slicing, architectural models also become easier to extend. First, slices remain stable in case of extensions that do not affect their ensembles/constraints. Second, affected slices are easier to identify. Finally, existing global constraints automatically apply to new ensembles.

Consider for illustration the following scenario that occurred during the evolution of Hibernate from version 1.0 to version 1.2.3. In this step, a new ensemble — called Metadata — to represent Hibernate's new support for metadata was introduced. This change was accommodated mostly incrementally. First, the specification as a whole was extended incrementally by introducing a new slice, referring to the ensembles that Metadata is allowed to use and be used from. Second, the set of existing slices that eventually required revision was restricted to those modeling the dependencies of ensembles referred to in the new Metadata slice. For example, the slice that defined constraints for DataTypes was refined to enable the usage by Metadata. Slices that modeled unrelated architectural decisions, e.g., those governing dependencies of ConnectionProvider (cf. Figure 3), did not require any reviewing. Yet, previously stated global constraints carry over to the new ensemble, ensuring e.g., that it does not unintentionally use SQLDialects (slice in Figure 2); the usage of non-constrained ensembles, e.g., Utilities, is also granted automatically.

The way the mapping between ensembles and source code is modeled has an effect on the stability of the model in face of evolution of the system. Here we hit a variant of the well-known "fragile pointcut problem". One way to mitigate this problem is by using stable abstractions in the source code in the queries. However, this is not always feasible; in the case study queries had to be adapted as the system evolved. Here the tool support provided by Vespucci offered some help to identify changes in the source code by: a) showing elements that do not belong to an ensemble, b) showing (sub-)queries with empty results; c) specifying that a list of ensembles should be non-overlapping (i.e., to prevent accidental matches). Even so we are aware that better source code query technology and tool support for it is needed; in this paper, we focus on the modularity mechanisms on top of the query language.

# 4. EVALUATION

In this section, we evaluate quantitatively the effectiveness of Vespucci's mechanisms to modularize the specification of a software's intended structure. This evaluation is performed from two complementary perspectives: (a) reduction of complexity, which is measured as the number of ensembles and constraints, and (b) facilitating architecture maintainability during system evolution. As a basis we use the re-engineered architecture of Hibernate (c.f. Sec. 2), which allows us to study an architecture of a size that is representative for mid- to large-scale projects. We also give a critical discussion of the broader applicability of our results and of threats to the validity of our study at the end of the section.

The goal of our evaluation is to assess the modularization mechanisms of Vespucci and not the accuracy of architectural violation control. Therefore, even though Vespucci is targeted at continuous architecture conformance checking, the identification of violations to the architecture is not the purpose of our quantitative evaluation. Nevertheless, it is important to highlight that in terms of enforcing conformance Vespucci is able to control violations in the source code similar to related approaches [25, 21, 28, 10, 32, 3, 8].

## 4.1 Scalability

We first analyze the reduction in complexity when reasoning about an architecture specification. This analysis was performed by comparing the architecture of Hibernate 1.0 modeled in the four principal ways schematically depicted in Figure 7 and outlined in the previous section. The model with both slices and hierarchies (Figure 7, D) was the primary model produced during our study of Hibernate. The other three models were produced to measure the complexity reduction for the different mechanisms (hierarchies, slices, combination of both).

*Scalability with regard to the number of ensembles.*

We first compare different mechanisms w.r.t. the number of ensembles referenced by isolated dependency rules. The baseline is a single monolithic specification with a total of 79 ensembles, modeled by following Figure 7 (A). The other three models Figure 7 (B-D) are quantified in the diagrams in Figure 8. The y-axis of all three diagrams denominates the number of ensembles referenced per architectural model.

The diagram on the left shows reduction in complexity for hierarchical structuring only. The model is a single specification, but high-level ensembles may be collapsed to reduce the overall number of ensembles to consider at once. The x-axis

denominates the number of collapsed ensembles ordered by the number of their sub-ensembles. The values on the y-axis show how many ensembles are referenced after collapsing an enclosing ensemble, i.e., the enclosing ensemble is referenced instead of all its children. The values are accumulated, since multiple ensembles can be collapsed together. For example, in a model with the top five most complex high-level ensembles collapsed, the architect has to consider 41 ensembles at once. When collapsing all ensembles in the hierarchy, we are left with 22 top level ensembles, hence hierarchical structuring reduces the number of ensembles to approx. 27% of the total (22 of 79).

The diagram in the middle of Figure 8 shows the number of ensembles per slice when using only slices (no hierarchies). The x-axis denominates the modeled slices in the decreasing complexity order (decreasing number of referenced ensembles). Almost all slices refer to less than 27% of the ensembles (12% on average). The exemption are the three first slices that capture rules for the following building blocks (of central importance) (i) persisting classes, (ii) persisting collections, and (iii) the interface to Hibernate's internal data types. The combination of both mechanisms (diagram on the right-hand side of Figure 8), yields a much smaller number of slices (x-axis), since it focuses on the top-level building blocks. In addition, the combined approach features slightly smaller slices; on average each slice references only 9% of the total number of ensembles.

*Scalability with regard to the number of constraints.*

In the following, we compare how much each mechanism reduces the number of constraints used in isolated dependency rules. The comparison is similar to the comparison regarding the number of ensembles and the numbers are shown in Figure 9. The x-axis is organized in the same manner as in Figure 8. The y-axis denominates the number of constraints that are referenced in each architectural model.

The y-axis for hierarchical structuring (left-hand side diagram in Figure 9) shows the total number of constraints after collapsing an enclosing ensemble. The number includes (i) constraints that are abstracted away, since they are internal to the enclosing ensemble (cf. Figure 7 B; 1&2) and (ii) constraints that are abstracted away, since several constraints at the low level are subsumed by a single constraint at the high level (cf. Figure 7 B; 1&2 to 4). Both internal and external constraints contribute approx. half of the reduction in constraints (external slightly outweighs internal). As in the evaluation for ensembles, the y-values for the hierarchical composition (B) are accumulated, since we can use several hierarchical groupings together. For the architectural models using slices (C,D) the number of constraints is simply the number of constraints modeled in one slice.

The baseline (A) consists of 705 constraints in a single specification. If we consider the hierarchical model and collapse all enclosing ensembles, approx. 2/3 of the constraints are removed (down to 214, last value in the left-hand side diagram in Figure 9)). In comparison, slices (diagram in the middle of Figure 9) show less than 5% of the total number of constraints and 1,3% on average (9 of 705) per slice. The combination of slices and hierarchical structuring (right-hand side diagram in Figure 9) features slightly smaller slices; on average 0.9% (6.5 constraints) of the total of 705 constraints modeled.

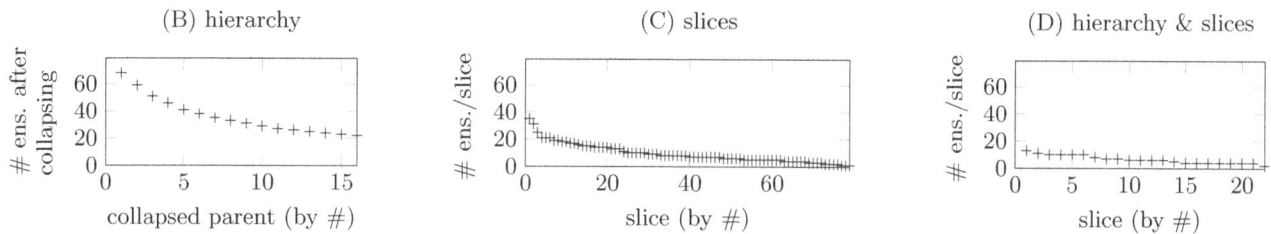

Figure 8: Comparison of ensemble reduction w.r.t. hierarchies and architectural slices (Hibernate 1.0)

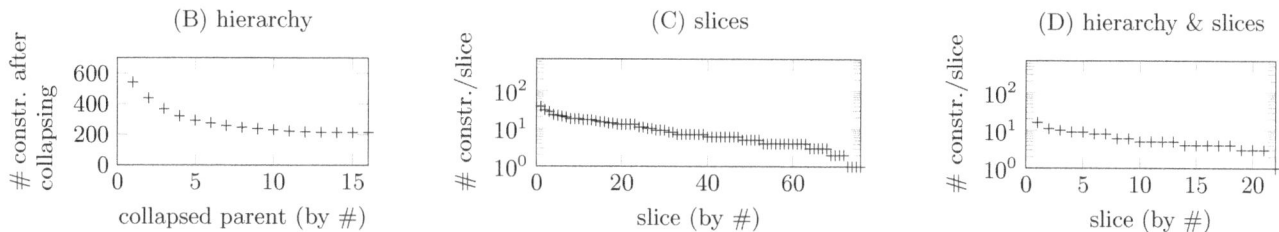

Figure 9: Comparison of constraint reduction w.r.t. hierarchies and architectural slices (Hibernate 1.0)

*Scalability with regard to the number of slices.*

To control the architecture of Hibernate we have modeled top-level slices comparable to Figure 7 (D) and slices for the internal constraints of the 16 ensembles that are further structured; totaling to 35 slices. Thus, the overall number of slices is smaller than the overall number of ensembles (79) and remains manageable. Note that in these models we do not use the total of the 705 constraints. First, three ensembles at the top level (SessionManagement, PersistenceManagement, and UserAPI) have no slice (and no constraints), for the reason of being used by and using almost all other ensembles. This is an inherent problem of the modularization of the software system and should be treated by refactoring the code base. Second, the modeled top-level constraints subsume several constraints on the internal ensembles. We found the control provided by the top-level constraints mostly sufficient during the evolution of Hibernate. Hence, we modeled detailed constraints only in few cases to further our understanding of the dependencies between selected ensembles.

*Summary.*

In this study the hierarchical structuring included 22 ensembles and 215 constraints (both approx. 1/3 of the total). Slices are much smaller; we have to collapse the first seven enclosing ensembles of the hierarchy to reduce the number of ensembles to 35, the number referenced in the most complex slice (persisting classes). Collapsing all ensembles still references 5 times more constraints than the number referenced in the slice for persisting classes. Hence, the modeling approach based on slices scales much better by reducing each slice to 9.5 ensembles and 9 constraints on average. The combination of both mechanisms produces the best results by reducing each slice to 7.1 ensembles and 6.5 constraints on average, which means that a typical slice in the Hibernate model had about 7 ensembles and 6 to 7 constraints. Thus especially the number of constraints that need to be reasoned over at once remains manageable and includes on average only 3% of the constraints of the model using hier-

archical structuring, with a maximum of 16 constraints, or 7% of the constraints in the single hierarchical model.

## 4.2 Evolvability

To evaluate the effectiveness of Vespucci in supporting architecture evolution we have compared a single model with hierarchies (Figure 7 B) to slices with hierarchies (Figure 7 D). The results are summarized in Table 2. The first three Columns show the analyzed version, its release year, and the number of LoC as an estimate for the size. Columns four and five characterize the architecture evolution in terms of ensembles and their queries. Overall, the number of ensembles has doubled. Column six shows the total number of slices in each version. We followed the methodology of one slice per ensemble – hence, the number of slices roughly follows the number of ensembles, with the exception of those ensembles that were not constrained (c.f. Sec. 4.1). Column seven shows that on average 33% of all slices (1/3 of the architecture specification) remained stable w.r.t. the previous version. The least stable revisions were the first and the last one. In the first revision, Hibernate was close to its inception phase, hence requiring more adaptations to its features. The last revision was the most extensive in terms of the timespan covered. The last three columns compare the complexity involved in performing the required updates of the architecture specifications. Columns eight and nine show the average, resp. maximal number of ensembles per slice, whose dependencies were updated, in the approach using slicing. The last column shows how many dependencies were updated in the single hierarchical model. On average only 4% to 6% of the number of dependencies updated in the single model were reviewed per slice (the maximum ranging between 7% and 15%). This reduction in complexity of the updates per slice is comparable with the reduction of the number of constraints between (B) and (D) in Figure 9.

The numbers indicate that the maintenance of individual slices is much easier than the evolution of the single architecture model and confirm what is qualitatively discussed in the previous section.

Table 2: Analysis of the Evolution of Hibernate's Architecture

| Vers. | Release Year | LoC | # Ens. (Top-Level) | Add./Rem. Ensembles | # Slices (Top-Level) | Stable Slices | Dependencies reviewed with | | |
|-------|--------------|-----|--------------------|----------------------|----------------------|---------------|------------------|------------------|--------------|
| | | | | | | | Slices (Avg.) | Slices (Max.) | Single Model |
| 1.0 | 2002 | 14703 | 22 | n/a | 19 | n/a | n/a | n/a | n/a |
| 1.2.3 | 2003 | 27020 | 26 | +5 / -1 | 23 | 4 (21%) | 2.4 | 6 | 61 |
| 2.0 | 2003 | 22876 | 28 | +5 / -3 | 25 | 12 (52%) | 1.9 | 6 | 40 |
| 2.1.6 | 2004 | 44404 | 30 | +2 / -0 | 27 | 9 (36%) | 2.6 | 6 | 38 |
| 3.0 | 2005 | 79248 | 36 | +9 / -3 | 33 | 8 (30%) | 4.5 | 8 | 118 |
| 3.6.6 | 2011 | 106133 | 39 | +3 / -0 | 36 | 9 (27%) | 5.0 | 11 | 87 |

## 4.3 Threats to Validity

We identify two threats to the *construct validity* of our study. First, the reverse engineering of Hibernate's architecture was primarily performed by this paper's authors, i.e., not by the original Hibernate developers. Hence, the resulting architecture design may not accurately reflect Hibernate's real/intended architecture, which may lead to inconsistencies in the results. To mitigate this threat, the architectural model was created by three people — one student, one PhD candidate and one post-doctoral researcher — that together have many years of experience on object-relational mapping frameworks. Further, we extensively studied the available documentation to make sure that the model is true to Hibernate's architecture. Yet, it is likely that a different group would reverse engineer a different architectural model. But, it is unlikely that the architecture would be such different that our evaluation would become invalid. A second threat to construct validity is that other architects may modularize the architecture specification differently, resulting in a different number and scope of slices. However, the approach that we followed — roughly creating one slice per top-level ensemble — has proven to be useful and can at least be considered as one reasonable approach.

Threats to *conclusion validity* in our study could be related to the number of ensembles and architectural constraints involved in our analysis. We tried to mitigate this threat by considering an architectural model of a significant complexity. Our analysis concerned an architectural model that involved 79 ensembles, more than 700 architectural constraints and 35 architectural slices for Hibernate 1.0.

The main issue that threatens the *external validity* of our study is that it involved a single software system. To mitigate this threat we have used a well-known medium-size framework, which has been designed by taking into consideration guidelines and good practices. These characteristics allow us to analyze the benefits of Vespucci when modeling architecture designs of well-modularized software systems. In addition, we have discussed the properties of Hibernate's architecture that influence the results and compared them to other studies. However, we are aware that more studies involving other systems should be performed in the future. All our findings should be further tested in repetitions or more controlled replications of our study.

## 5. RELATED WORK

Closely related to Vespucci are approaches that support checking the conformance between code and architectural constraints on static dependencies [25, 21, 28, 10, 32, 3, 8]. The key difference is that none of the above approaches (nor other related work) offers the ability to modularize the architecture description into arbitrary many slices. They rather require a self-contained monolithic specification of the architecture, which does not support the kind of black-box reasoning enabled by slices (cf. Sec. 3.5). In the following, we discuss the above approaches separately; a summary of their support for the features elaborated in Sec. 3 is presented in Table 3.

Reflexion Models (RM) [25] pioneered the idea of encoding the architecture via a declarative mapping to the source code. RM is an analytical approach that uses the modeled system architecture to generate deviations between source code and planned architecture, which is reviewed by the architect. The RM approach is not a constraint system, but rather requires the specification of the complete set of valid dependencies. Omission of dependencies is interpreted as "no dependency is allowed". Other approaches extend RM by (i) incorporating hierarchical organization [21], (ii) visual integration into the Eclipse IDE [20] and (iii) extending the process to continuously enforce compliance of structural dependencies between a planned architecture and the source code [27].

Sangal et al. [28] discuss the scalability issue of architecture descriptions and propose a hierarchical visualization method called design structure matrices(DSMs), which originates from the analysis of manufacturing processes. The key advantage is the notation (matrices) that facilitates identification of architectural layers via a predominance of dependencies in the lower triangular half of the matrix. DSM features a very verbose constraint system. For example, exemptions on lower level ensembles are encoded by the order in which rules are declared, e.g., by first allowing Persistence-Management to use SQLDialects and then disallowing the use of ConcreteDialects. While effective, this approach requires a carefully crafted sequences of constraints.

In previous work [10] we proposed an approach to continuous structural dependency checking; integrated into an incremental build process. As in Vespucci, we referred to conceptual building blocks as ensembles. However, the specification of architectural constraints has been completely revised for Vespucci. Previously we have defined LogEn; a first order logic DSL, that integrated query language and constraint specification. However, the meaning of a violation, i.e., a constraint, is defined by the end-user, which is complex in first order logic. Hence, we provided a visual notation (VisEn), which is less complex, but focuses on documenting the architecture and hence is not a constraint system, but requires explicit modeling of all dependencies. The focus of this work was on the efficient incrementalization

Table 3: Comparison with the State of the Art

| | Reflexion Models (RM) [25] | Hierarchical RM [21] | DSM [28] | LogEn/VisEn [10] | DCL [32] | Vespucci |
|---|---|---|---|---|---|---|
| Architectural slices | - | - | - | - | - | ✓ |
| Constraint system[1] | - | - | + | +++/-[2] | ++ | +++ |
| Hierarchies | - | ✓ | ✓ | ✓ | - | ✓ |
| Dependency Kinds | - | - | - | - | ✓ | ✓ |

[1] - (non existent) to +++ (very expressive)
[2] LogEn is very expressive; VisEn does not offer a constraint system

of the checking process, hence slicing architecture specifications into manageable modular units was not supported.

Terra et al. [32] propose a dependency constraint language (DCL) that facilitates constructive checking of constraints on dependencies; discrimination of dependencies by kind is also supported. DCL offers a textual DSL for specifying constraints. DCL's constraint system is closest to Vespucci's, and can express the not-allowed, expected and incoming constraints. Yet, it lacks outgoing constraints and a scoping mechanism such as global/local constraints, which goes hand in hand with the lack of support for slicing specifications into modular units. The language supports no inherent hierarchical structure in the architecture.

A number of commercial tools have been documented (c.f. [8]) for checking dependency among modules and classes using implementation artefacts, e.g., Hello2Morrow Sotograph [1] . However, the scope of these tools is limited; they are only able to expose violations of "certain" architectural constraints such as inter-module communication rules in a layered architecture. That is, they do not provide means for expressing system constraints.

In [3] the authors propose a technique for documenting a system's architecture in source code (based on annotations) and checking conformance of code with the intended architecture. The representation of the actual architecture in the source code is hierarchical, however, they do not support slicing of specifications in modular units and the modular architectural reasoning related to it.

Languages specialized on software constraints like SCL [17], LePUS3 [14], Intensional Views [23], PDL [24] and Semmle .QL [7] can be used to check detailed design rules e.g., related to design patterns [12]. However, they are not expressive enough for formulating architectural constraints in a way that allows to abstract over irrelevant constraints, when reasoning about a part of the architecture in isolation.

In [30] authors introduce a technique to identify modules in a program called concept analysis. A concept refers to a set of objects that deal with the same information. The authors observed that, in certain cases, there is an overlap among concept partitions. The notion of slice in Vespucci could be considered as conceptually close to the notion of concept overlapping since Vespucci supports the grouping of ensembles that are ruled by the same design decisions. Other than that slices and concepts are different in the way they are defined and used. Concepts emerge while slices are explicitly modeled. Moreover, use case slices [18] are also related to our notion of slices, but focus on the modularization of the scattered and tangled implementation of use cases.

In [13] the authors discuss foundations and tool support for software architecture evolution by means of evolution styles. Basically, an evolution style is a common pattern how software architectures evolve. This case study complements our work by helping to identify evolution styles w.r.t. a software's structural architecture. The evolvability of a software that is developed in a commercial context is also discussed by Breivold et al. [6]. They propose a model that — based on a software's architecture — evaluates the evolvability of the software. Based on our experience, the model also applies to open-source software, such as Hibernate. Aoyama [4] presents several metrics to analyze software architecture evolution. He made the general observation that discontinuous evolution emerges between certain periods of successive continuous evolution. Our case-study confirms this observation. We observed that some parts of Hibernate evolved continuously, while in other parts the evolution was disruptive. Using our modular architecture conformance checking approach architects can focus on continuous and disruptive slices individually.

## 6. SUMMARY AND FUTURE WORK

In this paper, we proposed and evaluated Vespucci, an approach to modular architectural modeling and conformance checking. The key distinguishing feature of Vespucci is that it enables to break down specification and checking into an arbitrary number of models, called architectural slices, each focussing on rules that govern the structural dependencies of subsets of architectural building blocks, while treating the rest of the architecture as a black box. Vespucci features an expressive constraint system to express architectural rules and also supports hierarchical structuring of architectural building blocks.

To evaluate our approach, we conducted an extensive study of the Hibernate framework, which we used as a foundation for a qualitative evaluation, highlighting the impact of Vespucci's mechanisms on managing architectural scalability and evolvability. We also quantified the degree to which Vespucci can (a) reduce the number of ensembles and constraints that need to be considered at once, and (b) facilitates architecture maintainability during system evolution. For this purpose, we played back the evolution of Hibernate's structure. The results confirm that Vespucci's is indeed effective in managing complexity and evolution of large-scale architecture specifications. However, given that we have only done one extensive case study so far, we need to carry out further case studies before final conclusions on the scalability of the approach can be made.

In future work, we will explore how IDE support can help to "virtually merge" slices into a virtual global architectural model and to automatically create "on-demand" slices to help architects to plan a software's evolution. Obviously, further empirical studies are needed to better understand the benefits and limitations of Vespucci. New studies need to be designed to asses the impact of the approach on architect's productivity and on software quality. In this respect it would be interesting to study the effect of modularization w.r.t. enlarged control, i.e., the modularization allows to efficiently maintain an architecture containing more ensembles, which provides tighter control over the source code.

# 7. REFERENCES

[1] Hello2Morrow Sotograph. http://www.hello2morrow.com/products/sotograph (accessed Oct. 2012).

[2] Vespucci. http://www.opal-project.de/vespucci_project.

[3] M. Abi-Antoun and J. Aldrich. Static extraction and conformance analysis of hierarchical runtime architectural structure using annotations. OOPSLA, 2009.

[4] M. Aoyama. Metrics and analysis of software architecture evolution with discontinuity. IWPSE, 2002.

[5] C. Bauer and G. King. *Hibernate in Action*. Manning Publications Co., 2004.

[6] H. Breivold, I. Crnkovic, and P. Eriksson. Analyzing software evolvability. COMPSAC, 2008.

[7] O. de Moor, D. Sereni, M. Verbaere, E. Hajiyev, P. Avgustinov, T. Ekman, N. Ongkingco, and J. Tibble. .QL: Object-Oriented Queries Made Easy. In *Generative and Transformational Techniques in Software Engineering II*. Springer-Verlag, 2008.

[8] L. de Silva and D. Balasubramaniam. Controlling software architecture erosion: A survey. *Journal of Systems and Software*, 85(1), 2012.

[9] M. Eichberg, M. Kahl, D. Saha, M. Mezini, and K. Ostermann. Automatic incrementalization of prolog based static analyses. PADL. 2007.

[10] M. Eichberg, S. Kloppenburg, K. Klose, and M. Mezini. Defining and continuous checking of structural program dependencies. ICSE, 2008.

[11] S. G. Eick, T. L. Graves, A. F. Karr, J. S. Marron, and A. Mockus. Does code decay? assessing the evidence from change management data. *IEEE Trans. Softw. Eng.*, 27(1), 2001.

[12] E. Gamma, R. Helm, R. E. Johnson, and J. Vlissides. *Design Patterns: Elements of Reusable Object-Oriented Software*. Addison-Wesley, 1995.

[13] D. Garlan, J. Barnes, B. Schmerl, and O. Celiku. Evolution styles: Foundations and tool support for software architecture evolution. WICSA/ECSA, 2009.

[14] E. Gasparis, J. Nicholson, and A. H. Eden. Lepus3: An object-oriented design description language. Diagrams, 2008.

[15] M. W. Godfrey and E. H. S. Lee. Secrets from the monster: Extracting mozilla's software architecture. *COSET*, 2000.

[16] S. Herold. Checking architectural compliance in component-based systems. SAC, 2010.

[17] D. Hou and H. J. Hoover. Using scl to specify and check design intent in source code. *IEEE Trans. Softw. Eng.*, 32(6), 2006.

[18] I. Jacobson and P.-W. Ng. *Aspect-Oriented Software Development with Use Cases*. Addison-Wesley, 2004.

[19] J. Knodel, D. Muthig, U. Haury, and G. Meier. Architecture compliance checking - experiences from successful technology transfer to industry. CSMR, 2008.

[20] J. Knodel, D. Muthig, M. Naab, and M. Lindvall. Static evaluation of software architectures. CSMR, 2006.

[21] R. Koschke and D. Simon. Hierarchical reflexion models. WCRE, 2003.

[22] A. MacCormack, J. Rusnak, and C. Y. Baldwin. Exploring the structure of complex software designs: An empirical study of open source and proprietary code. *Manage. Sci.*, 52, 2006.

[23] K. Mens, A. Kellens, F. Pluquet, and R. Wuyts. Co-evolving code and design with intensional views. *Comput. Lang. Syst. Struct.*, 32(2-3), 2006.

[24] C. Morgan, K. De Volder, and E. Wohlstadter. A static aspect language for checking design rules. AOSD, 2007.

[25] G. C. Murphy, D. Notkin, and K. Sullivan. Software reflexion models: bridging the gap between source and high-level models. *SIGSOFT Softw. Eng. Notes*, 20, 1995.

[26] D. E. Perry and A. L. Wolf. Foundations for the study of software architecture. *SIGSOFT Softw. Eng. Notes*, 17(4), Oct. 1992.

[27] J. Rosik, A. Le Gear, J. Buckley, and M. Ali Babar. An industrial case study of architecture conformance. ESEM, 2008.

[28] N. Sangal, E. Jordan, V. Sinha, and D. Jackson. Using dependency models to manage complex software architecture. OOPSLA, 2005.

[29] M. Shaw and D. Garlan. *Software Architecture: Perspectives on an Emerging Discipline*. Prentice Hall, Upper Saddle River, NJ, USA, 1996.

[30] M. Siff and T. Reps. Identifying modules via concept analysis. 25(6):749–768, 1999.

[31] H. A. Simon. The architecture of complexity. In *Proceedings of the APS*, 1962.

[32] R. Terra and M. T. Valente. A dependency constraint language to manage object-oriented software architectures. *Softw.: Practice and Experience*, 39(12), 2009.

[33] S. Wong, Y. Cai, M. Kim, and M. Dalton. Detecting software modularity violations. ICSE, 2011.

# Using Roles to Model Crosscutting Concerns

Fernando Sérgio Barbosa
Escola Superior de Tecnologia de Castelo Branco
Avenida do Empresário
6000-035 Castelo Branco, Portugal
+351 272339300

fsergio@ipcb.pt

Ademar Aguiar
INESC TEC and Faculdade de Engenharia,
Universidade do Porto
Rua Roberto Frias, 4200-465 Porto, Portugal
+351 225081400

ademar.aguiar@fe.up.pt

## ABSTRACT

In object oriented languages the problem of crosscutting concerns, due to limitations in the composition mechanisms, is recurrent. In order to reduce this problem we propose to use roles as a way of composing classes that extends the Object Oriented approach and can be used to model crosscutting concerns. To support our approach we developed a role language that extends Java, while being compatible with existing virtual machines. As validation we conducted a case study using three open source systems. We identified crosscutting concerns in the systems and then modeled them using our role approach. Results show that roles are a viable option for modeling crosscutting concerns.

## Categories and Subject Descriptors

D.1.5 [**Programming Techniques**]: Object-oriented Programming; D.2.2 [**Software Engineering**]: Design Tools and Techniques; D.2.13 [**Software Engineering**]: Reusable Software

## General Terms

Design, Experimentation, Languages.

## Keywords

Roles, Crosscutting Concerns, Code reuse, Modularity, Composition.

## 1. INTRODUCTION

A single decomposition strategy cannot capture all possible views of a system [51], so there are always concerns that cannot be adequately decomposed using only a single decomposition strategy, and end up scattered among the various modules. These are called the crosscutting concerns. Crosscutting concerns have a negative impact on modularization: it makes systems hard to maintain and evolve. When a change has to be made it affects several places. To prevent such consequences we need to use other decomposition techniques. Several proposals are available, like multiple inheritance [49][33], mixins [7][8], traits [43][15], features [1], aspects [27] and roles . We believe that if we explore the way roles can be used to compose classes we will find that roles can be used to model crosscutting concerns.

There are several definitions, examples and targets of roles [30] [40][41][42][46]. Most deal with dynamic aspects, i.e., the attachment/detachment of roles to objects at runtime. However we do not address this dynamic use of roles. We are interested in using roles for composing classes and thus eliminate crosscutting concerns that are present due to the lack of other compositional techniques. Namely we are interested in extending the Object Oriented paradigm using a new kind of construct.

Static roles have been used for modeling [39][41][42] and despite their benefits no programming language has appeared that deals with this static nature [11]. This is a gap we try to bridge with JavaStage, a Java extension that support roles as programming constructs. We will briefly describe JavaStage, so that examples are understandable.

With our approach, when a class plays a role the role methods are added to the class interface, so a class can be seen either as being composed from several roles or as an undivided entity.

We validate our work showing how roles can be used to model crosscutting concerns in three open source systems, namely JHotDraw, Open JDK compiler and Spring Framework. We used an aspect mining technique - clone detection - to identify crosscutting concerns and then proceed to model them using roles. The results obtained indicate that roles can be used to model a significant number of crosscutting concerns.

We can summarize our paper contributions as: an alternative way of modeling crosscutting concerns that is a natural extension of the Object Oriented paradigm; a role supporting language that extends Java but is compatible with existing Java Virtual Machines; JavaStage has features like a powerful renaming mechanism that allows methods names to reflect the context in which they are used, allow block renaming with a single configuration and allow multiple method versions; a case study showing how roles can be used to the proposed effect.

This paper is organized as follows. Section 2 presents roles and role modeling. Section 3 presents our proposal of composing classes from roles. In section 4 we present design and implementation decisions made when developing our approach. We study how roles are capable of handling crosscutting concerns in section 5 where we applied them to three systems. Related work is presented in section 6 and section 7 concludes the paper.

## 2. Roles

Perhaps the best known use of roles is to extend objects by attaching/detaching new behavior at runtime as described by Steinmann in [47]. This is helpful when we want to use objects in situations for which they were not developed. When in a new situation we write a new role for the object and attach it. This way the development of the class is focused on the basic concerns it models. The use of dynamic roles can include the definition of a context [2][23]. When the object enters the context the roles are automatically attached to the object. Normally this use of roles implies that roles are often developed for a concrete class and the reuse of role code is not a priority, even if they can be reused.

The use of roles that we propose does not fall into this category of dynamic roles at all. We want to use roles as static entities, that is, a role that is permanently attached to the objects of the class throughout their life cycle and cannot be detached. To stress this point we refer to the class that plays the role as the player of the role and not as the intrinsic as used in dynamic roles.

In JavaStage roles are meant to provide code for concerns that are not the player's primary concern. A role is not a superclass or a subclass of its players. Its methods are just added to the player interface via forwarding methods. Hence roles do not redefine class behavior nor classes redefine role behavior. Even so it is possible for roles to redefine the class inherited methods. A class cannot redefine a role method. If a class defines a method with the same signature of the role it just means that the role method is not added to the class interface and the class method is used.

We use the definition of role used by Riehle in [42] where "a role is an observable behavioral aspect of an object". Riehle also defines a role type as something "that defines the behavior of a role an object may play. It defines the operations and the state of the role, as well as the associated semantics". We use roles in the sense they were used in Riehle's work as well as in the OORam method [39]. In both works roles were used as a static modeling construct, neither as a dynamic entity nor a programming artifact.

## 2.1 Modeling with Roles

Role modeling using static roles was used as an integral part of the OORam method [39] and by Riehle in [42]. We took these modeling approaches into the programming level using roles as building blocks for classes. To support roles we developed the JavaStage language, which we will discuss in Section 3.

To exemplify role modeling we present in the left side of Figure 1 the class diagram of a simplified Figure Handling Framework, based on the JHotDraw Framework. The main concepts in the framework are the figures, so we created a class to model each figure. All figures share a common concept and may share structure and behavior, so an inheritance hierarchy was used. Some figures may contain other figures, like the group figure, so the Composite pattern [20] was used. Because the view has to draw the figures it must know when they have changed. We used the Observer pattern for this purpose.

Clients interested in knowing if a Figure is being handled may not want to interact with it in any other way. For these clients, a Figure assumes the role of a Subject in a Observer pattern and only those operations related to that role are of interest. The Parent role is responsible for managing a collection of children. Clients may set or read properties of the figure, for them the figure plays the role of a PropertyProvider. The mentioned roles are depicted on the upper right side of Figure 1, where we also show the role associations and the revised class diagram using roles.

## 2.2 Advantages of Role Modeling

Role modeling has several advantages in system comprehension, reuse, development and documentation [42]. Describing a class as a set of roles helps to separate the various ways in which a class is used. Documentation can be done in these terms as well. This helps clients to focus on whichever aspect they are interested in, providing a better understand and use of the class. Designing the class can also be done in role terms, thus developers are able to focus only on one aspect of the class. This enables independent development of a class with all its benefits in terms of reduced development time and complexity.

Class relationships are reduced to role relationships. Since roles focus on a particular view of a class we need not to understand the player in its whole. This eases the understanding and development of these relationships. Whenever needed the broader perspective

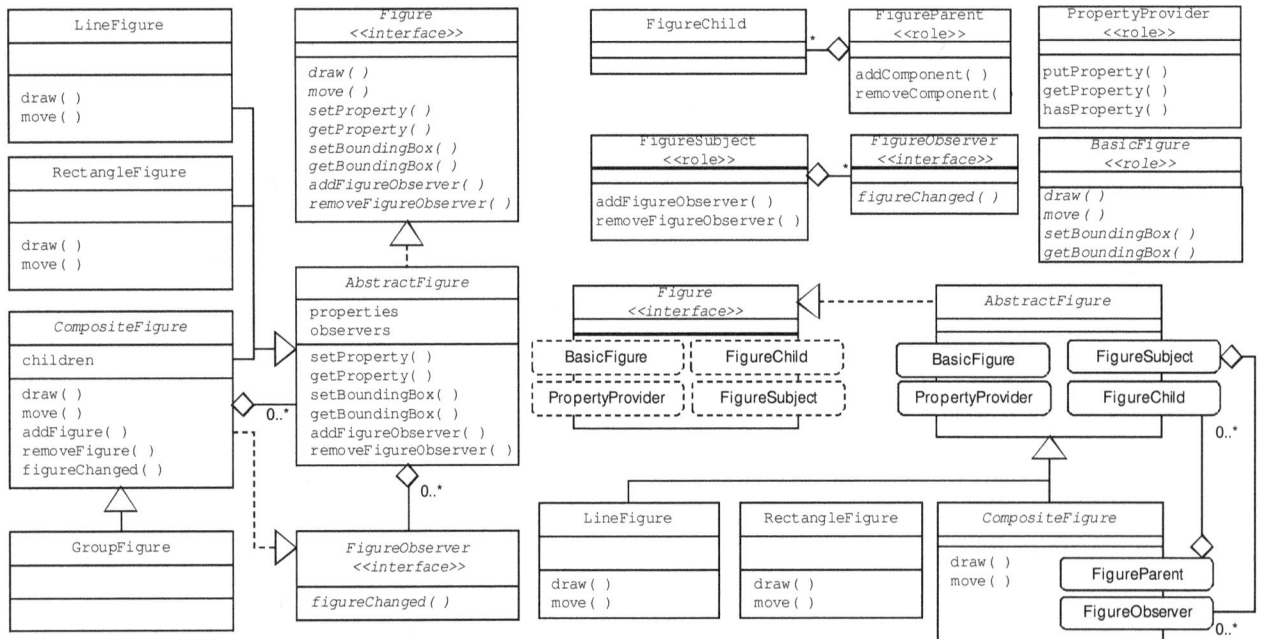

**Figure 1. Example of modeling with roles. On the left: class diagram of the Figure Framework. On the right: Roles and their relations in the Figure Framework, and the revised class diagram of the Figure Framework, now with roles. Rounded rectangles identify roles played by the class. Dashed round rectangles represent the interface provided by the role**

can also be used. Role modeling allows shifting between role level and class level without any information loss.

Role modeling also allows for better understanding using previous experiences. When a developer knows how to use roles that have a relationship in a system, then when he encounters different roles with similar relationships the past experience will allow a quicker understanding. One such example is the use of the Observer pattern. When experienced with a FigureSubject and how it works with a FigureObserver to use another instance of the pattern is much simpler and straightforward.

## 2.3 Modeling Crosscutting Concerns

Crosscutting concerns appear when several modules must deal with the same problem because one cannot find a single module responsible for it in the light of a decomposition strategy. This leads to scattered code. Its consequences are the opposite of the benefits of modularization. Since a module deals with a part of a problem that is spread over other modules, changes to that code may affect those modules. This affects independent development. Maintenance is impaired too because changes in the code needs to be done in all modules transversely.

Because a role is a smaller composition unit than a class we can put the crosscutting concern in a role, or a set of roles, and the classes that have the crosscutting concern play those roles. Any changes to the crosscutting concern are limited to the roles thus greatly improving maintenance and reducing change propagation.

Crosscutting concerns are present even in the simplified Figure framework. It is not the figure's main concern to act like a Subject but it has that role superimposed on it. With roles we are able to extract those concerns from the class and reduce code scattering. Furthermore, those roles are reusable whenever we need a class to address those concerns, even if it is not a Figure. We can also argue that being a PropertyProvider is not the figure's main concern. Assume that a property is of type Object and identified by a name of type String. For the Figure framework this would suffice but it would be more reusable if it used generics for the property type. We can also use generics to specify the value type - instead of type Object. After a closer look, the property provider is in fact a map that maps keys to values. With this configuration we can develop a role – Mapper - that could be used in several other situations. We could also reuse a map implementation if we inherited from a Map class, but that would be conceptually wrong. A Figure is not a map, it plays the role of a property map.

## 3. Composing Classes with Static roles

We propose to use roles as a basic construct from which we can compose classes. Roles provide the basic behavior for concerns that the classes must deal with but are not their main concern. This way we can better modularize the construction of classes.

To support roles we developed JavaStage, an extension to Java. We extended java but we believe that our approach also applies to other single-inheritance languages and can even be useful in multiple inheritance languages. Examples in this paper (we use the Figure framework as a running example) have been compiled in the JavaStage compiler we developed. We will not discuss JavaStage's syntax in full detail but will explain it so that examples are understandable.

When developing JavaStage we tried to introduce as few extensions to the language as possible and we only introduced 5

new keywords - role, plays, requires, performer and Performer - which are responsible for the definition of a role, declaring that a class plays a role, the requirements list and referencing the player. We also introduced a renaming mechanism that uses as the special character: #. The JavaStage syntax is presented in Figure 2.

```
type_decl ::= (role_decl | class_decl | … ) ";"
role_decl ::= {class_modifier} "role" identifier
           ["extends" class_name] "{" role_body "}"
role_body ::= ( [requires] class_body )
requires ::= "requires" type "implements"
           ( (type method_decl) | constructor_decl)
           [throws] ";"
class_body ::= ( [plays] … )
plays ::= {access_modifier} "plays" class_name
           [configs] identifier [role_params] ";"
configs ::= "(" config {"," config} ")"
config ::= identifier "=" identifier
role_params ::= "(" args_list ")"
```

**Figure 2. The extension of java syntax in JavaStage.**

## 3.1 Declaring roles

A role may define methods and fields, like a class, including access levels. For example, whenever a figure is changed the view must be updated. We can use an Observer [20] with Figure as a subject and View as an observer. Being a subject is not the Figure's main concern so we developed a subject role, see [20] and Figure 3, to capture it and figures just play it.

Our roles have state so the FigureSubject role doesn't need the player to supply a container for the observers. This way the subject concept is totally encapsulated within the role and the class only needs to call the fire method when it changes.

The role is independent of the player, so whenever a role calls a method it will be a role method not a player's method. To call player's methods the role must call them explicitly using the performer keyword, which is a reference to the object that plays the role. Inside the role the this reference always refers to the role

## 3.2 Classes play roles

A class can play any number of roles, and can even play the same role more than once. We refer to a class playing a role as a player of that role. When a class plays a role all the non private methods of the role are added to the class. To play a role the class uses a plays directive and gives the role an identity, as shown in Figure 3. To refer to the role the class uses its identity.

## 3.3 Role–player communication

A role may need to communicate with its player, so it may require the player to have specific methods. Those methods are stated in a requirement list. The role indicates the type that must supply the method and the method signature. To indicate that the supplier is

```
public role FigureSubject {
  private Vector<FigureObserver> observers =
               new Vector<FigureObserver>();
  public void addFigureObserver(FigureObserver o){
    observers.add( o );        }
  void removeFigureObserver(FigureObserver o){
    observers.remove( o );   }
  protected void fireFigureChanged( ){
   for( FigureObserver o : observers ) o.update();
   }
}
```

**Figure 3. A FigureSubject role for an observer pattern.**

```
public class AbstractFigure implements Figure {
 plays FigureSubject figureSbj;

 public void moveBy(int dx, int dy) {
   // ... code for moving the figure
   // firing change, using role identity
   figureSbj.fireFigureChanged();
 }
}
```

**Figure 4. A Figure class playing the subject role.**

the player we use the Performer keyword. Performer is used within a role as a place-holder for the player's type. This enables roles to declare fields and parameters of the type of the player. A role can impose restrictions on other types as well.

As an example suppose we want to expand the use of our subject role to other situations. Then we can use generics to specify the type of the observer. When developing the role the concrete type of the observer is unknown but using the requirement list we still can develop a full purpose role as shown in Figure 5.

In a class, when we call a method without using a role identity, it will always refer to a class method. The same is true for the role, if the role calls a method without the use of the performer keyword it always calls role methods. In fact for a role to call a method from the player the method must be present in the requirements list, so all player methods must be called explicitly.

## 3.4 Playing the same role more than once

A class can play a role more than once as long as there are differences between role instances. The GenericSubject role could be played more than once if the observer type is changed. If the Figure was a FigureObserver subject and a FigureHandleObserver subject it could play both roles using

```
plays GenericSubject<FigureObserver> figureSbj;
plays GenericSubject<FigureHandleObserver>
          figHandlerSbj;
```

## 3.5 Renaming methods

A method's name must clearly state its purpose but in generic roles such names are difficult to attain. Role methods may be tuned for a collaboration but in similar collaborations, where the same role can be reused, those names can be inadequate. We developed a renaming mechanism to enhance role reuse.

In the last example the figure played the GenericSubject twice, but the observers' management and fire methods had the same name. Methods like addFigureObserver and addFigureHandleObserver would be more adequate. The same applies to the other methods. We can also improve the ObserverType's update method if we use methods like figureChanged and figureHandleChanged.

Our renaming mechanism allows the easy configuration of those methods, with a simple configuration. Each name may have three parts: a configurable one and two fixed. Both fixed parts are optional so that the name of a method can be fully configurable by the class. The configurable part is bounded by # as shown next.

```
            fixed#configurable#fixed
```

The name configuration is done by the class playing the role in the plays clause as depicted in Figure 6. To play the role the class must define all configurable methods. In the example the AbstractFigure class would have the methods addFigureObserver and the fireFigurechanged that calls the figureChanged method of

```
public role GenericSubject<ObserverType> {
 requires ObserverType implements void update();

   private Vector<ObserverType> observers =
           new Vector<ObserverType>();
   public void addObserver( ObserverType o ){
     observers.add( o ); }
   public void removeObserver( ObserverType o ){
     observers.remove( o ); }
   protected void fireChanged( ){
     for( ObserverType o : observers )  o.update();
}
```

**Figure 5. A generic Subject role for the Observer pattern (requirements are in bold).**

```
public role GenericSubject<ObserverType> {
 requires ObserverType implements
                     void #Fire.update#( );
   public void add#Observer#( ObserverType o ){
     observers.add( o ); }
   protected void fire#Fire#( ){
     for( ObserverType o : observers)
       o.#Fire.update#();
   }
}
public class AbstractFigure implements Figure {
 plays GenericSubject<FigureObserver>
 (Observer = FigureObserver,
  Fire= FigureChanged, Fire.update= figureChanged
 ) figureSbj;

 plays GenericSubject<FigureHandleObserver>
 ( Observer = FigureHandleObserver,
   Fire = FigureHandleChanged,
   Fire.update = figureHandleChanged
 ) figHandleSbj;
}
```

**Figure 6. The generic subject role now with configurable methods (in bold). Also shown a class playing that role twice.**

the observer. The class also has the addFigureHandleObserver and fireFigureHandleChanged that calls the figureHandleChanged.

## 3.6 Providing multiple versions of a method

It's possible to declare several versions of a method using multiple definitions of the configurable name. This way, methods with the same structure are defined once.

For this feature to be used we must use a configurable called method inside a configurable role method. We must name the called method after the method it is called from. This is done using a dot name, where the configuration name before the dot is the configuration name of the outer method, as in the example:

```
    void role#Method#( ){
        performer.called#Method.inner#( );
    }
```

This way the compiler knows that both methods are to be used together and can check if one configuration name has the same number of configurations of the other and it also checks that they are defined sequentially in the plays clause:

For example we can expand FigureObserver to include several update methods, not just figureChanged. We can provide figureMoved, figureChanged, figureRemoved, etc, to specify which actual change occurred. We would write the plays clause as

```
plays GenericSubject<FigureObserver>
( Fire= FigureChanged, Fire.update= figureChanged,
  Fire= FigureMoved,   Fire.update= figureMoved,
```

```
Fire= FigureRemoved, Fire.update= figureRemoved,
Observer = FigureObserver ) figureSbj;
```

## 3.7 Roles playing or inheriting from roles

Roles can play roles but can also inherit from roles. When a role inherits from a role that has configurable methods it cannot define them. When a role plays another role it must define all its configurable methods, but can have its own configurable methods. Both situations are depicted in Figure 7.

Inside a role the super reference refers to the super role of the role, not the player's super class. This follows the view that roles are independent of their players.

For example managing observers is a part of a more general purpose concern that is to deal with collections. We can say that the subject role is an observer container and develop a generic container role and make the subject inherit from the container.

If the FigureSubject role can be played by several classes then we'll create a FigureSubject role based on GenericSubject. Because we need to rename the role methods the FigureSubject role must play the Generic role and define all its methods. It would use the configuration used in AbstractFigure (using Thing as name configurator instead of Observer). AbstractFigure would then use FigureSubject without any configuration.

## 3.8 Visibility control

Role members have all the visibility control available to classes. We extended the protected level to include the role-class relationship. A protected role member is accessible to its players and subroles. A protected class member is also accessible to roles. A class can reduce the visibility of the role members. If a class uses protected in the plays clause then all the public role methods are imported to the class as protected. This way a class can use roles to provide an interface for their subclasses only. A class cannot change a single member's visibility.

```
public role GenericContainer<ThingType> {
  private Vector<ThingType> ins =
                    new Vector<ThingType>();
  public void add#Thing#(ThingType t){ ins.add(t);}
  void insert#Thing#At( ThingType t, int idx ){
      ins.insertElementAt( t, idx );
  }
  protected Vector<ThingType> get#Thing#s(){
      return ins;
  }
}
public role GenericSubject<ObserverType> extends
                  GenericContainer<ObserverType>{
  requires ObserverType implements
                  void #Fire.update#();
  protected void fire#Fire#( ){
    for( ObserverType o : get#Thing#s() )
       o.#Fire.update#( );
  }
}
public role FigureSubject {
  plays GenericSubject<FigureObserver>
  (Fire= FigureChanged, Fire.update= figureChanged,
  Fire= FigureMoved,   Fire.update = figureMoved,
  Fire= FigureRemoved, Fire.update= figureRemoved,
  Thing = FigureObserver ) figureSbj;
}
public class AbstractFigure implements Figure {
  plays FigureSubject figureSbj;
}
```

**Figure 7. Role inheritance and role composition.**

## 3.9 Conflict resolution

Class defined methods always take precedence over role methods. Role methods override the class inherited methods. Conflicts may arise when a class plays roles that have methods with the same signature. When conflicts arise the compiler issues a warning. Developers can handle the conflict by redefining the method and calling the intended method. This is not mandatory because the compiler uses, by default, the method of the first role in the plays clause order. This may seem a fragile rule, but we believe that it will be enough for most situations. Even if a conflicting method is later added to a role the compiler does issue a warning so the class developer is aware of the situation and can solve it. The important part is that the class composer can solve the situation as he wishes and not as imposed by the role or superclass' developers.

## 3.10 Role constructors

We may need to parameterize roles. In our example we may want the container to be an ArrayList instead of a Vector. We support role constructors, so we can initialize role fields, but role constructors cannot be called directly. To initialize a role we use the plays clause like

```
plays FigureSubject figureSbj( new ArrayList() );
```

## 4. Role Implementation and Design Decisions

In this section we briefly describe how we implemented our version of roles and the design decisions we made along the way.

## 4.1 Implementation of JavaStage

To support roles we opted for a version using inner classes with forwarding methods. This supports all our options for the approach while maintaining the final code executable in existing virtual machines.

When a class plays a role the role code is copied to the class as an inner class. Figure 8 shows the translation of the code in Figure 7. We can see that a role is used as an object inside the class. This allowed roles to have constructors. However, no one can directly instantiate a role, as roles aren't meant to have instances.

The # in the inner class name guarantees that there isn't a name clash between synthetic classes and developer's code. The role identity in the class name guarantees that no conflicts arise when playing the same role twice. For example, the name of the class for the FigureObserver role is GenericSubject#figureSbj and for a FigureHandleObserver would be GenericSubject#figHandlerSbj.

Role methods are copied to the class interface and call the corresponding method on the role object (see addFigureObserver). This apparently introduces a redirection but it is easily solved by inlining the method. No performance loss is therefore introduced.

## 4.2 Role identity

To some authors roles have no identity [29] because they are not independent. Others argue that roles have an identity different from its intrinsic [35]. With role identity classes can have multiple instances of the same role. There are also those to whom roles share the same identity with its intrinsic and also have one that distinguishes it from other roles in the intrinsic [46].

```
public class AbstractFigure implements Figure {
    public void moveBy(int dx, int dy) {  figureSbj.fireFigureChanged(); }

    private class GenericContainer#figureSbj {
        private java.util.Vector<FigureObserver> ins = new java.util.Vector<FigureObserver>();

        public void addFigureObserver(FigureObserver t)                { ins.add( t ); }
        protected java.util.Vector<FigureObserver> getFigureObservers(){ return ins;   }
    }
    private class GenericSubject#figureSbj extends GenericContainer#figureSbj {
        protected void fireFigureChanged( ){
            for(FigureObserver o : getFigureObservers() )   o.figureChanged();
        }
        // … other fires
    }
    private GenericSubject#figureSbj figureSbj  =  new GenericSubject#figureSbj();
    public void addFigureObserver( FigureObserver t){ figureSbj.addFigureObserver( t ); }
    protected void fireFigureChanged( )                { figureSbj.fireFigureChanged();   }
}
```

**Figure 8. Excerpt of how the AbstractFigure class from Figure 7 would look after the role is added to the class.**

Our roles have an identity, associated with the player, given by the player in the plays clause. When the player accesses role members it uses this identity. If the role is public then its identity is accessed just like any class member. Class clients can then select the role they want. However, we consider that roles should not be public, for the same reasons fields should not be public.

Role identity is also used to distinguish between roles when resolving a conflicting method. We use the identity to specify which role we want to access. This is better than using class names, because we can change the role hierarchy and still be able to maintain the code unchanged.

Another reason to provide roles with an identity is to support state. Since each role field must be accessed using the role identity there never is a name conflict between fields of different roles.

### 4.3 The plays clause
Shouldn't the plays clause be considered equivalent to the extends or implements clauses and be placed accordingly? After all it does have impact in the class interface. There are in fact several reasons for not doing so. One is the role identity which, purposely, resembles an object declaration (see the implementation section). Yet another reason is the naming configuration, which would clutter that declaration.

### 4.4 Role as types
We could let roles define a type and write code that would work with any class that plays the role. Our renaming strategy, however, forbids this because the actual interface a role provides is configured by the class. Roles that do not use renaming could define a type. In our approach making a role a type is simple, however we still haven't explored if there are advantages in considering roles as types. We defer this decision to future work.

### 4.5 Aliases vs Method renaming
Some approaches, like Traits, use aliases to rename a method within a class to solve conflicts. Our renaming mechanism goal is enhancing role reusability. It allows: configuration of a role with meaningful method names, in the context of the player class; fast renaming of several methods; multiple versions of methods.

### 4.6 Requirements listing
Requirements in most role languages are made by a playedBy clause that states which classes can play it. In our case restrictions are imposed via the requirement list, using structural equivalence, rather than nominal equivalence. This allows us to provide a renaming mechanism that tailors the role to the class, which could not be done using playedBy. This also allowed us to impose restrictions on other types that are part of the interaction, like the observer type in the subject role.

### 4.7 Summary
Having shown in the previous sections our role approach we now make a summary of its features.

**Role support**. JavaStage is a role supporting language.

**State support**. Examples show how roles can have state and how it doesn't cause any conflict, mainly because roles have identity. Also the implementation using inner classes facilitates the introduction of this feature in a compatible and straightforward way with traditional java.

**No conflicting fields**. State cannot generate conflicts as each state variable is confined to a single class or role. The only conflicts that can occur are methods with the same signature either inherited or added from roles. To resolve them the class may redefine the conflicting method, usually calling the intended one. It can do that either using super to refer to the base class or by using the intended role identity.

**Accessing features from roles and superclasses**. We do not support multiple inheritance but we do support multiple role playing. We can distinguish between superclasses and role features using super and role identity as seen in the previous topic.

**Visibility control**. Roles have the same visibility control that classes have. We only extended the protected access to include role-class communication as discussed in section 3.8.

**Multiple role playing**. A class can play the same role multiple times, as long as there are some differences between instances like methods names or used types. We showed how we could play a subject role twice in the same class.

**Block renaming**. Our renaming mechanism allows renaming several methods in one go. It can also provide multiple versions of a method. We showed in the subject role how this can be used effectively to enhance role reuse.

**Multiple Method Versions**. Our renaming mechanism allows the defining of several structurally identical methods using a single method definition and the renaming mechanism. The subject role is an example on how this can be used to enhance role reuse.

# 5. Case Study

To explore how roles can model crosscutting concerns we conducted an experiment using three systems. In this section we present the target systems, the experiment setup and methodology.

## 5.1 Target Systems

For the experiment we used 3 open source systems: the JHotDraw framework, the OpenJDK compiler and the Spring Framework.

JHotDraw is a Java GUI framework for technical and structured Graphics. The JHotDraw framework defines the basic structure for a GUI-based editor with tools in a tool palette, different views, user-defined graphical figures, and support for saving, loading, and printing drawings.

The OpenJDK compiler (javac) is an open source java compiler, and other tools, that has support from Oracle. The JavaStage compiler was based on this compiler so it was a good candidate for the case study.

Spring is a layered Java application platform for building enterprise solutions. Spring framework provides a powerful and flexible collection of technologies to improve the development of enterprise Java applications and is claimed to be used by millions of developers.

## 5.2 Detecting Crosscutting Concerns

In order to identify crosscutting concerns we opted for an aspect mining technique that as been used for other studies [9]. The aspect mining technique used was the clone detection. Code duplication was considered a good indicator of crosscutting concerns in the source code of a system: because the crosscutting concerns could not be cleanly modularized, certain parts of the implementation show high levels of duplicated code.

To detect clones we used CCFinder [24] with its standard options. CCFinder is a token-based clone detection tool. We only considered clones that appeared in, at least, two files. This reduced the amount of clones removable with refactoring. Due to the great number of clones (416) reported in the Spring framework, and since we intended to inspect all clones, we removed, for this system, clones with less than 20 tokens.

We manually inspected all detected clones to identify their nature and the concern they dealt with. We grouped clones according to their concerns. Each role must focus on a concern so this grouping step helped us reasoning on which roles we should develop. In this phase we dismissed false clones. We also ignored clones that used deprecated code or code marked for substitution.

After the concern classification we inspected all clones to remove them with refactoring. However we could not remove all clones or felt the solution could be improved.

## 5.3 Case Study Results

After analysis of the concerns we counted the concerns that were addressed by roles and the ones that roles could not address. We also counted the lines of code used for each concern in the original system and in the system with roles. This way we can

determine if the role system is better than the original system or, at least, has fewer lines of code. Results are presented in table 1.

**Table 1. Clone Removal Results**

| System | JHotDraw | javac | Spring |
|---|---|---|---|
| Concerns | | | |
| Considered | 38 | 31 | 84 |
| Solved | 30 (79%) | 24 (77%) | 81 (96%) |
| Unsolved | 8 (21%) | 7 (23%) | 3 (4%) |
| Code Statistics | | | |
| Original Clone LOC | 1390 | 1571 | 4763 |
| Final Clone LOC | 883 | 986 | 3346 |
| Size reduction | 36,5% | 37,2% | 29,8% |
| Roles with less LOC | 27 | 20 | 65 |
| Roles with more LOC | 2 | 3 | 14 |
| Roles with same LOC | 1 | 1 | 2 |
| Concerns better with roles | 90,0% | 83,3% | 80,2% |

## 5.4 Discussion

The first observation we can make is that the number of unresolved concerns is low. The greater absolute value belongs to JHotDraw with 8 (21%). But they can be easily explained and derives from the fact that we considered as clones code that could be considered false clones. They are clones only in the structure and not on the code itself. A "creating undo activity" concern creates an undo activity object for each tool and command. Each tool class has an UndoActivity inner class hence all the constructors from these inner classes have similar structures and were marked as clones. Each tool class also has a method that creates the undo activity. Since this method is similar it was marked as a clone too. We opted to maintain this code as clones because they dealt with the same concern.

Another example from JHotDraw is persistence: figures are streamed so they all have a write and read methods with similar structures, but not quite identical code. One role, DisplayBoxed, considerably reduced this duplicated code, though. In another unresolved concern the cloned method overrides the superclass method for performance issues that we failed to understand.

Some concerns (3 from JHotDraw and 4 from javac) could be removed with roles but their configuration would be complex and since clones had, at most, 4 simple lines of code, we decided that the clone was a better solution.

The remaining javac unresolved concern was a declaration of static constants in different classes. The initialization was made in a static block so we could not put them into a role. Our roles also do not support public static variables.

Interestingly the system with more concerns was the one with the least (only 3) unresolved concerns. Those concerns could not be resolved because they needed features that JavaStage does not support. One concern had a clone with a static method in one class and a regular method in another class. Roles do not support changing the nature of a method. Another concern used static inner classes and the other was used in anonymous classes.

From this discussion we can see that roles could address most of the concerns and the ones it didn't address some were small (less than 4 lines), some could be counted as not real clones, and some had rarely used properties.

We can see that, for the majority of the resolved concerns, roles needed fewer lines of code, ranging from a 30% to 37% reduction in code size. This seems to indicate a smaller effort when developing a system with roles. One can argue that LOC count is not a good measure for the effort when using different languages, but since JavaStage is an extension to Java and the development of a role in JavaStage is very similar to that of a class in plain Java, we can apply it here with relative confidence. The concerns that had more LOC than the original code were concerns with few lines of code where the role requirements and configuration overhead did not overcome the replicated code.

We counted as LOC the requirements statements that roles must declare and the plays directive. Assume one concern that presents 8 lines of replicated code in each class which could be resolved with a simple role. We would expect this role to have the same 8 LOC. That is not so because we do not count the class declaration as a clone LOC (the class does other things) but count the role declaration as a solution LOC. Roles may also require methods, and these requirements are counted as LOC. Thus for the 8 LOC clone the role would have 1 more fixed, 1 more for each player and 1 more for each requirement. If the role requires 3 methods and the concern appears in two classes then the clone has 16 LOC and the role solution would count 14 LOC. This may not seem a great improvement but LOC do not account for the modularity and maintenance issues. Removing the crosscutting concern gives the system a great advantage in modularity terms.

## 5.5 Threats to Validity

### 5.5.1 Complexity of JavaStage
This study did not take into account the difficulty in learning the JavaStage language. Nevertheless we believe that the few extensions that JavaStage introduces are simple to understand and do not pose great difficulties. To address this issue we must conduct experiments involving developers.

### 5.5.2 System Comprehension and Evolution
Using roles to model crosscutting concerns does not mean that our solution is easier to understand and develop, even if we do believe that all roles we developed contributed to a better system. Some studies must be made to assess if systems with roles are easier to maintain/evolve than without roles. For this we intend to analyze the newer versions of these target systems and analyze the impact roles have on issues like maintenance and evolution.

There is also the work of Riehle [41] that shows that modeling a system with roles is easier and provides a better comprehension so we expect to find those advantages when they are used in programming as well.

### 5.5.3 Aspect Mining Technique Used
We used clone detection for detecting crosscutting concerns. This technique has some limitations, namely it cannot deal well with tangled code [26]. This means that crosscutting concerns that use different code in different places are not detected by this technique. The results also tend to have too much noise (false clones), and that was evident from our study, so the need to filter the results, like we did on the Spring framework, may lead to concerns being disregarded. If we used another aspect mining technique we could have different concerns and therefore different results. Nevertheless, we believe that the amount of concerns detected in our study provides a good code base from which to draw conclusions. This technique also does not detect all the code associated with a crosscutting concern so when analyzing the code we must not limit ourselves to the clone code but also to its surrounding and associated methods. This implies an additional effort, which we did in our case study.

### 5.5.4 Case Study Setup
The clone detecting settings can also affect the detected clones and that would lead to different concerns. That and the removal of clones from the same file could have removed important clones. However, we would need to reduce the amount of clone sets to a manageable number, otherwise there would be a greater number of false clones. We even went under the limit of 30 tokens recommended in [24] for the limitation of false clones.

### 5.5.5 Analyzed Systems
We analyzed 3 open source systems from different fields, each with its particularities. We believe that they make a good case study, but nevertheless results could be different if we used other systems and/or a bigger number of systems. The uniformity of results from these three very different systems is somewhat reassuring, though.

## 6. Related Work
We divided related work into three sections. Our solution is based on roles so we considered role related work even if some work is dedicated to dynamic roles. We do consider related work other forms of composition even if some are object composition rather than class composition. Because class extension can also relate to our approach some approaches are discussed. There are some features present in our approach has that none of the debated work has, namely: the multiple method version, the possibility to rename the constraints methods and block renaming. Further differences are discussed in each related work.

## 6.1 Role related work
Chernuchin and Dittrich [11] compared five approaches for role support in OO languages. They were multiple inheritance, the role object pattern, interface inheritance, object teams and roles as components of classes. They used criteria such as encapsulation, dependency, dynamicity, identity sharing and the ability to play the same role multiple times. Roles as components of classes compared fairly well and the only drawback, aside dynamicity, was the fact that there where no tools to supported it. With JavaStage that drawback is eliminated.

Chernuchin and Dittrich [10] use the notion of natural types and role types that we followed. They also described ways to deal with role dependencies which we didn't consider as it would introduce extra complexity to the role language. Their role model is similar to ours in that it is an extension to the object model. Even though they suggest programming constructs to support their approach no role language has emerged.

Riehle in [42] proved roles usefulness in the various challenges frameworks are faced with: documentation, comprehensibility, etc. He did not propose a role language, but simply described how roles could be used in several languages. We extended his work taking the advantages of roles all the way from modeling to code.

VanHilst and Notkin in [52] proposed to use roles in the C++ language. They did not extend the language to introduce role support, instead they proposed the use of template classes [49] to implement role like behavior. A role is implemented as a template class and defines their collaborators via other template types. To play the role a class inherits from the template role, defining each collaborator type, so roles are a supertype of the player. Since roles are template classes they can play other roles and inherit from other roles as in our approach. To support multiple role playing they use intermediate steps to avoid name clashes and other multiple inheritance problems. Our roles do not rely on inheritance. Our plays declaration also solves name clashes without the use of intermediate steps. C++ does not support method renaming, like JavaStage does, thus a role must provide generic methods, which can make it impossible to use the same role several times. The C++ template engine is very powerful but its use makes their approach limited to the C++ language. Our approach seems to require fewer changes. For example, it would be easier to apply our approach in the C++ language (which supports inner classes) than supporting templates in Java.

In [3] we discussed how roles can be used to address modularity issues, why we choose to adopt a static approach and how we could use roles to implement some of the GoF [20] patterns.

The remainder of this subsection is dedicated to programming languages that support roles, but on their dynamic nature. Our approach focus is on static class composition, so a direct comparison between them and our approach is not feasible.

Object Teams [23] is an extension to Java that uses dynamic roles. They introduce the notion of team. A team represents a context in which several classes collaborate to achieve a common goal. Roles are implemented as inner classes of a team. When an object is used by the team is gets attached one of the defined team roles. Whenever used inside the team the object has that role attached, whenever used outside the team the role is not considered (a process named as lifting/lowering). A role from a team cannot be reused by another, unrelated, team. Roles are limited to be played by a specific type, because of the playedBy directive.

EpsilonJ [50] is another java dynamic role extension that, like Object Teams, uses aspect technology. In EpsilonJ roles are also defined as inner classes of a context. Roles are assigned to an object via a bind directive. EpsilonJ uses a requires directive similar to ours, but, unlike ours, affects only the player. It also offers a replacing directive to rename methods names but that is done on an object by object basis when binding the role to the object, and does not allow block renaming. It also applies only to the names of the methods the role offers.

PowerJava [2] also supports dynamic roles. In PowerJava roles always belong to a so called institution. When an object wants to interact with that institution it must assume one of the roles the institution offers. To access specific roles of an object castings are needed. Roles are written for a particular institution, so we cannot reuse roles between unrelated institutions.

## 6.2 Composition Techniques

Traits [43][15] have first appeared in smalltalk but some attempts have been made into bringing traits in to java-like languages [38][45] and we consider them to be the most related to our approach. Traits can be seen as a set of methods that provide common behavior. When a class uses a trait the trait methods are added to the class. The class also provides glue code to compose the several traits it uses. Traits have a flattening property, i.e., a class can be seen indifferently as a collection of methods or as composed by traits, and trait methods can be seen as trait methods or as class methods. In our approach a class can also be seen as being composed from several roles or as an undivided entity. This is not to be confused with the traits' flattening property. A super reference in a trait refers to the superclass of the class, while in our approach a super reference in the role refers to the superrole.

Unlike our roles, traits cannot store state. State is maintained by the class that uses the trait. Traits access state through required methods. When modeling a concept we, often, need to express state. For example when modeling a container we need a structure for storage. Forcing the class to supply that structure is breaking the container's encapsulation or, at least, the class knows more about the container than it should. Proposals to solve this introduced a significant complexity to the trait model [5][13] and encapsulation problems. In [13] traits ceased to be static entities to become dynamic entities and lost the flattening property.

Traits have no visibility control. Freezable traits [16] compensate this by allowing classes to freeze/unfreeze methods, i.e., declare a method as private (freeze) or redeclaring it as public (defrost).

A class can do the same task multiple times with some minor differences. For example, we could use a container for two types of objects in the same class, using the container trait twice. Traits support aliases that could be used to distinguish between methods for each case but a class cannot declare to use the same trait twice, at least directly. It would need to do an intermediate composition.

One advantage traits have over mixins and single inheritance, is that the order of composition is irrelevant. In our approach the order of role playing is also irrelevant if roles have no conflicting methods. When there are conflicting methods, the method of the first played role is used. This behavior is designed to facilitate development but the compiler generates a warning so the developer can be aware of the situation.

The decomposition strategy of Feature Oriented Programming (FOP) is to decompose the system into features [1]. Features are the main abstractions in FOP during design and implementation. Features reflect user requirements and incrementally refine each other. Because features are the distinguishing product characteristics of Software Product Lines, FOP is mainly used for SPL and program generators.

FOP relies on a step-wise refinement of applications by adding new features or refining existing ones. To compose a system we just state which features it has. The composition is made automatically with tool support, like AHEAD [4]. This is a more powerful technique than JavaStage. AHEAD uses several tools for composing the code and extra files for configuring the composition step. JavaStage is a programming language that statically composes classes using only source code. AHEAD can be used to compose classes. For example, we can develop a class that defines the basic behavior of a class, undistinguishable from a normal Java class, except that it has a feature keyword indicating to which feature it is associated to. We can then construct several refinements to that class. Each refinement must indicate the associated feature and the class it refines. This is where we can use JavaStage together with AHEAD. Refinements may be similar between some classes and, in AHEAD, we would duplicate that code in all refinements. With roles we just place the refinement code in a role and then all refinements just play the role, thus preventing code replication. To avoid this duplication we could use Mixins [44], but Roles offer more ways of configurations and don't have mixins limitations like a linear composition order.

Multi-dimensional separation of concerns (MDSOC) claims that crosscutting concerns are a consequence of the "tyranny of the dominant decomposition" [51]. A key reason is that one needs different decompositions for different concerns at different times, but most languages and modularization approaches support only one "dominant" kind of modularization [36]. MDSOC allows the encapsulation of overlapping, interacting and crosscutting concerns, including features, aspects, variants, roles, components, frameworks, etc., simultaneously [14]. All concerns are first-class components that can be integrated flexibly. Concerns are placed in hyperslices that are composed together in hypermodules following a set of composing rules. Hyperslices may be used by many hypermodules. Hypermodules may be reused and can contain other hypermodules. MDSOC is a natural evolution of subject programming [22].

In the Figure framework a figure could be decomposed in several hyperslices, one for each concern. For example, we could have the properties hyperslice, the observer hyperslice, etc. The Figure hypermodule would compose the several hyperslices. Extensions to the application can be made by adding new hyperslices and composing them in a new hypermodule. For example, if we wish to save the figures in a SVG or any other format we need to add a hyperslice for each format. In an OO version we would need to add a saveXFormat method in each class or use a Visitor pattern [20]. A further advantage of hyperslices is that we can mix-and-match them. This means we could add/remove the support for one format simply by including/not including the respective hyperslice.

Nevertheless, code replication can still be found between hyperslices. Several hyperslices that could implement observers would have code replication. We could reuse our GenericSubject role in those hyperslices. In MDSOC we cannot develop a full purpose Subject role like we did in JavaStage as it does not support multiple method versions and method renaming. MDSOC does not exclude the use of roles as they can add yet another form of composition to its tools. MDSOC also relies heavily on tools and configuration files, which can be overkill, while our approach is code based only.

Package Templates (PT) [31] combine the use of packages and templates. Classes defined in these packages are directly available when the package is instantiated. When instantiated, classes can be tailored to the use context by getting additions, renaming elements and type parameters can be given actual types. This tailoring is similar to ours as we also support renaming and type parameters. PT may also impose restrictions on type parameters using a constraints declarations that resembles our requirement list. Classes in a PT can be merged with classes from other PT and can be used more than once in the same merging operation (like roles can be used multiple times). To avoid multiple inheritance problems PT imposes a series of restrictions on inheritance inside a PT. These, however, can be overcome by simple workarounds. Name clashes are solved via renaming because both fields and methods can be renamed. The renaming cannot be used on the constraints, which means that intermediate classes may be needed to provide these constraints and/or methods added to classes just to meet that requirement. The main differences between PT and roles are: PT relies on inheritance for the merging and roles rely on inner classes and forward methods, so JavaStage does not need the PT restriction on inheritance and still allows multiple role playing; JavaStage allows the renaming of required methods, so it does not need intermediate classes. In PT a class may be the result of the merge of several classes. To refer to one of those classes PT uses the construct super( MergedClassName ) while we use the role identity. The use of the class name is more fragile than the identity, because the identity is given by the class composer and the class name is given by the PT and not the instantiator.

Aspect-Oriented Programming as used in AspectJ [27] is another approach that tries to modularize crosscutting concerns. In [27] the authors define pointcuts to identify points in the executing program that may trigger a different execution path and advices that indicate the new execution path. While the modularization of crosscutting concerns is the flagship of AOP several authors disagree [48][37]. The effects of pointcuts and advices, especially when several aspects have similar pointcuts, may be unpredictable. A particular problem is the fragile pointcut [28]. This problem arises when simple changes made to a method code make a pointcut either miss or incorrectly capture a joint point thus incorrectly introducing or failing to introduce the necessary advice. Thus simple changes in the class code can have unsought effects [25].

The obliviousness feature [18] means that a class is aspect unaware so aspects can be plugged or unplugged as needed. This somewhat resembles dynamic roles and explain why aspects are used in dynamic role languages. But it introduces problems in comprehensibility [21]. To fully understand the system we must not only know the classes but also have to know the aspects that affect each class. This is a major drawback when maintaining a system, since the dependencies aren't always explicit and there isn't an explicit interface between both parts.

With our approach all dependencies are explicit and the system comprehensibility is increased when compared to the OO version [41]. We do not, however, have the obliviousness of AOP because the class knows and is aware of the roles it plays. Any changes to the class code are innocuous to the role, as long as the contract between them stays the same. As a final point we do not believe that our approach can replace AOP. They are different and approach different problems. We believe that for modeling static concerns our approach is more suitable while AOP is better suited for pluggable and unpluggable concerns. A comparison between the two approaches would be useful to ascertain this point.

Caesar [34] uses aspect technology to modularize crosscutting concerns and enhance reuse of aspects leading to a greater reduction of repeated code. Caesar uses an Aspect Collaboration Interface that decouples aspects binding and implementations by defining them in a separated module. It also uses Virtual Classes (see section 6.3). Caesar does not allow method renaming. We can compare our Subject role with the subject role in [34]: our role subject has fully configurable methods names while in Caesar all subjects must have a addObserver method. In Caesar if we want a class to be a subject for two different actions (e.g. FigureListener and FigureHandleListener) we must define a binding class for each action, while with our approach we can do the configuration in the class alone. Lastly in Caesar observers are limited to a single notification method, named notify. This limits the observer pattern to report only a change, whereas in ours we can define several notification methods, each with a meaningful name.

Jiazzi [32] is based on Units [19] and aims at building systems out of reusable components integrated with the language. Jiazzi has two types of units: Atoms (composed by java classes) and Compounds (composed by atoms or other compounds). Jiazzi supports the addition of features to classes without editing their source code. Jiazzi could use Roles to specify these new features. Furthermore, a role could be used to add the same behavior for different classes in the same unit (used as a way of emulating multiple inheritance within a unit), or for the same class but in different units whenever those units are not composable together.

Open classes as used by MultiJava [12] allow external methods to be added to a class, without changing the class. They also support multi-method dispatching. So they are used to extend a class interface like our approach but from a different perspective: we focus on constructing the class and open classes in adding methods to existing classes. They are complementary to our approach because roles can be used to compose the class and MultiJava can later add new methods. Roles could also be used to add those new methods. Also MultiJava cannot redefine class methods, and while roles methods do not redefine the class's defined methods they can redefine the class's inherited methods.

## 6.3 Class extensibility

These approaches deal with the extensibility of classes. Our work's goal is not to extend existing classes but to express concerns that otherwise would be replicated in several classes. These techniques can be combined with ours so we consider them as complementary and not as alternatives.

In Classboxes [6] classes are defined within a kind of module, or unit of scoping. In each classbox we can define classes but can also import classes from other classboxes or refine other classes. Refinements may consist in adding or redefining new behavior or state from an imported class. Since these refinements are only visible within the classbox or classboxes that import from it, existing clients from the refined class are not affected. Roles could be used in conjunction with classboxes as a way to introduce refinements. Supposing that some, related or unrelated, classes need a refinement with the behavior common to all, we can use a role to model that behavior and the refinement would be each class playing that role.

Virtual classes [17] are used by Caeser. A class can define nested classes. Nested classes are the virtual classes and they can be redefined by subclasses of the enclosing class. Each virtual class has therefore an enclosing object - the object of the container class or one of its subclasses. At run time, the inner class to use depends on the type of the outer object. Again roles could be used with Virtual classes to add the new behavior to virtual classes.

## 7. Conclusion and Future work

We presented a way of handling crosscutting concerns code that has a finer grain than classes. For that we compose classes from roles. Roles can model specific concerns and classes play those roles. To better model concerns roles support state and visibility control. To better adapt to the context, roles support renaming methods. We supported roles and the mentioned features by using JavaStage, a proposed extension to the Java language.

We showed how roles can be used to reduce crosscutting concerns in a system, using three systems in a case study and results showed that roles could be used for almost all of the identified crosscutting concerns.

For future work we intend to conduct experiments with developers to assess the ease of use and productivity of the language. We also intend to make studies in whether roles have an impact on system's evolution by using the tests mentioned in the threats to validity section (section 5.5.25.5).

## 8. Acknowledgments

We would like to thank Erik Ernst and the anonymous reviewers for their comments which greatly helped in improving the paper.

## 9. References

[1] Apel, S., Kästner, C. (2009): An Overview of Feature-Oriented Software Development, in Journal of Object Technology, vol. 8, no. 5, July–August 2009,pages 49–84

[2] Baldoni, M., Boella, G. van der Torre, L., (2007): Interaction between Objects in power-Java, Journal of Object Technologies 6, 7 - 12.

[3] Barbosa, S. and Aguiar, A. (2011). Generic roles, a test with patterns In 18th Conference on Pattern Languages of Programs, PloP 2011, Portland, OR, USA.

[4] Batory, D., Sarvela, J. N. and Rauschmayer, A., Scaling Step-Wise Refinement. IEEE TSE, 30(6), 2004.

[5] Bergel, A., Ducasse, S., Nierstrasz, O. and Wuyts, R. (2008): Stateful traits and their formalization. Journal of Computer Languages, Systems and Structures, 2008.

[6] Bergel, A, Ducasse, S. and Nierstrasz, O. (2005). Classbox/J: controlling the scope of change in Java. In Proc. Ann. ACM SIGPLAN Conf. OOPSLA.

[7] Bracha, G. and Cook, W. (1990): Mixin-Based Inheritance. In Proceedings of the OOPSLA/ Proceedings of the ECOOP, Ottawa, Canada. ACM Press.

[8] Bracha, G. (1992): The programming language jigsaw: mixins, modularity and multiple inheritance. PhD thesis, University of Utah.

[9] Ceccato, M., Marin, M., Mens, K., Moonen, L, Tonella, P. and Tourwe, T. A qualitative comparison of three aspect mining techniques, Proc. of the 13th InternationalWorkshop on Program Comprehension (Washington, DC, USA), 2005

[10] Chernuchin, D., and Dittrich, G. 2005. Role Types and their Dependencies as Components of Natural Types. In AAAI Fall Symposium: Roles, an interdisciplinary perspective.

[11] Chernuchin, D., Lazar, O. S., and Dittrich, G., Comparison of Object-Oriented Approaches for Roles in Programming Languages, Papers from the 2005 Fall Symposium. Technical Report FS-05-08. American Association for Artificial Intelligence, Menlo Park, California.

[12] Clifton, C., Leavens, G., Chambers, C., and Millstein, T. (2000): MultiJava: Modular open classes and symmetric multiple dispatch for Java. In Proc. of OOPSLA, Oct..

[13] Cutsem, T., Bergel, A., Ducasse, S. and Meuter, W. (2009): Adding State and Visibility Control to Traits Using Lexical Nesting, Proceedings of the 23rd ECOOP 2009, Italy

[14] Devanbu, P., Balzer, B., Batory, D., Kiczales, G., Launchbury, J.; Parnas, D., Tarr, P., Modularity in the New Millennium: A Panel Summary, International Conference On Software Engineering (2003), Vol. 25, pages 723-725

[15] Ducasse, S., Schaerli, N., Nierstrasz, O., Wuyts, R. and Black, A. (2004): Traits: A mechanism for fine-grained reuse In Transactions on Programming Languages and Systems.

[16] Ducasse, S., Wuyts, R., Bergel, A., and Nierstrasz, O. (2007): User-changeable visibility: Resolving unanticipated name clashes in traits. In Proceedings of 22nd International Conference OOPSLA'07, New York, NY.

[17] E. Ernst, K. Ostermann, and W. R. Cook. A virtual class calculus (2006). In POPL'06: Conference record of the 33rd Symposium on Principles of Programming Languages.

[18] Filman, R.E., Friedman, D.P.: Aspect-oriented programming is quantification and obliviousness. In: Workshop on Advanced Separation of Concerns at OOPSLA (2000)

[19] Flatt, M. and Felleisen, M. (1998) Units: Cool modules for HOT languages. In Proc. of PLDI, pages 236–248, May

[20] Gamma, E., Helm, R., Johnson, R. and Vlissides, J., (1995): Design Patterns: Elements of Reusable Object-Oriented Software, Addison-Wesley.

[21] Griswold, W.G., Sullivan, K., Song, Y., Shonle, M., Tewari, N., Cai, Y., Rajan, H., 2006: Modular Software Design with Crosscutting Interfaces. IEEE Software 23(1), 51–60 (2006)

[22] Harrison, W. and Ossher, H. 1993. Subject-oriented programming: a critique of pure objects. In Proceedings of the Eighth Annual Conference OOPSLA '93, 411-428

[23] Herrmann, S., (2005): Programming with Roles in ObjectTeams/Java. AAAI Fall Symposium: "Roles, An Interdisciplinary Perspective".

[24] Kamiya, T., Kusumoto, S. and Inoue, K. (2002), Ccfinder: a multilinguistic tokenbased code clone detection system for large scale source code, IEEE Trans. Softw. Eng. 28, no. 7.

[25] Kästner, C., Apel, S., Batory, D., 2007: A Case Study Implementing Features using AspectJ. In:11th International Conference of Software Product Line Conference, Kyoto

[26] Kellens, A., Mens, K., Tonella, P., A survey of automated code-level aspect mining techniques, Transactions on aspect-oriented software development IV, Springer-Verlag, Berlin, Heidelberg, 2007

[27] Kiczales, G., Hilsdale, E., Hugunin, J., Kersten, M., Palm, J., Griswold. W. G., (2001): An overview of AspectJ. In proceedings of ECOOP 2001, Budapest, Hungary

[28] Koppen, C., Störzer, M.: PCDiff, 2004: Attacking the fragile pointcut problem. In: European Interactive Workshop on Aspects in Software, Berlin, Germany.

[29] Kristensen, B. B., (1995): Object-oriented modeling with roles, in Proceedings of the 2nd International Conference on Object-Oriented Information Systems, Springer-Verlag.

[30] Kristensen, B. B. and Østerbye, K.. (1996): Roles: Conceptual abstraction theory & practical language issues, Theory and Practice of Object Systems 2(3): 143–160.

[31] Krogdahl, S., Møller-Pedersen, B., Sørensen, F.: "Exploring the use of Package Templates for flexible re-use of Collections of related Classes", in Journal of Object Technology, vol. 8, no. 7, November - December 2005

[32] S. McDirmid, M. Flatt, and W.C. Hsieh, "Jiazzi: new-Age Components for Old-Fashioned Java", OOPSLA 2001.

[33] Meyer, B. (1988): Eiffel: the Language. Prentice-Hall.

[34] Mezini, M. and Ostermann, K. 2003. Conquering Aspects with Caesar. In Proc. of AOSD 2003, pages 90 – 99

[35] Odberg, E. (1995): Category classes: Flexible classification and evolution, in object-oriented databases, Proceedings of Advanced Information Systems Engineering, Springer-Verlag, London, UK, pp. 406–420.

[36] Ossher, H. and Tarr, P. (2000). Multi-dimensional separation of concerns and the hyperspace approach, Proceedings of the Symposium on Software Architectures and Component Technology, Kluwer.

[37] Przybyłek, A.(2001). Systems Evolution and Software Reuse in Object-Oriented Programming and Aspect-Oriented Programming , J. Bishop and A. Vallecillo (Eds.): TOOLS 2011, LNCS 6705, pp. 163–178.

[38] Quitslund, P. and Black, A. (2004): Java with traits - improving opportunities for reuse. In Proceedings of the 3rd International Workshop on Mechanisms for Specialization, Generalization and inheritance (ECOOP 2004), 2004.

[39] Reengskaug, T., Wold, P. and Lehne, O. A. (1996), Working with objects - the OOram software engineering method. Manning.

[40] Reengskaug, T. 2007: Roles and classes in object oriented programming. In: Roles 2007. Proceedings of the 2nd Workshop on Roles and Relationship in Object Oriented Programming, Multiagent Systems, and Ontologies

[41] Riehle, D. and Gross, T. 1998. Role Model Based Framework Design and Integration." In Proceedings OOPSLA '98. ACM Press

[42] Riehle, D. 2000. Framework Design: A Role Modeling Approach, Ph. D. Thesis, Swiss Federal Institute of technology, Zurich.

[43] Scharli, N., Ducasse, S., Nierstrasz, O. and Black, A. (2003): Traits: Composable units of behavior. In Proceedings of ECOOP 2003, volume 2743 of Lecture Notes in Computer Science. Springer.

[44] Smaragdakis, Y. and Batory, D., Mixin Layers (2002): An Object-Oriented Implementation Technique for Refinements and Collaboration-Based Designs. ACM TOSEM, 11(2).

[45] Smith, C. and Drossopoulou, S. (2005):Chai: Traits for Java-like languages. In Proceedings of the 19th European Conference on Object-Oriented Programming.

[46] Sousa, P., Silva, A. R. and Marques, J. A. (1995): Object identifiers and identity: A naming issue, Proceedings of the 4th International Workshop on Object-Orientation in Operating Systems, USA, p. 127.

[47] Steimann, F., (2000): On the representation of roles in object-oriented and conceptual modeling. Data & Knowledge Engineering 35(1):83–106.

[48] Steimann, F. (2006), The paradoxical success of aspect-oriented programming", in Proceedings of the 21st Annual Conference OOPSLA'06

[49] Stroustrup, B. (1986): The C++ Programming Language. Addison Wesley, Reading, Mass.

[50] Tamai, T., Ubayashi, N., and Ichiyama, R., (2007): Objects as Actors Assuming Roles in the Environment, in Software Engineering For Multi-Agent Systems V: Research Issues and Practical Applications, Lecture Notes In Computer Science, vol. 4408. Springer-Verlag,

[51] Tarr, P., Ossher, H., Harrison, W. and Sutton Jr., S. M. (1999) , N degrees of separation: multi-dimensional separation of concerns, Proceedings of the 21st International Conference on Software Engineering, New York, NY, USA

[52] VanHilst, M., and D. Notkin, D., (1996). Using Role Components to Implement Collaboration-Based Designs. In OOPSLA '96, pages 359–369, 1996.

# Aggregation for Implicit Invocations*

Sebastian Frischbier
Technische Universität
Darmstadt
frischbier@dvs.tu-
darmstadt.de

Alessandro Margara
Vrije Universiteit Amsterdam
a.margara@vu.nl

Tobias Freudenreich
Technische Universität
Darmstadt
freudenreich@dvs.tu-
darmstadt.de

Patrick Eugster
Purdue University
p@cs.purdue.edu

David Eyers
University of Otago
dme@cs.otago.ac.nz

Peter Pietzuch
Imperial College London
prp@doc.ic.ac.uk

## ABSTRACT

Implicit invocations are a popular mechanism for exchanging information between software components without binding these strongly. This decoupling is particularly important in distributed systems when interacting components are not known until runtime. In most realistic distributed systems though, components require some information about each other, be it only about their presence or their number. Runtime systems for implicit invocations—so-called publish/subscribe systems—are thus often combined with other systems providing such information.

Given the variety of requirements for information about interacting components across applications, this paper proposes a generic augmentation of implicit invocations: rather than extending a given publish/subscribe API and system in order to convey a particular type of information across interacting components, we describe domain-specific joinpoints that can be used to advise application-level invocation routers—so-called brokers—used by publish/subscribe systems. This enables aggregation of application-specific information to and from components in a scalable manner.

After presenting our domain-specific joinpoint model, we describe its implementation inside the REDS publish/subscribe middleware. The empirical evaluation of our approach shows that: (a) it outperforms external aggregation systems, by collecting and distributing information with a limited overhead; (b) the deployment of new functionalities has virtually no overhead, even if it occurs while the publish/subscribe system is running.

*Supported by German BMBF Software-Cluster EMERGENT (01IC10S01) and (01|C12S01V), LOEWE Dynamo PLV, Dutch national program COMMIT, NSF grant 0644013, Alexander von Humboldt Foundation, DARPA grant N11AP20014. The authors assume responsibility for the content.

## Categories and Subject Descriptors

C.2.4 [**Computer-Communication Networks**]: Distributed Systems—*Distributed applications*; D.3.3 [**Programming Languages**]: Language Constructs and Features

## General Terms

Design, Management, Measurement

## Keywords

implicit invocation; publish/subscribe; event; broker; aspect; joinpoint

## 1. INTRODUCTION

Many distributed software systems are designed according to the paradigm of *implicit invocations* [39, 43]: software components announce *events* which are conveyed to target components that registered interest in such events, instead of invoking the target components directly. By shifting communication away from direct invocations via component references to asynchronous consumption of events of interest produced anonymously by components, application components can be added (and removed) at runtime. This increases flexibility through decoupling—event sources need not manage their targets explicitly—and supports scalability by allowing for asynchronous, streamlined communication.

### 1.1 Publish/Subscribe Systems

*Publish/subscribe* systems [35] provide the communication backbone for software systems based on implicit invocations. They transmit events produced, potentially at high rates, by *many* publishing components *to many* subscribed components. For years, research on publish/subscribe systems has focused on scalable and flexible architectures [4, 9] that support fine-grained *content-based* addressing models, in which published events are routed as *messages* based on the subscribers' interests in the *content* of messages. Such architectures commonly use so-called *brokers* to form decentralized *overlay networks* of application-level routers that interconnect publishers and subscribers. Brokers perform efficient en-route filtering and selective forwarding of messages based on their content. While extremely scalable and flexible, these systems have only seen limited adoption in practice. In particular, they have not dethroned simpler systems based on centralized brokers and ones that route messages directly based on namespaces, i.e., by associating messages with single *topic* (or *subject*) names.

One reason for the poor adoption of content-based systems, particularly decentralized publish/subscribe systems, is that simpler systems assigning topics to individual single brokers can more easily provide additional information on the usage of these topics. Many applications that utilize implicit invocations for scalable communication, however, require such information, such as the number of publishers and subscribers connected to a given topic or the the set of subscribers that might have received published events. Concrete scenarios are presented in the remainder of this section.

Currently, these applications must use less scalable, *centralized*, publish/subscribe systems or group communication systems that can provide membership information albeit at high overhead. This reduces the performance benefits of implicit invocations. Moreover, these systems do not provide broader feedback information efficiently, such as the relevance of a message based on the number of potentially interested subscribers, unless using out-of-band communication which can hamper scalability.

We present two scenarios that motivate the need for additional information exchange in a publish/subscribe system, without sacrificing scalability, when implementing implicit invocations. In the first scenario—systems monitoring—subscribers seek additional information about publishers, while in the second scenario—online advertising—publishers receive information about subscriptions.

## 1.2 Example Scenarios

The *monitoring of networked systems* is a common application of publish/subscribe-style communication. Publishers may be servers in a data center that are monitored for load spikes and failures. The subscribers may be monitoring systems that determine the "health" of the data center, and assist with root-cause analysis of failures. The overall rate of monitoring data from servers can become high [30]: for example, probes may provide measurement data about utilization of server resources with sample rates of many times a second. When a client subscribes to all events that may be of interest, it can become easily overwhelmed. Similarly, by subscribing only to very specific events, a client may miss relevant ones, which are causal antecedents of observed symptoms. This prevents root cause discovery, thus making the diagnosis of failures difficult. Fig. 1 sketches the implementation of a monitor client based on the Java Message Service (JMS) [36] API. The API provides no way before or after the `publish` call to know the number of potential or actual receivers for the event. There may be no corresponding subscribers at all, in which case the client might want to use another communication band or raise an alarm.

Thus it would be desirable to aggregate performance characteristics, for example, by summarizing key metrics as statistical distributions. If aggregation happens in the broker network, clients can receive information about the general health of a large number of data center components without being overloaded by low-level events. As specific problems in the data center develop, subscriptions can be refined to focus on more detailed events for root cause analysis.

In *online advertising on the web* [20], advertising agents act as publishers of advertising messages (*ads*), and users are subscribers to particular ads. Subscriptions reflect user

```
public class JMSMonitoringClient {
    ...
    TopicSession t = ...;
    TopicPublisher tp = t.createPublisher(...);
    ObjectMessage om = t.createObjectMessage();
    Long loadLevel = ...;
    ... // how many interested subscribers?
    om.setObject(loadLevel);
    tp.publish(om);
    ... // how many receivers?
}
```

**Figure 1: Example of system monitoring client based on JMS [36] pushing load level readings regardless of subscribers. The client has no indication of the number of potential or actual event receivers.**

interests, potentially inferred from browsing activities. For targeted online advertising to be most successful, however, advertisers want to have information about potential consumers who subscribe to their advertisements. This goes beyond a basic publish/subscribe model, in which publishers and subscribers remain anonymous—in that case, subscribers cannot gauge the number of potentially matching publishers and vice-versa.

An aggregation mechanism provided within the broker network would satisfy the requirements of both parties in this scenario. An aggregation function can ensure that the feedback sent to the publishers only includes statistical distributions of subscribers' interests, and thus would preserve anonymity. Subscribers can safely use fine-grained subscriptions, and thus ensure that they receive only ads that might be of interest to them.

## 1.3 Application-Specific Integrated Aggregation

In order to support the above scenarios, we propose to augment publish/subscribe systems with *application-specific integrated aggregation* (ASIA), which offers the ability to convey additional information, besides published messages, as required by communicating components in an application.

Compared to the use of an external aggregation system, the integration of aggregation with the publish/subscribe system has benefits in terms of efficiency (e.g., combining network communication for aggregation with actual messages) and expressiveness (giving access to broker internals).

However, it raises an immediate question: *how to express aggregation logic in a generic manner that is suitable for decentralized evaluation?* In particular, the precise nature of aggregation remains application-specific. In the above monitoring scenario, for example, monitoring clients should have the flexibility to provide their own aggregation functions, potentially at runtime.

The core idea behind ASIA is inspired by aspect-oriented programming (AOP) [29]: we define a set of joinpoints that are *specific to the filtering and routing of messages in decentralized publish/subscribe systems*. These joinpoints can be advised in order to perform aggregation of relevant information by altering and augmenting the default broker code. Our joinpoints can also be used to support more general behavior change in brokers, such as protocol extensions.

## 1.4 Contributions

This paper reports on our experience with the design and implementation of a "reflective" publish/subscribe system according to the ASIA model. More specifically, we:

- present domain-specific joinpoints for decentralized implicit invocation (publish/subscribe) messaging systems;

- describe an implementation of this model within the Java-based REDS publish/subscribe system [10];

- demonstrate the benefits of the ASIA model by showing the performance improvements of integrated aggregation over the use of a separate aggregation system; and

- evaluate a mechanism for deploying new aggregation functions to be attached to joinpoints in brokers. We also discuss the supported flexibility in terms of the different forms of aggregation that are implementable in ASIA without changes to the main broker code.

## 2. BACKGROUND

After outlining the generic publish/subscribe model that we assume in this paper, we derive the requirements of a mechanism for additional information flow.

### 2.1 Publish/Subscribe Model

Several publish/subscribe models have been proposed in the literature [15] (cf. §6). We adopt a mixed *topic/content-based* publish/subscribe model, which provides a good compromise between performance and expressiveness in practice. It enables ASIA to be applied both to pure topic-based and pure content-based publish/subscribe systems.

Subscribers issue *subscriptions* to express interest in a topic, and may additionally use a *predicate filter* to further narrow their subscriptions to specific content. Publishers send *advertisements* to indicate that they will publish messages to a specific topic. *Publication* messages are then routed from publishers to interested subscribers.

We consider clients and brokers to be interlinked to form an *overlay network*, as shown in Figure 2(a). For ease of presentation, we assume the overlay network to be free of cycles. The brokers ($b_1$ to $b_5$) provide the publish/subscribe service to clients. Brokers directly connected to clients ($b_{\{1,3,5\}}$) are termed *edge brokers*. We further assume each client to be connected to one and only one broker. Each *broker state* includes information (i.e., advertisements and subscriptions) on directly connected clients. Furthermore, the broker state stores information on neighboring broker nodes and their respective (transitive) advertisements and subscriptions.

Broker nodes propagate events—each of which pertains to a topic $\tau$—from publishing nodes (i.e., publishers of $\tau$ events) to client nodes that have expressed interest in receiving $\tau$ events (i.e., subscribers to $\tau$ with content-based filters). A broker $b_i$ propagates events for each topic $\tau$ along its neighbor links to other brokers or clients via lower-level protocol *messages* (event messages); analogously a broker propagates advertisements (advertisement messages) and subscriptions (subscription messages).

To illustrate the routing of messages in the broker network, Figure 2(a) shows a sample sequence, in which client $c_1$ first advertises publications for a topic (step 1); clients $c_3$ and $c_4$ subscribe to the advertised topic (steps 2 and 3); and finally client $c_1$ publishes an event of that topic (step 4).

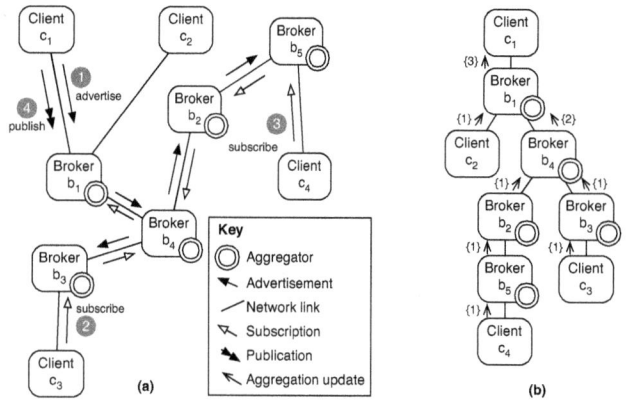

**Figure 2: Distributed publish/subscribe system with messages**

**Algorithm 1** A basic publish/subscribe client algorithm. The client is instantiated with an edge broker

Publish/subscribe client algorithm. Executed by client $c_i$
```
 1: init
 2:     b                                          {Edge broker}
 3:     for all published topics τ do
 4:         SEND(AD,τ) to b

 5: to PUBLISH(e) on topic τ do               {Publish new event}
 6:     SEND(PUB,τ,e) to b

 7: to SUBSCRIBE(φ) to topic τ do                   {Subscribe}
 8:     SEND(SUB,τ, φ) to b

 9: to UNSUBSCRIBE(φ) from topic τ do            {Unsubscribe}
10:     SEND(UNSUB,τ, φ) to b

11: upon RECV(PUB, τ, e) do                     {Receive event}
12:     DELIVER(e)
```

### 2.2 Broker Routing Algorithms

The routing of messages in a broker network is conducted according to *routing algorithms*. We now briefly introduce how these algorithms work before presenting our corresponding integrated aggregation mechanisms in the following sections.

Algorithm 1 shows the API utilized by clients and the corresponding broker routing algorithm is shown in Algorithm 2. The **to ... do** clauses describe how to achieve a desired function (e.g., publish a message) using lower layers in the network stack. In contrast, the **upon ... do** clauses indicate the operations that occur when lower layers in the network stack react to messages they have received. In both algorithms, we assume that nodes communicate by pair-wise reliable channels offering primitives SEND and RECV. For simplicity, a node $p$ acts either as a client $c$ or as broker $b$, and advertisements are only on topics and do not include value ranges. Unadvertisements are elided for brevity.

In our algorithm listings operations such as insertion of a predicate filter (INSERT()), removal (DELETE()) or matching (MATCH()) on partially ordered sets—which store subscriptions using *subsumption* [4]—are just shown as procedure calls. Subscribers (*subs*), subscriptions (partially ordered set $\mathcal{P}$), and publishers (*pubs*) are stored by topic $\tau$; *links* is the array of neighbor brokers.

We assume that brokers act as publishers and subscribers towards their respective *downstream* and *upstream* neigh-

**Algorithm 2** Algorithm for publish/subscribe event processing with subscription summarization. $\mathcal{P}[\tau]$ is the predicate poset ordered by $\preceq_\Phi$. $pubs[\tau]$ stores the advertising peers. $subs[\tau][\phi]$ stores peers that subscribe with $\phi$. $subs[\tau][\phi]$ avoids the need to duplicate $\phi$ in $\mathcal{P}[\tau]$, if more than one peer subscribes with $\phi$.

---

Publish/subscribe broker algorithm. Executed by broker $b_i$

```
 1: init
 2:     pubs[]                          {Process sets indexed by topic τ}
 3:     P[]                             {Posets indexed by topic τ}
 4:     subs[][]                        {Process sets indexed by τ & φ}
 5:     links                           {Neighbor brokers pⱼ}

 6: upon RECV(AD, τ) from pⱼ do         {Receiving advertisement}
 7:     if pⱼ ∉ pubs[τ] then
 8:         for all bₖ ∈ links\{pⱼ} do
 9:             SEND(AD, τ) to all bₖ
10:         SEND(SUB, τ, LUB(P[τ])) to pⱼ
11:         pubs[τ] ← pubs[τ] ∪ {pⱼ}

12: upon RECV(SUB, τ, φ) from pⱼ do      {Receiving subscription}
13:     subs[τ][φ] ← subs[τ][φ] ∪ {pⱼ}
14:     φᵒˡᵈ ← LUB(P[τ])
15:     if |subs[τ][φ]| = 1 then
16:         INSERT(P[τ], φ)
17:     UPD(τ, φᵒˡᵈ, LUB(P[τ]), pⱼ)

18: upon RECV(PUB, τ, e) from pⱼ do      {Receiving event}
19:     matches ← MATCH(P[τ], e)
20:     for all φ ∈ matches do
21:         for all pₖ ∈ subs[τ][φ]\{pⱼ} do
22:             SEND(PUB, τ, e) to pₖ

23: upon RECV(UNSUB, τ, φ) from pⱼ do {Receiving unsubscription}
24:     subs[τ][φ] ← subs[τ][φ] \ {pⱼ}
25:     φᵒˡᵈ ← LUB(P[τ])
26:     if |subs[τ][φ]| = 0 then
27:         DELETE(P[τ], φ)
28:     UPD(τ, φᵒˡᵈ, LUB(P[τ]), pⱼ)

29: procedure UPD(τ, φᵒˡᵈ, φⁿᵉʷ, pⱼ) {Update poset and neighbors}
30:     if φᵒˡᵈ ≠ φⁿᵉʷ then
31:         for all bₖ ∈ pubs[τ]\{pⱼ} do
32:             SEND(SUB, τ, φⁿᵉʷ) to bₖ
33:             SEND(UNSUB, τ, φᵒˡᵈ) to bₖ
```

---

bors. Since subscriptions involve predicates $\phi$ and several subscribers can have the same predicate, subscribers are also indexed by their predicate $\phi$. Advertisements (and unadvertisements) are handled by adding or removing corresponding publishers.

If a broker receives an advertisement for a new topic (line 6 in Algorithm 2), updates need to be performed recursively. When evaluated for a given event $e$, MATCH() returns the set of matching subscriptions; the corresponding subscribers are resolved via $subs$ (line 21). Procedure UPD (line 29) factors out common parts of reactions to addition and removal of subscriptions: in case the poset's least upper bound (LUB; the predicate covering all subscription predicates according to subsumption) was changed by either of these operations, the broker needs to replace its subscription towards upstream brokers by canceling its previous one and issuing a new one (line 30).

# 3. AN ASPECT-ORIENTED PUBLISH/SUBSCRIBE BROKER

In this section we present an aspect-oriented event broker algorithm for publish/subscribe systems that supports reflection by exposing joinpoints.

## 3.1 Overview

Algorithm 3 outlines a reflective broker algorithm exposing joinpoints as an evolution of that of Algorithm 2. The joinpoints refer to points in the execution, at which we may want to customize the algorithm's behavior. Execution "jumps" from a given joinpoint $X$-$Y$ to a corresponding advice ADVS$^{X\text{-}Y}$, which is said to advise joinpoint $X$-$Y$.

In ASIA, this advice is used to invoke code that effects the computation of an aggregation function within a given broker—a simple example would be a function that maintains a count of subscribers to a particular topic—down a subtree reachable from this broker. We discuss the aggregation computations in more detail in §3.3 shortly. Every joinpoint involves a specific set of arguments passed to its advice, which represent volatile computational state of the broker for the current action being performed.

## 3.2 Joinpoints

Table 1 summarizes the joinpoints used in ASIA and their arguments. Besides arguments, code for advice ADVS$^{X\text{-}Y}$ is constrained by a type of value that must be returned. Corresponding advice are thus able to substitute some of the volatile current computational broker state. For every joinpoint/advice, there is a default behavior, which corresponds to the logic of the basic, non-reflective, broker algorithm. In a custom advice ADVS$^{X\text{-}Y}$, this code can be optionally invoked explicitly via PROCD$^{X\text{-}Y}$. The primitive PROCD$^{X\text{-}Y}$ has the same formal argument and return types as the corresponding advice ADVS$^{X\text{-}Y}$. In ASIA, if no advice is specified for a given joinpoint $X$-$Y$, a call to PROCD$^{X\text{-}Y}$ is automatically performed (ADVS$^{X\text{-}Y}(args)$ = PROCD$^{X\text{-}Y}(args)$). Thus in the absence of custom advice, Algorithm 3 has the same semantics as Algorithm 2.

In addition to joinpoints, ASIA uses subtype polymorphism (subtyping) for easy expression of code for aggregation. Attributes of broker protocol messages can be augmented by appending to them (denoted $\oplus v$ to append a value $v$) for piggybacking information. This operation refers to an implicit subtype of the original attribute type, augmenting its super-type with an attribute of the type of $v$. We also allow aggregation features to add their own specific reactions (**upon** clauses) or even **tasks**, as well as custom message types issued and consumed via SEND and RECV, respectively. Due to space limitations, we elide the presentation of aggregated information to clients.

## 3.3 ASIA Model

While joinpoints and advice capture which code to inject into brokers for the sake of aggregation and where, they do not exactly say what such code should perform. This section thus provides an overview of the ASIA aggregation model, before presenting a concrete example in §3.4.

An aggregation function $f$ is evaluated by aggregators that operate at each broker in the distributed publish/subscribe system. Aggregators are pieces of code that are installed using advice. They can access the state at each broker, for example to include current aggregation computations, or to store their own state. By adding some restrictions on the type of aggregation function that can be computed, we can ensure that the results of the aggregation computations at each broker can be combined. Thus we can compute a global aggregation result by repeatedly combining the partial results generated within each broker.

| Joinpoint description ($X$-$Y$) | Arguments | Return value |
|---|---|---|
| Advertiser addition (AD-ADD) | $\tau$, new advertiser $p_j$ | - |
| Subscriber addition (SUB-ADD) | $\tau$, new subscription $\phi$, new subscriber $p_j$ | old LUB $\phi_{old}$ |
| Advertiser removal (AD-DEL) | $\tau$, old advertiser $p_j$ | - |
| Subscriber removal (SUB-DEL) | $\tau$, old subscription $\phi$, old subscriber $p_j$ | old LUB $\phi_{old}$ |
| Publication matching (PUB-MATCH) | $\tau$, publication $e$ | matching subscrs. |
| Publication sending (PUB-SEND) | $\tau$, publication $e$, upstream publisher $p_j$ | - |
| Subscription sending (SUB-SEND) | $\tau$, new own subscr. $\phi$, new subscriber $p_j$ | - |
| Unsubscr. sending (UNSUB-SEND) | $\tau$, old own subscr. $\phi$, old subscriber $p_j$ | - |
| Advertisement sending (AD-SEND) | $\tau$, new own advertisement $p_j$ | - |
| Unadvert. sending (UNAD-SEND) | $\tau$, new own advertisement $p_j$ | - |

**Table 1: Joinpoints $X$-$Y$.**

In this paper we focus on aggregators using additive functions $f$, including counting and rate measurements. However, any associative and commutative function can be used within an aggregator, including multiplication, set operations such as union and intersection, bitwise operations, maximum, minimum, and composite functions using other functions on this list, such as the arithmetic mean.

A key feature for scalability is that we include a notion of adjustable *imprecision* into the aggregation functions: a distance $d$ sets the maximum imprecision that will be tolerated by the aggregator at a particular broker. The imprecision factor $d$ allows for a tradeoff between the number of messages flowing through the distributed system (ideally low), and the precision of the aggregation computations (ideally high).

We illustrate the ASIA aggregation model using the online advertising scenario presented in §1.1. Consider the broker $b_1$ in Figure 2(a), and its directly connected subscriber $c_2$. Now assume an online advertiser, $c_1$, wants to discover the number of subscribers to messages within a topic $\tau$ that $c_1$ publishes to. Broker $b_1$ is able to determine the number of subscribers to topic $\tau$, because $b_1$ maintains the state necessary to deliver messages to subscribers that have subscribed to topic $\tau$.

Using an ASIA API, advertiser $c_1$ indicates its interest in an aggregation, providing a callback function that is invoked when the result of $f$ changes—e.g., if $c_2$ were to unsubscribe.

### 3.3.1 Distributed Aggregation

In a distributed broker network, a client that registers interest in an aggregation function at a broker $b_r$ causes the formation of a spanning tree rooted at $b_r$. The tree, which we call the *aggregation tree*, contains all of the other brokers $b_i$ that can contribute relevant aggregation results. To maintain aggregation of data, brokers exchange messages along the aggregation tree. For efficiency, these messages may be piggybacked onto existing publish/subscribe messages rather than being sent separately. Brokers that provide a broker $b_i$ with data to aggregate are considered *below* $b_i$. A broker that receives more aggregated data is considered to be *above* $b_i$, and is closer to the root of the tree.

Consider our online advertising scenario from before. Figure 2(b) shows the aggregation tree formed when $c_1$ requests the count of subscribers from $b_1$. From the perspective of $b_4$, the subscriber count is 2—i.e., $b_4$'s descendants $c_3$ and $c_4$. $b_4$ can determine this value by learning the counts from its immediate children, $b_3$ and $b_2$, namely one each, and then applying $f$ to these counts. The contents of the messages indicating updates to the aggregation value are shown on Figure 2(b) as sets propagating up the aggregation tree.

As a consequence, computing an overall $f$ result can be broken down into computations of $f$ at individual brokers $b_i$ within the distributed publish/subscribe system. To enable this distributed computation in our model, every broker $b_i$ stores the most recent evaluation of $f$ so as to track the $f$ that will be passed to its parent in the aggregation tree: we denote this cached value $v_i$. In addition, every broker stores the value of $v_k$ for its direct children in the aggregation tree. All of these $v$ values are part of the broker state.

When $v_i$ changes, a message is sent by broker $b_i$ to the broker above it in the tree, triggering an update of its $v$ value. So in Figure 2(b), if client $c_3$ were to unsubscribe from $\tau$, then $b_3$ would recompute $v_3$, and would need to inform $b_4$ in order for $v_4$ to be recomputed. This would continue up the tree to the root broker $b_1$, which would invoke client $c_1$'s registered aggregation callback. $c_1$ would thus learn the number of subscribers as requested—and in particular that this number had recently changed.

### 3.3.2 Precision of Distributed Aggregation

In this above approach, *every* change in the aggregation value $v_i$ at any broker causes aggregation value update messages to be sent up to the root of the aggregation tree, which limits scalability. Therefore, similar to other scalable aggregation models [26], we relax the precision of $v_i$ that is propagated upwards in the aggregation tree by having each broker $b_i$ maintain $v_i$ privately, which we term $\overline{v}_i$. Updates to $\overline{v}_i$ are sent upwards in the tree only if the previous values of $\overline{v}_i$ sent up the tree would otherwise be more than some *bound* away from $b_i$'s most recent $v_i$. The bound on the precision with which $\overline{v}_i$ approximates the true evaluation of $f$ (i.e., $v_i$) is denoted by $\hat{v}_i \in \mathbb{R}$. If $R$ is the range of an aggregation function $f$, we require $d : R \times R \to \mathbb{R}$ to be a distance metric on $R$. For our example scenario, $d$ is simply the positive integer difference between any two numbers of subscribers.

Clients control the quality (i.e., precision) of the aggregation data, by setting the $\hat{v}$ parameter when they register interest in an aggregation function. When a broker $b_r$ first constructs an aggregation tree, it divides the $\hat{v}_r$ that it has been given between itself and the subtrees, at which its immediate children are roots. For a broker with $n$ children, a simple heuristic is to set $\hat{v}_r = f^{n+1}(\hat{v}_k)$. For aggregators that perform summation, this means setting $\hat{v}_k = \frac{\hat{v}_r}{n+1}$ for each child $k$, and using the same bound over its directly connected clients.

To illustrate this approach using our running example,

**Algorithm 3** A reflective broker algorithm. Advice $\text{ADVS}^{X\text{-}Y}$ abstract core broker functionalities. When not implemented by an aggregation feature, $\text{ADVS}^{X\text{-}Y}(args)$ is implemented by a direct call to $\text{PROCD}^{X\text{-}Y}(args)$. When installing a custom $\text{ADVS}^{X\text{-}Y}$ the corresponding default behavior in $\text{PROCD}^{X\text{-}Y}$ can be invoked explicitly.

---

Reflective publish/subscribe broker algorithm. Executed by broker $b_i$

```
 1: init
 2:    pubs[]                          {Process sets indexed by topic τ}
 3:    P[]                             {Posets indexed by topic τ}
 4:    subs[][]                        {Process sets indexed by τ & φ}
 5:    links                           {Neighbor brokers pⱼ}

 6: upon RECV(AD, τ) from pⱼ do         {Receiving advertisement}
 7:    if pⱼ ∉ pubs[τ] then
 8:       ADVS^{AD-SEND}(τ, pⱼ)
 9:       ADVS^{SUB-SEND}(τ, LUB(P[τ]), pⱼ)
10:       ADVS^{AD-ADD}(τ, pⱼ)

11: upon RECV(PUB, τ, e) from pⱼ do      {Receiving event}
12:    match ← ADVS^{PUB-MATCH}(τ, e)
13:    for all φ ∈ match do
14:       ADVS^{PUB-SEND}(τ, e, pⱼ)

15: upon RECV(SUB, τ, φ) from pⱼ do      {Receiving subscription}
16:    φ^{old} ← ADVS^{SUB-ADD}(τ, φ, pⱼ)
17:    UPD(τ, φ^{old}, LUB(P[τ]), pⱼ)

18: upon RECV(UNSUB, τ, φ) from pⱼ do    {Receiving unsubscription}
19:    φ^{old} ← ADVS^{SUB-DEL}(τ, φ, pⱼ)
20:    UPD(τ, φ^{old}, LUB(P[τ]), pⱼ)

21: procedure UPD(τ, φ^{old}, φ^{new}, pⱼ) {Update poset and neighbors}
22:    if φ^{old} ≠ φ^{new} then
23:       ADVS^{SUB-SEND}(τ, φ^{new}, pⱼ)
24:       ADVS^{UNSUB-SEND}(τ, φ^{old}, pⱼ)

25: procedure PROCD^{AD-ADD}(τ, pⱼ)       {Default advice for new ad}
26:    pubs[τ] ← pubs[τ] ∪ {pⱼ}

27: function PROCD^{SUB-ADD}(τ, φ, pⱼ)    {Def. adv. for new subscript.}
28:    subs[τ][φ] ← subs[τ][φ] ∪ {pⱼ}
29:    φ^{old} ← LUB(P[τ])
30:    if |subs[τ][φ]| = 1 then
31:       INSERT(P[τ], φ)
32:    return φ^{old}

33: function PROCD^{SUB-DEL}(τ, φ, pⱼ)    {Def. adv. for unsubscription}
34:    subs[τ][φ] ← subs[τ][φ]\{pⱼ}
35:    φ^{old} ← LUB(P[τ])
36:    if |subs[τ][φ]| = 0 then
37:       DELETE(P[τ], φ)
38:    return φ^{old}

39: function PROCD^{PUB-MATCH}(τ, e) {Def. adv. for event forwarding}
40:    return MATCH(P[τ], e)

41: procedure PROCD^{PUB-SEND}(τ, e, pⱼ)  {Def. advice for new event}
42:    for all pₖ ∈ ⋃_φ subs[τ][φ]\{pⱼ} do
43:       SEND(PUB, τ, e) to pₖ

44: procedure PROCD^{X-SEND}(τ, φ, pⱼ) | X ∈ {SUB, UNSUB}  {Def.}
45:    for all pₖ ∈ pubs[τ]\{pⱼ} do      {adv. for (un)subscription}
46:       SEND(X, τ, φ) to pₖ            {forwarding}

47: procedure PROCD^{X-SEND}(τ, pⱼ) | X ∈ {AD, UNAD}   {Def. adv.}
48:    for all bₖ ∈ links\{pⱼ} do         {for (un)advertisement}
49:       SEND(X, τ) to bₖ                {forwarding}
```

---

consider $\hat{v}_1 = 2$ and broker $b_4$ having set $\hat{v}_2 = 1$ and $\hat{v}_3 = 1$. Because of subscriber $c_3$, $v_3 = 1$. If $c_3$ unsubscribes, now $v_3 = 0$. Broker $b_3$ knows that $b_4$ has previously stored $\overline{v}_3 = 1$. However, $\hat{v}_3$ is 1, so $b_3$ does not need to send an aggregation value update message, as $b_4$'s $\overline{v}_3$ is still within the $\hat{v}_3$ bound of the true value of $v_3$. This mechanism avoids sending updates for small changes in the aggregation value. As an example, assume a new client $c_5$ subscribes to broker $b_3$, $v_3$ returns to 1. $c_1$ still has an estimate of the number of

subscribers to within 2 (i.e., $\hat{v}_1$), and the system has avoided sending two aggregation value update messages.

## 3.4 Illustration: Subscriber Counts

Algorithm 4 illustrates a simple subscriber count aggregator implemented in ASIA. When a client registers interest in an aggregation function, the advice shown will need to be integrated into the brokers.

---

**Algorithm 4** Example of advice instantiation for implementing subscriber counts. Advice not listed are directly implemented by the respective default $\text{PROCD}^{X\text{-}Y}$ semantics as mentioned. We abbreviate UPDATEAGGREGATEVALUE to UAV. Summation operates over tuples as $\langle a, b \rangle + \langle c, d \rangle = \langle a + c, b + d \rangle$, and $\preceq$ indicates containment of ranges.

---

Subscriber count advice for reflective publish/subscribe broker algorithm. Executed by $p_i$

```
 1: init
 2:    cr[][]                    {Sub. count ranges, by τ & nghbr}

 3: function ADVS^{SUB-ADD}(τ, φ ⊕ ⟨c_min, c_max⟩, pⱼ)  {Advice for new}
 4:    cr[τ][pⱼ] ← ⟨c_min, c_max⟩                        {subscription}
 5:    φ^{old} ← PROCD^{SUB-ADD}(τ, φ, pⱼ)
 6:    if φ^{old} = LUB(P[τ]) then
 7:       ⟨c'_min, c'_max⟩ ← Σ_{pₖ ∈ subs[τ][φ]} cr[τ][pₖ]
 8:       if ∃pₖ ≠ pⱼ | ⟨c'_min, c'_max⟩ ⋠ cr[τ][pₖ] then
 9:          SEND(UAV, τ, ⟨c'_min, c'_max⟩) to pₖ
10:    return φ^{old}

11: procedure ADVS^{SUB-SEND}(τ, φ, pⱼ)   {Advice for neighbor update}
12:    ⟨c'_min, c'_max⟩ ← Σ_{pₖ ∈ subs[τ][φ]} cr[τ][pₖ]
13:    PROCD^{SUB-SEND}(τ, φ ⊕ ⟨c'_min, c'_max⟩)

14: function ADVS^{SUB-DEL}(τ, φ, pⱼ)     {Advice for unsubscription}
15:    cr[τ][pⱼ] ← ⟨0, 0⟩
16:    φ^{old} ← PROCD^{SUB-DEL}(τ, φ, pⱼ)
17:    if φ^{old} = LUB(P[τ]) then
18:       ⟨c'_min, c'_max⟩ ← Σ_{pₖ ∈ subs[τ][φ]} cr[τ][pₖ]
19:       if ∃pₖ ≠ pⱼ | ⟨c'_min, c'_max⟩ ⋠ cr[τ][pₖ]
           then
20:          SEND(UAV, τ, ⟨c'_min, c'_max⟩) to pₖ
21:    return φ^{old}

22: upon RECV(UAV, τ, ⟨c_min, c_max⟩) from pⱼ do  {Handler for}
23:    cr[τ][pⱼ] ← ⟨c_min, c_max⟩                  {aggregation updates}
24:    ⟨c'_min, c'_max⟩ ← Σ_{pₖ ∈ subs[τ][φ]} cr[τ][pₖ]
25:    if ∃pₖ ≠ pⱼ | ⟨c'_min, c'_max⟩ ⋠ cr[τ][pₖ] then
26:       SEND(UAV, τ, ⟨c'_min, c'_max⟩) to pₖ
```

---

For our advertising scenario, we require maintenance of a subscriber count—an internal variable $cr$ (line 2), which stores subscription count ranges for subscribers (i.e., $\hat{v}_j$ for each broker $b_j$).

Subscriptions are augmented ($\oplus$) with count ranges when new subscriptions are issued. Upon reception of a subscription, the call to $\text{ADVS}^{\text{SUB-ADD}}$ (line 16 in Algorithm 3) triggers the code at line 3 in Algorithm 4. Thereafter, the default behaviour is invoked via $\text{PROCD}^{\text{SUB-ADD}}$ at line 5. Subtype polymorphism allows us to append an actual *count* tuple (representing $\overline{v}$) to any such subscription. The appending of this value is illustrated on line 13 in Algorithm 4.

Now, if the LUB has not changed, but the subscriber count—which necessarily changed—has moved outside the count range held by the broker $p_k$ (as calculated by the implementation of $f$ at line 7), we send to appropriate neighboring brokers a message to update the aggregate value (abbreviated here as UAV) at line 9 before returning the result of $\text{PROCD}^{\text{SUB-ADD}}$. In case the LUB has changed, the advice for subscriptions invoked in UPD propagates the new count range along with the new subscription. Advising of unsubscription message reception proceeds analogously, except that unsubscriptions need not be augmented with counts as the information is present at respective brokers.

If a broker receives a message requesting it to update an aggregate value (line 22), it updates the local count ranges and recursively updates the neighbor brokers' count information, if their current count ranges do not cover the new count information. As mentioned, we must also advise the sending of new subscriptions in the context of UPDs, as subscriptions must be augmented with respective counts (line 13).

## 3.5 Dynamic Aggregation Function Changes

The capability to change broker behaviour at runtime is powerful, but potentially unsafe, if the change of broker behaviour leads to inconsistent processing of events that are in transit.

While we assume the correctness of the implementation of the new behaviour and do not explore generic mechanisms for ensuring safe modification of broker behaviour, we provide an intuition of how the problem can be tackled for the specific case of distributed evaluation of aggregation functions. As already described, given a function $f$ evaluated over an aggregation tree $t$, each broker $b_i$ receives aggregate updates from its children in $t$ and forwards updates towards the root of $t$. By propagating a new aggregation function $f_{new}$ from the root of an aggregation tree, down to the leaves, we ensure that a generic broker $b_i$ has obtained $f_{new}$ before it starts to receive updates for $f_{new}$ from brokers below it. Indeed, $b_i$ can receive updates only from its children in the tree, which obtain $f_{new}$ from (and, thus, *after*) $b_i$.

While this ensures safe deployment of new, independent functions, the generic case in which functionalities are substituted or modified is more complex. It potentially requires every broker $b$ to (i) modify its internal state and adapt it to new functionalities, and (ii) support transition phases in which $b$ receives updates from its children that are generated by an old version of the communication protocol. We leave this for future work.

## 4. IMPLEMENTATION

This section describes our prototype implementation of ASIA in the REDS publish/subscribe system [10]. REDS was chosen due to its modularity, which gives us the flexibility needed to implement our joinpoint model.

## 4.1 Overview of REDS

REDS provides a framework of Java classes and defines the architecture of a generic broker through a set of components with well-defined interfaces. The current release of REDS offers concrete implementations for each component, thus limiting our implementation effort to a small number of classes. Fig. 3 shows the REDS architecture. REDS is composed of two layers: an *overlay* layer and a *routing* layer. The former manages the connections with other brokers and has mechanisms for sending and receiving packets. The latter defines the routing strategy. The routing layer communicates with the overlay layer through the Overlay component, which relies on a TopologyManager component to establish and monitor connections with other brokers, and on a Transport component to receive and send packets.

REDS packets are composed of a Subject and a Payload (any Serializable object). Developers can register different traffic classes on the Overlay: for each traffic class, the

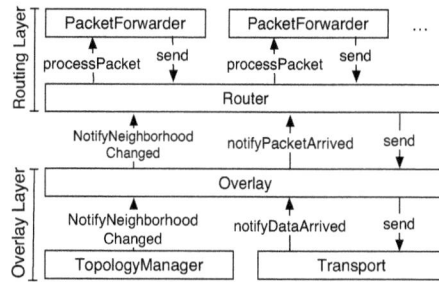

**Figure 3: Overview of REDS architecture**

Overlay creates a separate queue for storing incoming packets coming from the Transport. Developers can specify the maximum size of each queue, and how packets are associated with traffic classes based on their subjects.

The Overlay delivers packets to the Router, using the method notifyDataArrived. Based on the packet's subject, the Router selects the PacketForwarder in charge of processing it (if any). The actual routing strategy is implemented within the various PacketForwarder components installed on top of the Router. In processing an incoming packet, a PacketForwarder can: (i) modify its payload; (ii) forward it to one or more brokers; or (iii) generate and send new packets.

## 4.2 Implementing ASIA in REDS

Our implementation of ASIA in REDS requires fewer than 9,000 new lines of code, and no changes to existing code. We exploit the standard REDS Overlay layer, which adopts TCP for communication between brokers. In the routing layer, we use the standard Router component, limiting the scope of our implementation to the definition of two new PacketForwarder components: a TreeBuilder, as well as an ASIAPacketForwarder.

The TreeBuilder creates one forwarding tree on top of the existing overlay network. ASIA assumes an acyclic topology: the TreeBuilder creates it by electing a leader node $n_\ell$ and then running a protocol that creates the shortest path spanning tree rooted at $n_\ell$. The ASIAPacketForwarder is responsible for processing both packets for event propagation, and packets for aggregation. With reference to §3, the ASIAPacketForwarder implements both the protocols for offering a publish/subscribe service (e.g., protocols for advertisements, subscriptions, and publications forwarding), and the aggregators for evaluating aggregation functions.

Besides implementing a new PacketForwarder, we define new types of packets for propagating aggregation requests and updates, and modify the packets used for publish/subscribe communication so that they can piggyback aggregation data, using subtype polymorphism (see §3.2).

### 4.2.1 Aggregation Mechanisms

For each aggregation function $f$ (see §3.3), an aggregator in each broker needs to perform the following steps:

- it creates a data structure to store (i) the local view of the value of the aggregation function $\overline{v}$; (ii) the last value sent to each neighbor $k$ ($\overline{v}_k$); (iii) the maximum imprecision tolerated by each neighbor $k$ ($\hat{v}_k$);

- it starts monitoring the information that may change its local view of $\bar{v}$ (e.g., a new subscription);

- every time $\bar{v}$ changes, it checks, if it needs to send updates to its neighbors by applying $f$ and determining if the result falls beyond the boundary value for imprecision—i.e., a range check for additive aggregators;

- it determines if the updates can be piggybacked using existing messages.

Within this approach, defining and deploying a new aggregator is straightforward. Developers need to specify (i) what triggers an update in the value of the aggregation function, i.e., which specific data has to be monitored; and (ii) how to compute the value of the aggregation function. Functions currently accept and return simple Java `Objects`.

To evaluate ASIA, we implemented several aggregators, including subscriber and publisher counts, rate of subscriptions and publications (over a time-based sliding window), and active publishers (in a time-based window). We present some of the results based on these aggregators in the following section.

### 4.2.2 Runtime Distribution of Aggregation Functions

To allow for maximum flexibility, we support distributing new aggregation functions at runtime through clients supplying additional advice. Clients that want to supply a new, custom aggregation function write a corresponding class (implementing the common interface for all aggregation functions) and compile it locally. They then disseminate the compiled bytecode over the broker network. We make use of the existing publish/subscribe system for dissemination of code across all brokers.

Once a broker receives a message with a new aggregation function, it takes the bytecode from that message and uses a custom classloader to load that class at runtime. Classes are then instantiated using reflection, adding the new aggregation function to the respective joinpoints. While there is a risk of naming conflicts—two clients could use the same class name—this is no different to potential name collisions when using multiple libraries of any sorts, and is easily avoided by following Java naming conventions.

Weaving is simplified by assuming that different advice (represented by different classes) are independent. If this does not hold, programmers need to deal with the interaction explicitly knowing that different advice for a same joinpoint are executed in a non-deterministic order. Piggybacked information is unambiguously assigned to advice by conveying all such information inside a hashmap indexed by advice name rather than performing nested wrapping of messages and (possibly inconsistent) unwrapping.

## 5. EVALUATION

Our experimental evaluation has two goals:

1. We investigate the costs and benefits of introducing the ASIA model within a publish/subscribe middleware. For this purpose, we (i) compare ASIA with the REDS publish/subscribe system and measure the overhead introduced by the aggregation mechanism; (ii) compare ASIA with REDS and an external aggregation system running side by side.

2. We measure the cost of dynamically deploying new aggregate functions in the broker network at run-time.

| Parameter | Value |
|---|---|
| Number of brokers | 16 |
| Average subscriptions per client | 5 |
| Number of connections | 3 |
| Number of different topics | 500 |
| Average link latency | 0.2ms |
| Subscription/unsubscription ratio | 50% / 50% |
| Average publication rate (per client) | 1 ev/s |
| Average subscription rate (per client) | 0.1 subsc./s |
| Number of clients per broker | 100 |
| Number of clients (delay tests) | 8 |
| Average aggregation requests per client | 3 |
| Maximum imprecision | 0 |

**Table 2: Parameters used in the reference scenario**

We monitor the behavior of brokers while injecting new functions supplied by the clients.

### 5.1 Experiment Setup

For our experiments, we define a reference scenario with the parameters shown in Table 2. We consider a network of 16 brokers, each connected to 3 other brokers and serving 100 clients. Each client publishes 1 event per second on average and is subscribed to 5 topics out of 500. Since subscriptions select events based on their topics only, each client receives 1% of the published events on average. Topics of publications and subscriptions are normally distributed, and each client issues one subscription (or unsubscription) every 10 seconds.

We use 32 Intel Core i7 nodes, each with 8 cores at 3.4 GHz and 8 GiB of RAM, running Linux version 3.0.3. Each broker is deployed on a separate node. An additional 16 nodes host the clients that produce network traffic; all clients connected to a given broker are hosted on the same node. When measuring the delay of packets, we deploy all the clients on a single physical node, but connected to different brokers. This allows us to measure time in a precise way, based on the unique clock of the host machine. We also reduce the overall number of clients running concurrently to 8 to have sufficient resources to run them in parallel.

Depending on the specific test, we adopt the subscriber count or the publisher count functions. Each client requests subscriber (publisher) count for three event topics. When stressing the system, we configure it towards a maximum aggregation imprecision of zero, i.e., updates are always propagated. In §5.2.2 we explore the effect of increasing the imprecision. We use REDS with two different traffic classes, one for delivering events and one other to distribute subscriptions, aggregate updates, and new functionalities in terms of deployable code.

All the experiments presented below have been repeated five times, with different seeds for generating events and subscriptions. Each point plots the average value measured.

### 5.2 Overhead and Benefits

We evaluated the overhead of ASIA compared to a publish/subscribe-only middleware and studied the advantages of integrating the aggregation services with the publish/subscribe system.

### 5.2.1 Investigation of the Costs of Aggregation

We investigate the costs of the aggregation mechanisms introduced in ASIA. To do so, we compare it against the

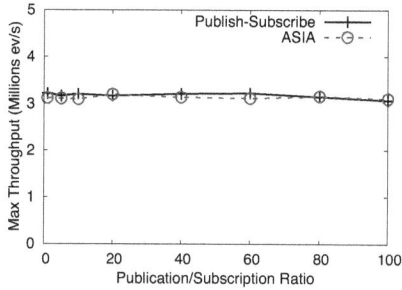

Figure 4: Throughput of ASIA compared to the REDS publish/subscribe system. Computing subscriber counts and changing the publication/subscription ratio

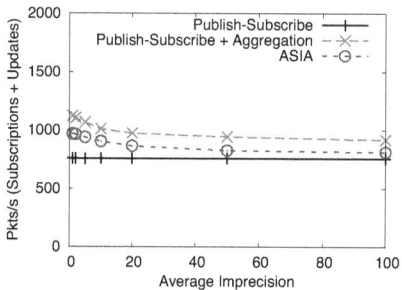

Figure 5: Traffic overhead of ASIA compared to the REDS publish/subscribe system and to an external aggregation system

REDS middleware, which only provides a publish/subscribe service with no support for aggregation. Both systems are configured to build and use the same overlay topology, and the same protocol for distributing subscriptions and events.

In this experiment, we configured all clients to publish events at their maximum rate, and we measure the overall number of events received by subscribers per second. In ASIA, we also configured publishers to ask for subscriber counts. Since this value is affected by the number of subscriptions generated by clients, we repeated the experiment while changing the number of subscriptions generated for every publication.

Figure 4 shows the maximum throughput we measured for the two systems. As the results show, the overhead of aggregate update computation and distribution is negligible: ASIA exhibits a maximum throughput that is almost identical to the unmodified publish/subscribe system. This is true even when we push the system to the extreme case in which subscriptions/unsubscriptions are generated at the same rate as publications.

### 5.2.2 Advantages of Integration

We studied the advantages of integrating the publish/subscribe and aggregation services, as proposed in the ASIA model. To do so, we compare ASIA with a solution that adopts two separate systems to provide the publish/subscribe and aggregation services, deployed side by side on the same brokers. We refer to this solution as *PS+Agg*.

Figure 5 compares the overall traffic generated by ASIA and PS+Agg when changing the maximum imprecision accepted by clients. As a baseline, we also plot the traffic gen-

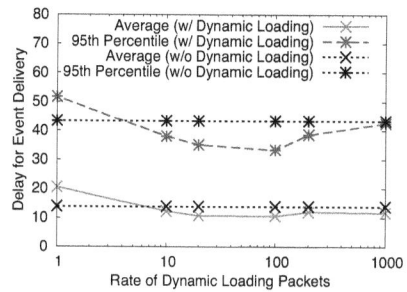

Figure 6: Delay for delivering events, changing the rate of dynamic loading packets

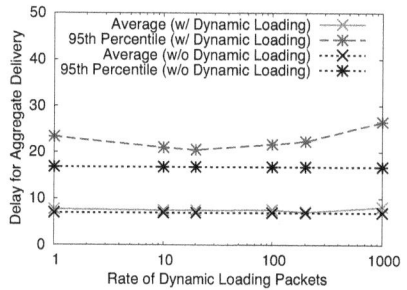

Figure 7: Delay for delivering aggregation updates, changing the rate of dynamic loading packets

erated by a traditional publish/subscribe system that does not implement any aggregation mechanism. Notice that we do not consider the traffic of events, which is identical for all the systems under analysis.

As Figure 5 shows, the overhead introduced by the aggregation mechanisms of ASIA is limited, and decreases with the maximum imprecision tolerated by clients. At the same time, Figure 5 highlights the advantages of ASIA with respect to the PS+Agg solution. By piggybacking aggregation updates, ASIA is capable of significantly reducing the network traffic.

We conducted several other experiments that we omit due to space constraints. Interested readers can refer to the ASIA project webpage,[1] where we investigate in greater detail the performance of ASIA in terms of throughput, network traffic, and delay for delivering events and aggregate updates.

Based on these results, we conclude that the ASIA aggregation model is a practical augmentation of distributed publish/subscribe middleware to provide components interacting via implicit invocations with aggregated crucial information that can be computed in a cheap, distributed manner.

## 5.3 Dynamic Function Loading

Having demonstrated the benefits of an integrated aggregation mechanism for publish/subscribe systems, we now investigate the cost of real-time deployment of new (aggregation) functions into the broker network.

We configure clients to periodically send special *DL Packets* (*Dynamic Loading Packets*), which include the Java byte-

---

[1] http://www.dvs.tu-darmstadt.de/research/events/asia/

117

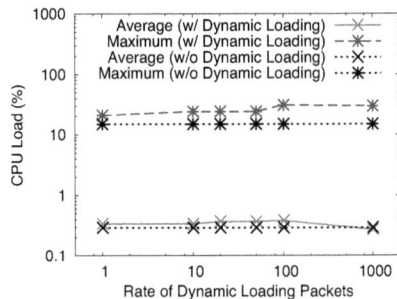

**Figure 8: CPU load of brokers, changing the rate of dynamic loading packets**

code of new functions. When a broker receives a DL Packet, it extracts the bytecode, loads the new functionality, and starts executing it, thus weaving the new code into the broker as outlined in Section 4.2.2. It then forwards the packet to other brokers through the overlay network. This happens until all brokers receive the new function. For our experiments, each new function implements some mathematical computation that is executed only once at each broker.

In realistic settings, we expect the deployment of new functions to occur rarely. To stress the system, however, we performed all our tests in the extreme situation in which new functions are deployed at high rate. We investigate the overhead of function deployment when changing the generation rate of DL Packets from 1 to 1000 per second.

Initially, we measure how the presence of dynamic loading impacts on the delay for delivering events. During our experiments, clients use ASIA both as a publish/subscribe system and as an aggregation system, to collect information about the total count of publishers.

Figure 6 shows the average delay measured by clients and its $95^{th}$ percentile. We compare the results we collect with a baseline system that does not offer dynamic loading. Interestingly, both the average delay and the $95^{th}$ percentile oscillate around the values collected from the baseline system. We do not measure any significant overhead. Moreover, increasing the rate of DL Packets does not significantly increase the measured delay.

Considering the same experiment, Figure 7 shows the delay for propagating aggregate updates. While our implementation uses a separate traffic class for events, aggregate updates and DL Packets share the same class. We thus expect a higher impact of dynamic loading. Indeed, as shown in Figure 7, the $95^{th}$ percentile is higher when dynamic loading is introduced. Moreover, it slightly increases when the rate of DL Packets becomes extremely high (over 100 DL Packets per second). Interestingly, however, even when dynamic loading occurs at high rate, the impact on the average delay remains negligible.

As a final experiment, to better understand the computational effort of dynamic loading, we monitored the CPU load of the machines hosting the brokers. The results are shown in Figure 8. Independently from the rate at which DL Packets are generated, the average CPU load remains almost constant and well below 1%. For this reason, we also plot the maximum load. In all cases, it never reaches 100%; moreover, when considering dynamic loading, this value only increases marginally.

# 6. RELATED WORK

*Implicit Invocations.*

Several programming languages support implicit invocations inherently through asynchronous "event" methods or similar constructs. Examples include ECO [21], Java$_{PS}$ [13], EventJava [14], Ptolemy [38], and EScala [19]. None of these provide any other features than *propagation* of implicit invocations; focusing on centralized deployments, Ptolemy or EScala could easily provide information on the number of publishers per subscriber or vice-versa. In contrast, languages based on Actors (e.g., Erlang [12]) and inspired by the Join Calculus [18] (e.g., Polyphonic C# [2]—now integrated with C#), or combining the two (e.g., Scala Actors [22], Erlang Joins [44, 37]) focus on asynchronous invocations on single destinations, and thus do not support implicit invocations.

*Distributed Aspects.*

In general, AOP is sometimes viewed itself as a form of event-based programming which is *implicit*, as opposed to the explicit events mentioned above. Several authors have proposed extensions to AOP specifically for asynchronous distributed systems, including *event-based aspect-oriented programming* (EAOP) [11], *aspects with explicit distribution* (AWED) [34], and *distributed aspects for distributed objects* (DADO) [46]. These focus on advising one-to-one remote object invocations and thus on a different distributed programming abstraction than implicit invocations, which are specifically designed for multicast (one-to-many) interaction. Together with a design pattern, for instance based on the use of dummy proxy objects, one could certainly use any of these approaches to implement some form of implicit invocations. Their applicability for implementing the aspects presented herein is less obvious.

The specific contribution of EAOP, AWED, or DADO is to be able to advise remote joinpoints in application components. In contrast, the joinpoints introduced herein are focused on broker logic which is internal to a publish/subscribe-based application; broker networks are precisely introduced to streamline propagated information via aggregation, and thus to avoid references and dependencies between application components. Similarly, *conspects* [24, 25] focus on advising application components—in order to add contextual arguments to implicit invocations—rather than on the components constituting the infrastructure implementing the implicit invocations.

*Publish/Subscribe & Group Communication Systems.*

Centralized publish/subscribe middleware systems, such as ActiveMQ [42], are used as a foundation for distributed applications that involve many participants, potentially joining and leaving at runtime [23]. To gain scalability, distributed publish/subscribe middleware such as Padres [17] have been developed, in which publishers and subscribers are decoupled in time, space and flow [15] by a network of brokers.

Researchers have considered additional information that can be provided by publish/subscribe middleware. Behnel et al. [1] describe various metrics that are of interest in such systems. For example, ActiveMQ publishes *advisory messages* providing information about the state of the server.

Traditional *group communication* systems have a membership service that manages a list of active *members* for each group [16, 32, 40]. Group communication systems typically provide strong delivery guarantees [6] that are aligned with changes in membership views. However, in contrast to ASIA, these approaches do not scale to large system sizes.

*Distributed Aggregation.*

Efforts on distributed aggregation for scalable monitoring deal with challenges in terms of robustness, scalability, and dynamism. The spectrum of aggregated information ranges from state information [45, 33, 5] to a quantification of system stability [26, 8, 27, 41].

Astrolabe [45] is an early system for monitoring the state of distributed resources and provides summarization aggregations based on user-defined aggregation functions. Astrolabe uses a single logical aggregation tree on top of an unstructured peer-to-peer gossip protocol. The authors describe how it could support a topic-based publish/subscribe model but, in contrast to ASIA, it is unclear how aggregation would be affected by changes in the topology.

Yalagandula and Dahlin [47] extend distributed hash tables (DHTs) into a *scalable distributed information management system* (SDIMS). SDIMS performs hierarchical aggregation based on attribute types and names through aggregation functions associated with a certain attribute type. Jain et al. [27] introduce *network imprecision* (NI), a consistency metric for large-scale distributed systems. NI allows a system to quantify its stability in terms of currently (un)reachable nodes and number of updates that might have been repeatedly processed due to network failures. The STAR [26] protocol adaptively sets the precision constraints for processing aggregate queries and is used by NI. However, such generic aggregation systems cannot leverage specific properties of publish/subscribe systems, such as overlay topologies or exchanged messages. Consequently aggregation trees do not necessarily match routing trees, resulting in inefficiency and delayed adaptation to system changes.

Other proposed aggregation solutions specialize on certain goals. For example, Cheung and Jacobsen [5] propose an algorithm that probabilistically traces publication messages through replies with data aggregation in order to find the best broker for connecting a new publisher. Migliavacca and Cugola [8] provide an approach for handling replies in publish/subscribe communication. REMO [33] builds many optimized routing trees for different monitoring tasks, taking available resources into account and allowing for efficient data aggregation towards the root of the tree. Adam2 [41] estimates the distribution of a value by using a gossip-based algorithm, in which nodes exchange aggregation instances. All of these systems cannot support the application-specific aggregation requirements of generic applications.

The paradigm of *stream processing* (SP) bears certain resemblances with aggregation as described herein. While SP systems also make use of distributed overlay networks, they pursue a different goal: every node performs a different operation, and the end result is not inherently multicast.

*Reflective Middleware.*

Middleware systems have long been designed with reflective features in mind. *Object Request Brokers* (ORBs) have been a particular early target of such design. ORBs are analogous to our event brokers, except that they focus on remotely accessible objects and explicit invocations thereon whereas the latter implement implicit invocations. Examples of reflective ORBs include the DynamicTao [31] and Open ORB [3] systems. Similar to more recent and generic work on adaptive middleware and composable systems relying on reflection features [7], these approaches aim at improving the internal design and other non-functional characteristics (e.g., adaptivity, extensibility, and QoS) of a software system. Preservation of external interfaces is a declared goal. In contrast the goal of ASIA is the addition of features for individual applications or families of applications, thus extending the external interfaces of a system.

# 7. CONCLUSIONS

Publish/subscribe systems for implicit invocations enable decoupling among the communication components in a distributed system. In practice, components often require some information on others, and rely on ad-hoc middleware services to obtain such information. Progressing from this premise, we introduced ASIA, a model for augmenting a distributed publish/subscribe system with integrated mechanisms to collect aggregated information.

Since different applications may require different information about interacting components, our solution does not provide a predefined set of aggregated functions. Rather, it provides domain-specific joinpoints which developers can program advice for that compute application-specific aggregation functions.

In terms of future work, we are investigating a more formal definition of the conditions required for safe modification of broker behavior at run-time. In addition, we are working on an implementation of ASIA for the ActiveMQ [42] publish/subscribe system. Last but not least, we are considering leveraging the well-matured AspectJ [28] tool chain, by compiling ASIA advice to AspectJ advice, and exploiting existing dynamic weaving support for runtime deployment of aggregation functions.

# 8. REFERENCES

[1] S. Behnel, L. Fiege, and G. Mühl. On Quality-of-Service and Publish-Subscribe. In *ICDCSW'06*.

[2] N. Benton, L. Cardelli, and C. Fournet. Modern Concurrency Abstractions for C#. *ACM TOPLAS*, 26(5):769–804, 2004.

[3] G. S. Blair, G. Coulson, M. Clarke, and N. Parlavantzas. Performance and Integrity in the OpenORB Reflective Middleware. In *Reflection 2001*

[4] A. Carzaniga, D. Rosenblum, and A. Wolf. Design and Evaluation of a Wide Area Event Notification Service. *ACM TOCS*, 19(3):332–383, Aug. 2001.

[5] A. K. Y. Cheung and H. Jacobsen. Publisher Placement Algorithms in Content-based Publish/Subscribe. In *ICDCS'10*.

[6] G. Chockler, I. Keidar, and R. Vitenberg. Group Communication Specifications: a Comprehensive Study. *ACM CSUR*, 33(4):427–469, 2001.

[7] G. Coulson, G. Blair, P. Grace, F. Taiani, A. Joolia, K. Lee, J. Ueyama, and T. Sivaharan. A Generic Component Model for Building Systems Software. *ACM TOCS*, 26(1):1–42, Mar. 2008.

[8] G. Cugola, M. Migliavacca, and A. Monguzzi. On Adding Replies to Publish-Subscribe. In *DEBS'07*.

[9] G. Cugola, E. D. Nitto, and A. Fuggetta. The JEDI Event-Based Infrastructure and Its Application to the Development of the OPSS WFMS. *IEEE TSE*, 27(9):827–850, 2001.

[10] G. Cugola and G. P. Picco. REDS: a Reconfigurable Dispatching System. In *SEM'06*.

[11] R. Douence, P. Fradet, and M. Südholt. Composition, Reuse and Interaction Analysis of Stateful Aspects. In *AOSD'04*.

[12] Ericsson Computer Science Laboratory. *The Erlang Pogramming Language.* http://www.erlang.org.

[13] P. Eugster. Type-based Publish/Subscribe: Concepts and Experiences. *TOPLAS*, 29(1), 2007.

[14] P. Eugster and K. Jayaram. EventJava: An Extension of Java for Event Correlation. In *ECOOP'09*.

[15] P. T. Eugster, P. A. Felber, R. Guerraoui, and A.-M. Kermarrec. The Many Faces of Publish/Subscribe. *ACM CSUR*, 35:114–131, 2003.

[16] P. Felber. Lightweight Fault Tolerance in CORBA. In *DOA'01*.

[17] E. Fidler, H. Jacobsen, G. Li, and S. Mankovski. The PADRES Distributed Publish/Subscribe System. *Feature Interactions in Telecommunications and Software Systems, VIII*, 2005.

[18] C. Fournet and C. Gonthier. The Reflexive Chemical Abstract Machine and the Join Calculus. In *POPL'96*.

[19] V. Gasiunas, L. Satabin, M. Mezini, A. Núñez, and J. Noyé. EScala: Modular Event-driven Object Interactions in Scala. In *AOSD 2011*.

[20] S. Guha, A. Reznichenko, K. Tang, H. Haddadi, and P. Francis. Serving Ads from Localhost for Performance, Privacy, and Profit. In *HotNets'09*.

[21] M. Haahr, R. Meier, P. Nixon, V. Cahill, and E. Jul. Filtering and Scalability in the ECO Distributed Event Model. In *PDSE'00*.

[22] P. Haller and T. Van Cutsem. Implementing Joins using Extensible Pattern Matching. In *COORDINATION'08*.

[23] A. Hinze, K. Sachs, and A. Buchmann. Event-based Applications and Enabling Technologies. In *DEBS'09*.

[24] A. Holzer, L. Ziarek, K. Jayaram, and P. Eugster. Putting Events in Context: Aspects for Event-based Distributed Programming. In *AOSD'11*.

[25] A. Holzer, L. Ziarek, K. Jayaram, and P. Eugster. Abstracting Context in Event-based Software. In *TAOSD'12*.

[26] N. Jain, D. Kit, P. Mahajan, P. Yalagandula, M. Dahlin, and Y. Zhang. STAR: Self-tuning Aggregation for Scalable Monitoring. In *VLDB'07*.

[27] N. Jain, P. Mahajan, D. Kit, P. Yalagandula, M. Dahlin, and Y. Zhang. Network Imprecision: a New Consistency Metric for Scalable Monitoring. In *OSDI'08*.

[28] G. Kiczales, E. Hilsdale, J. Hugunin, M. Kersten, J. Palm, and W. Griswold. An Overview of AspectJ. In *ECOOP 2001*.

[29] G. Kiczales, J. Lamping, A. Mendhekar, C. Maeda, C. Videira Lopes, J.-M. Loingtier, and J. Irwin. Aspect-Oriented Programming. In *ECOOP'97*.

[30] R. R. Kompella, J. Yates, A. Greenberg, and A. C. Snoeren. IP Fault Localization via Risk Modeling. In *NSDI'05*.

[31] F. Kon, M. Román, P. Liu, J. Mao, T. Yamane, L. C. Magalhães, and R. H. Campbell. Monitoring, Security, and Dynamic Configuration with the *dynamicTAO* Reflective ORB. In *Middleware 2000*.

[32] S. Landis and S. Maffeis. Building Reliable Distributed Systems with CORBA. *Theory and Practice of Object Systems*, 3(1):31–43, 1997.

[33] S. Meng, S. Kashyap, C. Venkatramani, and L. Liu. Remo: Resource-Aware Application State Monitoring for Large-scale Distributed Systems. In *ICDCS'09*.

[34] L. Navarro, M. Südholt, W. Vanderperren, B. D. Fraine, and D. Suvée. Explicitly Distributed AOP using AWED. In *AOSD'06*.

[35] B. Oki, M. Pfluegl, A. Siegel, and D. Skeen. The Information Bus - An Architecture for Extensible Distributed Systems. In *SOSP'93*.

[36] Oracle Co. Java Message Service - Specification, version 1.1. http://www.oracle.com/technetwork/java/jms/index.html, 2008

[37] H. Plociniczak and S. Eisenbach. JErlang: Erlang with Joins. In *COORDINATION'10*.

[38] H. Rajan and G. Leavens. Ptolemy: A Language with Quantified, Typed Events. In *ECOOP'08*.

[39] S. P. Reiss. Connecting Tools Using Message Passing in the Field Environment. *IEEE Software*, 7(4):57–66, 1990.

[40] M. K. Reiter. A Secure Group Membership Protocol. *IEEE TSE*, 22(1):31–42, 1996.

[41] J. Sacha, J. Napper, C. Stratan, and G. Pierre. Adam2: Reliable Distribution Estimation in Decentralised Environments. In *ICDCS'10*.

[42] B. Snyder, D. Bosanac, and R. Davies. *ActiveMQ in Action*. Manning Publications Co., 2011.

[43] K. Sullivan and D. Notkin. Reconciling Environment Integration and Software Evolution. *ACM TOSEM*, 1(3):229–268, July 1992.

[44] M. Sulzmann, E. Lam, and P. V. Weert. Actors with Multi-headed Message Receive Patterns. In *COORDINATION'08*.

[45] R. Van Renesse, K. P. Birman, and W. Vogels. Astrolabe: A robust and Scalable Technology for Distributed System Monitoring, Management, and Data Mining. *ACM TOCS*, 21:164–206, May 2003.

[46] E. Wohlstadter, S. Jackson, and P. Devanbu. DADO: Enhancing Middleware to Support Crosscutting Features in Distributed, Heterogeneous Systems. In *ICSE'03*.

[47] P. Yalagandula and M. Dahlin. A Scalable Distributed Information Management System. In *SICOMM'04*.

# Specification and Verification of Event Detectors and Responses

Cynthia Disenfeld        Shmuel Katz
Department of Computer Science
Technion - Israel Institute of Technology
{cdisenfe,katz}@cs.technion.ac.il

## ABSTRACT

Events and aspects that respond to them can and should be defined, specified, and verified in a modular way, as an aid in understanding and guaranteeing the correctness of each on its own. However, finding the appropriate interfaces and abstractions and expressing them precisely is not an easy task. Moreover, formally verifying large models is often unfeasible for existing model-checking tools.

We present an abstraction refinement scheme to verify aspects and to define and correct both aspect and event specifications. This allows considering smaller models and learning the needed event guarantees at each step. In addition, this technique can be used to find sound abstractions to check event reachability. Moreover, the technique is applied for detecting interference in systems where there are responses to complex events and aspects may be activated within the execution of other aspects.

## Categories and Subject Descriptors

D.2.1 [**Software Engineering**]: Requirements/Specifications; D.2.4 [**Software Engineering**]: Software/Program Verification—*Correctness proofs, Model checking*

## Keywords

Aspects; Events; Specification; Verification; Composition

## 1. INTRODUCTION

In reactive systems, events occur and responses are executed when necessary. Aspect-oriented programming (AOP) [23] represents crosscutting concerns with aspects, and responses with advice defined within the aspects. Advice is applied when its pointcut designator matches a joinpoint such as a method call, an exception thrown, and others. Complex Event Processing (CEP) [15, 24], a separate programming paradigm, analyzes the necessary processing of events in order to generate new events or notify event consumers such as services or business processes. Events can be filtered, transformed, aggregated in different time/location windows in order to generate new events. While one approach focuses on

crosscutting concerns and the other on event processing, both capture main principles of reactive systems.

In AOP, pointcut designators as presented in AspectJ [22] and other aspect languages are very dependent on how the code was written and the names chosen for the different classes, methods, and fields. Moreover, if an aspect responds to some complex event, the internal fields to record needed information about the event and the processing to detect it are tangled with the response to its detection. Any small change to the underlying system may lead to the fragile pointcut problem [31], where previously captured joinpoints are no longer matched, or new unintended joinpoints are now matched. Moreover, pointcut composition is restricted to boolean operators or knowing if it is in the execution flow of another pointcut.

Some work [2, 5, 28] has considered extending pointcut expressive power to regular expressions or patterns. However pointcuts are still limited to the restricted expressive power of the language presented.

In [4], events were introduced in the context of aspect-oriented programming, distinguishing more clearly between when an aspect should be activated, and what it must do. A joinpoint is then an event occurrence, and pointcuts are low-level event detectors. More complex event detectors can be defined, by writing pieces of code that resemble advice and may trigger new events. Examples of events are: a commit request to a control version system repository without having the tests run and succeed, insufficient purchases of a certain product during a certain period, or identifying a series of messages as a potential denial of service attack. Event evaluation and detection does not have any external side-effect besides those caused by the event being triggered. This allows applying optimizations and distinguishing clearly between when an event is being detected and what the responses to detection are that may affect the entire system.

Considering events as in [4] in the context of AOP aids in reducing the gap between systems considered in this paradigm and the CEP paradigm where events are announced by several event producers, these events are then processed (filtered, transformed, etc.) and different consumers can be notified so that they can apply the appropriate response. Events can be added with the syntactic sugar presented in [4] or using coding conventions with, e.g., AspectJ, to express event detectors as usual spectative aspects (also called observers [21, 11]). We will call the use of events within aspect-oriented programming, whether explicit or using coding conventions, *Event AOP*.

Previous work has considered how to express formal specification of events [13] and aspects [16, 20]. Such specification provides a better understanding of the system, and moreover, allows applying formal verification techniques based on model checking to guarantee the correctness of each modularly or detect interference among aspects [16, 14].

However, when applying verification to a library containing both events and aspects, new practical issues arise. In particular, aspect and event specifications are difficult to write precisely, as needed for formal verification. Especially, given the complexity of a hierarchy of events, where one depends on numerous others, and considering that model checking suffers from the state explosion problem, a sound, practical, and feasible technique must be used. In this work we present an abstraction refinement technique, inspired by CEGAR [10] (Counterexample-Guided Abstraction Refinement) which allows checking smaller models, and only refining and considering bigger models when necessary. Basically, an abstraction of the model to be checked is used, and if a counterexample is found, then the concrete model is checked automatically to find whether the error is real or spurious (and due to overabstraction of the actual system). If the counterexample found represents a spurious error, the abstract model is automatically refined and with new information from the concrete version, a repeated attempt is made to verify the new abstract model relative to its specification.

We adapt this idea to verify smaller models when applying aspect verification, exploiting our knowledge of the relationship between detecting events and responding to them in aspects in order to propose generic abstractions that often are adequate for verification, and provide a basis for refinement when they are not. Using the abstraction refinement scheme to be presented, the task of writing specifications often becomes much simpler. The abstraction refinement mechanism can be used for finding appropriate abstractions to write event and aspect specifications, and/or for correcting existing specifications. Moreover, we take advantage of these ideas both to find sound abstractions for verifying event reachability (i.e. determining whether an event may be eventually detected) and to analyze possible *interference* in a library including both events and aspects. Here *interference* means that on their own events and aspects are correct, but weaving them together may cause one or more of them not to achieve their expected behavior.

Therefore, the main contributions of this work are to:

- Introduce to aspect specification the aspects' assumption about underlying events.

- Present an abstraction refinement technique for verifying a library of events and aspects. This method allows verifying smaller models in most cases, preventing the state explosion problem, thus making model checking feasible in practice for many systems.

- Aid in the difficult task of defining and correcting precise formal specifications. In particular, the necessary abstractions are detected iteratively, through interaction with the user.

- Extend the approach to detect interference now that aspects respond to complex events.

- Show that the approach provides sound abstractions, i.e., when verification succeeds, the property indeed holds in the actual system, and can also be used to check reachability of complex events.

In the next section we provide the necessary background for understanding event and aspect specification and verification. In Section 3 we present our abstraction refinement technique for aspect verification and in Section 4 the algorithm to define events and correct event specifications is presented. A combination of these two algorithms is presented in Section 5. Correctness and termination of the algorithms are presented in Section 6. Next, in Section 7 we introduce how these ideas can be used for interference detection among aspects that respond to complex events, including the case of aspects within aspects. In Section 8, the issue of event reachability is addressed, and finally in Section 9 we conclude.

## 2. BACKGROUND

### 2.1 Linear Temporal Logic

We use Linear Temporal Logic (LTL) [25] for aspect specification. Since LTL describes temporal properties that must hold along every path, several interesting properties can be expressed with this logic. Any atomic proposition $p$ is an LTL formula which expresses that in the current state $p$ must be satisfied. In this work, atomic propositions represent method calls, fields set, variable values and others. In particular, we may assume that for method calls and events (to be described in Sec. 2.2) detected there is an atomic proposition with their name indicating their occurrence.

Some of the temporal operators are:

- $G\varphi$: (Globally) at every state starting from the current one $\varphi$ holds.

- $F\varphi$: (Future or Eventually) eventually $\varphi$ will hold.

- $X\varphi$: (Next) in the next state $\varphi$ will hold. $X^k$ will be used to denote $\underbrace{X \ldots X}_{k \text{ times}}$

Past operators can be used too:

- $O\varphi$: (Once) at some state in the past $\varphi$ was satisfied.

- $Y\varphi$: (Previous) in the previous state $\varphi$ was satisfied.

- $H\varphi$: (Historically) always in the past $\varphi$ was satisfied.

### 2.2 Events

Event detectors [4] are aspect-like modules that react to lower-level events by either gathering information in locally defined fields, or processing the gathered information to trigger a detection announcement that can be used either by other event detectors, or by aspects that respond to event detection and may change the underlying system. Evaluation of event detectors can change local fields only and does not otherwise affect the underlying system besides the consequences of being detected. They can be hierarchically composed, thus obtaining complex events and abstracting from syntactic joinpoint matching.

EXAMPLE 1. *LowActivity (Fig. 1) is an event detected whenever a period of examining sales of a product ends and there have not been enough purchases. It is possible to see that this event declaration relies on RelevantPurchase, a lower-level event which exposes the purchases being considered. A possible response to this event could be an aspect that applies a discount to the price of the product, in order to encourage future sales.*

Given that events are used to define other events, and aspects respond to them, it is important to understand what the event assumes and what it guarantees. In [13] the issue of event specification and verification was addressed. Basically, the specification of an event $E$ consists of what it assumes about the underlying system, and what is guaranteed about the augmented system, that is, the underlying system with the event detection. Event specification can be expressed in LTL, PSL [1], state machines, regular expressions and

```
event LowActivity(P product, int purchases)
 int UPPER_BOUND = 100;
 Info purchaseInfo = new Info();
 after(Purchase purchase):RelevantPurchase(purchase){
   purchaseInfo.increase(purchase.product()):
 }
 when(P product):call(P.timeDone()) && target(product){
  if (purchaseInfo.count(product) <= UPPER_BOUND)
   trigger(product, purchaseInfo.count(product));
  purchaseInfo.reset(product):
 }
```

**Figure 1: `LowActivity` event**

others. In most cases, indicating exactly when an event is detected is most naturally expressed by a state machine where all the valid underlying event sequences are represented. If the event guarantee is given by a formula or a regular expression, the equivalent automaton can be obtained with well-known techniques.

EXAMPLE 2. *The guarantee of LowActivity can be given by a state machine maintaining a counter (cnt) of how many relevant purchases have occurred in the current period and a set of additional guarantees such as:*

1. $G(LowAct \Rightarrow TmDone)$ : *If LowActivity is detected then $TmDone$ must have occurred.*

2. $G(\#RelPur \leq B \Leftrightarrow (cnt \leq B))$ : *The predicate $\#RelPur \leq B$ holds if and only if the counter is lower than the upper bound.*

3. $G((TmDone \wedge \#RelPur \leq B) \Leftrightarrow LowAct)$ : *LowActivity is detected if and only if $TmDone$ has occurred and the predicate $\#RelPur \leq B$ holds.*

*Note that these properties represent abstractions and conclusions about the event implementation.*

## 2.3 Aspects

In order to apply modular verification, aspect specification also consists of an assumption and a guarantee. In particular, for the verification of an aspect $A$, it has been shown [16, 14] that the specification needs to include what is assumed about the underlying system ($A$'s *external* assumption), what is assumed about any aspect that may execute during $A$ ($A$'s *internal* assumption) and what is guaranteed about the augmented system (base system with $A$ woven).

Then, aspect verification can be applied as follows:

1. Build the *tableau* of the assumption. The *tableau* of a formula $\varphi$ is a state machine which accepts precisely every possible path satisfying $\varphi$. There exist automatic mechanisms that, given a formula $\varphi$, build the tableau.

2. Weave into the tableau of the assumption the state machine model derived from the aspect advice code. The state machine model can be obtained automatically using existing tools such as Bandera [12].

3. Verify whether the obtained model satisfies the aspect guarantee.

This technique guarantees that if the verification above succeeds, for any system satisfying the assumption of the aspect, when weaving the aspect on its own to the system, the resulting model satisfies the aspect guarantee. The main idea is that the tableau–state machine–of the assumption contains every possible path that satisfies the assumption. Therefore, if a system satisfies the aspect assumption, every path of the system is included in the tableau. And the aspect woven at every necessary place in the tableau in particular includes the paths where the aspect would be woven in any such system.

## 2.4 Interference

Verifying each aspect in a library is usually not enough to guarantee that a set of the aspects in a library satisfies its expected behavior. One aspect may affect the behavior of another aspect, for instance when both change the same variables. In previous work on interference detection [14, 16], it has been shown that checking the conditions presented below in addition to aspect verification guarantees interference-freedom:

- Every aspect $A$ is verified assuming every other aspect $B$ satisfy $A$'s internal assumptions. ($A$'s assumption about aspects that may be activated within $A$'s execution).

- Every aspect $A$ preserves the assumption and guarantee of every other aspect $B$. This proof obligation is called pairwise aspect verification. Note that as proven in [19], showing these properties for every pair of aspects is sufficient to guarantee that any subset of the aspects included in a particular system are interference-free.

- Every other aspect $B$ possibly activated within $A$ satisfies $A$'s internal assumption.

## 2.5 CEGAR

Counterexample-Guided Abstraction Refinement (CEGAR) [10] is a technique that allows verifying properties of smaller (abstract) models and refining only when necessary. All the steps involved in CEGAR–verifying an abstract model, analyzing if a counterexample is spurious, refining the abstract model–are fully automatic.

CEGAR can be combined with various techniques. For example [10] presents spuriousness checking and refinement using BDDs [6]; BLAST [3] obtains a sequence of interpolants [27] and applies lazy abstraction [17] (refining only where needed with the interpolant found–a minimal formula to prove a conjunction unsatisfiable); SATABS [9] translates the program to a boolean program and uses a SAT Solver to verify and find refinements; and MAGIC [7] represents concurrent components as labelled transition systems (LTS) and verifies and finds refinements compositionally.

Although all the related work applies CEGAR ideas, the use of a counterexample guided abstraction refinement scheme has not yet been considered for aspects. In particular, taking advantage of the knowledge about the interaction between events and aspects, a compositional technique will be presented.

## 3. VERIFYING ASPECTS RESPONDING TO COMPLEX EVENTS

The problem to address now is aspect specification and verification when aspects respond to complex events.

In most previous work it was assumed that aspects could only respond to code-level events. Considering that aspects may respond to, add, or remove complex events introduces new questions such as: How is aspect verification applied now? What do aspects need to assume about events?

Given that aspects respond to events, we could naively include all relevant event guarantees as part of the assumption of the aspect and verify aspect correctness. However, the event hierarchy–where

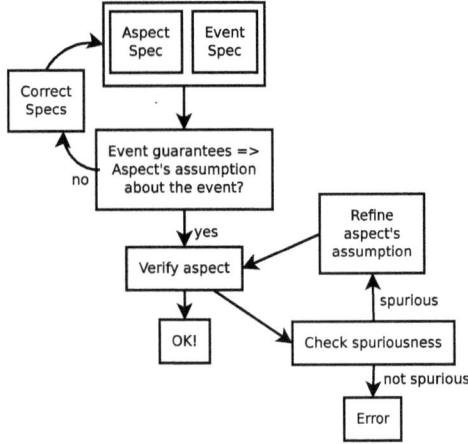

Figure 2: Aspect verification

```
aspect Discount
  after(Product p): LowActivity(p)
    applyDiscount(p);
```

Figure 3: **Discount** aspect

one event detector depends on many simpler event detectors–may include many events, and the model would turn out to be unreasonably large, making model checking unfeasible. Instead, we would like to have the aspect make only the necessary assumption about the events on which it depends–an *abstraction* of the full properties of the events, so that the needed guarantee of the aspect can still be shown.

Therefore, the aspect assumption should include only necessary assumptions about the events it relies on to show a potential guarantee. This provides better understanding of what the aspect needs for each guarantee, and if an event definition changes, the affected aspects can be easily identified.

Now, considering their assumptions about events, we present an abstraction refinement scheme to verify aspects that respond to complex events.

In this section, we assume all events have already been specified and verified with a maximal (strongest possible) guarantee. In the next section we consider the more general case where event guarantees also need to be developed in stages, along with the aspect assumptions and guarantees.

The mechanism to find a weak aspect assumption about the events needed for verification of the aspect guarantees is presented in Fig. 2.

When verification is to be applied, the first step is to check that the aspect assumption about the underlying events is correct (and thus this assumption is an *overapproximation* of the event guarantee). This is done by either deriving the assumption from the already verified event guarantees, or checking whether the guarantees of the events imply the aspect assumption.

Then, each iteration in the algorithm consists of verifying the aspect, checking if the counterexample is spurious and refining when necessary. In the next sections we will describe in more detail the steps applied at each iteration.

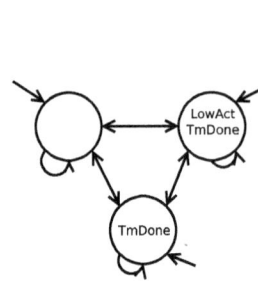

Figure 4: Assumption      Figure 5: Augmented model

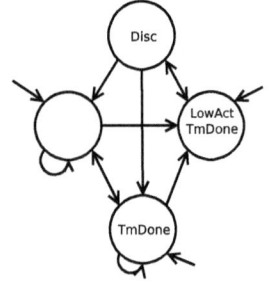

1. $LowActivity$ causes the discount to be applied at the next state:
$$G(LowAct \Rightarrow X\ Disc)$$

2. If $TmDone$ has occurred without enough purchases, the discount is applied:
$$G((TmDone \wedge \#RelPur \le B) \Rightarrow F\ Disc)$$

3. If there have not been any purchases ($Pur$), an occurrence of $TmDone$ causes the discount to be applied.
$$G((H\neg Pur) \Rightarrow (TmDone \Rightarrow F\ Disc))$$

Figure 6: Example: $Discount$ **guarantees**

## 3.1 Verifying the aspect

The aspect is verified as presented in Section 2.3. In previous work, the external aspect assumption included both general properties about the underlying system (such as liveness properties) and particular properties about the dynamic behavior given now by event detectors. In this work, we distinguish between these two kinds of assumptions and allow the aspect assumption about the underlying events to be incomplete.

Basically, the user provides the general aspect assumption and the aspect guarantee property to be checked, the specifications of relevant events, and a first version of the aspect assumption about the underlying events. This first assumption could be obtained by static code analysis [29].

EXAMPLE 3. *The Discount aspect in Fig. 3 applies a discount to the products with LowActivity. We can easily see from the code of LowActivity in Fig. 1 that if the event is detected then a call to TimeDone has occurred: $G(LowAct \Rightarrow TmDone)$. Here LowAct is the proposition indicating that LowActivity has been detected, while TmDone represents the call to TimeDone. Therefore, we can define Discount's initial assumption as $P_{Disc} \equiv G(LowAct \Rightarrow TmDone)$.*

To apply verification the tableau of the assumption is built, the aspect is woven, and the aspect guarantee is checked in the augmented model (see Sec. 2.3).

If the methodology succeeds, the aspect is correct. If the obtained model does not satisfy the property, it might be the case that the error is caused because of the overapproximation used.

EXAMPLE 4. *Continuing with the Discount aspect example, in Fig. 4 the tableau of the assumption of Discount is presented (always LowAct implies TmDone). A state not containing an atomic proposition is a state where that atomic proposition does not hold. For instance, the empty state represents that neither TmDone nor LowAct holds. In Fig. 5 the augmented model is* obtained (*whenever LowAct occurs, a Disc state is injected: the discount is applied*). *The aspect guarantees in Fig. 6 are checked against this augmented model and it can be seen that $P_{Disc}$ is enough to prove only the first guarantee.*

## 3.2 Checking spuriousness

In this section we explain in more detail which event guarantees should be considered and how to detect whether a counterexample is spurious.

Let $E$ be an event. The hierarchy tree $T_E$ of the events involved in $E$'s detection can be easily obtained. In this tree, $E_1$ is a parent of $E_2$ if $E_1$ directly uses $E_2$. Then, the events involved in the verification of an aspect $A$ are all those belonging to $T_E$ where $E$ is either an event the aspect responds to, an event appearing in the aspect guarantee, or an event possibly invoked within the aspect execution.

As explained earlier, when an aspect guarantee is not verified with a general assumption, it does not imply that the aspect is not correct. Since the current assumption about the events represents an overapproximation of the event detection, failure to verify may mean that the aspect specification needs to be refined, i.e., more information about the event detection should be added.

EXAMPLE 5. *Recall guarantee 2 of Fig. 6*

$$G((TmDone \land \#RelPur \leq B) \Rightarrow F\ Disc)$$

*expressing that whenever there is a call to $TimeDone$ without enough relevant purchases, the discount should be applied. A counterexample (1) is found when checking this guarantee since a key state–where $TmDone$ is true but $LowAct$ is not–is an initial state (because we have made no assumptions so far about the number of relevant purchases).*

$$\{TmDone \land \neg LowAct\}, \{TmDone \land \neg LowAct\}, \dots \quad (1)$$

*This counterexample contradicts that the discount should be applied when $TimeDone$ occurs and no relevant purchases have been registered in the period. The problem is that neither the relevant purchases nor their influence on the counter are represented in the aspect assumption about the event. Therefore, the counterexample represents the abstract path where $TmDone$ occurs but the event is not detected at every state since it does not know that initially the counter is 0 (no relevant purchases have been detected in the current period), and that only $RelPur$ can increase the counter.*

Now, when checking whether a counterexample $\pi$ is spurious, we must check whether $\pi$ is consistent with the guarantees of the events involved.

Verifying if a counterexample is consistent with an event guarantee is done using model checking. Without loss of generality, we can assume the event guarantee given as a state machine. The main idea is that we build a formula from the counterexample and check whether the counterexample is a path (or prefix) of the event guarantee state machine.

The counterexample presented in equation (1) is not consistent with the guarantee of $LowActivity$: in the guarantee of $LowActivity$ every path satisfying $TmDone$ at the initial state causes the event to be detected.

## 3.3 Refinement

An adaptation of known techniques can be used to find an appropriate refinement for a spurious counterexample automatically. As a first step, we can check whether there is a single formula that by adding it to the aspect assumption, the previous counterexample

can not arise in a new attempt of model checking. If there is not, since we assume the event guarantee is complete, there is for sure a subset of formulas of the state machine representation such that the counterexample is avoided. A small subset of formulas sufficient to avoid the counterexample can be found using an SMT solver (automatic satisfiability solver of fragments of first order logic). SMT solvers can compute a core of conjuncts that make a formula unsatisfiable.

Refinement is done automatically by adding this core to the assumption of the aspect, thereby eliminating the problematic behavior.

Following, we present an example where only one event is included in the refinement, and another example with more than one event involved.

EXAMPLE 6. *As shown in the previous section, a counterexample is obtained when the second guarantee of Fig. 6 is verified under the initial assumption. Given $LowActivity$'s guarantee (Example 2), our technique finds automatically that this counterexample can be avoided by adding $G((TmDone \land \#RelPur \leq B) \Leftrightarrow LowAct)$. The next iteration of aspect verification succeeds.*

EXAMPLE 7. *Recall guarantee 3 of Fig. 6*

$$G((H \neg Pur) \Rightarrow (TmDone \Rightarrow F\ Disc))$$

*expressing that if no purchases have been detected, the end of a period ($TmDone$) causes the discount to be applied. This guarantee fails to be verified with the initial aspect assumption. We can now apply CEGAR for aspects. Although adding a single formula is not enough a subset of the event guarantees–inferred from the state machine–can be added to show that the guarantee holds.*

*The information added is:*

- *($LowAct$) $LowActivity$ is detected if and only if there is a call to $TmDone$ with less than $B$ purchases.*

- *($LowAct$) Initially the counter is set to 0.*

- *($LowAct$) $TmDone$ resets the counter to 0.*

- *($LowAct$) $RelPur$ with less than $B$ relevant purchases detected in the current period increases the counter.*

- *($RelPur$) $RelPur$ detected implies a purchase $Pur$ has occurred.*

*By running the tool, we find that the behavior of $RelPur$ with more than $B$ relevant purchases does not affect this guarantee.*

## 3.4 Summary of the algorithm

Assuming that event specifications imply the initial aspect assumption about the events, CEGAR for aspects is applied. The main components involved have been explained in the previous subsections:

- The aspect is verified as in previous work but considering only a partial assumption about underlying events.

- On failure, the counterexample is checked along with the event guarantees. If the counterexample is consistent with all event guarantees then the error is real.

- Otherwise, the counterexample is spurious and an appropriate refinement is added. The appropriate refinement is such that the counterexample will not be found in a future iteration (using model checking or SMT solvers).

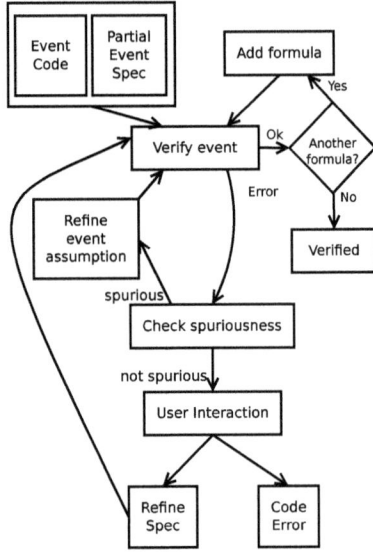

**Figure 7: Abstraction-Refinement for Events**

- The aspect is verified again, starting a new iteration.

Note that if the guarantee holds, but a partial aspect assumption about the events is considered, the technique finds the necessary refinements automatically. If the guarantee does not hold, there will be an iteration where verification will fail and the counterexample will be real providing the user an explanation of the failure.

## 4. ABSTRACTION-REFINEMENT FOR EVENTS

The next issue to consider is how a similar abstraction refinement scheme can be used as well to refine event specifications. We now assume that events have already been implemented but there is only a partial specification available.

An initial event specification may completely ignore the contents of unbounded variables, or the assumption about the underlying events and have a very general guarantee (perhaps the guarantee obtained by static code analysis).

The algorithm for refining event specifications is presented in Fig. 7. At each iteration, the user enters an LTL formula $f$ about the event behavior. If $f$ holds (i.e. $f$ is implied by the current event specification and implementation), a new iteration starts. Otherwise, the counterexample obtained may be spurious due to an over-approximation of the underlying events or due to having a partial guarantee of those underlying events. If the assumption is the one needing refinement, applying CEGAR as in the previous section may find the appropriate refinements. Otherwise, interaction with the user is required to identify if the error is real, or the code or the event guarantee needs to be changed, or the underlying events need additional guarantees.

In the next subsections we describe the algorithm in more detail.

### 4.1 Input formula

Part of the input of the algorithm is the formula to be verified. This formula has (in many cases) one of the following forms:

1. Properties describing execution paths from the initial state

   - $\text{PREFIX}_k \Rightarrow X^k \text{ EVENT}$
   - $\text{PREFIX}_k \Rightarrow X^k \neg \text{ EVENT}$

2. Properties describing subsequences within the execution (starting at any state)

   - $G(\text{SUBPATH}_k \Rightarrow X^k \text{ EVENT})$
   - $G(\text{SUBPATH}_k \Rightarrow X^k \neg \text{ EVENT})$

Notation:

- $\text{PREFIX}_k$ represents a computation path prefix until the $k^{th}$ successor state.

- $\text{SUBPATH}_k$ represents a finite subsequence within the computation such that from its start point reaches until the $k^{th}$ successor state.

Recall that event guarantees should include information of both when the event should be detected and when it should not [13] . Therefore, the above guarantees are natural ways of describing how underlying event sequences affect event detection.

EXAMPLE 8. *The following event guarantees are expressed using the patterns above:*

1. *Property* (2) *considers the paths that start without a relevant purchase, and in the next state there is a call to* $TimeDone$. *Then* $LowActivity$ *is detected.*

$$\underbrace{(\neg RelPur \wedge XTmDone)}_{PREFIX_1} \Rightarrow XLowAct \qquad (2)$$

2. *The subformula in* (3) *represents the subsequence in the computation path where there have been* $B+1$ *relevant purchases in a row followed by* $TmDone$

$$\begin{aligned} SUBPATH_{B+1} \quad = \quad & RelPur \wedge \cdots \wedge X^B RelPur \\ & \wedge X^{B+1} TmDone \qquad (3) \end{aligned}$$

*Then, Property* (4) *represents that whenever this sequence occurs, the event should not be detected.*

$$G(SUBPATH_{B+1} \Rightarrow X^{B+1} \neg LowAct) \qquad (4)$$

3. *These patterns can include any lower-level event, for example the subformula in* (5) *is very similar to the one in* (3) *but now refering to lower-level events (purchases). In* $SUBPATH\_LOWER$, *deploySite represents the moment when the site is deployed. Purchases since that state are considered relevant until the site is undeployed.*

$$SUBPATH\_LOWER_{B+1} = deploySite \wedge$$
$$(Pur \wedge \neg undeploySite) \wedge \cdots \wedge X^B(Pur \wedge \neg undeploySite)$$
$$\wedge X^{B+1} TmDone \qquad (5)$$

*Then, Property* (6) *represents that whenever this sequence occurs, the event should not be detected.*

$$G(SUBPATH\_LOWER_{B+1} \Rightarrow X^{B+1} \neg LowAct) \qquad (6)$$

*Note that the properties* (2) *and* (4) *can be verified without further refinement of the specification. However, Property* (6) *will need to be refined to include the relation between* $RelPur$, $Pur$, *and site deployment.*

```
event RelevantPurchase(Purchase)
boolean isRelevant = false;
after():deploySite(){
  isRelevant = true;
}
after():undeploySite(){
  isRelevant = false;
}
when(Purchase pur):Purchase(pur){
  if (isRelevant) trigger(pur);
}
```

**Figure 8: RelevantPurchase event**

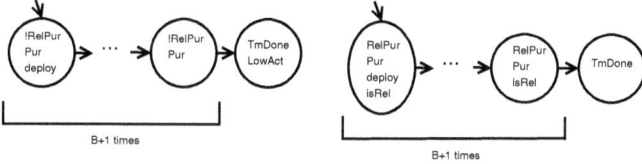

**Figure 9:** $B+1$ relevant purchases counterexample

**Figure 10:** $B + 1$ relevant purchases counterexample (with counter)

## 4.2 Automatic Refinement

Given a counterexample, the abstraction refinement scheme explained in the previous section can also be applied for event specifications. The counterexample is checked against the lower-level events. If found spurious, an appropriate refinement is added. Otherwise, the user should be involved to identify whether the counterexample is real or information is missing.

EXAMPLE 9. *Consider that LowActivity does not assume anything about lower-level events, and the formula being checked about LowActivity is "If there have not been any purchases, and $TmDone$ occurs, then there is $LowActiviy$": $G(H\neg Pur \Rightarrow (TmDone \Rightarrow LowAct))$. In this case, it is enough to add $G(RelPur \Rightarrow Pur)$ to satisfy the guarantee. This is very similar to Example 7. However, we are now verifying LowActivity, therefore the only additional information needed is the one about RelevantPurchase.*

## 4.3 User-related Refinement

Sometimes, CEGAR is not enough, and user-interaction is required to find suitable refinements.

EXAMPLE 10. *Consider again LowActivity with no assumptions about lower-level events, and the formula being checked is Property (6) from Example 8 "If there are $B + 1$ states such that there purchases considered relevant (the site has been deployed), followed by an occurrence of $TmDone$, then $LowActiviy$ should not be detected". This property is not satisfied by the current definition and specification of LowActivity: the counterexample in Fig. 9 is obtained.*

*We should now check whether the counterexample is spurious or real using the same ideas presented in Section 3.2. The added difficulty now is that underlying events may not be completely specified, therefore a counterexample may not contradict existing lower-level event guarantees even when spurious. Hence, user intervention is required.*

When the user knows the counterexample is spurious but the counterexample is not found spurious (because event specification

is incomplete), the user should refine the specification to avoid the counterexample found. This refinement can be adding the representation of a variable, or adding assumptions / guarantees. When some variable is missing, the user should provide as well an appropriate abstraction for that variable so that the obtained model is still finite. There exist several abstractions usually applied such as symbolic abstraction (a symbol represents all instances of a class), predicate abstraction (a predicate among variables), allocation sites (an abstract instance representing all objects allocated at the same place). The user may choose among them, or others. Once the necessary abstractions have been added, the necessary refinements are added as well.

EXAMPLE 11. *Given the previous counterexample, we notice that we want to detect $RelPur$ if and only if there is a purchase and the site has been deployed ($G(RelPur \Leftrightarrow (Pur \wedge isRel))$). We can add the predicate $isRel$ which holds if and only if purchases at the current state are relevant (i.e. the site has been deployed).*

*From the code of RelevantPurchase as in Fig. 8, we notice that the event relies on the lower-level events deploySite, undeploySite, and Purchase. We can define the guarantees about $isRel$ with respect to these lower-level events:*

- *If deploySite occurs then $isRel$ is set to true.*
- *If undeploySite occurs then $isRel$ is set to false.*
- *If the site is neither deployed nor undeployed $isRel$ preserves the previous state (or FALSE if it is the initial state).*

*Since we can (in general) describe a variable by its initial value and by how it is updated for different lower-level events, writing the necessary refinements for an added abstractions turns to be quite systematic.*

After the specification is refined, we should check whether the lower-level event is correct with respect to its new specification (In our example, $RelevantPurchase$ is verified with respect to its refined specification).

A new iteration starts with the refined lower-level event specifications.

EXAMPLE 12. *Once the specification of RelevantPurchase has been refined with the predicate $isRel$, applying specification of Property (6) may provide the counterexample in Fig. 10. Applying CEGAR as in the previous section, we find that the refinement should be $G(RelPur \Leftrightarrow (Pur \wedge isRel))$ with the information related to $isRel$.*

The assumptions of the technique presented can be relaxed even more by allowing that we do not have yet a complete event implementation. In this case, having the complete transition definition as presented above (as in Example 11), we can combine the previous ideas with synthesis as presented in [26] and obtain the complete definition of the event.

## 5. COMBINATION OF THE ALGORITHMS

A useful mechanism is obtained when the algorithms in Figures 2 and 7 are combined, obtaining a technique where the events are defined and corrected and the aspect is verified using the obtained event definitions. Recall that the first step in the algorithm of Fig. 2 is to check whether the event guarantees imply the aspect assumption about the events. If this assumption is not implied then it is not a proper overapproximation. Let $\varphi$ be the subformula within the aspect assumption not implied. This formula ($\varphi$) can be given as input to the algorithm for defining and correcting event definitions, and once the assumption can be shown to be an overapproximation of the event guarantees, a new attempt to verify the aspect is made.

For example, if $LowActivity$'s guarantee does not refer to the detection on $B$ relevant purchases and the aspect assumes that the event is detected in that case, the algorithm in Fig. 7 is used, the event guarantee is corrected, and the procedure for aspect verification is applied again.

On the other hand, it may occur that the event specification implies the aspect assumption about events but the aspect guarantee cannot be proven, either because the event specification is incomplete or there is an error in the code. User intervention is required to discern which of these causes the aspect guarantee not to hold. If the problem is due to the incomplete specification, the algorithm in Fig. 7 is applied with the missing event behavior as input. Then, the aspect guarantee can be checked again.

# 6. CORRECTNESS

In this section we will consider the following issues:

- Why is it sound to consider an overapproximation of the event detection?

- Do the algorithms presented in Figures 2 and 7 terminate?

- Why is it sound to check spuriousness and find the necessary refinements analyzing each event individually?

## 6.1 Soundness of the overapproximation

The aspect assumption about the underlying events usually represents an overapproximation formula $\varphi$ of the event detection. This means that every sequence of underlying events where the event is detected in any concrete model is included in the set of sequences of events where the formula $\varphi$ indicates detection, and the same occurs for every sequence of underlying events where the event is explicitly not detected by $\varphi$. In addition, the overapproximation may be satisfied by additional sequences where the event detection does not coincide with the actual behavior. That is, all the actual behaviors are included, but additional behaviors (representing the overapproximation) are also allowed.

It is possible to see that $G(LowAct \Rightarrow TmDone)$ represents an overapproximation of the event detection. Every sequence of lower-level events in the guarantee of $LowActivity$ satisfies this formula. However, additional sequences are also included such as those where the event is detected because of $TimeDone$ even if it shouldn't be.

We may ask ourselves why verifying properties using an overapproximation of event detection is sound. Following is the lemma proving that assuming the event guarantee we have so far is correct relative to the event detection code, then considering an overapproximation of the event guarantee yields sound results.

LEMMA 1. *Checking whether any LTL formula holds in the augmented model of an aspect considering an overapproximation of the event detection is sound.*

PROOF. Assuming the procedure is not sound, then there exists an LTL formula $\varphi$, and an augmented model $M'$ considering an overapproximation $Guar'_E$ of the guarantee of the event such that $\varphi$ **holds in** $M'$, but does not hold in the augmented model $M$ which detects the event exactly where it should. Because we are checking an LTL formula, there must exist a path $\pi$ in $M$ where the formula does not hold. However, given that $M'$ includes an

overapproximation of the detection of $E$, in particular it includes the path $\pi$ (possibly abstracted to the relevant variables). Then $\varphi$ **does not hold in** $M'$ as well, contradicting the assumption. □

Note that every refinement applied in the algorithm in Fig. 2 only restricts existing paths therefore, any property already proven correct continues being satisfied.

## 6.2 Termination

We can also consider the termination of the suggested refinement algorithms. The algorithm for aspect verification (Fig. 2) will at most apply a finite number of refinements since there is a finite number of events involved and a finite number of LTL formulas describing their guarantees.

The algorithm for event definition (Fig. 7) continues as long as there are new formulas the user would like to check. We assume that the user will consider only a bounded number of formulas. For each formula every step in the algorithm ends, and assuming that eventually the correct event specification is introduced by the user in step 3) the procedure halts.

## 6.3 Modularity

In this section we consider why is it sound to consider each event individually when checking spuriousness and searching for appropriate refinements. Note that this property of the method is central for preserving modularity of the verification technique.

Since events do not affect the underlying system, their detection can be assumed to be atomic. If we consider events as CSP [18] processes, the composition of a library of events is then equivalent to the alphabetized parallel composition of their CSP representation. For example, given an event $E_{lower}$ and two event definitions $E_1$ and $E_2$ such that both respond to $E_{lower}$, then both do the necessary processing of $E_{lower}$ concurrently. In the formal model and taking advantage that events do not affect the underlying system this assumption is sound. Now, we can use the distributivity of $traces$ [30] and check if a counterexample is consistent with every event in the composition. If there is at least one event such that the counterexample is not consistent (i.e, the counterexample trace does not belong to its CSP representation), then the counterexample is found spurious.

Since the specification of events is modular, and the event that causes a counterexample to be spurious can be individually identified, the refinement can be obtained from the event found.

# 7. INTERFERENCE DETECTION

As mentioned above, there is previous work on interference detection when aspects respond to low-level code events (i.e, traditional joinpoints), including when aspects may execute within other aspects.

The issues we address now are:

1. How to check interference because of an aspect activated within another aspect?

2. How to apply pairwise verification (assumption and guarantees must be preserved) now that aspects respond to complex events?

The difficulty arises from the use of complex events, which having a possibly more complex behavior, need to be represented consistently with the aspects' representation (e.g. same atomic propositions for the same method calls, same name for the same variables). Moreover, there is a hierarchy of events to be considered.

Both questions can be solved assisted by the abstraction refinement scheme presented. For example, we may first check the necessary conditions for interference-freedom (Section 2.4) considering only a general assumption about the events. If these checks succeed, since an overapproximation is applied, there is no interference among the actual aspects and events. Otherwise, the error must be analyzed and the model must be possibly refined.

In the next sections we refer in more detail to how to answer these two questions when complex events are involved.

## 7.1 Aspects within aspects

We first consider the question about interference due to aspects activated within aspects. In this section, we will consider two aspects $A$ and $B$ such that $B$ responds to the detection of the event $E_B$. Detecting interference when aspects can be activated within other aspects has been addressed in previous work [14]. Every aspect $A$ has an internal assumption about every other aspect $B$ that may be activated within $A$. In that work, it was easy to see if an aspect $B$ may be activated within $A$ since simple pointcuts were used. However, now aspects respond to complex events, and we may want to get more precise information about $B$ being possibly activated within $A$, that is whether $E_B$ is reachable within $A$.

We can first check whether executing the advice of $B$ could violate $A$'s internal assumption, i.e. $A$'s assumption about aspects that may be activated during $A$'s execution. If $B$ may violate $A$, it then is crucial to check whether $E_B$ could in fact be detected within $A$ since it is irrelevant whether $B$ satisfies $A$'s internal assumption if $E_B$ is unreachable within $A$. It should be taken into account that the guarantee of $E_B$ and $A$ may work on different abstractions, therefore the ideas presented in Section 4 to define/correct events are convenient in this case to detect sound abstractions. Once all the necessary abstractions have been found and the event guarantee built, we check whether $E_B$ is reachable within $A$. If so, then there is potential interference between $A$ and $B$.

Thus, the steps are:

**Internal Assumption Preserved** Verify whether $B$ may interfere with $A$ considering a general assumption about the event detection. This step consists of checking whether $B$ satisfies $A$'s internal assumption.

**Adequate Abstractions** If $B$ may interfere with $A$, obtain the guarantee of the event being abstracted consistently with the aspect definition, i.e. same atomic propositions representing the same methods, same abstractions for variables of the same type, etc. Here, we can apply the algorithm of Fig. 7 to aid in finding appropriate abstractions.

**Event Reachable Within Aspect** Check whether $E_B$ is reachable within $A$ by checking whether $G \neg (inAspect \wedge E_B)$ holds in the augmented model including the current assumption about the underlying events. If the formula holds, then the event is not reachable within the aspect (there is no state in the aspect where the event is detected). Otherwise, there is potential interference, the assumption should be refined and verification applied again.

The abstraction refinement scheme considers now the path where the event is reachable, and checks spuriousness and finds refinements as in Sections 3.2 and 3.3. This scheme is applied until $E_B$ is shown reachable with a path consistent with all the events involved, or $E_B$ is shown unreachable with the current assumption about the events.

```
aspect Auth
before(): doTrans()
  Usr u = requestUsr();
  Pwd p = requestPwd();
  authed = isAllowed?(u, p);


event NotBothEntered
boolean usrSet = false;
boolean pwdSet = false;
after(): requestUsr()
  usrSet = true;
after(): requestPwd()
  pwdSet = true;
when(): isAllowed?(Usr, Pwd)
  if (!usrSet || !pwdSet) trigger();
  usrSet = false;
  pwdSet = false;


aspect ThrowNotBothEntered
before(): NotBothEntered()
  throw new Exception("Both usr and pwd must be entered")
```

**Figure 11: Interference example**

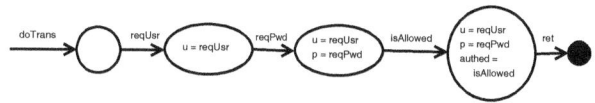

**Figure 12:** $Auth$ **aspect**

If $B$ does not satisfy $A$'s internal assumption, and $E_B$ is reachable within $A$, there may be interference. Otherwise, $B$ is guaranteed not to interfere with $A$'s internal assumption.

EXAMPLE 13. *We illustrate these ideas with the aspects presented in Fig. 11.*

*Auth authenticates transactions in a website. Before any transaction, the aspect requests a user $u$ and a password $p$ to check whether the user-password is allowed to perform the transaction. The variable* authed *keeps authentication information. The event* $NotBothEntered$ *checks whether the method* isAllowed? *is being called without having entered both a user and password. When* $NotBothEntered$ *is detected,* $ThrowNotBothEntered$ *throws an exception.*

*In the example, Auth has an internal assumption that every aspect that executes within Auth returns without throwing exceptions, and does not call to the methods* requestUsr, requestPwd, *or* isAllowed? *(to guarantee the correctness of Auth).*

*It is possible to see that if* $ThrowNotBothEntered$ *was activated within Auth, the internal assumption would not be satisfied (an exception is thrown). However, the event* $NotBothEntered$ *is*

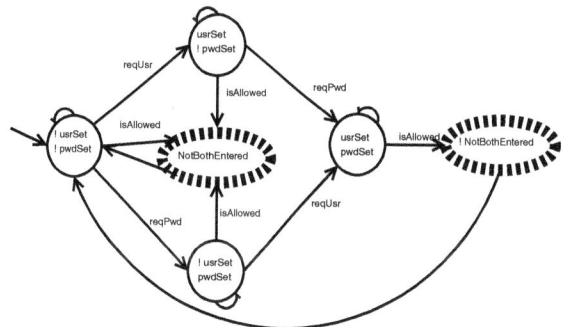

**Figure 13:** $NotBothEntered$ **event**

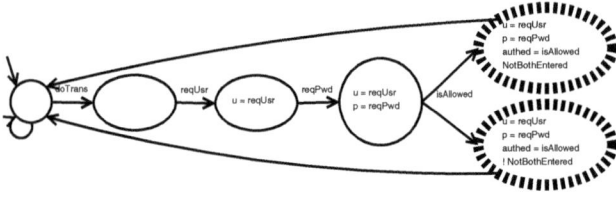

**Figure 14:** *Auth* **restricted to** $G(NotBothEntered \Rightarrow isAllowed)$

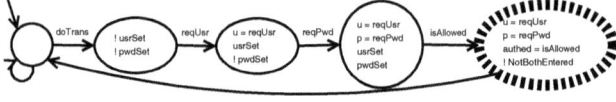

**Figure 15:** *Auth* **restricted to** $NotBothEntered$

*not detected within Auth and therefore ThrowNotBothEntered does not execute within Auth.*

*Applying the steps explained above we observe the following:*

**Internal Assumption Preserved** *Since an exception might be thrown during the execution of ThrowNotBothEntered, Auth's internal assumption (no exception is thrown) is not satisfied.*

**Adequate Abstractions** *Given the model of Auth (Fig. 12), the guarantee of NotBothEntered should be found such that consistent abstractions/atomic propositions are used. The event guarantee is presented in Fig. 13. Note that the atomic propositions $\{reqUsr, reqPwd, isAllowed\}$ belong to Auth's model and to NotBothEntered's guarantee.*

**Event Reachable Within Aspect** *If Auth's initial assumption about the event NotBothEntered is $G(NotBothEntered \Rightarrow isAllowed)$, Auth's augmented model including this assumption is presented in Fig.14. The only state where the event could be detected is where isAllowed is called. This overapproximation still causes NotBothEntered to be reachable, but the path where the event is reachable is spurious: the values of usrSet and pwdSet of the event are missing. When refining (adding the values of these variables), the model in Fig. 15 is obtained. Since both the user and the password are set, the event is not detected. Therefore, ThrowNotBothEntered is not reachable within Auth.*

The steps presented need not be executed in the order above. An alternative order is first checking whether the event might be detected within $A$ (applying steps **Adequate Abstractions** and **Event Reachable Within Aspect**) and only in case it might, checking whether $B$ interferes with $A$. Which order to apply usually depends on how hard it is to prove a property. If the internal assumption can be easily checked then the first order presented should be used. Otherwise, if the event detector $E_B$ is very simple then the second option may make verification easier.

*Observations.*

In step **Adequate Abstractions**, to find a precise enough overapproximation of $E_B$, the event refinement scheme (see Sec. 4) can be applied as before.

In step **Event Reachable Within Aspect** the abstraction refinement scheme for aspects (see Sec. 3) can be applied to find appropriate refinements.

Soundness of the method: the internal assumption only adds paths. Then, if the event is shown unreachable for an overapproximation of the event detection, then it is in particular unreachable for the event detection.

If the event is shown reachable for a path consistent with all the events, then there will be no refinement obtained from the events that will cause the path to be avoided.

## 7.2 Pairwise interference verification

We will now consider interference detected while applying pairwise verification. We recall from the definition of events [4] that events do not have any side-effects besides those of being triggered. We may think that events do not interfere at all since they do not affect the system besides being detected. However, the specification of other events and aspects may influence their detection. For example, some aspect may assume that under certain conditions an event is never detected (to prevent another aspect from executing), or that certain lower-level events are detected in a certain order. Hence, in order to check whether an event may interfere with another event or aspect $A$, we should first apply check (7).

$$VarsDetection(E) \cap Vars(\varphi_A) \stackrel{?}{=} \emptyset \qquad (7)$$

In the equation above, $VarsDetection(E)$ returns the set containing all the atomic propositions representing the event detection and parameters. Any other variable is not considered as it is known to be unmodified during the event evaluation. $\varphi_A$ in $Vars(\varphi_A)$ represents either the assumption or the guarantee of $A$. That is, we check whether the atomic propositions of the event detection appear in the assumption or the guarantee of $A$. If the intersection in (7) is empty then the event is guaranteed not to interfere with the aspect, otherwise non-interference between the event and its dependencies, and the aspect should be proven.

Considering an initial overapproximation about the events including in its alphabet (at least) all the atomic propositions of the intersection $VarsDetection(E) \cap Vars(\varphi_A)$, the technique presented can be applied to refine the event and check for interference. If there is no interference considering an overapproximation, then there will not be interference considering the actual event detection. If there is interference, the procedure considers the problematic path as a counterexample, analyzes its spuriousness and find an appropriate refinement if necessary. This methodology is sound for the same reasons as in Sec. 6.1.

Considering pairwise verification between two aspects that respond to complex events, when weaving each aspect an overapproximation of the event detection can be used and refined only when necessary. CEGAR as in Section 3 is applied until the aspects are shown interference-free, or real interference (consistent with all event detectors) is found.

EXAMPLE 14. *Consider Auth as in Fig. 11 and Extr an aspect that under particular conditions (such as a payment to be done) extracts money from the current user's account. Money extraction is considered as part of a transaction, therefore the user should be authenticated before the transaction is done.*

*Extr may respond to a complex event $E_{extr}$ which in lower-level event terms is related to the user having been authenticated, and then some sort of payment being made. If initially nothing is assumed about $E_{extr}$ pairwise verification will indicate interference: There could be a transaction (Extr) applied such that the user has not yet be authenticated. However, when the assumption about the event is refined and we know that $E_{extr}$ is applied only when the user has already been authenticated, interference-freedom can be proven.*

```
event >200Purchases(P product,int purchases)
 int LOWER_BOUND = 200;
 Info purchaseInfo = new Info();
 after(Purchase purchase):RelevantPurchase(purchase)
  purchaseInfo.increase(purchase.product());

 when(P product): call(P.timeDone()) && target(product)
  if (purchaseInfo.count(product) > LOWER_BOUND)
   trigger(product,purchaseInfo.count(product));

 after(P product): call(P.timeDone()) && target(product)
  purchaseInfo.reset(product);

aspect SpecialDiscount
 after(P product): LowActivity(p,int) && >200Purchases(p,int)
  applySpecialDiscount(p);
```

**Figure 16:** $LowActivity \land\ >200Purchases$

# 8. REACHABILITY OF COMPLEX EVENTS

Finally, we consider the issue of *complex event reachability*. This is relevant for all the previous sections since we can always have an aspect that responds to the detection of an event and we want to see if that event is eventually detected.

Given the whole hierarchy of events it may be hard to see whether there are no contradictions. The problem is by itself undecidable but in some cases certain checks can be applied.

As a first step, we must check that there are not any contradictions among the assumptions of the events, that is, the tableau considering both assumptions is not empty. As a second step, we check that the guarantees do not yield an empty model, and that the events are reachable.

One possibility is to benefit from event modularity and check whether the lower-level events are reachable at some path i.e. checking $G\neg$event or $G\neg$(event sequence) yields a counterexample. For example, we first verify that $TimeDone$ is reachable with less than UPPER_BOUND relevant purchases and assuming that we check whether $LowActivity$ may be reachable.

Another possibility is to check reachability of the conjunction of a set of event detectors (i.e. $E_1 \land \cdots \land E_n$). In several situations event detectors are expressed as the conjunction of lower-level events, thus it may help checking whether the conjunction may be reachable.

As occurred in previous sections, it is important that the abstractions used to represent the different guarantees involved are consistent among the events being considered. Then, all the guarantees of the events in the conjunction are considered together and the property $\varphi = G\neg(E_1 \land \cdots \land E_n)$. If $\varphi$ is satisfied, the conjunction of the events is unreachable.

In the example of Fig. 16 the aspect $SpecialDiscount$ applies a special discount when $LowActivity$ is detected together with at least 200 purchases. We would like to check whether the aspect might be applied at some place, that is, whether the conjunction of the events $LowActivity$ and $>200Purchases$ is reachable.

Note that both events should have the same atomic propositions to represent lower-level events ($RelPur$ and $TmDone$). In order to obtain consistent abstractions for the guarantee of the events involved in the conjunction, the algorithm in Fig. 7 is used again. The only difference is that the abstractions should be applied simultaneously to both events to guarantee that they are consistent.

Then, it is possible to see that equation (8) is satisfied (and since no assumption was made, it is in particular valid).

$$G\neg\ (LowActivity \land\ >200Purchases) \qquad (8)$$

This equation expresses that the two events never occur together.

Since the property is satisfied, we learn that the aspect $SpecialDiscount$ will never be applied. The user should then decide how to handle this situation.

This method guarantees that under sound abstractions, if shown unreachable, then the combination of events is in fact unreachable for any computation.

# 9. CONCLUSIONS

Finding the right specification for events and aspects is well-known to be a difficult task. Nevertheless, providing a specification yields better modular understanding, and avoids the need of reading the code and understanding every internal field and advice every time the event detector or aspect is to be reused. Moreover, the specification, being an abstraction, is less susceptible to change, while the code suffers from more variability (such as applying new naming standards, added information, or others).

The first algorithm allows automatically identifying a weak assumption of an aspect about events. This aids in understanding how a change in the event definition affects the aspects in the system. Moreover, using event detection overapproximations allows applying model checking over smaller models, making it more feasible in practice.

The refinement is done automatically to eliminate spurious counterexamples, and therefore we obtain only a necessary (weak) assumption about the event. Different aspects may assume different things about events. For example, there may be aspects that do not care about the events' exposed information.

As mentioned before, capturing a weak assumption of an aspect about events is particularly useful for understanding how events affect aspects or which aspects would be affected if some change is made to an event.

The second algorithm relaxes the condition of events being completely specified and aids in finding appropriate abstractions that allow building and completing the event specification.

Then, the two algorithms can be combined to obtain a more sophisticated technique where both the events are corrected and the aspect is verified (with user-intervention where necessary).

These ideas are not only relevant for events as defined in [4], but also for code-level events and other pointcut language extensions such as tracematches [2], pointcuts for distributed aspects [28], and others. Moreover, these ideas are relevant for interference analysis and checking event reachability.

Therefore, the algorithms not only provide aspect verification, but present an iterative method based on CEGAR (counterexample guided abstraction refinement) for verifying smaller models, and allowing defining and correcting event specifications.

A partial implementation of the ideas presented above has been achieved, making use of some already existing tools and applying the necessary processing as explained above. The model checker used is NuSMV [8] and modular verification of aspects is achieved using MAVEN [16]. The tool provides the connection between aspect verification and the event guarantee in order to apply refinement when necessary using the information obtained from the model checker.

The main advantage of the tool is that the technical details are hidden from the user, whose only role is to take care of the user input as explained in the sections above. The technical issues the tool takes care on its own include applying verification, determining whether a counterexample is spurious, obtaining automatically the formula representing the counterexample, and refining the assumptions by adding the formula for spurious counterexamples.

As the tool is further developed, applying verification to aspects and defining events should become a much easier task, obtaining a verified library of events and aspects, where every event and aspect is specified modularly.

By taking advantage of the interaction among events and aspects a compositional technique has been presented where the behavior of the events is initially abstracted and only refined when necessary. We have shown how these ideas are useful for aspect and event verification, interference detection and event reachability. By taking advantage of event modular specification, spuriousness checking and refinement finding are done modularly. Since the necessary assumptions learnt for verifying a property are only dependent on the property being checked, applying this technique allows leaving aspect specification simple.

Although other work on compositional CEGAR [10, 7] can be viewed as refining assumptions about different parts of a parallel composition, in our case we just refine the aspect assumption and only then weave the aspect. That is, we find the necessary assumption the underlying system must satisfy such that the augmented system will satisfy the guarantee. This allows preserving aspect and event modularity in the verification methodology.

## 10. REFERENCES

[1] IEEE standard for property specification language (PSL). IEEE Std 1850-2005.

[2] Chris Allan, Pavel Avgustinov, Aske Simon Christensen, Laurie Hendren, Sascha Kuzins, Ondřej Lhoták, Oege de Moor, Damien Sereni, Ganesh Sittampalam, and Julian Tibble. Adding trace matching with free variables to AspectJ. *SIGPLAN Not.*, 40, 2005.

[3] Dirk Beyer, Thomas A. Henzinger, Ranjit Jhala, and Rupak Majumdar. The software model checker blast: Applications to software engineering. *Int. J. Softw. Tools Technol. Transf.*, 9(5), 2007.

[4] Christoph Bockisch, Somayeh Malakuti, Mehmet Akşit, and Shmuel Katz. Making aspects natural: events and composition. In *Proc. of the Tenth Intl. Conf. on Aspect-Oriented Software Development*, 2011.

[5] Eric Bodden and Volker Stolz. Tracechecks: Defining semantic interfaces with temporal logic. In *Software Composition*, 2006.

[6] Randal E. Bryant. Graph-based algorithms for boolean function manipulation. *IEEE Trans. Comput.*, 35(8), 1986.

[7] S. Chaki, E. Clarke, A. Groce, S. Jha, and H. Veith. Modular verification of software components in C. In *Software Engineering, 2003. Proc. 25th Intl. Conf. on*, 2003.

[8] Alessandro Cimatti, Edmund M. Clarke, Enrico Giunchiglia, Fausto Giunchiglia, Marco Pistore, Marco Roveri, Roberto Sebastiani, and Armando Tacchella. NuSMV 2: An opensource tool for symbolic model checking. In *Proc. of the 14th Intl. Conf. on Computer Aided Verification*. Springer-Verlag, 2002.

[9] Edmund Clarke, Daniel Kroening, Natasha Sharygina, and Karen Yorav. Satabs: Sat-based predicate abstraction for ansi-c. In *In TACAS, volume 3440 of LNCS*. Springer, 2005.

[10] Edmund M. Clarke, Orna Grumberg, Somesh Jha, Yuan Lu, and Helmut Veith. Counterexample-guided abstraction refinement. In *Proc. of the 12th Intl. Conf. on Computer Aided Verification*. Springer-Verlag, 2000.

[11] Curtis Clifton and Gary T. Leavens. Observers and assistants: A proposal for modular aspect-oriented reasoning. In *Proc. of the 10th Intl. Workshop on Foundations of Aspect-Oriented Languages*, 2002.

[12] James C. Corbett, Matthew B. Dwyer, John Hatcliff, Shawn Laubach, Corina S. Păsăreanu, Robby, and Hongjun Zheng. Bandera: extracting finite-state models from Java source code. In *Proc. of the 22nd Intl. Conf. on Software Engineering*, 2000.

[13] Cynthia Disenfeld and Shmuel Katz. Compositional verification of events and observers: (summary). In *Proc. of the 10th Intl. Workshop on Foundations of Aspect-Oriented Languages*, 2011.

[14] Cynthia Disenfeld and Shmuel Katz. A closer look at aspect interference and cooperation. In *Proc. of the 11th Annual Intl. Conf. on Aspect-Oriented Software Development*, 2012.

[15] Opher Etzion and Peter Niblett. *Event Processing in Action*. Manning Press, 2010.

[16] Max Goldman, Emilia Katz, and Shmuel Katz. MAVEN: modular aspect verification and interference analysis. *Form. Methods Syst. Des.*, 37, November 2010.

[17] Thomas A. Henzinger, Ranjit Jhala, Rupak Majumdar, and Grégoire Sutre. Lazy abstraction. POPL '02. ACM, 2002.

[18] C. A. R. Hoare. Communicating sequential processes. *Commun. ACM*, 21(8), 1978.

[19] Emilia Katz and Shmuel Katz. Incremental analysis of interference among aspects. In *Proc. of the 7th Workshop on Foundations of Aspect-Oriented Languages*, 2008.

[20] Emilia Katz and Shmuel Katz. Modular verification of strongly invasive aspects: summary. In *Proc. of the 2009 Workshop on Foundations of Aspect-Oriented Languages*, 2009.

[21] Shmuel Katz. Aspect categories and classes of temporal properties. *T. Aspect-Oriented Software Development I*, 2006.

[22] Gregor Kiczales, Erik Hilsdale, Jim Hugunin, Mik Kersten, Jeffrey Palm, and William G. Griswold. An overview of AspectJ. In *ECOOP*, 2001.

[23] Gregor Kiczales and Mira Mezini. Aspect-oriented programming and modular reasoning. In *Proc. of the 27th Intl. Conf. on Software Engineering*, 2005.

[24] David C. Luckham. *The Power of Events: An Introduction to Complex Event Processing in Distributed Enterprise Systems*. Addison-Wesley Longman Publishing Co., Inc., 2001.

[25] Zohar Manna and Amir Pnueli. *The temporal logic of reactive and concurrent systems*. Springer-Verlag, 1992.

[26] Shahar Maoz and Yaniv Sa'ar. AspectLTL: an aspect language for LTL specifications. In *Proc. of the Tenth Intl. Conf. on Aspect-Oriented Software Development*, 2011.

[27] K. L. Mcmillan. Interpolation and SAT-Based Model Checking Financial Cryptography. volume 2742 of *Lecture Notes in Computer Science*. Springer, 2003.

[28] Luis Daniel Benavides Navarro, Mario Südholt, Wim Vanderperren, Bruno De Fraine, and Davy Suvée. Explicitly distributed AOP using AWED. In *Proc. of the 5th Intl. Conf. on Aspect-Oriented Software Development*, 2006.

[29] Flemming Nielson, Hanne R. Nielson, and Chris Hankin. *Principles of Program Analysis*. Springer-Verlag, Secaucus, NJ, USA, 1999.

[30] A. W. Roscoe, C. A. R. Hoare, and Richard Bird. *The Theory and Practice of Concurrency*. Prentice Hall PTR, Upper Saddle River, NJ, USA, 1997.

[31] M. Störzer and C. Koppen. Pcdiff: Attacking the fragile pointcut problem. Technical report, 2004.

# Past Expression: Encapsulating Pre-states at Post-conditions by Means of AOP

Jooyong Yi[†]      Robby[‡]      Xianghua Deng[‡,*]      Abhik Roychoudhury[†]

[†]School of Computing, National University of Singapore      [‡]Kansas State University
[†]{jooyong,abhik}@comp.nus.edu.sg, [‡]{robby,deng}@ksu.edu

## ABSTRACT

Providing a pair of pre and post-condition for a method or a procedure is a typical way of program specification. When specifying a post-condition, it is often necessary to compare the post-state value of a variable with its pre-state value. To access a pre-sate value at a post-condition, most contract languages such as Eiffel and JML provide an old expression; old(x) returns a pre-state value of variable x. However, old expressions pose several problems, most notably the lack of encapsulation; old(x) does not encapsulate an object graph rooted from the pre-state value of x. Thus, method-call expressions like x.equals(old(x)) should generally not be used, and instead each field of x should be compared individually as in x.f1==old(x.f1) && x.f2==old(x.f2). In this paper, we first describe this lack of encapsulation and other problems of old expressions in more detail. Then, to address those problems, we propose our novel past expression along with its formal semantics. We also describe how our past expression can be supported during runtime assertion checking. We explain the involved problems, and show how we solve them. We implement our solution by means of AOP where we exploit various primitive pointcuts including our custom branch pointcut.

## Categories and Subject Descriptors

D.3.1 [**Programming Languages**]: Formal Definitions and Theory—*Semantics, Syntax*; D.2.4 [**Software Engineering**]: Software/Program Verification—*Programming by contract*

## Keywords

Contract; Encapsulation; Old Expression; Past Expression; Branch Pointcut; Runtime Assertion Checking (RAC)

## 1  INTRODUCTION

Modern programming languages and tools are not only able to construct and run a program, they can also specify

---

*This author moved to Google Inc.

and check what programmers really intend to do with a program. One promising way of doing such program-level specification is to use general assertions such as pre-conditions and post-conditions. Programming languages supporting such general assertions include Eiffel [14], JML [5], Spec# [3] and SPARK [2], to name but a few. Those languages are often called design-by-contract languages (contract languages in short). Those contract languages extend base languages, such as Java, C# and a subset of Ada, with specification-related features such as pre-conditions, post-conditions, and specification-purpose expressions.

The specification-purpose expression that is most commonly used is an old expression. An old expression is used at a post-condition to compare a post-state value (i.e., the value at the end of a method or a procedure) with a pre-state value (i.e., the value at the entry of a method or a procedure). More specifically, an old expression, old($E$), returns a pre-state value of expression $E$. In fact, an old expression is the only means of retrieving pre-state values in most contract languages.

However, old expressions of contract languages pose several problems, most notably the lack of encapsulation; old(x) does not encapsulate an object graph rooted from the pre-state value of x. For this reason, non-destructive method-call expressions such as x.equals(old(x)) in general should not be used, and instead each field of x should be compared one by one as in x.f1==old(x.f1) && x.f2==old(x.f2). In the next section, we describe this and other problems in more detail before proposing our solution.

Our major contributions of this paper are as follows:

1. We identify three problems of old expressions that have been largely neglected, most notably the lack of encapsulation.

2. To address the identified problems, we suggest a past expression as an alternative to an old expression, and provide its formal semantics.

3. We identify the obstacles in supporting past expressions for runtime assertion checking, and show how they can be overcome through various aspect-oriented techniques.

## 2  OVERVIEW

In this section, we identify the three problems of existing old expressions of contract languages before proposing our solution by a past expression. We also compare our solution with more traditional formal specification languages such as Z [19] and VDM [9]. Lastly, we provide an overview of the obstacles in supporting our past expression in a runtime

```
public class PatientSet {
  private Patient[] patients;  private int size;
  //@ invariant (\exists int i; 0 <= i && i < size; patients[i] != null);

  //@ requires contains(p);
  //@ ensures size == \old(size) && (\forall int i; 0 <= i && i < size; (\exists int j; 0 <= j && j < size;
  //@   patients[i].name.equals(\old(patients[j].name)) && patients[i].height == \old(patients[j].height) &&
  //@   patients[i].weight == \old(patients[j].weight) && patients[i].birthDate.year == \old(patients[j].birthDate.year) &&
  //@   patients[i].birthDate.day == \old(patients[j].birthDate.day) && patients[i].birthDate.year == \old(patients[j].birthDate.year)));
  //@ also
  //@ requires !contains(p);
  //@ ensures size == \old(size)+1 && contains(p) && (\forall int i; 0 <= i && i < \old(size); (\exists int j; 0 <= j && j < \old(size);
  //@   patients[i].name.equals(\old(patients[j].name)) && patients[i].height == \old(patients[j].height) &&
  //@   patients[i].weight == \old(patients[j].weight) && patients[i].birthDate.year == \old(patients[j].birthDate.year) &&
  //@   patients[i].birthDate.day == \old(patients[j].birthDate.day) && patients[i].birthDate.year == \old(patients[j].birthDate.year)));
  public void add(/*@ non_null @*/ Patient p) { /* omitted */ }

  public /*@ pure @*/ boolean contains(Patient p) { for (int i = 0; i < size; i++) {if (patients[i].equals(p)) return true;} return false; }
  public /*@ pure @*/ boolean containsAll(/*@ non_null @*/ PatientSet set)
  { for (int i = 0; i < set.size(); i++) {if (!contain(set.get(i))) return false;} return true; }
  public /*@ pure @*/ int size() { return size; }    public /*@ pure @*/ Patient get(int i) { return patients[i]; }
}
```

(a) A **PatientSet** Java-class stub annotated with JML specifications

```
public class Patient {
  /*@ non_null @*/ String name; float height, weight;
  /*@ non_null @*/ Date birthDate;

  public boolean equals(Object o) {
    return (o instanceof Patient) && ((Patient) o).name.equals(name)
      && ((Patient) o).height == height && ((Patient) o).weight==weight
      && ((Patient) o).birthDate.equals(birthDate); }
}
```

```
public class Date {
  short year, month, day;

  public boolean equals(Object o) {
    return (o instanceof Date) &&
    ((Date) o).year == year && ((Date) o).month == month
    && ((Date) o).day == day; }
}
```

(b) Java classes **Patient** and **Date**

```
public class Set<T> {
  private T[] arr;    private int size;

  public void add(T p) { /* omitted */ }
  public boolean contains(T p) { /* omitted */ }
  public boolean containsAll(Set<T> set) { /* omitted */ }
}
```

```
public class Set {
  private ISetElem[] arr;    private int size;

  public void add(ISetElem p) { /* omitted */ }
  public boolean contains(ISetElem p) { /* omitted */ }
  public boolean containsAll(Set set) { /* omitted */ }
}
```

(c) A **Set** stub with a type variable **T**               (d) A **Set** stub with an interface **ISetElem**

Figure 1: **PatientSet** example (above the line) and generalized set examples (below the line)

assertion checker. Later sections explain how those obstacles are addressed (§ 4) and implemented through AOP (§ 5) after providing formal semantics of a past expression (§ 3). In § 6, we provide related work.

## 2.1 Three problems of an old expression

**Problem I.** To see the first problem, consider a Java class PatientSet of Figure 1(a). An instance of class PatientSet stores patient information in its array, patients. To add new patient information, PatientSet has a method add, and its specification is provided above the method. The specification says in a nutshell: (i) if array patients already contains a Patient instance equivalent to the one given through parameter p, then after exiting add, the patients array should only contain all the Patient information that existed before the add method is called, and (ii) otherwise, not only that but also the new patient information should be added to patients.

The first problem of an old expression is that it tends to cause a specification to be lengthy. Notice the long ensures clauses of method add. To compare i-th patient information, patients[i], to its pre-state value, each of the final fields reachable from patients[i] is compared to its corresponding pre-state value obtained using an old expression. These long ensures clauses are in contrast to the short requires clauses of

the same method that use method contains for the containment check.

While it is tempting to use method containsAll in the ensures clauses, using containsAll(\old(this)) will result in a misleading specification due to the semantics of an old expression that can be described informally as follows. Given an old expression, old(x), the value of x is stored in a fresh variable y before executing a method. When old(x) appears in a post-condition, say through old(x).f, old(x) is replaced by y resulting in y.f. It is important to notice that the resulting y.f is executed at the post-state because this expression is inside a post-condition. Thus, going back to the aforementioned tempting solution, i.e., containsAll(\old(this)), even if some fields of a Patient instance accessible through this are modified by method add, the call to containsAll returns, to the surprise of those who are not familiar with contract languages, true. In general, it is risky to pass to a method a reference value returned from an old expression unless the object pointed to by that reference is immutable as is the case of Java's String.

The described problem reflects the lack of encapsulation of an old expression; \old(this) does not represent the PatientSet object that existed at the time when method add is entered; it only represents the reference to that object. This is the

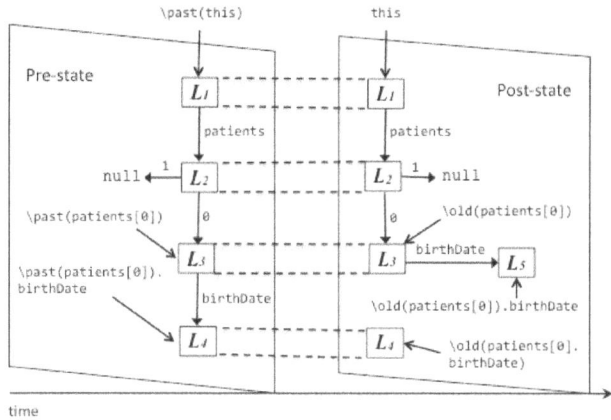

Figure 2: A possible pair of pre/post-states of the add method of class PatientSet

high-level reason why expressions like containsAll(\old(this)) cannot be used in the way one would wish. As a result, one has to end up writing a long specification such as the one in Figure 1(a) where all the fields are revealed. It is unfortunate that the built-in encapsulation capability of the underlying OOP language is lost when using an old expression.

**Problem II.** The second problem of an old expression is even more serious in terms of expressibility. To see it, consider the two styles of generalized class Set shown in the bottom part of Figure 1. Notice how different add methods are from before. The first one takes as its parameter a generic-type value, and the second one an interface-type value. For these kinds of modern programs, the previous verbose approach is not even applicable because the data structure of the passed parameter cannot be known a priori.

One possible workaround is to confine the type of a set element to a specific one. For example, method add of Figure 1(c) can be specified as follows assuming the Patient type for set elements. In the below, \typeof(p) returns the runtime type of p.

```
//@ requires \typeof(p) <: Patient && contains(p);
//@ ensures size==\old(size) && (\forall int i; 0 <= i && i < size;
//@    (\exists int j; 0 <= j && j < size;
//@    ((Patient) arr[i]).name.equals(\old(((Patient) arr[j]).name))
//@    && ((Patient) arr[i]).height==\old(arr[j].height)
//@ /* rest of them omitted */
public void add(T p) { /* omitted */ }
```

The above approach, however, not only makes the specification even lengthier (because similar specification should be given to each potential type of set elements), but it also does not match the nature of generic programs. Similarly, programs using interfaces (e.g., Figure 1(d)) suffer from the same problem. Overall, conventional old expressions are not expressive enough to handle modern programming languages.

**Problem III.** Lastly, we also point out that the old expression's copy-based semantics is not memory efficient. Each instance of an old expression takes up an additional variable. Note that one old expression can cause many instances of it if it appears in a quantified expression. Figure 1(a)'s old-expression-based specification in effect deep-clones the patients array. Such semantics of old expressions causes exponential increase in memory usage as the size of PatientSet increases as will be shown in § 5.2.

```
public class PatientSet {
  private Patient[] patients;  private int size;
  //@ invariant (\exists int i; 0 <= i && i < size; patients[i] != null);

  //@ requires contains(p);
  //@ ensures size == \past(size) && containsAll(\past(this));
  //@ also
  //@ requires !contains(p);
  //@ ensures size==\past(size)+1 && contains(p) && containsAll(\past(this));
  public void add(Patient p) { /* omitted */ }

  public /*@ pure @*/ boolean contains(Patient p) { /* omitted */ }
  public /*@ pure @*/ boolean containsAll(PatientSet set) { /* omitted */ }
}
```

Figure 3: A PatientSet stub specified with past expressions

## 2.2 Our solution with a past expression

The root cause of the identified problems is that only a single object of the pre-state is accessible from one old expression instance. In order to use an expression like patient.equals(old(patient)) with proper meaning, an object graph of the pre-state rooted from patient should be available to equals. To achieve this, we suggest in this paper an alternative to an existing old expression, i.e., a past expression. The central goal of our past expression is to provide a user means to make an access to not only a single object but also an object graph so that an expression such as patient.equals(past(patient)) can be used with proper meaning. In more general terms, we want to make encapsulated accesses to pre-state objects; one clear benefit of it is the ability to apply a non-destructive (i.e., pure [13]) method [1] to pre-state objects at post-conditions.

Before providing the formal semantics of a past expression in the next section, let us first explain the overall idea of a past expression with Figure 2. The two panes of the figure show one possible pair of pre-state and post-state of method add shown earlier. The left pane shows that before entering add, the current instance of PatientSet points to location $L_1$; its patients field points to an array-type location $L_2$; its first two array elements are location $L_3$ of type Patient and null, respectively; lastly, $L_3$ points to $L_4$ through field birthDate. Similarly, the right pane shows the post-state of method add. Most of the locations of the pre-state remain the same as depicted through a set of two parallel dashed lines. For expositional purposes, however, we assume that patients[0].birthDate is modified to a fresh location $L_5$.

Notice in the figure that \old(patients[0]).birthDate refers to $L_5$ while \past(patients[0]).birthDate refers to $L_4$. Below, we explain the reason for that difference. Since the location of patients[0] is not modified over the method execution, \old(patients[0]) returns the value of patients[0], i.e., $L_3$. The same is true for \past(patients[0]). Meanwhile, \old(patients[0]).birthDate accesses field birthDate at the post-state as explained earlier, and returns the modified field value $L_5$. In contrast, \past(patients[0]).birthDate accesses the same field at the pre-state, and returns the pre-state value $L_4$. This is because \past(patients[0]) represents not only the pre-state value of patients[0] but also the object graph rooted from patients[0].

Being equipped with a past expression, it becomes easier to compare values of the post-state with their pre-state counterparts. Compare the specification shown in Figure 3 to the original specification using old expressions; the new specifica-

---

[1] I.e., a method that always terminates and whose execution does not change the program state of its caller.

135

tion is shorter and more comprehensible; the previously used long quantified expression is now replaced with a simpler and more understandable expression, containsAll(\past(this)).

The same benefit of a past expression also goes to generic programs and interface-oriented programs. For example, the add methods of Figure 1(c) and Figure 1(d) can be specified identically to Figure 3. Without having to reveal the details of the Set class which may be unknown a priori, a specification can be written only with method calls and abstract data including those provided by past expressions.

Our past expressions can be supported in a memory efficient way. However, before moving into tool issues, let us first finish the issue of expressibility by comparing our past expression to old expressions of traditional formal specification languages.

## 2.3 Traditional formal specification languages

Contract languages such as Eiffel and JML were greatly influenced by more traditional formal specification languages such as Z [19] and VDM [9]. A number of specification expressions of contract languages, including an old expression, were originated from those formal specification languages.

For example, Figure 4 shows a PatientSet expressed in VDM notations. Special attention needs to be paid to the post clause where two implies clauses are listed; those two clauses correspond to the two pre/post-condition pairs of the previous PatientSet examples.[2] Notations size~ and patients~ refer to the pre-state values of size and patients, respectively. Notice that the patients variable is defined as a mathematical sequence (i.e., patients: seq of Patient). Thus, patients~ means the sequence of patients that existed at the pre-state, not the memory address that pointed to that sequence at the pre-state. Thus, patients~(n) refers to the n-th element of the pre-state patients sequence. In the same regard, patients = patients~ means the mathematical equality of two sequences, patients and patients~. Similarly, patients = patients~^[p] means the equality of the patients sequence and the pre-state patients sequence extended with a Patient p at its tail.

As can be seen in this example, traditional formal specification languages in general do not suffer from the lack of encapsulation seen in modern contract languages. This difference between the old expressions of modern contract languages and traditional formal specification languages stems from the the fact that the latter treats mathematical data types such as a sequence atomically as encapsulated data types. Meanwhile, programming languages like Java provide such mathematical data types in a form of classes. While those classes themselves can be nicely encapsulated, the accesses to their pre-state instances through old expressions are not encapsulated. Our past expression fills this missing gap. But of course, there is no reason to restrict the use of a past expression to only the classes for mathematical data types. Our past expression can be used with instances of any classes.

## 2.4 Tool support

Contract languages are usually shipped with tools to check specifications that are performed through either static checking or runtime assertion checking (RAC). According to the Chalin's survey of over 200 programmers in industry [6], 97% of respondents use specifications for RAC. Meanwhile, only 20% of respondents use specifications for static check-

---

[2]VDM does not allow multiple pre/post-condition pairs.

```
class PatientSet
 types
   Date:: year: nat1 month: nat1 day: nat1;
   Patient:: name: seq of char  height: real  weight: real  birthDate: Date
 instance variables
   private patients: seq of Patient;   private size: nat
 operations
   add : Patient ==> ()
   add(p) == /* body omitted */
   post (exists n : nat1 & n <= size~ and patients~(n) = p)
          => (size = size~ and patients = patients~)
        and
        (not (exists n : nat1 & n <= size~ and patients~(n) = p))
          => (size = size~ + 1 and patients = patients~ ^ [p])
end PatientSet
```
Figure 4: PatientSet in VDM notations

ing. Thus, it seems logical to conclude from that survey result that support for RAC is practically more urgent than support for static checking. In this paper, we show how past expressions can be supported for RAC of Java programs. Our approaches should be applicable to other similar languages.

In fact, it is quite challenging to support past expressions for RAC. This is because the language's standard execution semantics should be respected during RAC unless a custom execution environment is available. RAC is usually performed by instrumenting the original source code with assertion code and then running that instrumented code on a standard execution environment such as the HotSpot JVM (Java Virtual Machine) in the case of Java.

The first problem encountered when supporting RAC of Java programs is that a standard JVM does not use two separate heaps while one of the key elements of the formal semantics of a past expression is to maintain and manipulate two separate heaps independent of each other. A naive workaround would be to use cloning. However, not only is cloning resource-consuming but also its naive uses cause a serious semantic flaw. The reference to a cloned object is always different from the reference to the original object. Thus, if cloning was used, expressions such as \past(x) == x would wrongly return false even if x points to the same object before and after a method execution. Although such a flaw can be avoided by relating cloned objects with their original objects at the expense of more memory, problems still remain because not every Java class is cloneable.

There is a more challenging obstacle caused by Java's call-by-value semantics. When using an expression like x.equals (\past(x)), we need to inform the body of equals that its parameter is passed as a past expression. This is because, depending on whether the parameter of equals is past-ed at a call site, a field of that parameter should be accessed either in the pre-state or the post-state. Java's call-by-value semantics, however, makes it difficult to distinguish whether or not a parameter value is returned from a past expression.

In the latter part of this paper, we show how the above obstacles in supporting RAC can be tackled. We will also show how our solution is implemented through AOP.

## 3  FORMAL SEMANTICS

In this section, we provide formal semantics of our past expression. For efficiency of discussion, we define formal semantics on a minimal language. A past expression can be added to any contract languages that can accommodate our minimal language.

## 3.1 Programming language

We use a typical imperative procedural language shown in Figure 5(a) that can manipulate integers, booleans, and records. To manipulate records, our language has field-access expressions $E.f$ and field-update commands $E.f := E$. We omit arrays because they can be dealt with similarly to records by treating array indices as if they are record fields.

A past expression, like an old expression, can be used only in a post-condition. In our minimal language, a post-condition appears as an assert command at the end of a procedure. We assume that every procedure ends with a single assert command. We do not consider a pre-condition because a past expression cannot be used there.

An assert command takes as its argument a boolean expression $E^b$ such as $E==E$ and $E>E$. Note that $E>E$ can be used only with integer-type expressions. Record-type expressions can be compared only with ==. Although our minimal language does not allow a call to a boolean function such as equals, extension towards it is straightforward.

Only the boolean expressions used in an assert command can have past expressions $\past(E)$ as their sub-expressions. The commands of a procedure preceding its only assert command cannot use a past expression.

A past expression cannot be nested inside another past expression. For example, we disallow $\past(\past(x))$ and $\past(\past(x).f)$. This coincides with the fact that the conventional specification does not consider the pre-state of a pre-state at a post-condition.

For simplicity, procedures of our minimal language do not take a parameter. We assume that every variable of a program is declared global. We also assume that every variable is initialized to a certain value. An extension to a more sophisticated language supporting multiple scopes is straightforward.

## 3.2 Semantic rules

Figure 6 shows the operational semantic rules for the critical expressions in a big-step style. As mentioned, we restrict past expressions to be used only in the single assert command located at the end of a procedure. Such restriction is denoted with the assertion-context notation "assert ⊢" in the rules. Semantic reductions described behind this notation should occur in an assertion context, i.e., inside an assert command. We only describe the semantics under an assertion context while assuming the standard semantics under a non-assertion context.

An assertion context is introduced when a boolean expression of an assert command starts to be evaluated. And it exits when that boolean expression is finished to be evaluated. Figure 6(a) shows the rules that introduce an assertion context. Note that a past expression, $\past(E)$, is as a whole considered a base expression in an assertion context. Therefore, when encountered with assert $\past(x) > 0$, an assertion context should be introduced by $\past(x)$, not by $x$.[3]

In an assertion context, an expression is evaluated in an extended state $(\sigma_1, h_1, \sigma_2, h_2)$. Its first two components, a store $\sigma_1$ and a heap $h_1$, constitute the pre-state, and similarly, $\sigma_2$ and $h_2$ the post-state. Such meanings of those symbols are assigned through $\langle C, (\sigma_1, h_1) \rangle \Downarrow_c (\sigma_2, h_2)$ in the premises of the rules. This command-reduction relation $\Downarrow_c$

---

[3]Although the rules allow to enter an assertion context through $x$, there is no rule to interpret the remaining $\past$ afterwards.

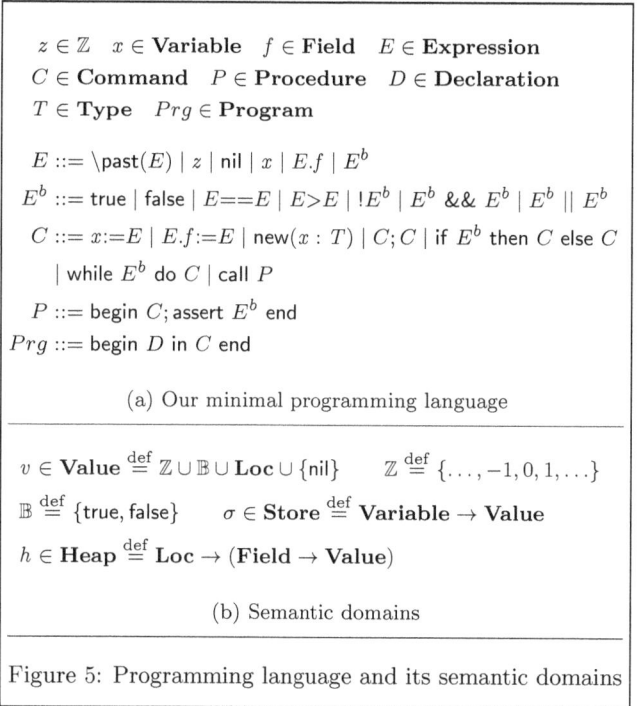

$z \in \mathbb{Z}$   $x \in$ **Variable**   $f \in$ **Field**   $E \in$ **Expression**

$C \in$ **Command**   $P \in$ **Procedure**   $D \in$ **Declaration**

$T \in$ **Type**   $Prg \in$ **Program**

$E ::= \past(E) \mid z \mid \mathsf{nil} \mid x \mid E.f \mid E^b$

$E^b ::= \mathsf{true} \mid \mathsf{false} \mid E{==}E \mid E{>}E \mid {!}E^b \mid E^b \text{ \&\& } E^b \mid E^b \mathbin{||} E^b$

$C ::= x{:=}E \mid E.f{:=}E \mid \mathsf{new}(x:T) \mid C;C \mid \mathsf{if}\ E^b\ \mathsf{then}\ C\ \mathsf{else}\ C$
$\quad \mid \mathsf{while}\ E^b\ \mathsf{do}\ C \mid \mathsf{call}\ P$

$P ::= \mathsf{begin}\ C; \mathsf{assert}\ E^b\ \mathsf{end}$

$Prg ::= \mathsf{begin}\ D\ \mathsf{in}\ C\ \mathsf{end}$

(a) Our minimal programming language

$v \in$ **Value** $\stackrel{\text{def}}{=} \mathbb{Z} \cup \mathbb{B} \cup$ **Loc** $\cup \{\mathsf{nil}\}$    $\mathbb{Z} \stackrel{\text{def}}{=} \{\dots, -1, 0, 1, \dots\}$

$\mathbb{B} \stackrel{\text{def}}{=} \{\mathsf{true}, \mathsf{false}\}$    $\sigma \in$ **Store** $\stackrel{\text{def}}{=}$ **Variable** $\to$ **Value**

$h \in$ **Heap** $\stackrel{\text{def}}{=}$ **Loc** $\to ($**Field** $\to$ **Value**$)$

(b) Semantic domains

Figure 5: Programming language and its semantic domains

between configuration $\langle C, (\sigma_1, h_1) \rangle$ and state $(\sigma_2, h_2)$ captures the fact that the program state changes from $(\sigma_1, h_1)$ to $(\sigma_2, h_2)$ by executing command $C$. Since $C$ represents the whole command preceding the sole assertion at the end of a procedure, $(\sigma_1, h_1)$ and $(\sigma_2, h_2)$ are interpreted as the pre-state and the post-state of a procedure, respectively.

Our past expressions, $\past(E)$, are evaluated using the two rules in the upper row of Figure 6(a). Its sub-expression $E$ should be evaluated with the pre-state, $(\sigma_1, h_1)$, for an obvious reason. Thus, those rules have in common an expression-reduction relation, $\langle E, (\sigma_1, h_1) \rangle \Downarrow_e v$, in their premises.

The uniqueness of a past expression is revealed in the consequent part of the left-hand-side rule of the upper row; $\past(E)$ reduces to not a value $v$ but a pair $(v, h_1)$. As will be described in detail shortly, the second component $h_1$ indicates the heap in which fields of the record represented by $v$ are accessed. Since $h_1$ is the heap of the pre-state, field accesses such as $\past(E).f$ will be made in the pre-state as desired. On the contrary, when reducing non-past base expressions $x$, the value of $x$ is paired with $h_2$, i.e., the heap of the post-state. Such semantics is described in the left-hand-side rule of the lower row of Figure 6(a).

However, not every base expression should be reduced to a pair. If a base-expression value $v$ represents not a record but an integer or a boolean value, there is no need to pair that value $v$ with a heap because no further field access from $v$ is possible. Such a semantic difference is captured in the two right-hand-side rules of Figure 6(a). In case where $v$ is nil, we still pair $v$ with a heap despite that a further field access from nil is impossible too. As will be shown shortly, this slight compromise reduces the number of necessary semantic rules for equality expressions.

As mentioned, such paired heaps are looked up when a field is accessed subsequently. The two upper-row rules of Figure 6(b) show such a usage of a paired heap. Given an expression $E.f$, the owner expression $E$ is first reduced to a pair $(v, h)$. The subsequent access to the field $f$ is made

When $P ::= \text{begin } C; \text{assert } E^b \text{ end}$,

$$\dfrac{\langle C, (\sigma_1, h_1)\rangle \Downarrow_c (\sigma_2, h_2) \qquad \langle E, (\sigma_1, h_1)\rangle \Downarrow_e v \qquad v \in \mathbf{Loc} \cup \{\text{nil}\}}{\text{assert} \vdash \langle \backslash\text{past}(E), (\sigma_1, h_1, \sigma_2, h_2)\rangle \Downarrow_e (v, h_1)} \qquad \dfrac{\langle C, (\sigma_1, h_1)\rangle \Downarrow_c (\sigma_2, h_2) \qquad \langle E, (\sigma_1, h_1)\rangle \Downarrow_e v \qquad v \in \mathbb{Z} \cup \mathbb{B}}{\text{assert} \vdash \langle \backslash\text{past}(E), (\sigma_1, h_1, \sigma_2, h_2)\rangle \Downarrow_e v}$$

$$\dfrac{\langle C, (\sigma_1, h_1)\rangle \Downarrow_c (\sigma_2, h_2) \qquad \langle x, (\sigma_2, h_2)\rangle \Downarrow_e v \qquad v \in \mathbf{Loc} \cup \{\text{nil}\}}{\text{assert} \vdash \langle x, (\sigma_1, h_1, \sigma_2, h_2)\rangle \Downarrow_e (v, h_2)} \qquad \dfrac{\langle C, (\sigma_1, h_1)\rangle \Downarrow_c (\sigma_2, h_2) \qquad \langle x, (\sigma_2, h_2)\rangle \Downarrow_e v \qquad v \in \mathbb{Z} \cup \mathbb{B}}{\text{assert} \vdash \langle x, (\sigma_1, h_1, \sigma_2, h_2)\rangle \Downarrow_e v}$$

(a) Semantic rules for base expressions (i.e., $\backslash\text{past}(E)$ and $x$) under an assertion context

---

When $S = (\sigma_1, h_1, \sigma_2, h_2)$,

$$\dfrac{\text{assert} \vdash \langle E, S\rangle \Downarrow_e (v, h) \qquad h(v)(f) = v' \qquad v' \in \mathbf{Loc} \cup \{\text{nil}\}}{\text{assert} \vdash \langle E.f, S\rangle \Downarrow_e (v', h)} \qquad \dfrac{\text{assert} \vdash \langle E, S\rangle \Downarrow_e (v, h) \qquad h(v)(f) = v' \qquad v' \in \mathbb{Z} \cup \mathbb{B}}{\text{assert} \vdash \langle E.f, S\rangle \Downarrow_e v'}$$

$$\dfrac{\text{assert} \vdash \langle E_1, S\rangle \Downarrow_e (v, h) \qquad \text{assert} \vdash \langle E_2, S\rangle \Downarrow_e (v, h')}{\text{assert} \vdash \langle E_1 == E_2, S\rangle \Downarrow_e \text{true}} \qquad \dfrac{\text{assert} \vdash \langle E_1, S\rangle \Downarrow_e (v_1, h) \qquad \text{assert} \vdash \langle E_2, S\rangle \Downarrow_e (v_2, h') \qquad v_1 \neq v_2}{\text{assert} \vdash \langle E_1 == E_2, S\rangle \Downarrow_e \text{false}}$$

(b) Semantic rules for field access expressions and equality expressions under an assertion context

---

When $P ::= \text{begin } C; \text{assert } E^b \text{ end}$,

$$\dfrac{\langle C, (\sigma_1, h_1)\rangle \Downarrow_c (\sigma_2, h_2) \qquad \langle E, (\sigma_1, h_1)\rangle \Downarrow_e v \qquad v \in \mathbf{Loc} \cup \{\text{nil}\}}{\text{assert} \vdash \langle \backslash\text{old}(E), (\sigma_1, h_1, \sigma_2, h_2)\rangle \Downarrow_e (v, h_2)} \qquad \dfrac{\langle C, (\sigma_1, h_1)\rangle \Downarrow_c (\sigma_2, h_2) \qquad \langle E, (\sigma_1, h_1)\rangle \Downarrow_e v \qquad v \in \mathbb{Z} \cup \mathbb{B}}{\text{assert} \vdash \langle \backslash\text{old}(E), (\sigma_1, h_1, \sigma_2, h_2)\rangle \Downarrow_e v}$$

(c) Semantic rules for \old expressions under an assertion context (shown for comparison with \past expressions)

Figure 6: Semantic rules for past and other critical expressions under an assertion context

using this pair. That is, the value of the field $f$ is obtained by $h(v)(f)$. Recall that the previous rules of Figure 6(a) pass the pre-state heap when evaluating a past expression, and the post-state heap when evaluating a non-past expression. By the combination of the rules in Figure 6(a) and Figure 6(b), only a field access followed by a past expression is looked up in the pre-state as desired.

Once the value of $E.f$ is obtained using a passed heap $h$, the rules for $E.f$ continuously pass the same heap $h$ by pairing it with the obtained field value. This way, subsequent field accesses can be made in the same heap. Of course, there is no need to continue to pass the heap when the obtained field value is an integer or a boolean.

Although our semantic rules reduce a field access expression $E.f$ to a pair $(v, h)$, only $v$ should be considered a value. Thus, when comparing $(v, h)$ with $(v', h')$, the actual comparison is made only between $v$ and $v'$ while ignoring $h$ and $h'$. Such semantics of comparison is described in the two lower-row rules of Figure 6(b). Recall that we earlier chose to pair nil with a heap. Thank to that choice, we do not need additional rules to handle the comparison involving nil.

Our semantic rules are sound in the sense of the following. The proof is trivial and omitted.

THEOREM 1. *Given a procedure* " $\text{begin } C; \text{assert } E^b \text{ end}$ " *and a legitimate reduction relation* $\langle C, (\sigma_1, h_1)\rangle \Downarrow_c (\sigma_2, h_2)$, *let us define an extended state* $S$ *as* $(\sigma_1, h_1, \sigma_2, h_2)$. *Then the following holds true: If* $\text{assert} \vdash \langle E^b, S\rangle \Downarrow_e \text{true}$, *then it holds that* $S \models E^b$, *which means that* $E^b$ *is true at extended state* $S$. *Also conversely, if* $\text{assert} \vdash \langle E^b, S\rangle \Downarrow_e \text{false}$, *then* $S \models \neg E^b$.

**Comparison to old expressions.** While the conventional old expression, $\backslash\text{old}(E)$, is not part of our language, we describe in Figure 6(c) its semantics for the comparison with our past expression. While the sub-expression $E$ is evaluated in the pre-state as in a past expression, its reduction result is paired with $h_2$, i.e., the heap of the post-state, unlike in a past expression, if $E$ represents a record.

# 4  SUPPORT FOR RAC

As been alluded to earlier, the semantic rules of Figure 6 cannot be naturally extended for runtime assertion checking (RAC) for Java programs. The problems are twofold. First, the conventional Java virtual machine does not provide pre-states. Second, Java's call-by-value semantics does not pass heap information used extensively in the semantic rules. That is, only $v$ instead of $(v, h)$ is passed around. Recall that the second component, $h$, indicates the heap to consult when accessing the fields of $v$. We address each of the problems by the techniques we call difference heap and proxy, respectively. We describe our solutions with the same minimal language we used before, and refine the original semantics to the one in Figure 9. It is straightforward to extend the given solutions to Java language as empirically evidenced by our prototype. We describe Java-specific issues we found interesting in the last part of this section.

## 4.1  Difference heap

Given a procedure, "$\text{begin } C; \text{assert } E^b \text{ end}$", only a small portion of the program state is likely to be modified while executing $C$. Recall that we consider the state before and after executing $C$ as the pre-state and the post-state of a given procedure, respectively. If those differences between the pre-state and the post-state are known, the pre-state can be retrieved from the post-state and can be used to evaluate past expressions. While a conventional execution environment does not make pre-states readily available, it is relatively easy to maintain difference heaps at run time by instrumenting the original code.

To maintain such state differences, we use a difference heap for each procedure call. As its name indicates, a difference heap maintains differences occurring to the heap. Note that there is no need to do the same for the store $\sigma$ because for an arbitrary variable, its pre-state value can be stored in an extra fresh variable without too much cost. Meanwhile, it is impractical to use the same technique for the heap given the complexity and the size of the heap.

Whenever a procedure is called, a fresh empty difference heap is associated with that procedure call. Being equipped with difference heaps, we reformulate a program state to a tuple of a store $\sigma$, a heap $h$, and a stack of difference heaps $\delta$. The top of the difference-heap stack is the difference heap of the current procedure. In the semantic rules, we denote a state with triple $(\sigma, h, \delta)$ for store $\sigma$, heap $h$, and the difference heap $\delta$ of the current procedure. Unlike before, we do not distinguish an assertion context, and use this triple throughout the all semantic rules. Recall that in our first formal semantics, we use a state $(\sigma_1, h_1)$ under a non-assertion context, and an extended state $(\sigma_1, h_1, \sigma_2, h_2)$ under an assertion context.

We mentioned that a fresh empty difference heap is associated with each procedure call. An empty difference heap satisfies the following property: for every location $l \in \mathbf{Loc}$ and field $f \in \mathbf{Field}$, $\delta(l)(f) = \bot$. The difference heap may be updated during the execution of $C$ given a procedure, "begin $C$; assert $E^b$ end". More specifically, whenever a field $f$ of a location $l$ that was already in use in the pre-state is modified for the first time during the execution of $C$, the original field value is stored in the difference heap of the procedure.

In our minimal language, such a field update can be made only through field update commands, $E_1.f := E_2$. The corresponding semantic rules are shown in Figure 7(a). Note that in the above semantic rules, $\langle E, (\sigma, h, \delta) \rangle$ reduces to its value $v$ regardless of difference heap $\delta$ because a past expression cannot appear in $E$. Notice in the first rule that the regular heap $h$ and the difference heap $\delta$ are updated differently. Given an owner location $v_1$ and a field $f$, $h(v_1)(f)$ is updated to an assigning value $v_2$; i.e., $h[(v_1, f) \mapsto v_2]$. Meanwhile, $\delta(v_1)(f)$ is updated to the original value of the field, $v_f$; i.e., $\delta[(v_1, f) \mapsto v_f]$.

Such an update of the difference heap can take place only if the three conditions listed in the second line of the first rule hold. These three conditions are that (i) the owner location $v_1$ of the field $f$ is already in use in the pre-state – in other words, $v_1$ is not freshly allocated during the current procedure, i.e., $v_1 \notin Fresh$ where $Fresh$ represents a set of locations allocated during the current procedure, (ii) the assigning field value $v_2$ is different from the original field value $v_f$, and (iii) the field $f$ of $v_1$ has not been updated before at this instance of procedure call. Such a first-time-update condition can be checked by looking at the value of $\delta(v_1)(f)$ before updating the difference heap. Only if $\delta(v_1)(f)$ remains undefined (i.e., $\delta(v_1)(f) = \bot$), the field update is consid-

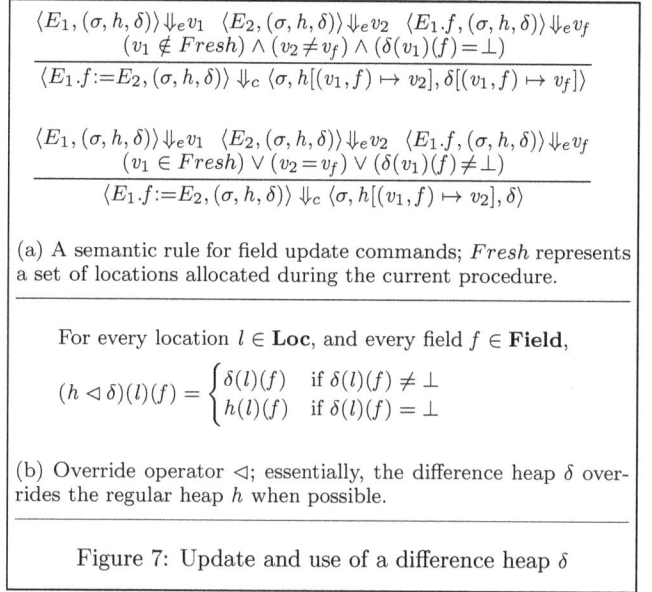

$$\frac{\langle E_1, (\sigma, h, \delta) \rangle \Downarrow_e v_1 \quad \langle E_2, (\sigma, h, \delta) \rangle \Downarrow_e v_2 \quad \langle E_1.f, (\sigma, h, \delta) \rangle \Downarrow_e v_f}{(v_1 \notin Fresh) \wedge (v_2 \neq v_f) \wedge (\delta(v_1)(f) = \bot)}{\langle E_1.f := E_2, (\sigma, h, \delta) \rangle \Downarrow_c \langle \sigma, h[(v_1, f) \mapsto v_2], \delta[(v_1, f) \mapsto v_f] \rangle}$$

$$\frac{\langle E_1, (\sigma, h, \delta) \rangle \Downarrow_e v_1 \quad \langle E_2, (\sigma, h, \delta) \rangle \Downarrow_e v_2 \quad \langle E_1.f, (\sigma, h, \delta) \rangle \Downarrow_e v_f}{(v_1 \in Fresh) \vee (v_2 = v_f) \vee (\delta(v_1)(f) \neq \bot)}{\langle E_1.f := E_2, (\sigma, h, \delta) \rangle \Downarrow_c \langle \sigma, h[(v_1, f) \mapsto v_2], \delta \rangle}$$

(a) A semantic rule for field update commands; $Fresh$ represents a set of locations allocated during the current procedure.

For every location $l \in \mathbf{Loc}$, and every field $f \in \mathbf{Field}$,

$$(h \lhd \delta)(l)(f) = \begin{cases} \delta(l)(f) & \text{if } \delta(l)(f) \neq \bot \\ h(l)(f) & \text{if } \delta(l)(f) = \bot \end{cases}$$

(b) Override operator $\lhd$; essentially, the difference heap $\delta$ overrides the regular heap $h$ when possible.

Figure 7: Update and use of a difference heap $\delta$

ered the first-time update. The third condition is necessary because we are interested only in the pre-state value, and mid-state values are unnecessary for our purposes. If one of those three conditions is not met, the difference heap $\delta$ is not modified as shown in the second rule of Figure 7(a).

When a procedure is about to exit, items in the difference heap of the exiting procedure are transferred into the difference heap of the caller. When this happens, it is unnecessary to move all items; we move only the items $(l, f, v)$[4] whose location elements $l$ are in use in the pre-state of the caller and $\delta(l)(f)$ is not $\bot$. Recall that the first condition can be checked by looking up $l$ in the $Fresh$ set of the caller. In addition to migrating difference heaps, we also move all locations contained in the $Fresh$ set of the exiting procedure into the $Fresh$ set of the caller to ignore subsequent updates made on the fields of those locations during the remaining execution of the caller.

The semantic rules of Figure 7(a) guarantee the following property. The proof is trivial and omitted.

PROPERTY 1. *Given a procedure, "begin $C$; assert $E^b$ end", suppose that for an arbitrary location $l$ and field $f$, it holds that $h_1(l)(f) = v$ before executing $C$ for the then heap $h_1$. Then, the following holds true after executing $C$. In the below, $\delta$ and $h_2$ represent, respectively, the difference heap and the regular heap existing after executing $C$.*

$$h_1(l)(f) = v \Rightarrow (\delta(l)(f) = v \vee (\delta(l)(f) = \bot \wedge h_2(l)(f) = v))$$

Now it is possible to retrieve pre-state values from the difference heap and the post-state heap. The difference heap can be used to retrieve the pre-state values when those values are overwritten while executing a procedure. If the difference heap returns $\bot$ for a location $l$ and a field $f$, that means that that value was not overwritten with a different value while executing the procedure. Thus, the post-state heap can be used to retrieve the original value. Formally, we define an override operator $\lhd$ as in Figure 7(b), and use it in the new semantic rules that will be shown shortly.

---

[4]Recall that $\mathbf{Heap} \stackrel{\text{def}}{=} \mathbf{Loc} \to (\mathbf{Field} \to \mathbf{Value})$.

## 4.2 Proxy record

While pre-state values can be retrieved using the difference heap, the difference heap alone is not enough to support past expressions at run time. It is in addition necessary to be able to decide when a pre-state value should be retrieved. For example, given a field access expression $E.f$, it is necessary to be able to distinguish the case where $E$ is \past(x) from the case where $E$ is x.

To make such a distinction, we in our first formal semantics paired a value with a heap that should be used subsequently. However, call-by-value semantics used in languages like Java generally cannot accommodate such pairing unless the language itself is modified to allow such pairs of a location and a heap as part of program values. For this reason, the previous pairing-based semantics is not suitable for runtime assertion checking that uses the execution environment of a language as it is.

To see the problem caused by call-by-value semantics, consider x.equals(\past(x)). Following the call-by-value semantics, \past(x) is first evaluated before method equals receives that evaluated value as its parameter. Thus, from the perspective of equals, it cannot distinguish whether its parameter was evaluated from a past expression or a non-past expression. This causes a problem when accessing a field through that parameter. There is no clue about in which heap a field should be accessed.

To address the above problem, we use a proxy record (a proxy in the sequel), i.e., a proxy for a non-scalar pre-state value such as a pre-state location. As the name indicates, every non-scalar pre-state value is accessed through a proxy. In the above example of x.equals(\past(x)), equals receives as its parameter a proxy for the pre-state value of x. Thus, when encountered with $E.f$, we can decide the heap in which the field should be accessed depending on whether or not the evaluation result of an owner expression $E$ is a proxy. Only if an owner expression returns a proxy, we look up a subsequent field in the pre-state heap restored using the difference heap and the post-state heap.

Let us take a concrete example. Figure 8 shows a post-state $(\sigma, h, \delta)$ for a store $\sigma$, a heap $h$, and a difference heap $\delta$ of the current method. This post-state satisfies the following three conditions: $\sigma(\text{x\_old}) = l_2$, $h(l_2)(f) = l_4$, and $\delta(l_2)(f) = l_3$. A special variable x_old has location $l_2$ as its value equivalent to the pre-state value of the original program variable x. A field f of $l_2$ points to location $l_4$ at the given post-state (i.e., $h(l_2)(f) = l_4$). However, at the pre-state, the same field pointed to another location $l_3$ as shown with the difference heap $\delta$ (i.e., $\delta(l_2)(f) = l_3$). To distinguish those two distinct field accesses between $h$ and $\delta$, notations $l_2 \xrightarrow{f}_h l_4$ and $l_2 \xrightarrow{f}_\delta l_3$ are used in the figure.

Notice in the figure that \past(x) does not directly refer to x's pre-state value $l_2$. Instead, it refers to $l_1$, the proxy for $l_2$. Every proxy record has a special field, actual, through which the "actual" value of a proxy can be accessed.

Now that \past(x) returns a proxy, we can choose the pre-state heap when accessing a subsequent field. To get the value of \past(x).f, the actual value of \past(x) is first retrieved and then the difference heap is consulted to obtain the pre-state field value $l_3$. However, $l_3$ is not directly returned. Its proxy $l_5$ is created at runtime while being linked to $l_3$ through the actual field, and returned. This way, subsequent field accesses occurring after \past(x).f can still be looked up in the pre-state heap.

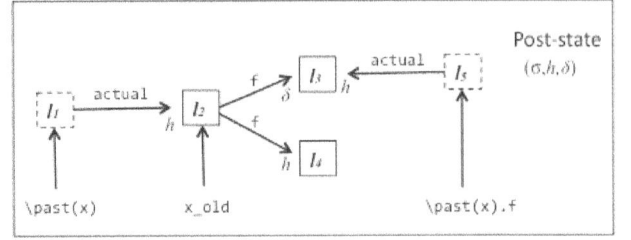

Figure 8: A post-state where the locations for proxy records are in dashed boxes; at this post state, it holds that $\sigma(\text{x\_old}) = l_2$, $h(l_2)(f) = l_4$, and $\delta(l_2)(f) = l_3$.

Non past-expressions are evaluated ordinarily without using a proxy. For example, consider the case where it holds that $\sigma(x) = l_2$ in the above figure. The evaluation result of x is directly $l_2$. Similarly, the evaluation result of x.f is $l_4$.

Note that using proxies barely increases memory usage in general. Although multiple proxies can be generated while evaluating expressions such as $\past(x).f_1.f_2.f_3$, most of them soon become garbages. Also, the fact that proxies are generated on-demand basis has a positive impact on memory maintenance.

## 4.3 Refined semantics

Now that we explained a difference heap and a proxy record, we can show in Figure 9 the refined semantics of the original one. Unlike in the original semantics, we do not distinguish an assertion context. Notice that notation assert ⊢ is not used in the refined semantics. This is because we uniformly use the same state configuration $(\sigma, h, \delta)$ across a whole program.

As shown in the left-hand-side rule of Figure 9(a), $\past(E)$ now reduces to a fresh proxy $v_p$ for the pre-state value $v$ of $E$ if $v$ is non-scalar, i.e., $v \in \mathbf{Loc} \cup \{\text{nil}\}$. We use a predicate notation "$v_p : \mathsf{Proxy}$" to denote that $v_p$ is a proxy. Notice that after applying the rule, the current heap $h_2$ is updated to reflect that $v_p$'s actual field points to a pre-state value $v$. The store and the difference heap are not modified. Meanwhile, if the pre-state value of $E$ is a scalar value $v$ (i.e., $v \in \mathbb{Z} \cup \mathbb{B}$), then $v$ is directly returned as the right-hand-side rule shows.

Figure 9(b) shows how to evaluate a field access expression $E.f$ when its owner expression $E$ reduces to a proxy. As informally explained earlier, in such cases, the actual value $v_a$ of the proxy $v_p$ is first retrieved from the current heap $h'$. See $h'(v_p)(\text{actual}) = v_a$. Then, the field value $v_f$ is obtained from the restored pre-state heap. See $(h' \lhd \delta)(v_a)(f) = v_f$. If $v_f$ is non-scalar, a fresh proxy $v'_p$ that points to $v_f$ through the actual field is returned (the left-hand-side rule). Otherwise, $v_f$ is directly returned (the right-hand-side rule). We omit to show the rules for the rest of the cases where an owner expression reduces to a non-proxy value. They are handled in a standard way.

Lastly, Figure 9(c) shows some of the rules for equality expressions $E_1{==}E_2$ while the rest of the cases can be handled similarly. As usual, we assume that $E_1$ is evaluated first before $E_2$. The rules show that if a sub-expression reduces to a proxy, then its actual value should be used for comparison.

The semantics presented in this section is a refinement of the original semantics of § 3 in the following sense. The proofs for them are not difficult and omitted for the lack of space.

When $P ::=$ begin $C$; assert $E^b$ end,

$$\frac{\langle C, (\sigma_1, h_1, \delta_1) \rangle \Downarrow_c (\sigma_2, h_2, \delta_2) \quad \langle E, (\sigma_1, h_1, \delta_1) \rangle \Downarrow_e v}{\langle \backslash \mathsf{past}(E), (\sigma_2, h_2, \delta_2) \rangle \Downarrow_e \langle v_p, (\sigma_2, h_2[(v_p, \mathsf{actual}) \mapsto v], \delta_2) \rangle} \quad \frac{\langle C, (\sigma_1, h_1, \delta_1) \rangle \Downarrow_c (\sigma_2, h_2, \delta_2) \quad \langle E, (\sigma_1, h_1, \delta_1) \rangle \Downarrow_e v}{\langle \backslash \mathsf{past}(E), (\sigma_2, h_2, \delta_2) \rangle \Downarrow_e \langle v, (\sigma_2, h_2, \delta_2) \rangle}$$

(with side conditions $v \in \mathbf{Loc} \cup \{\mathsf{nil}\}$, $v_p : \mathsf{Proxy}$, $v_p \notin \mathit{dom}\ h_2$ on the left, and $v \in \mathbb{Z} \cup \mathbb{B}$ on the right)

(a) Semantic rules for $\backslash \mathsf{past}(E)$

$$\frac{\langle E, (\sigma, h, \delta) \rangle \Downarrow_e \langle v_p, (\sigma, h', \delta) \rangle \quad v_p : \mathsf{Proxy} \quad h'(v_p)(\mathsf{actual}) = v_a \quad (h' \lhd \delta)(v_a)(f) = v_f \quad v_f \in \mathbf{Loc} \cup \{\mathsf{nil}\} \quad v_p' : \mathsf{Proxy} \quad v_p' \notin \mathit{dom}\ h'}{\langle E.f, (\sigma, h, \delta) \rangle \Downarrow_e \langle v_p', (\sigma, h'[(v_p', \mathsf{actual}) \mapsto v_f], \delta) \rangle} \quad \frac{\langle E, (\sigma, h, \delta) \rangle \Downarrow_e \langle v_p, (\sigma, h', \delta) \rangle \quad v_p : \mathsf{Proxy} \quad h'(v_p)(\mathsf{actual}) = v_a \quad (h' \lhd \delta)(v_a)(f) = v_f \quad v_f \in \mathbb{Z} \cup \mathbb{B}}{\langle E.f, (\sigma, h, \delta) \rangle \Downarrow_e \langle v_f, (\sigma, h', \delta) \rangle}$$

(b) Semantic rules for field access expressions (partial)

$$\frac{\langle E_1, (\sigma, h, \delta) \rangle \Downarrow_e \langle v_p, (\sigma, h', \delta) \rangle \quad v_p : \mathsf{Proxy} \quad h'(v_p)(\mathsf{actual}) = v_a \quad \langle E_2, (\sigma, h', \delta) \rangle \Downarrow_e \langle v_p', (\sigma, h'', \delta) \rangle \quad v_p' : \mathsf{Proxy} \quad h''(v_p')(\mathsf{actual}) = v_a}{\langle E_1 == E_2, (\sigma, h, \delta) \rangle \Downarrow_e \mathsf{true}} \quad \frac{\langle E_1, (\sigma, h, \delta) \rangle \Downarrow_e \langle v, (\sigma, h, \delta) \rangle \quad \neg(v : \mathsf{Proxy}) \quad \langle E_2, (\sigma, h, \delta) \rangle \Downarrow_e \langle v_p, (\sigma, h', \delta) \rangle \quad v_p : \mathsf{Proxy} \quad h'(v_p)(\mathsf{actual}) = v}{\langle E_1 == E_2, (\sigma, h, \delta) \rangle \Downarrow_e \mathsf{true}}$$

(c) Semantic rules for equality expressions (partial)

Figure 9: Refined semantic rules friendly to runtime assertion checking (RAC)

THEOREM 2. *Given a procedure "begin $C$; assert $E^b$ end", if one can obtain $\langle C, (\sigma_1, h_1, \delta_1) \rangle \Downarrow_c (\sigma_2, h_2, \delta_2)$ using the refined semantics, then $\langle C, (\sigma_1, h_1) \rangle \Downarrow_c (\sigma_2, h_2)$ can be obtained in the original semantics. And also subsequently, if one can obtain $\langle E^b, (\sigma_2, h_2, \delta_2) \rangle \Downarrow_e \mathsf{true}$ using the refined semantics, then assert $\vdash \langle E^b, (\sigma_1, h_1, \sigma_2, h_2) \rangle \Downarrow_e \mathsf{true}$ can be obtained in the original semantics. A similar implication also holds for the $\mathsf{false}$ case.*

## 4.4 Java-specific issues

**Arrays.** While our minimal language does not have arrays, our prototype tool supports arrays. We treat arrays similarly to records. For example, $\backslash \mathsf{past}(\mathsf{a}[0])$ returns a proxy for the pre-state value of $\mathsf{a}[0]$. Similarly, $\backslash \mathsf{past}(\mathsf{a})$ refers to a proxy array for a pre-state array $\mathsf{a}$. The length of such a proxy array is set to zero to minimize memory usage. When the length of a pre-state array is queried through a proxy array, the actual array of the given proxy array is first retrieved and its length is returned.

However, a proxy array cannot have an $\mathsf{actual}$ field unlike a proxy record. To address this issue, we maintain a global map that associates proxy arrays with their actual arrays.

**Equality with this.** Semantic rules for equality expressions shown in Figure 9(c) enforce equality between values regardless of whether they belong to the pre- or the post-state. Thus, an expression such as $\backslash \mathsf{past}(\mathsf{o}) == \mathsf{o}$ for a reference variable $\mathsf{o}$ returns true if $\mathsf{o}$ points to the same object in the pre and the post-state. Such equality across the time is usually desirable because comparison methods such as $\mathsf{equals}$ often compare between two references of some fields.

However, there is an important exception to consider. A number of Java classes have $\mathsf{equals}$ methods that in common start with the following if-statement for an efficiency reason:

```
public boolean equals(Object o) {if (this == o) return true;
```

Thus, an expression such as $\mathsf{o.equals}(\backslash \mathsf{past}(\mathsf{o}))$ would return true even if some fields of $\mathsf{o}$ are modified during method execution and the $\mathsf{equals}$ method of $\mathsf{o}$ is defined to compare those modified fields between the receiver and the parameter.

A solution for this problem is debatable. Currently, our prototype, when faced with expressions such as this $==$ o or o $==$ this, returns false if o represents a pre-state value. If this and o point to the same object, a warning is issued. Disequality expressions are handled similarly; true is returned from this != o if o represents a pre-state value, and a warning is issued if this and o point to the same object.

**Type of proxies.** Consider x.equals($\backslash \mathsf{past}$(x)) again. Typical $\mathsf{equals}$ method returns false if its parameter is not a subtype of the enclosing class $C$. Substituting a proxy for the parameter of $\mathsf{equals}$ should not deceive $\mathsf{equals}$ into returning false despite that the type of the actual value the proxy represents is a subtype of $C$. More generally speaking, substituting a proxy for its actual value should not fool the type system. To achieve that, we assign a proxy the type of its actual value. This can be done by dynamically generating a proxy class as a subtype of the runtime type of its actual value. Simple set of bytecode engineering of non-proxy classes, which includes inserting absent default constructors and removing final modifiers, is necessary to instantiate proxy classes. Such a proxy class also implements the IProxy interface to manifest itself as a proxy.

**Dynamic dispatch.** Not only x.equals($\backslash \mathsf{past}$(x)) but also $\backslash \mathsf{past}$(x).equals(x) is valid while resulting in the same result. This is because $\backslash \mathsf{past}$(x) is assigned the type as specific as the dynamic type of x.

## 5 IMPLEMENTATION THROUGH AOP

In implementing our prototype of an OpenJML [16]-based RAC system for Java programs that supports past expressions, we exploited AOP. Figure 10 shows the part of the aspect that captures our RAC solution for past expressions. In the figure, we list the critical pointcuts and advices used to implement our prototype.[5] Auxiliary parts and the part to handle the migration of difference heaps and fresh sets at method boundaries are omitted.

---

[5]Slight simplification is done for the presentation.

```
pointcut fieldWrite(Object obj): set(* *.*) && target(obj) && ... ;

pointcut fieldRead(Object obj): get(* *.*) && target(obj) && ... ;

pointcut arrElemWrite(Object[] arr, int idx):
  arrayset() && target(arr) && args(idx) && ... ;

pointcut arrElemRead(Object[] arr, int idx):
  arrayget() && target(arr) && args(idx) && ... ;

pointcut eq(Object thiz, Object o1, Object o2):
  this(thiz) && branch(Object == Object) && args(o1,o2) && ... ;

pointcut diseq(Object thiz,Object o1, Object o2):
  this(thiz) && branch(Object != Object) && args(o1,o2) && ... ;
```

(a) Pointcuts (partial); omissions are indicated by ellipses

```
// 1.updates of difference heap due to field writes
before(Object obj) : fieldWrite(obj) {
  updateDiffHeap(obj,(FieldSignature) thisJoinPoint.getSignature());
}

// 2.updates of difference heap due to array element writes
before(Object[] arr, int idx) : arrElemWrite(arr, idx)
{ updateDiffHeap(arr, idx); }

// 3.field accesses
Object around(Object obj) : fieldRead(obj) {
  if (obj instanceof IProxy) {
    try { return fieldVal(actual(obj),
          (FieldSignature) thisJoinPoint.getSignature());
    } catch (NoFieldCase e) { return proxy(proceed(actual(obj))); }
  } else { return proceed(obj); }
}

//4. array element accesses
Object around(Object[] arr,int idx): arrElemRead(arr,idx) {
  if (isProxyArray(arr)) {
    try { return arrElem(actual(arr), idx);
    } catch (NoArrayElemCase e) {
      return proxy(proceed(actual(arr), idx));
    }
  } else { return proceed(arr,idx); }
}

//5. equality evaluation
boolean around(Object thiz,Object o1,Object o2):eq(thiz,o1,o2) {
  if ((o1 == thiz && o2 instanceof IProxy) ||
      (o2 == thiz && o1 instanceof IProxy)) {
    println("Warning"); return false;
  }
  return proceed(thiz,actual(o1),actual(o2));
}

//6. disequality evaluation
boolean around(Object thiz,Object o1,Object o2):diseq(thiz,o1,o2) {
  if ((o1 == thiz && o2 instanceof IProxy) ||
      (o2 == thiz && o1 instanceof IProxy)) {
    println("Warning"); return true;
  }
  return proceed(thiz, actual(o1), actual(o2));
}
```

(b) Advices (partial)

Figure 10: An aspect to support RAC of past expressions

In defining pointcuts of Figure 10(a), various primitive pointcuts are used. In addition to some standard primitive pointcuts of AspectJ [10] such as set and get, we also used non-AspectJ pointcuts provided by *abc* [1] such as arrayset and arrayget. In addition, we extended *abc* with our custom primitive pointcut branch to accommodate our special needs.

In the sequel, we explain how each of advices of Figure 10(b) matches a specific need of RAC. The first advice updates the difference heap when a field-write takes place in the target program. The body of this advice calls updateDiffHeap with two parameters, i.e., (i) the target object of the current field update and (ii) the field information.

The second advice similarly updates the difference heap when an array-element-write takes place. To make necessary interventions, we use *abc*'s custom pointcut, arrayset, in defining the pointcut of the advice.

The third advice looks up the proper value of the field that is currently being read in either the pre-state or the post-state of the currently running method. If the target object obj is a proxy, the pre-state field value is first looked up in the difference heap by calling fieldVal along with the actual object of the proxy (obtained by actual(obj)) and the field information. This difference-heap-lookup-method fieldVal returns a proxy for the pre-state field value of the given object at its normal termination. If the field under consideration was not modified during method execution, fieldVal raises a NoFieldCase exception because the pre-state field value is not in the difference heap; instead, it is in the current heap. In this case, proceed is called to obtain the unmodified pre-state field value from the current heap, and subsequently method proxy is called to return a fresh proxy for the obtained pre-state value.

The fourth advice is a synonym of the previous one that deals with array element accesses. Instead of an object and field information, an array and an index are used. We use *abc*'s custom pointcut, arrayget, for this advice.

The last two advices handle (dis)equality expressions. To handle those expressions, we need to intervene when two values are compared to each other with operator == or !=. We, however, could not find an appropriate pointcut for our need neither in AspectJ nor *abc*, and extended *abc* with our custom branch pointcut.

## 5.1 Branch pointcut

Our branch pointcut reveals as join points the binary comparison expressions satisfying the comparison pattern given as the parameter of this pointcut. The grammar of the branch pointcut is "branch($Type_1$ $op$ $Type_2$)" where $op$ represents a binary comparison operator such as == and !=. A comparison pattern is deemed met if the comparison expression of a program consists of the matching comparison operator in the middle, the left-hand-side sub-expression that is an instance of $Type_1$, and the right-hand-side sub-expression that is an instance of $Type_2$. As usual, the compared values can be exposed to an advice through an additional accompanying args pointcut.

The last two advices of Figure 10(b) use branch pointcuts so that actual values of proxies can be first retrieved before performing comparison by calling proceed. Notice that the call to proceed located at the last line of each advice takes as its parameters actual(o1) and actual(o2) to pass actual values of proxies. We define actual(o) to return o itself if o is not a proxy.

We explained in §4.4 that we treat (dis)equalities with this conservatively; e.g., for the equality case, we consider this and \past(this) to be different from each other. Such conservatism is programmed in the last two advices as an if-statement before the proceed call. The conditional expression checks if one of comparison operands is the same as this and the other operand is a proxy. If that is a case, the same conservative boolean value is returned regardless of the actual value of the proxy.

142

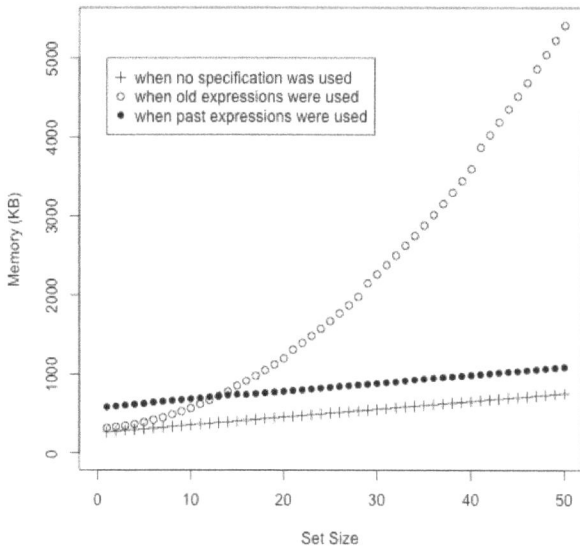

Figure 11: Comparison of memory usage

```
//@ model import org.jmlspecs.models.JML_Elem_Set;
public class PatientSet {
  private Patient[] patients;   private int size;
  //@ public model JML_Elem_Set set;

  //@ private represents set <- abs();
  /*@ private model pure JML_Elem_Set abs() {
    @   JML_Elem_Set ret = new JML_Elem_Set();
    @   for (Patient p : patients) ret = ret.insert(p.clone());
    @   return ret; }
    @*/

  //@ requires set.has(p);
  //@ ensures size == \old(size) && \old(set).isSubset(set);
  //@ also
  //@ requires !set.has(p);
  //@ ensures size==\old(size)+1 && set.has(p) && \old(set).isSubset(set);
  public void add(Patient p) { /* omitted */ }
}
```

Figure 12: A PatientSet stub described using a modeling type

We originally developed the branch pointcut to support symbolic execution [11]. The path condition of symbolic execution is determined depending on the execution result of the comparison expression of an encountered branch statement. Our branch pointcut can be used to construct the path condition on the fly.

A similar pointcut to our branch pointcut was used in [17] to define the criteria for measuring code coverage such as branch coverage of C# programs. We point out three subtle differences from our branch pointcut, which are mainly caused by different needs. First, not comparison expressions but branch statements are revealed as join points. Second, types of comparison operands are not exposed as if the usage of the pointcut is "branch()". Third, various kinds of branch statements such as if-statements, switch-statements and while-statements are distinguished from one another.

A more recent AOP language such as LogicAJ2 [18] can flexibly define a number of pointcuts including our branch pointcut only with a few primitive pointcuts. However, it does not yet support dynamic information available in AspectJ through this and target pointcuts, and cannot be used for our needs.

## 5.2  Evaluation

Use of past expressions improves not only comprehensibility and expressiveness of specifications, but also memory usage during RAC as compared to when using old expressions. To compare the memory usage, we measured the memory used when executing the add method of class PatientSet introduced in § 2. Recall that an instance of PatientSet represents a set of Patient instances. We observed how memory usage increased as we added Patient instances one by one to increase the set size. To make the changes of memory usage more easily visible, we padded out Patient with an extra array of bytes. We observed memory usage increase in three cases where method add is either (i) specified using old expressions similarly to Figure 1(a), (ii) specified using past expressions similarly to Figure 3, or (iii) not specified at all.

Figure 11 shows the result of our experiment. The graph shows that memory usage increases in different rates depending on whether past or old expressions are used. While mem-

ory usage increased exponentially when old expressions were used for the reason we explained in § 2, only linear growth was observed when past expressions were used. When no specification was used, a similar linear growth was observed as expected.

The reason why past expressions do not cause exponential memory usage growth should be evident by now; unlike an old expression, a past expression does not make a copy of the pre-state value.

In the figure, observe the two parallel lines representing respectively the memory usage when past expressions were used and when no specification was used. The gap between these two parallel lines corresponds to the memory overhead of using past expressions. For our prototype, we did not employ optimized collection data structures and simply used Java's standard collections to implement the difference heap. Replacing Java's standard collections with more memory efficient collections will shrink the gap between the two lines.

## 6  RELATED WORK

We mentioned in § 2 that benefits of our past expressions include encapsulated accesses to pre-state values representing mathematical data types. There is another approach that provides similar benefit by using a library of modeling types [7]. Each modeling type of this library corresponds to a mathematical data type such as a sequence and a set.

For example, Figure 12 shows a PatientSet that uses a modeling type JML_Elem_Set for its specification. The set field of this modeling type is used to specify method add unlike in Figure 1(a) where the patients field was directly used. The fact that set is a specification-only field is indicated with modifier model. The relation between the two fields, set and patients, needs to be made, and this is done through a model method abs and a represents clause in the example; the notation "represents set <- abs()" of the figure means that the value of set is assigned by calling abs. Notice in the definition of abs that a fresh set of a modeling type JML_Elem_Set is returned as a result whose set elements are the clones of the patients-array elements. [6] Note that set and \old(set) of add's ensures clause point to two different instances of JML_Elem_Set because their values are assigned through two different calls of

---

[6]We assume that class Patient implements interface Cloneable with an appropriate clone method being provided.

abs at the pre-state and the post-state of add, respectively. Also, those two sets contain the elements that are not shared between the sets because of the aforementioned cloning. In this special case, the subset relation, \old(set) ⊆ set, can be validly checked through a method call \old(set).isSubset(set).

The main potential benefit of using the modeling types is that program data such as an array (e.g., patients) can be treated in a specification as a mathematical data such as a set (e.g., set). However, the use of modeling types often entails extra effort to relate modeling-type data to programming-type data. Our past expressions can be used as an alternative that is simpler and arguably more programmer-friendly; programmers of an object-oriented language are already familiar with the concept of encapsulation.

While formal semantics of past expressions is the novelty of this paper, informal descriptions of similar concepts, that are all based on cloning unlike in past expressions, can be found in the literature. Jass [4], another contract language for Java, requires its old expression Old($E$) to be used only with expressions $E$ of interface Cloneable. This is because the value of Old($E$) is obtained by calling the clone method of $E$. We pointed out in § 2 the problems of cloning such as performance penalty and semantic flaws. In addition, it seems too intrusive to require a class to be a Cloneable one whenever a variable of that class needs to be used in an old expression.

In our experience in building a prototype, AOP proved to be handy for supporting past expressions during RAC. Similarly, AspectJ was used to support OCL (Object Constraint Language) [15]'s pre expression by Kosiuczenko [12]. While OCL is originally a specification language for a modeling language (i.e., UML), and thus independent of a specific programming language, tools such as ocl2j [8] use OCL to monitor behaviors of Java programs. The semantics of OCL's pre expressions is generally identical to the one of old expressions. Kosiuczenko used AspectJ to store pre-state field values at their modification sites. While being similar to our difference heap, his approach is distributed unlike our centralized difference heap; for each field f of class C whose pre-state value needs to be stored, a fresh field fHIST is inserted into C to keep track of the history of f. Also, a fresh method fATPre() is added to C to be used when getting the pre-state value of f. Our implementation of the difference heap is less intrusive; we do not add extra fields and methods to the original program. In addition, we used AOP not only to implement the difference heap but also to access fields in desired time contexts and compare objects that may reside in different time contexts.

## 7 CONCLUSION AND DISCUSSION

In this paper, we have (i) pointed out the problems of existing old expressions of contract languages, most notably the lack of encapsulation, (ii) suggested a past expression as an alternative, and (iii) showed how past expressions can be supported during RAC through AOP in a memory-efficient way.

While we focused on RAC in this paper in terms of tool support, this does not mean that static checking with past expressions is impossible. In fact, it seems, as compared to RAC, straightforward to implement static checking following the formal semantics we provide in this paper. Indeed, our prototype tool can also support static checking at the rudimentary level. However, static checking poses its own challenges. For example, it is not yet clear to us how the add method of a generic class Set<T> can be statically checked against a specification where past expressions are used. We leave static-checking support for past expressions as future work.

## Acknowledgements

We thank Bruno Oliveira for his valuable comments. This work was partially supported by a Ministry of Education research grant MOE2010-T2-2-073 (R-252- 000-456-112 and R-252-100-456-112) from Singapore, a US National Science Foundation grant (# 0709169 and # 0644288), and a US Air Force Office of Scientific Research grant (contract FA9550-09-1-0138).

## References

[1] P. Avgustinov, A. S. Christensen, L. Hendren, S. Kuzins, J. Lhoták, O. Lhoták, O. de Moor, D. Sereni, G. Sittampalam, and J. Tibble. *abc*: an extensible AspectJ compiler. In *AOSD*, pages 87–98, 2005.

[2] J. Barnes. *High Integrity Software: The SPARK Approach to Safety and Security*. Addison-Wesley, 2003.

[3] M. Barnett, K. R. M. Leino, and W. Schulte. The Spec# programming system: An overview. In *CASSIS*, pages 49–69, 2004.

[4] D. Bartetzko, C. Fischer, M. Möller, and H. Wehrheim. Jass - Java with assertions. *ENTCS*, 55(2):103–117, October 2001.

[5] L. Burdy, Y. Cheon, D. R. Cok, M. D. Ernst, J. R. Kiniry, G. T. Leavens, K. R. M. Leino, and E. Poll. An overview of JML tools and applications. *STTT*, 7(3):212–232, 2005.

[6] P. Chalin. Logical foundations of program assertions: what do practitioners want? In *SEFM*, pages 383–392, 2005.

[7] Y. Cheon, G. T. Leavens, M. Sitaraman, and S. Edwards. Model variables: cleanly supporting abstraction in design by contract. *SPE*, 35(6):583–599, May 2005.

[8] W. J. Dzidek, L. C. Briand, and Y. Labiche. Lessons learned from developing a dynamic OCL constraint enforcement tool for Java. In *MoDELS*, pages 10–19, 2006.

[9] C. B. Jones. *Systematic Software Development using VDM*. Prentice-Hall, 1990.

[10] G. Kiczales, E. Hilsdale, J. Hugunin, M. Kersten, J. Palm, and W. Griswold. An overview of AspectJ. In J. Knudsen, editor, *ECOOP*, pages 327–354, 2001.

[11] J. C. King. Symbolic execution and program testing. *Communications of the ACM*, 19(7):385–394, 1976.

[12] P. Kosiuczenko. On the implementation of @pre. In *FASE*, pages 246–261, 2009.

[13] G. T. Leavens, A. L. Baker, and C. Ruby. Preliminary design of JML: a behavioral interface specification language for Java. *ACM SIGSOFT Software Engineering Notes*, 31(3):1–38, May 2006.

[14] B. Meyer. *EIFFEL: The language and environment*. Prentice Hall, 1992.

[15] OMG. *Object Constraint Language - version 2.2*, 2010.

[16] OpenJML. http://jmlspecs.sourceforge.net/.

[17] H. Rajan and K. Sullivan. Aspect language features for concern coverage profiling. In *AOSD*, pages 181–191, 2005.

[18] T. Rho, G. Kniesel, and M. Appeltauer. Fine-grained generic aspects. In *FOAL*, pages 29–35, 2006.

[19] J. M. Spivey. An introduction to Z and formal specifications. *Software Engineering Journal*, 4(1):40–50, January 1989.

# A Pointcut Language for Setting Advanced Breakpoints

Haihan Yin, Christoph Bockisch, Mehmet Akşit
Software Engineering group, University of Twente, 7500 AE Enschede, the Netherlands
{h.yin, c.m.bockisch, m.aksit}@cs.utwente.nl

## ABSTRACT

In interactive debugging, it is an essential task to set breakpoints specifying where a program should be suspended at runtime to allow interaction. A debugging session may use multiple logically related breakpoints so that the sequence of their (de)activations leads to the expected suspension with the least irrelevant suspensions. A (de)activation is sometimes decided by some runtime context values related to that breakpoint. However, existing breakpoints, which are mainly based on line locations, are not expressive enough to describe the logic and the collaboration. Programmers have to manually perform some repeated tasks, thus debugging efficiency is decreased.

In this paper, we identify five frequently encountered debugging scenarios that require to use multiple breakpoints. For such scenarios, it is often easier than using the traditional debugger to write pointcuts in an aspect-oriented language, and to suspend the execution at the selected join points. However, existing languages cannot handle the scenarios neatly and uniformly. Therefore, we design and implement a breakpoint language that uses pointcuts to select suspension times in the program. Our language allows programmers to use comprehensible source-level abstractions to define breakpoints. Also, multiple breakpoints can be freely composed to express their collaboration. In this way, an expected suspension can be expressively programmed and reached with less or even no irrelevant suspensions.

## Categories and Subject Descriptors

D.2.5 [**Testing and Debugging**]: Debugging aids; D.3.2 [**Language Classifications**]: Very high-level languages; D.2.2 [**Design Tools and Techniques**]: User interfaces

## Keywords

Debugging, Advanced breakpoint, Pointcut language

## 1. INTRODUCTION

Software is kept being maintained from its delivery until its end. Maintenance takes the majority of effort spent in developing software and a significant portion of maintenance is carried out for debugging [14, 11]. An important step, which is called fault localization, of debugging is finding out the root cause based on some observed symptoms. The root cause always happens before unexpected symptoms appear. Eisenstadt [9] studied 59 bug anecdotes and he concluded that over 50% of the difficulties resulted from this temporal or spatial chasm between the root cause and the symptom, or from inapplicable debugging tools.

In interactive debugging, programmers use breakpoints to mark places in the source code where the program should be suspended at runtime. When the debuggee program is suspended, programmers can inspect the program state, observe the program behavior, or perform other debugging tasks by using a debugger. How breakpoints are set can significantly affect the efficiency of debugging. An inexperienced programmer may set too many breakpoints; redundant ones distract her attention from those revealing the root cause. Or she may set too few, which results in passing the root cause.

Breakpoints should not be arbitrarily set, because each suspension has a cost. At least, programmers need to decide whether the suspension is relevant. Observing a symptom, programmers usually first roughly choose a program slice that is most likely to cause the symptom, and then set breakpoints to observe the slice. For example, if a field stores a wrong value, breakpoints are set at places where the field is modified. Thus, logically, breakpoints are grouped by programmers according to what they do instead of where they are. However, the traditional breakpoints are mainly based on source lines, which do not embrace any logic of why breakpoints are placed there [7]. Programmers have to mentally sketch the logic relationships between these breakpoints.

Also, breakpoints are independent from each other. At runtime, each breakpoint has its states such as being activated, and contexts such as the value of a variable. Sometimes, a desired suspension requires multiple breakpoints and their states which sequentially form a path leading to the suspension. This may require non-trivial manual effort, such as recording the state or context at one suspension and using it at another. Limitations of the traditional breakpoints often require programmers to perform many repeated debugging steps. Thus, debugging efficiency is decreased.

We identify five frequently encountered scenarios that the traditional breakpoints cannot handle well. The identified scenarios require breakpoints to be set at places sharing some common characteristics, such as similar syntax. The concept of *pointcut* perfectly fits in this context. These scenarios show that using pointcut-advices to construct places for setting breakpoints is a more convenient and efficient approach than traditional debugging. However, current AOP languages are not specifically designed for solving these scenarios. Thus, pointcut-advices are too verbose. Furthermore, pointcut-advices cannot be used to set a breakpoint to specific advice compositions, which is one of our identified scenarios. Also, programs that are added for debugging may accidentally stay in the project. This will introduce unnecessary maintenance effort in the future.

Therefore, we propose a declarative breakpoint language (BPL). By building on AspectJ, BPL can be used to debug Java or AspectJ programs. The breakpoint, that is the core concept of BPL, is a first-class value. A breakpoint can be defined by AspectJ-like pointcuts which use source-level abstractions. This makes the description of breakpoints more comprehensible than line breakpoints. We extend and improve existing AspectJ pointcut designators with seven novel ones. In BPL, breakpoints are named and can be used to compose higher-level breakpoints. The composition level can be infinite because we treat the primitive and the composed breakpoints in a uniform way. BPL is the first approach providing a pointcut for selecting a specific action composition at runtime.

This paper is structured as follows. Section 2 presents five debugging scenarios and describes how debugging processes are performed in different approaches. Section 3 gives a detailed introduction to the new features introduced in BPL. Section 4 highlights several implementation considerations in our prototype. Section 5 describes two debugging examples by using the traditional debugging, the program solution, and our solution respectively. Section 6 and 7 describe related work and conclude this paper respectively.

## 2. PROBLEM STATEMENTS

Debugging is a cognitive process and how it is performed significantly depends on the programmer's observations and experience. A programmer tends to give the same treatment when she observes the same symptom, such as a certain exception being thrown. In this section, we select several debugging scenarios that are frequently encountered. For each scenario, we elaborate the way of using the traditional debugger. We tag debugging steps in the description like "a.1", in which the letter represents a debugging process and the digit represents the step order of that process.

Each scenario requires multiple breakpoints or suspensions at different locations, which share some common properties, such as similar syntax, relation to the same variable, etc. In AspectJ, a pointcut is used to select places with common properties. This has inspired us to use AspectJ programs during debugging, where pointcuts select the join points at which we want to suspend the execution and where we set a breakpoint in the otherwise empty advice body. In this section, we also demonstrate this approach for the identified scenarios. Actually, this approach is a variant of program instrumentation.

The program solution serves two purposes. First, the program can describe the scenario in a more succinct and clear

way than instructions for manual debugging given in natural language. Take pointcut **call(void** Shape.set∗(..)) for example, it can be seen as two debugging tasks in this context: finding all places calling methods which satisfy the pattern "void Shape.set∗(..)", and then setting line breakpoints there. Second, the programs will be compared with our solution, which is an AspectJ-like language.

We have two basic criteria for the program solution. First, the program should be in a separate module. AspectJ modularizes scattering concerns and we want to keep this principle in the program solution. Second, the program should be simple. Effort spent on writing the program should be comparable to that spent on the traditional debugger. In most cases, writing a lengthy or a sophisticated program for setting a breakpoint is not desirable.

### 2.1 Scenario 1: Selecting Multiple Locations

Sometimes, it is difficult to decide which specific location is executed at runtime. For example, to debug unexpected behavior in a system, which the programmer is not familiar with, she may deduce roughly which function is executed by matching names of the function with the observed runtime behavior. A function can be implemented as a set of overloading constructors or methods. However, to know which specific one is executed at runtime, she may need to set a breakpoint to each implementation.

The difficulties of using the traditional debugging in this scenario mainly come from finding locations for setting breakpoints and managing breakpoints as logic units.

*Finding locations.*

Suppose the programmer observes that a field stores an unexpected value, she needs to monitor the runtime states of this fields. The first option is setting a watchpoint to this field (**a.1**). When the watchpoint is hit, the programmer needs to perform one "step over" to inspect the field value after the modification (**a.2**).

The second option requires the programmer to manually find out the last assignment (**b.1**) to this field before the unexpected value is observed. This assignment can be in any constructor or method modifying this field. She needs to set a line breakpoint to each found place (**b.2**) and specify a condition to check whether the expression on the right-hand side of the assignment equals the unexpected value (**b.3**).

*Managing breakpoints.*

Using the *Breakpoint* view provided in modern IDEs, such as Eclipse, the programmer can organize the breakpoints she set. The view can group the breakpoints according to their types, such as line breakpoint or watchpoint, or their locations, such as files or projects. Breakpoints can be (de)activated and deleted at the granularity of groups. However, there is no approach provided to group breakpoints as logic units. Thus, a debugging task applied to a logic operation will possibly required repeating steps.

Listing 1 shows how an AspectJ program monitors unexpected assignments to the field Clazz.var. AspectJ can access the value assigned to a field by using **args()**. The pointcut describes the desired places for suspensions and a breakpoint is set in the body of the **before** advice. When the program is suspended in the advice, the programmer can locate the root cause by using the stack trace. Usually, the second top frame in the stack trace points to the root cause, because the top frame represents the execution of the advice.

```
1  public aspect Scenario1Aspect {
2    before(int val) : set(int Clazz.var) && args(val) &&
3      if(val==/*Unexpected value*/) {
4      // set a breakpoint on this line
5    }
6  }
```

**Listing 1: An AspectJ program monitoring assignments to a field**

## 2.2 Scenario 2: Monitoring Updates on a Field

Listing 2 shows a program slice updating a field, which is a HashMap. On lines 4-8, we use ellipsis to indicate that the separated statements may reside in different methods and their execution order is not the same as the lexical order.

```
1  class Scenario2 {
2    private HashMap map1;
3
4    map1.put(key1, value1);
5    ...
6    map1.get(key2);    //returns a null value
7    ...
8    map1.put(key2, value2);
9  }
```

**Listing 2: Multiple places updating a field**

When a value retrieved from the HashMap is wrong, as line 6 shows, the potential root causes are places updating this HashMap, such as lines 4 and 8.

To debug this with a traditional debugger, the programmer needs to find all the updating locations (**c.1**), set breakpoints there (**c.2**), and evaluate the values of the expressions for updates at runtime (**c.3**). A watchpoint for the field is not helpful in this scenario, because it can only suspend the program when the field is accessed or modified instead of being updated. Setting a breakpoint to the called method HashMap.put() may result in redundant suspensions, because there may be other HashMaps in the program.

Listing 3 gives an AspectJ solution for this scenario. It is a *privileged* aspect which can access protected members of other classes. On line 6, the advice checks whether the callee object t is same as the value stored in the expected field, such as the private field map1 in Listing 2.

```
1  public privileged aspect Scenario2Aspect {
2    before(Scenario2 caller, HashMap t, String s) :
3      call(public Object HashMap.put(..)) && this(caller) &&
4      target(t) && args(s, *) &&
5      if(s.equals(/*value of key2*/)) {
6      if(caller.map1 == t) {
7        // set a breakpoint on this line
8      }
9  }}
```

**Listing 3: An AspectJ program monitoring updates on the object referenced by a specific field**

## 2.3 Scenario 3: Finding Null Pointer Dereferences

The dot operator dereferences an object pointer to access a member from that object. A line of code may contain multiple dereference operations as in the following listing:

```
1  total.getObjects().addAll(current.getObjects());
```

If a *NullPointerException* occurs on this line, the error message only tells the line number where the exception occurs instead of the specific operation. For debugging this scenario, the programmer needs to place a breakpoint at the line where the exception occurs (**d.1**). When the program is suspended, she needs to repeatedly perform "step into" and then "step return" to check each dereference operation until the exception occurs (**d.2**). Meanwhile, she has to manually note which dereference operation the debugger reaches (**d.3**).

Another option is to change the layout of the code so that there is a dereference operation per line, like the following listing shows (**e.1**). After rerunning the program (**e.2**), the error message can accurately tell which line throws the exception. Since this option requires rewriting the source code, it is also not generally applicable.

```
1  total.getObjects()
2    .addAll(
3      current.getObjects());
```

Listing 4 presents an AspectJ program corresponding to this scenario. The pointcut is only satisfied if the receiver of a dereference operation is **null** (see line 5). The advice body further restricts the line number on line 7. If a breakpoint is set at line 8, it can suspend the execution before the dereference operation, which ends up with a *NullPointerException*, is about to occur.

```
1  public aspect Scenarios3Aspect {
2    before(Object receiver) :
3      (call(* *.*(..)) || get(* *.*)) && target(receiver) &&
4      withincode(/*a method pattern*/)
5      && if(receiver == null) {
6      int line = thisJoinPoint.getSourceLocation().getLine();
7      if(line == /*expectedLine*/) {
8        // set a breakpoint on this line
9      }
10   }
11 }
```

**Listing 4: An AspectJ program checking *null* receivers on a source line**

## 2.4 Scenario 4 : Recording Execution History

Listing 5 shows program slices related to operations on two stream objects. An exception would be thrown when line 6 is executed, because it tries to read data from a closed Stream.

```
1  InputStream s1 = new FileInputStream(...);
2  InputStream s2 = new FileInputStream(...);
3  s1.close();
4  s2.read();
5  s2.close();
6  s1.read();   // An exception is thrown.
```

**Listing 5: A program performing operations on stream objects**

An execution path can lead to unexpected behavior, such as first close then read. Programmers need to track the cause backwards from the point where the symptom is observed. However, most traditional debuggers do not provide backtracking. Using breakpoints, the programmer is likely to suspend the program either before or after the cause. If the cause is passed, the programmer needs to restart a new debugging session. Moreover, debugging Listing 5 requires that all events of the path refer to the same object. The programmer has to manually note corresponding information with the traditional debugger.

Tracematch [1] is an AspectJ extension designed for observing execution traces. Therefore, we choose Tracematch as the alternative debugging solution for this scenario. Listing 6 shows a Tracematch program. Lines 2 and 3 define two events named close and read. Both of them bind the **target** value to the parameter of the tracematch (line 1). Events with different **target**s are not recorded by the same **tracematch** instance. Line 4 declares the expected, but undesired execution path with the two names. When the path is matched on the same Stream, the instruction represented by line 5 is executed.

```
1 tracematch(Stream s) {
2   sym close before : call(* Stream.close(..)) && target(s);
3   sym read before : call(* Stream.read(..)) && target(s);
4   close read {
5     // set a breakpoint on this line
6   }
7 }
```

Listing 6: A tracematch specifying an undesired execution path

## 2.5 Scenario 5: Exploring a Program Composition

The execution of an advice can alter the flow of its base program to any extent. Many works [17, 8, 13, 15, 12] have identified the problem of aspect (or advice) interference. An incorrect composition, which can be either between advices or between advices and the base program, at a join point causes unexpected runtime behavior. Therefore, the programmer needs to inspect the execution of the composition at such a join point. What complicates this task is that pointcuts can include dynamic tests. Thus, when advice share a join point shadow, these advice are not necessarily executed together.

To debug this scenario, the programmer first needs to find join point shadows (JPS) affected by all advices of the expected composition (**f.1**) and then set breakpoints to these shared JPSs (**f.2**). Because in AspectJ, whether an advice is applied can only be seen when it is actually executed, the programmer needs to execute the program once (**f.3**), manually perform bookkeeping of the program composition at each join point (**f.4**) and record the hit count of the line which contains the desired join point (**f.5**). In the second debugging session, setting line breakpoints with the recorded hit counts (**f.6**) leads to suspensions at the expected join point before any advice is executed.

For this case, AspectJ cannot provide a clean way for putting debugging code in a separate module. There is no pointcut which can uniquely identify the execution of an advice, because advices are unnamed in AspectJ. Furthermore, an advice using the pointcut **adviceexecution**() cannot easily obtain information about the join point triggering the execution of the advice. Without this information, it is impossible to know whether different advice executions are composed at the same join point. Though there are works, such as Oarta [16] and dependent advice [5], supporting named

advices, none of them can use advice names to specify an expected runtime composition.

## 2.6 Summary

In this section, we have described five debugging scenarios that require non-trivial manual tasks such as setting breakpoints, repeating steps, and recording past states. For these scenarios, the program solutions that describe the suspension conditions in a declarative way show their strength and potential.

However, some debugging programs are verbose. In Listing 3, comparing the field value and the current target object is always required in debugging scenario 2. In Listing 4, most of the parts are generic except the location information. These programs can be more reusable if the configurable parts are parameterized. Besides, there is no solution that treats these scenarios in a uniform way. Take scenarios 4 and 5 for example, Tracematch can easily specify a sequential execution of operations $a$ and $b$. However, it is impossible to reuse the previous declarations to express that the operation $a$ should also be advised by the advice $c$.

Last but not least, we do not encourage to add code, which is not part of the main functionality, to the source program. The added code may introduce unnecessary maintenance effort if the programmer forgets to remove it after fixing a bug. Even though source control management systems, such as subversion, can be used to tell the differences between two versions, programmers need to distinguish the added debugging programs and the fixed parts.

## 3. BREAKPOINT LANGUAGE

Based on the observations described in section 2, we have designed and developed a breakpoint language (BPL) for setting advanced breakpoints. BPL reuses many features of AspectJ and Tracematch. Additionally, it has its own unique functionalities.

The debuggee programs of BPL are Java programs or AspectJ programs. During the debugging, breakpoints specified by the BPL suspend the debuggee program at join points where they are satisfied.

Listing 7 shows the grammar rule of a breakpoint declaration. A breakpoint declaration has a name, a parameter list, and a pointcut expression. The rule for PointcutExpr extends the AspectJ pointcut with seven designators. We describe these designators and their usages in the following subsections.

```
1 BreakpointDeclaration :
2   Name '(' FormalParameterList? ')' ':' PointcutExpr ';' ;
```

Listing 7: The grammar rule of a breakpoint declaration

## 3.1 The Pointcut call()on()

The pointcut **call**()**on**() is derived from the pointcut **call**(). It matches join points where a method is called on an object referenced by a specific field. The **call**() and **on**() parts take the method and the field specifications respectively. This pointcut can be used at any place where **call**() is applicable. It should be noted that the **on**() part matches based on the referential identity of the values, i.e., it also matches alias of the specified field.

Listing 8 shows an example of using **call**()**on**(). Compared to Listing 3 in **Scenario 2**, it implicitly constrains the callee object of the method.

```
1 bp(String s) :
2   call(public Object HashMap.put(..))on(Scenario2.map1) &&
3   args(s, *) && if(s.equals("key2"));
```

**Listing 8: A breakpoint declaration using pointcut call()on()**

## 3.2 The Pointcuts location() and checkNPE()

As the following listing shows, the pointcut **location**() takes three parameters which represent the file path, the file name, and the line number in the file respectively. The third parameter takes a list of line numbers or line ranges, e.g., [97, 100..102]. This pointcut can be used jointly with other pointcuts to restrict locations of JPSs, for example **call**(...) && **location**(...).

```
1 bp() : location("Spacewar", "SpaceObject.java", [97]);
```

The pointcut **checkNPE**() matches dereference operations where the receiver is *null*. A breakpoint using **checkNPE**() suspends the program just before the satisfied dereference operation is performed, and thus the suspension happens before a *NullPointerException* is thrown. The first breakpoint declaration shown in Listing 9 checks whether a line contains a null pointer dereference. The second breakpoint declaration does the same for the specified method body.

Compared to Listing 4 in **Scenario 3**, **checkNPE**() omits the fixed parts specifying the cause of *NullPointerException*; only the source location is left to be configured.

```
1 bp1() : checkNPE() &&
2     location("Spacewar", "SpaceObject.java", [97]);
3 bp2() : checkNPE() && withincode(/*a method pattern*/);
```

**Listing 9: Breakpoint declarations using pointcuts location() and checkNPE()**

## 3.3 The Pointcuts path() and bind()

The pointcut **path**() matches a specific execution path existing in the history. It takes a path expression, which consists of breakpoint references, as the parameter. We use a blank space to represent the sequential order and rectangular brackets to represent the exact expected hit count. For example, **path**(a[2] c) expects that breakpoints a and c are hit in the sequence "aac". Expression c is shortened from c[1]. Besides, the "+" sign, which means 1 or more, and "*", which means 0 or more, can be appended to a breakpoint reference. The **path**() expression is satisfied when all referenced breakpoints are satisfied in the sequence specified as the path expression.

The pointcut **bind**() is used to bind context values exposed by lower-level breakpoints to the higher-level breakpoint. Listing 10 shows our solution for **Scenario 4**, which is about recording execution history.

Lines 1 and 2 declare two breakpoints for read and close operations respectively. Lines 3–5 declare a composite breakpoint. Line 4 describes an expected execution path which requires that breakpoints close and read are hit sequentially. Line 5 binds values from lower-level breakpoints to the parameter declared on line 3. A binding relies on the name of

a parameter in the composite breakpoint and the position of a parameter in the lower-level breakpoint. For example, read(s) binds the first parameter of the breakpoint read to the parameter named s of the composite breakpoint closeRead. Wildcards can be used to skip parameters that are not relevant to the breakpoint declarations. For example, read(*, s) and read(.., s) bind the second and the last parameter respectively. In Listing 10, both bindings bind values to the same parameter s and this implies that the bound values must refer to the same object.

```
1 read(Stream t) : call(public * Stream.read()) && target(t);
2 close(Stream t) : call(public * Stream.close()) && target(t);
3 closeRead(Stream s) :
4     path(close read) &&
5     bind(read(s), close(s));
```

**Listing 10: Declaration of a composite breakpoint**

It is also possible to bind values to different parameters, as Listing 11 shows. The equality of bound values is specified explicitly in the **if**() expression on line 4. Both breakpoints closeRead and closeRead_if suspend the program at the same times. The former is more succinct and the latter is more flexible with restricting the bound values.

```
1 closeRead_if(Stream rStream, Stream cStream) :
2     path(close read) &&
3     bind(read(rStream), close(cStream)) &&
4     if(cStream == rStream);
```

**Listing 11: A composite breakpoint using if()**

Our solution for the path expression is greatly inspired by Tracematch, but there are two fundamental distinctions. In the view of the structure, a primitive event declared in one tracematch cannot be referred to by other tracematches. In BPL, primitive breakpoints are more reusable, because they can be referred in any number of composite breakpoints. In the view of the join point model, Tracematch is interested in event kinds, such as *before* and *around*. BPL runs with an interactive debugger that provides only forward execution. Therefore, it only suspends the program *before* executions of the satisfied join points.

## 3.4 The Pointcuts adviceexecution() and composition()

In AspectJ, **adviceexecution**() does not take any parameter and it cannot select the executions of a specific advice. BPL provides a backwards-compatible extension of **adviceexecution**(), which can take the fully qualified name of an advice as the parameter. As an example, pointcut **adviceexecution**(GameInfo.guiInitiation) selects the execution of the advice declared in the aspect "GameInfo" and named with "guiInitiation". Section 4.3 describes how advices are named.

We use the term "action" to refer to an advice, a method or constructor call, a field access, etc. The **composition**() pointcut designator selects join points with an action composition where actions have the specified relationship. To use this pointcut, the programmer first needs to declare breakpoints that suspend the program at the executions of the desired actions. Then, she can use the names of the declared breakpoints to specify a *composition pattern*. Last, the pattern is used as the parameter of **composition**().

We provide two types of composition pattern. Suppose beforeExe and afterExe are two breakpoints that both use **adviceexecution**(). A breakpoint using **composition**() suspends the program at a join point where the composition satisfies the specified pattern.

**Existence** - the actions referenced in the specified pattern should exist in the composition. We use commas to list breakpoints corresponding to desired actions, e.g., **composition**(afterExe, beforeExe).

**Exclusion** - the actions referenced in the pattern should not occur in the composition. We use an exclamation mark for this relationship, e.g., **composition**(!afterExe).

# 4. IMPLEMENTATION CONSIDERATIONS

In earlier work [19], we have developed a debugger for AO programs on top of the execution environment NOIRIn from the ALIA4J language-implementation architecture [3]. In ALIA4J and thus in NOIRIn, aspect-oriented concepts, such as join point and pointcut evaluation, are modelled as first class objects. The AO debugger complements a Java debugger with functionalities for debugging AO features. It allows programmers to inspect the context values, the composition, etc., at a join point. We modified the AspectBench compiler [2] to generate an intermediate representation of AspectJ programs as required by the ALIA4J approach, which preserves the full source locations of AO entities and makes them accessible at runtime.

BPL is implemented to work together with the AO debugger. When a breakpoint is hit at a join point, the programmer can use the AO debugger to observe the suspended program.

At runtime, the breakpoint declarations are sent to NOIRIn. They are evaluated in the context at a join point along with the execution of rest of the program.

## 4.1 Evaluation of Breakpoints

Figure 1 shows a diagram of the classes that are used in our implementation to represent breakpoints in BPL in the execution environment. *AdvancedBreakpoint* represents the breakpoint and it is managed by a *BreakpointManager*. Each breakpoint has a *Condition* specifying in what condition the breakpoint can be hit. The figure includes only the conditions related to the pointcut designators introduced in section 3.

When the program reaches a join point at runtime, NOIRIn first analyzes the call context, then computes the action composition performed at this join point, and finally executes actions in the composition. Breakpoints are evaluated when an action is about to be executed after all context information, including the call context, the composition, and the executing action, is prepared.

For all primitive conditions except *AdviceExecutionCondition*, it is enough to be evaluated once at a join point. To distinguish this different evaluation frequency, we put a flag to *Condition* and its subtypes. The flag has two values, which are composition-level (*c-level*) and action-level (*a-level*). A breakpoint with a c-level condition is evaluated once at a join point and one with an a-level condition is evaluated at every action in the composition. A binary condition such as *AndCondition* is c-level if and only if its two operators are c-level.

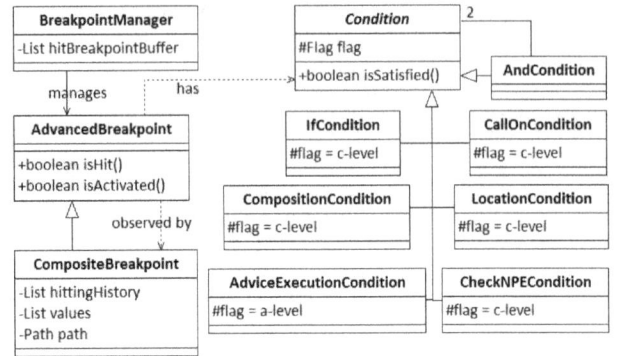

**Figure 1: A class diagram of classes related to the breakpoint in BPL.**

Multiple breakpoints may suspend the program at the same join point. We use a buffer *hitBreakpointBuffer* to store the hit breakpoints at a join point. A hit breakpoint sends a message with required debugging information to the buffer. When the evaluations of all breakpoints are finished, the *BreakpointManager* checks whether there is any message in the buffer. If the buffer is not empty, the manager emits a suspending request and releases all messages stored in the buffer. The buffer is cleared when the next evaluation process starts.

## 4.2 Evaluation of Composite Breakpoints

Using a **path**() expression, a composite breakpoint can be composed of primitive breakpoints or other composite breakpoints. It may use **bind**() to access values from lower-level breakpoints and further restrict its suspending condition by using **if**().

In Figure 1, there are two lists in class CompositeBreakpoint. The list *hittingHistory* records the hit history of the lower-level breakpoints. The list *values* records values bound to the parameters. A *CompositeBreakpoint* is an observer of all its lower-level breakpoints. Whenever one of its lower-level breakpoints hits, the evaluation of the composite breakpoint starts. The evaluation first updates lists *hittingHistory* and *values*. Then, it explores the *hittingHistory* list and tries to find an expected path. If an expected path exists, the composite breakpoint starts to evaluate the *if()* condition. If the *if()* condition is true, the composite breakpoint is hit and it produces a message containing all the debugging information, such as locations and bound values, of its lower-level breakpoints.

For illustration, take Listing 5 as the debuggee program, Listing 11 as the breakpoint declarations. Figure 2 shows a complete evaluation process of the breakpoint closeRead_if. Column *Code* lists code where the primitive breakpoints hit. Columns *Hit History* and *Values* describe the runtime states of the lists *hittingHistory* and *values* respectively. Column *Evaluation* has two sub-columns which represent the two-stage evaluation respectively. Sub-column *path* represents matches on the execution path and sub-column *If expr.* represents the test of the **if**() expression. "T" and "F" stand for true and false. When a path is found, a "T" is put in the sub-column *path*. A list representing the indexes of the hit history is put after "T", as in like T{0,1}.

| Code | Hit History | | Values | | Evaluation | |
|---|---|---|---|---|---|---|
| | Index | Bp. ref. | Index | Map | path | If expr. |
| s1.close( ) | 0 | close | 0 | "cStream" →s1 | F | - |
| s2.read( ) | 1 | read | 1 | "rStream" →s2 | T{0,1} | F |
| s2.close( ) | 2 | close | 2 | "cStream" →s2 | F | - |
| s1.read( ) | 3 | read | 3 | "rStream" →s1 | T{2,3} T{0,3} | F T |

**Figure 2: A table showing how a composite breakpoint stores the hit history and the bound values**

The breakpoints are hit sequentially from top to bottom in the table and the evaluation is performed accordingly. When the program reaches s2.read(), a path is found with indexes {0,1}. Then, the composite breakpoint uses the indexes of the path to retrieve values, which are required by the **if**() condition, from the *values* list. However, the condition "cStream==rStream" does not hold. When the program reaches s1.read(), the composite breakpoint finds a path with indexes {2,3} but the **if**() condition again does not hold. Then, another path with indexes {0,3} is found and satisfies the **if**() condition. The composite breakpoint closeRead_if is hit and it produces a suspension message.

## 4.3 Named Advices

Bodden et al. [5], as well as Marot and Wuyts [16] proposed named advices for the purpose of uniquely identifying an advice. They extended the AspectJ syntax to achieve this goal. We name advices for referring to them in point-cut **adviceexecution**(). Besides, we do not intend to extend the syntax of the debuggee program, because the effort integrating it with the rest of our debugging infrastructure [19] is not trivial. Annotations are not supported in the *abc* compiler, which is part of our tool chain. Therefore, we choose to use comments as the approach for naming advices.

Listing 12 shows a **before** advice with a comment naming the advice as firstBefore. The programmer can refer to this advice in the breakpoint declaration. For example, **adviceexecution**(Azpect.firstBefore). During compilation, methods with unique identifiers as method names are created. For simplicity, we do not use the specified name as method name, but let the compiler decide the name as usual. Instead, we keep a mapping between the advices' *virtual names* as specified in the comment and their *compiled name*, as Listing 13 shows. During compilation, a validator checks whether there are ambiguous virtual names and prints an error message, if so.

```
1 aspect Azpect() {
2   /**
3    * @advicename=firstBefore
4    */
5   before() : call(...) {}
6 }
```

**Listing 12: A named advice**

```
1 <map>
2   <entry>
3     <virtual>firstBefore</virtual>
4     <compiled>before$1</compiled>
5   </entry>
6 </map>
```

**Listing 13: A naming map**

When a breakpoint declaration using **adviceexecution**() is sent to NOIRIn, which reads the virtual advice name and replaces it with the compiled name by using the naming map. When this breakpoint is hit, a hitting message is produced. The creation of the message replaces the compiled name with the virtual name.

## 4.4 User Interface

We implemented a dedicated user interface to manage and set breakpoints. The snapshots are given in Figure 3. The breakpoint view (Figure 3(a)) lists all the breakpoints and it has five columns, which are for (de)activation, presenting the breakpoint names, giving the complete declarations, showing whether a breakpoint is hit, and counting the hits respectively.

The panel shown in Figure 3(c) is for creating or editing a breakpoint. From top to bottom, it consists of four parts. The *message* part is for showing error or warning messages. The *script* part is for writing the breakpoint declaration. The *hit count* part is for specifying the expected hit count. The *graph* part is for presenting the reference relationships to other breakpoints. The panel in Figure 3(c) constructs a breakpoint which refers to breakpoints *close* and *read* in its expected path. The corresponding graph shows their hierarchical relationship. The declaration detail is shown when hovering the mouse over a label in the graph. Each label has a check box where programmers can (de)activate the corresponding breakpoint. If the declaration refers to breakpoint names which do not exist, as shown in Figure 3(d), an error message is shown on the *message* part and names in the graph are labeled with "invalid name".

Moreover, we provide functionalities to ease the burden of typing static information, such as signatures and line locations. By using the context menu (Figure 3(b)) in the editor, text can be generated according to where the cursor is. For example, if the programmer selects a method name, the signature of the method can be generated. The generated text is appended to the *script* part of the breakpoint panel. Therefore, programmers do not have to manually type such information and can further customize the generated texts, such as replacing part of the text with wildcards.

## 4.5 Runtime Interactivity

We allow programmers to add, delete, and update breakpoints at editing time and during the execution of the debuggee program. The addition, deletion, or update of a breakpoint does not only changes itself but may also propagate the effect to other breakpoints. A breakpoint is invalid if its declaration contains invalid reference names. When a composite breakpoint becomes invalid, it does not make sense to keep its recorded history. Therefore, the runtime changes may alter the behavior of breakpoints. Other operations, such as additions and deactivations, do not affect the validity of other breakpoints. In the following list, we discuss these effects in detail.

- If a breakpoint is deleted, all breakpoints directly or indirectly referring to it become invalid. An invalid breakpoint discards the information it has recorded. This effect propagates until no more valid breakpoint can become invalid.

- The addition of a breakpoint triggers a re-compilation of all invalid breakpoints. If the added breakpoint

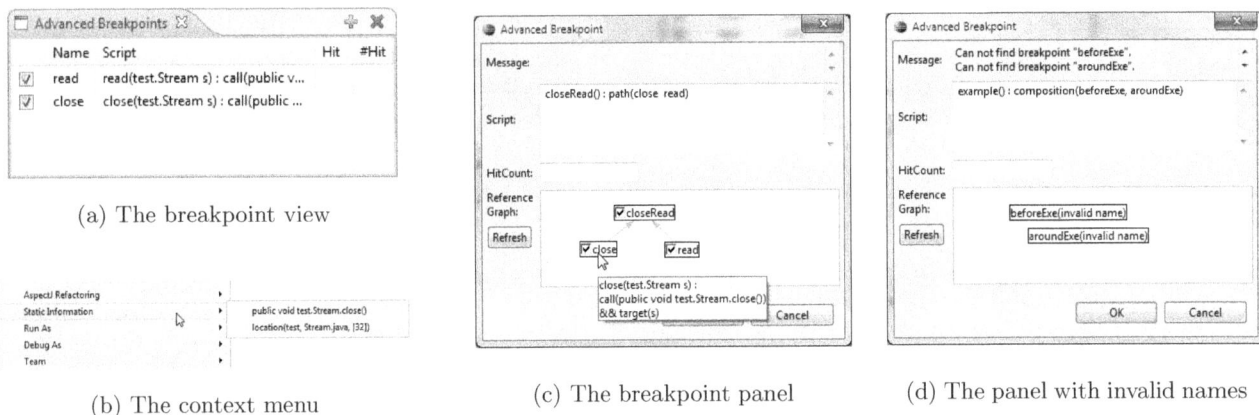

(a) The breakpoint view

(b) The context menu

(c) The breakpoint panel

(d) The panel with invalid names

**Figure 3: Snapshots of the user interface**

matches the missing part in previously invalid breakpoints, the invalid breakpoints become valid. This effect propagates until no more invalid breakpoint can become valid.

- If the script of a breakpoint is updated, this can be deemed as a deletion followed by an addition.

## 4.6 Performance

To get an indication of the runtime overhead imposed by our proposed debugging approach, we performed preliminary micro benchmarks. For this purpose, we use an application performing *bubble sort* and added a dummy aspect to be able to use our **adviceexecution** and **composition** pointcuts. We first executed this selected program on a plain Java Virtual Machine; second, we added breakpoints which use all the new features introduced in BPL and executed this on NOIRIn+BPL. The program uses fields instead of local variables for storing all temporary values, therefore the join points are dense in the program execution, and NOIRIn+BPL performs evaluation of breakpoints at each join point. This represents *the worst case* of using our approach. To avoid measuring the time spent during the suspensions, we change the infrastructural code to let breakpoints only print a message when they are hit. Though breakpoints do not cause any suspension throughout the whole execution, the evaluations whether each breakpoint is hit are actually performed. Comparing the execution times on the plain Java Virtual Machine and on NOIRIn+BPL with the breakpoints shows a slow-down of 20 times. For very short running debug sessions such a slow-down is already acceptable. Since we have focused on the semantics rather than on an efficient implementation, currently many breakpoint evaluations are redundant. In future work, we will also consider optimizations to avoid these redundant evaluations.

We have not yet analyzed the memory overhead which may especially be imposed by the **path**() pointcuts. Based on preliminary experiments, nevertheless, we do not expect this memory overhead to be limiting in typical, short-running debugging sessions.

## 5. EXAMPLES

## 5.1 Debugging Action Compositions

In this example, we add a new requirement to the **Scenario 4**, which is about recording execution history. A closed stream can be re-opened before it reads data in the context of the vital parts of the program. The **around** advice (lines 3–11) in Listing 14 implements this new requirement. Lines 13–17 contain a bug, which is reading data from a closed stream. The ellipses mean that statements may not be close to each other. We use the italic font for the variables *stream* to indicate that they refer to the same object but may use different names. The symptom of the bug is observed at line 17 but the root cause is the premature close() at line 15. The purpose of debugging is to intentionally suspend the program at the root cause. In the following paragraphs, we describe how a typical debugging process is performed in different solutions.

```
1  aspect SafeStream {
2    pointcut VitalPart() : ...;
3    Object around(Stream s) :
4        call(* Stream.read()) && target(s) &&
5        cflow(VitalPart()) {
6      if(s.isClosed()) {
7        s.open();
8      }
9      Object result = proceed();
10     return result;
11   }
12 }
13 Stream stream;
14 ...
15 stream.close();   // the root cause
16 ...
17 stream.read();  // an exception is thrown.
```

**Listing 14: An aspect with an advice that checks whether a Stream is closed and a program slice that performs operations on a Stream**

*The traditional debugging*

Before starting the debugging, the programmer should consider where to put the breakpoints. Reading the exception message, she knows that the root cause must be an invocation of close(). A natural thought is setting a breakpoint to the call site of close(). However, there may be many invocations of close() in the program and setting breakpoints to each of them is troublesome. Therefore, putting the breakpoint to the body of close() and then tracing back to its call sites is more feasible.

**1.** Set two breakpoints to the body of methods close() and read(). Start debugging.
**2.** When the program is suspended, note the hit count of the breakpoint and the object identity of the Stream object. Resume debugging.
**3.** Repeat step 2 until the exception occurs. The first debugging session terminates.
**4.** Check the note and find out where the problematic Stream was closed by comparing the object identity. Record the corresponding hit count of the breakpoint set in close().
**5.** Delete or deactivate the breakpoint set in read(). Use the recorded hit count to restrain the breakpoint set in close(). Start debugging again.
**6.** When the program is suspended, it is in the execution of close() invoked by the root cause. The source, which is located at the second top stack frame in the stack trace, is the root cause.

Besides finding appropriate places to set breakpoints, this process spends significant effort on manually noting record and searching history.

### The program solution

In this solution, the programmer realizes that using the program in Listing 6 will result in many false positives. The **around** advice interrupts the matching of the trace pattern "close read" because it calls Stream.open(). We see two ways of handling this. First, Stream.open() can be considered in the pattern. Second, and more straightforward, calls to Stream.read() which are advised by the **around** advice can be excluded. Suppose the programmer selects the latter way, the desired condition should contain the negation of the pointcut **cflow**(VitalPart()) (Listing 14, line 5).

**1.** Add the following aspect and tracematch to the program. Replace the ellipsis on line 11 with the definition of the pointcut VitalPart() (Listing 14, line 2). Set a breakpoint at line 13. Start debugging.

```
1  Aspect debugging {
2    private int hitcount=0;
3    before() : execution(* Stream.close()) {
4      hitcount++;
5      print(hitcount);
6    }
7  }
8  tracematch(Stream s) {
9    sym close before : call(* Stream.close()) && this(s);
10   sym read before : call(* Stream.read()) && this(s)
11     && !(cflow(...));
12   close read {
13     // set a breakpoint on this line
14   }
15 }
```

**2.** When the program is suspended, read the printed hit count produced by line 5. The first debugging session terminates.

**3.** Delete or deactivate the set breakpoint. Use the recorded hit count to set a breakpoint in the body of Stream.close(). Start debugging again.
**4.** When the program is suspended, it is in the execution of close() invoked by the root cause. The method execution, which is located at the second top stack frame in the stack trace, is the root cause.

This process automates the manual work in the previous process. The most effort spent concentrates on writing the program, especially the condition that excludes the **around** advice. Designing such a correct condition may be non-trivial. For example, a condition requires to exclude or include multiple advices.

### The BPL solution

The programmer has the same flow of thought as she does in the program solution. She needs to exclude the calls to Stream.read() where the **around** advice is applied.

**1.** Name the **around** advice as "aroundAdvice" and define the following five breakpoints. Line 6 shows how the **around** advice is excluded. Method signature such as Stream.close() can be generated by using the user interface. Activate only close_readNoAround. Start debugging.

```
1  close(Stream s) : call(* Stream.close()) && target(s);
2  read(Stream s) : call(* Stream.read()) && target(s);
3  aroundAdvice() :
4    adviceexecution(SafeConnection.aroundAdvice);
5  read_NoAround(Stream s) :
6    composition(read, !aroundAdvice) &&
7    bind(read(s));
8  close_readNoAround(Stream s) :
9    path(close read_NoAround) &&
10   bind(read_NoAround(s), close(s));
```

**2.** When the program is suspended, close_readNoAround prints the following information on the console. The first debugging session terminates.

```
1  *************** close_readNoAround ***************
2  matched path (close read_NoAround)
3  close examples\StreamTest.java(line 10, hitcount 3)
4  read_NoAround examples\MyLogger.java(line 30, hitcount 1)
5  ——read examples\MyLogger.java(line 30, hitcount 2)
```

**3.** Delete or deactivate close_readNoAround. According to the printed information, activate the breakpoint close and configure the hit count with 3. Start debugging again.
**4.** When the program is suspended, the root cause is found.

This process overcomes the shortcomings of the previous two processes. It not only automates manual works but also constructs the condition in a straightforward way. The **composition**() expression is an intuitive way for expressing a certain action composition and it does not require a sophisticated analysis.

## 5.2 Debugging Dereference Operations

*JabRef* is an open source bibliography reference manager. We have scanned the commits in its subversion repository to find all the revisions with reports containing the keyword "bug". Revision #25 reports a fixed null pointer bug. Listing 15 and 16 show the buggy program and the revised program. Line 4, which tests whether frame.basePanel() is

```
1  class SearchManager {
2    public void actionPerformed(ActionEvent e) {
3      if (e.getSource() == escape)
4
5        frame.basePanel().stopShowingSearchResults();
6    ...
7    }
8  }
```

**Listing 15: A buggy program**

```
1  class SearchManager {
2    public void actionPerformed(ActionEvent e) {
3      if (e.getSource() == escape)
4        if (frame.basePanel() != null)
5          frame.basePanel().stopShowingSearchResults();
6    ...
7    }
8  }
```

**Listing 16: A revised program**

null, is added in the revised program. By reverse engineering, we can deduce that a *NullPointerException* is thrown in the execution of line 5 of the buggy program. There are two possible root causes, one is the field frame and another is the expression frame.basePanel(). The exception occurs on the line where frame is accessed the first time in method actionPerformed. Therefore, the possibility that frame stores a null value cannot be excluded. The programmer needs to check both dereference operations during debugging.

### The traditional debugging

**1.** Set a breakpoint on line 5. Start debugging.
**2.** When the program is suspended, inspect the value of the field frame.
**3.** The frame is not null. Deduce that the expression frame.basePanel() returns a null value.
**4.** The root cause is found. Terminate debugging.

In this process, most effort concentrates on finding the expression that returns null. This effort increases with the number of dereference operations on the same line. The programmer has to inspect each dereference operation until she can decide which one is the root cause. There are usually two ways of inspection. One is to copy and evaluate an expression. Another is to repeatedly perform "step into" and "step return" and meanwhile note which dereference operation the debugger comes to.

Another solution is putting each problematic dereference operation in a separate line and rerun the program. Then, the line number from the error message indicates that the dereference operation on that line is the cause. However, the chopped format of code is not as readable as it was. After the bug is fixed, the code fragments need to be put back together. Similar to the other traditional solution, it does not scale well when the number of dereference operations increases. Besides, this option requires changing the source code, which is not generally possible.

### The program solution

**1.** Manually code the following aspect, add it to the debuggee project, and set a breakpoint at line 9 in the added program. Start debugging.

```
1  public aspect ProgramSolutionAspect {
2    before(Object receiver) :
3      (call(* *.*(..)) || get(* *.*))
4      && target(receiver) && withincode(public void
5        SearchManager.actionPerformed(ActionEvent))
6      && if(receiver == null) {
7      int line = thisJoinPoint.getSourceLocation().getLine();
8      if(line == 5) {
9        // set a breakpoint on this line
10     }
11   }
12 }
```

**2.** When the program is suspended, inspect the variable **thisJoinPoint** and find out the signature of the method call or the field access.
**3.** The signature contains stopShowingSearchResults(). Deduce that the expression frame.basePanel() returns a null value.
**4.** The root cause is found. Terminate debugging.

This process automates the task of finding the root cause but complicates the way of setting breakpoints. The program describes the task and constructs a place for setting the breakpoint. It needs non-trivial designing and implementation effort.

### The BPL solution

**1.** Set a breakpoint with the following breakpoint declaration using the user interface to generate the **location()** expression. Start debugging.

```
1  bp() : checkNPE() && location(
2    "net\sf\jabref", "SearchManager.java", [5]);
```

**2.** When the program is suspended, the breakpoint prints the following information on the console. Deduce that the expression frame.basePanel() returns a null value.

```
1  *************** bp ***************
2  Panel.stopShowingSearchResults() has a null receiver.
```

**3.** The root cause is found. Terminate debugging.

This process combines the advantages of the previous two processes. It not only automatically detects the root cause, but also requires only trivial effort for setting the breakpoint. This comparison highlights the great convenience and efficiency of using BPL.

## 6. RELATED WORK

We categorize the related work into three groups, which are breakpoints, debuggers, and pointcut languages.

### 6.1 Breakpoints

Modern IDEs have developed some advanced breakpoints. The IntelliJ Java debugger supports temporal dependency between two breakpoints: If a breakpoint $A$ depends on another breakpoint $B$, then $A$ cannot be hit until $B$ is hit. Visual Studio allows setting breakpoints on a specific call to a function by using the stack trace. Such a breakpoint suspends the program when the call stack is exactly the same as the one it was set on. Nevertheless, these advanced breakpoints are developed based on line breakpoints, which hardly show why breakpoints are placed there. BPL uses programs to specify breakpoints and the intention of using the breakpoints, like suspending the program at method calls or field accesses, are explicit.

Chern and De Volder [6] proposed the control-flow breakpoint to suspend the program according to the state of the stack trace. The control-flow breakpoint can specify that an event should or should not occur in the control-flow of another event. The breakpoint specification can be gradually refined at runtime until only the expected suspensions occur. In our approach, control flow refinements result in a breakpoint declaration which consists of multiple *cflow* expressions. In addition to the control flow, we also support sequential execution pattern.

The stateful breakpoint [4] allows programmers to suspend the program when some line breakpoints are hit in an expected order and certain values at those hits are coincident. A stateful breakpoint consists of three parts: a set of named line breakpoints, variables bound by the line breakpoints, and an execution trace composed by the names of the line breakpoints. Our composite breakpoint has two main differences to the stateful breakpoint. First, we provide more flexible ways specifying conditions on the bound variables by explicitly using if(). Second, a primitive breakpoint in BPL, once it has been defined, can be referenced by multiple composite ones. A primitive breakpoint is not reusable in stateful breakpoints.

## 6.2 Debuggers

Bugdel [18] is an AO debugging system in which programmers can set AO breakpoints by using dedicated graphical user interfaces. It can insert statements at a breakpoint to specify what to do when the breakpoint is hit. Its breakpoint model is join-point-shadow based, which is more fine-grained than line breakpoints. However, breakpoints defined in Bugdel are independent from each other. Thus, they cannot be used to compose higher-level breakpoints.

JavaDD [10] is a declarative debugger on which programmers can perform queries over the recorded execution history. Like other query-based debuggers, JavaDD records all the salient events, such as method calls, or field assignments, at runtime. In our approach, the breakpoint declarations are similar to queries but they are written before and applied to the following execution. Our approach records only the interesting values and, thus, programmers cannot query values which were not recorded.

Ducassé [7] complained that line-based breakpoints do not have semantics and, therefore, proposed Coca, which is a debugger using only events related to source abstractions as queries. Our approach does not throw away line-based breakpoints, because they are sometimes easier to be specified and more straightforward than breakpoints using source abstractions. Besides, Coca uses Prolog, which is completely different from the language of the debuggee programs, as the query language. Learning Prolog increases the cost of using debugging facilities. Our approach aims at minimizing the learning effort by using a pointcut language using abstractions that are natural to programmers of object-oriented and aspect-oriented programs, namely one similar to AspectJ.

## 6.3 Pointcut Languages

Tracematch [1] uses regular expressions consisting of references to primitive pointcuts to specify an expected execution trace. It supports free variables in specifying a trace. Therefore, a matched trace depends on not only the order of events but also the associated variables of the events. The design of the composite breakpoint in BPL is greatly inspired by

Tracematch. However, there are two fundamental differences. First, a *tracematch* is a standalone unit and other *tracematch*es cannot reuse its members including definitions of primitive events. In BPL, a primitive breakpoint can be referred by any number of other breakpoints. Second, there are several event kinds, such as *before* and *after returning*, in Tracematch. BPL is only interested in *before* because we want to suspend the program before the interesting events occur.

Oarta [16] extends AspectJ with features, which are similar to some of BPL. It supports *named advices* by putting a name in the declaration of an advice. It also allows declaring precedence at the advice level. Our approach does not change the syntax of AspectJ and it names advices by using Java comments. Besides, our composition specification targets finding advice compositions at runtime instead of defining precedence rules for weaving.

## 7. CONCLUSION AND FUTURE WORK

In this paper, we identified five scenarios of using breakpoints. These scenarios are frequently encountered but not supported sufficiently by existing breakpoints. Programmers need to manually perform some repetitive tasks, thus debugging efficiency is decreased.

Targeting all scenarios, we proposed a breakpoint language (BPL) which models the breakpoint as a first-class values. Breakpoints are named and they are defined by AspectJ-like pointcuts which use comprehensible source-level abstractions. We devised five completely new pointcut designators and improved two or AspectJ's pointcut designators. In our language, primitive and composite breakpoints are treated uniformly and the composition level can be infinite. It is the first language to support selecting join points with a specific advice composition.

We illustrate the usage of our approach by means of two example walkthroughs. The examples show that BPL has the following advantages over the traditional debugging and the approach using other languages.

- It allows programmers to describe the logic relationship between multiple breakpoints with succinct code.

- It allows using pointcut **composition**() to express an advice composition in a straightforward way.

- It automatically records and prints information, such as the source location and the hit count, of a hit breakpoint and its referenced breakpoints. The information is helpful for localizing the root cause in an additional debugging sessions.

The pointcut **composition**() can also be used for other purposes. For example, it can verify whether a certain advice composition exists or not in the program. Another example is handling the fragile pointcut problem. Sometimes, changes to a pointcut may unexpectedly exclude or include some join points. Therefore, behavior occurring at these join points becomes undesired. To find the join-point differences, the following breakpoint declarations can be used.

```
1 oldOne() : adviceexecution(someAzpect.oldAdvice);
2 newOne() : adviceexecution(someAzpect.newAdvice);
3 inOldNotInNew() : composition(oldOne, !newOne);
4 inNewNotInOld() : composition(newOne, !oldOne);
```

Breakpoint inOldNotInNew is hit on the excluded join points and breakpoint inNewNotInOld is hit on the newly included ones. In this way, the program execution is suspended at the join points which are (not) advised in both the old and the new versions of the program. The two breakpoints narrow down the potential places causing the undesired behavior.

The BPL is motivated by some ad hoc scenarios based on our past experiences and observations. To standardize their usage, a systematic requirement analysis is needed to enhance BPL's generality. Some language features can be explored more in the future. For example, the **path()** expression should support more types of path patterns, the runtime interactivity should be fault-tolerant in case of an accidental update, etc. Also, we plan to build an omniscient debugger for advanced-dispatching languages and our BPL can be reused as part of the built-in queries.

# 8. ACKNOWLEDGEMENTS

This work is partly funded by the Chinese Scholarship Council (CSC Scholarship No.2008613009).

# 9. REFERENCES

[1] C. Allan, P. Avgustinov, A. S. Christensen, L. Hendren, S. Kuzins, O. Lhoták, O. de Moor, D. Sereni, G. Sittampalam, and J. Tibble. Adding trace matching with free variables to AspectJ. *SIGPLAN Not.*, 40(10):345–364, Oct. 2005.

[2] P. Avgustinov, A. S. Christensen, L. Hendren, S. Kuzins, J. Lhoták, O. Lhoták, O. de Moor, D. Sereni, G. Sittampalam, and J. Tibble. abc: an extensible AspectJ compiler. In *Proceedings of the 4th AOSD*, pages 87–98, New York, NY, USA, 2005. ACM.

[3] C. Bockisch, A. Sewe, H. Yin, M. Mezini, and M. Aksit. An in-depth look at alia4j. *Journal of Object Technology*, 11(1):7:1–28, Apr. 2012.

[4] E. Bodden. Stateful breakpoints: a practical approach to defining parameterized runtime monitors. In *Proceedings of the 19th ACM SIGSOFT symposium and the 13th European conference on Foundations of software engineering*, ESEC/FSE '11, pages 492–495, New York, NY, USA, 2011. ACM.

[5] E. Bodden, F. Chen, and G. Rosu. Dependent advice: a general approach to optimizing history-based aspects. In *Proceedings of the 8th ACM international conference on Aspect-oriented software development*, AOSD '09, New York, NY, USA, 2009. ACM.

[6] R. Chern and K. De Volder. Debugging with control-flow breakpoints. In *Proceedings of the 6th international conference on Aspect-oriented software development*, AOSD '07, pages 96–106, New York, NY, USA, 2007. ACM.

[7] M. Ducassé. Coca: an automated debugger for C. In *Proceedings of the 21st international conference on Software engineering*, ICSE '99, pages 504–513, New York, NY, USA, 1999. ACM.

[8] P. E. A. Dürr. *Resource-based verification for robust composition of aspects*. PhD thesis, Enschede, June 2008.

[9] M. Eisenstadt. My hairiest bug war stories. *Commun. ACM*, 40(4):30–37, Apr. 1997.

[10] H. Z. Girgis and B. Jayaraman. JavaDD: a declarative debugger for java. Technical report, 2006.

[11] B. Hailpern and P. Santhanam. Software debugging, testing, and verification. *IBM Syst. J.*, Jan. 2002.

[12] A. Hannousse, R. Douence, and G. Ardourel. Static analysis of aspect interaction and composition in component models. In *Proceedings of the 10th ACM international conference on Generative programming and component engineering*, GPCE '11, pages 43–52, New York, NY, USA, 2011. ACM.

[13] E. Katz and S. Katz. Incremental analysis of interference among aspects. In *Proceedings of the 7th workshop on Foundations of aspect-oriented languages*, FOAL '08, New York, NY, USA, 2008. ACM.

[14] B. P. Lientz and E. B. Swanson. *Software Maintenance Management*. Addison-Wesley Longman Publishing Co., Inc., Boston, MA, USA, 1980.

[15] A. Marot and R. Wuyts. Detecting unanticipated aspect interferences at runtime with compositional intentions. In *Proceedings of the Workshop on AOP and Meta-Data for Software Evolution*, RAM-SE '09, New York, NY, USA, 2009. ACM.

[16] A. Marot and R. Wuyts. Composing aspects with aspects. In *Proceedings of the 9th International Conference on Aspect-Oriented Software Development*, AOSD '10, New York, NY, USA, 2010. ACM.

[17] I. Nagy. *On the design of aspect-oriented composition models for software evolution*. PhD thesis, Enschede, June 2006.

[18] Y. Usui and S. Chiba. Bugdel: An aspect-oriented debugging system. In *Proceedings of the 12th Asia-Pacific Software Engineering Conference*, APSEC '05, pages 790–795, Washington, DC, USA, 2005. IEEE Computer Society.

[19] H. Yin, C. Bockisch, and M. Aksit. A fine-grained debugger for aspect-oriented programming. In *Proceedings of the 11th annual international conference on Aspect-oriented Software Development*, AOSD '12, New York, NY, USA, 2012. ACM.

# Secure and Modular Access Control with Aspects

Rodolfo Toledo     Éric Tanter[*]

PLEIAD Laboratory
Computer Science Department (DCC)
University of Chile – Santiago, Chile
http://www.pleiad.cl

## ABSTRACT

Can access control be fully modularized as an aspect? Most proposals for aspect-oriented access control are limited to factoring out access control checks, still relying on a non-modular and ad hoc infrastructure for permission checking. Recently, we proposed an approach for modular access control, called ModAC. ModAC successfully modularizes both the use of and the support for access control by means of restriction aspects and scoping strategies. However, ModAC is only informally described and therefore does not provide any formal guarantee with respect to its effectiveness. In addition, like in many other proposals for aspect-oriented access control, the presence of untrusted aspects is not at all considered, thereby jeopardizing the practical applicability of such approaches. This paper demonstrates that it is possible to fully modularize aspect control, even in the presence of untrusted aspects. It does so by describing a self-protecting aspect that secures ModAC. We validate this result by describing a core calculus for AspectScript, an aspect-oriented extension of JavaScript, and using this calculus to prove effectiveness and non-interference properties of ModAC. Beyond being an important validation for AOP itself, fully modularizing access control with aspects allows access control to be added to other aspect languages, without requiring ad hoc support.

## Categories and Subject Descriptors

D.3.3 [**Programming Languages**]: Language Constructs and Features; D.3.1 [**Formal Definitions and Theory**]: Semantics

## General Terms

Languages, Design

## Keywords

Access control, aspect-oriented programming, restriction aspects, scoping strategies.

---

[*]Partially funded by FONDECYT project 1110051.

# 1. INTRODUCTION

Access control [24] is a cornerstone of every security architecture: it is the component in charge of ensuring that sensitive resources are accessed only by the entities authorized to do so. In modern runtime environments such as the JVM [16] and the CLR [4], access control architectures rely on a fine-grained specification based on permissions. Permissions represent the ability to access and use a particular resource (*e.g.* a file) in a certain manner (*e.g.* read-only or read-write). Fine-grained access control in these architectures allows one to assign different sets of permissions to different entities. Furthermore, stack inspection [15] is used to dynamically examine if a sensitive operation can be performed or not. This is known as *basic permission checking.*

The Java access control architecture also includes two other mechanisms: *privileged execution* and *first-class permission contexts*. Privileged execution allows a trusted entity to take responsibility for a certain action. This makes it possible for untrusted entities to access sensitive resources—such as the screen—in a controlled manner. First-class permission contexts allow the programmer to capture the set of permissions at a certain point and restore it later on, for instance to incrementally perform a long task—such as classloading—in different threads safely.

While these three mechanisms together provide a very powerful access control system, they also introduce modularity issues. Indeed, using basic permission checking is a crosscutting concern: in order to trigger stack inspection, explicit calls to the access control architecture are necessary. As a consequence, code related to permission checking ends up scattered at each and every place where sensitive resources are accessed, tangled with other concerns. In addition to the crosscutting nature of the *use* of access control, the *implementation* of access control is itself non-modular in the sense that it does not only lie in standard libraries, but depends on native support from the runtime environment. For instance, the Java VM provides specific support for reifying the stack and permission contexts. This native support in the VM is specific to (and can only be used for) access control enforcement. This tends to suggest that access control needs to be supported as a primitive in the language, and that therefore, access control is not something that can be plugged into an existing language without having to modify its semantics.

Considering the fact that security has long been considered a typical aspect, this work addresses the following research question:

*Can access control be fully modularized as an aspect?*

Here, we are concerned not only with modularizing basic permission checking—a somewhat easy and well-explored problem [9, 19, 21, 23, 25, 34, 35]. We want to express the *whole* access control infrastructure as an aspect, including the support for advanced features ignored in the literature, namely privileged execution and capturable permission contexts. Also, we aim at answering the question: *is it possible to leave the programming language semantics completely oblivious to the presence of access control?* If so, can we ensure that malicious code, including other aspects, do not interfere with the access control aspect, and how? What are the requirements on the underlying general-purpose aspect language?

Importantly, a positive answer to these questions should also contribute the formulation of a general-purpose aspect model that can be used to add access control to languages that do not include any (or very limited) support for it, like JavaScript. Indeed, in previous work [31], we have explored how it is possible to aspectize stack-based access control with support for privileged execution and capturable permission contexts. The approach, called ModAC (for Modular Access Control) consists of expressing access control using *restriction aspects* scoped with an appropriate *scoping strategy* [26]. Restriction aspects modularize the use of access control whereas scoping strategies replace the need for a native VM mechanism specific to access control, making it possible to modularly provide basic permission checking, privileged execution, and capturable permission contexts.

The ModAC approach was instantiated in AspectScript, an aspect-oriented extension of JavaScript that supports scoping strategies [30]. The resulting implementation (hereafter called ModAC/AS) was used to provide an extensible access control library for JavaScript, called ZAC [33]. However, previous work on ModAC answers part of the above research question. First, the formulation of ModAC is informal; its actual effectiveness in controlling accesses to sensitive resources has not been proven. Second, it leaves open the possibility for untrusted aspects to interfere with access control aspects, thereby ruining its effectiveness.

**Contribution.** This work extends previous work on ModAC [31] with three contributions:

- We show that it is possible to fully modularize access control as an aspect, even in the presence of untrusted aspects, thanks to a *self-protecting* restriction aspect that impedes untrusted aspects to interfere with critical access control components (Sect. 3).

- We develop $\lambda_{AS}$, a core calculus for AspectScript based on $\lambda_{JS}$ [17] for modeling JavaScript, and a variation of the semantics of LAScheme [28] for aspect weaving (Sect. 4).

- We state and prove the effectiveness and non-interference properties of an instantiation of ModAC in $\lambda_{AS}$, ModAC/$\lambda_{AS}$. The formulation of these results is detailed in Sect. 5; the proofs are available online [32]. We discuss the extension of the results to ModAC/AS and other aspect languages in Sect. 6.

Section 2 briefly introduces access control, and aspect-oriented approaches to it, in particular ModAC. Section 7 describes related work and Section 8 concludes.

**Implementation.** This work is implemented in the ZAC library for AspectScript. Also, the executable semantics of $\lambda_{AS}$ are implemented in PLT Redex [13]. Both artifacts are available online [32].

## 2. BACKGROUND & MOTIVATION

We briefly introduce stack-based access control, illustrating its main features (Section 2.1). We then describe aspect-oriented approaches to access control, including ModAC (Section 2.2). Section 2.3 classifies various threats to modular access control.

## 2.1 Access control by example

In this section we describe the three access control features based on stack inspection: basic permission checking, privileged execution, and permission contexts. We illustrate each one with real-world examples from the JavaScript realm.

### Basic permission checking.

When a sensitive resource is about to be accessed, a call to the access control infrastructure triggers a stack inspection algorithm [15], which *checks* whether all the entities in the current stack of execution (starting from the top of the stack) possess the necessary permission to access the resource. If not, an exception is thrown. Stack inspection is triggered by calling SecurityManager.checkPermission in Java, passing the required permission; in C#, this is done by invoking Demand() on a permission object. In both systems, the entities to which permissions are assigned to are classes. In the following examples, permissions are assigned to individual objects, since JavaScript is prototype based.

This basic behavior prevents the confused deputy problem [18] from happening: an untrusted entity cannot lead a trusted one to access a sensitive resource on its behalf by simply invoking a method, because the stack inspection algorithm will eventually notice the presence of the untrusted entity on the stack. This is exemplified in the following piece of code, in which accessing a sensitive resource—the network—is forbidden:

When m is executed, the untrusted object invokes newRequest on trusted to create a new XMLHttpRequest object. Assuming that the stack inspection algorithm is triggered as in Java with a call to checkPermission (signaled by the CP gray square in the figure above), the instantiation is prevented by throwing an exception. This is so because the stack inspection algorithm eventually checks the permissions of untrusted and discovers that it does not hold the necessary permission to access the network.

### Privileged execution.

In some scenarios, it is necessary for an entity to access a sensitive resource on behalf of another—possibly untrusted—entity. For this, the JVM supports *privileged execution*. For instance, suppose that we want to provide a netService object that allows any client to access the network, provided that the target site pertains to a list of known sites. In this case, the creation of an XMLHttpRequest object should be allowed even when there are untrusted objects participating in the current call stack.

158

A self call to doPrivileged initiates a privileged action[1]. Consequently, stack inspection only considers the permissions of objects on the stack corresponding to the dynamic extent of the privileged action, *including* the initiator of the action; *i.e.* the stack inspection algorithm stops at the frame of the initiator of the call to doPrivileged.

### Permission contexts.

When accessing a sensitive resource, it can be necessary for an entity to use the permissions present at another point in the execution of the application. The JVM provides built-in means to capture a *permission context* and restore it later on.

For instance, this can be used in JavaScript to capture the permission context at the time a network connection is initiated, and reinstall it when the response from the server is received (asynchronously). This way, the response processing is performed with the same permissions as the call, similarly to a synchronous communication.

## 2.2 Access control with aspects

Due to its inherently crosscutting nature, access control has been a repeated target for applying aspects. We briefly explain these approaches in the following, and then dive into a recent proposal for fully modularizing access control.

### Permission aspects.

The most obvious source of crosscutting due to access control is the necessity of explicitly triggering stack inspection upon access to sensitive resources. Many approaches based on aspects have been proposed in order to factor out these calls into advices [9, 19, 21, 23, 25, 34, 35]. In all these approaches, aspects follow the same pattern: their pointcuts match accesses to sensitive resources, and their advice triggers access control. For example, the following aspect, declared in AspectScript [30], guards the accesses to the network:

```
var netPermission = {
    pointcut: function(jp){ return jp.kind == NEW &&
                            jp.fun === XMLHttpRequest; },
    advice: function(jp){
        checkPermission(new Permission(NETWORK)); //triggers stack inspection
        return jp.proceed();
    } };
```

This aspect[2] successfully modularizes the triggering of basic permission checking for network accesses. Aspects following this pattern are classified as *permissions aspects* due to their use of the permissions infrastructure and the stack inspection algorithm [31].

### Restriction aspects.

While permission aspects modularize calls to check if the necessary permissions are available, they do not fully modularize access control, because they rely on native support from the runtime environment in order to perform stack inspection. Recently, we described an approach for fully modular access control, ModAC [31], based on restriction aspects and scoping strategies.

In contrast to permission aspects, *restriction aspects* do not rely on any permission infrastructure or stack inspection algorithm. Instead, the *scoping mechanism* of the aspect language is used to ensure proper resource protection. A restriction aspect works by adhering to a different, dual pattern: the pointcut selects accesses to

a sensitive resource (just like a permission aspect), but the advice immediately aborts the access by not proceeding with the primitive operation; scoping strategies are used to ensure that the aspect only *sees* forbidden accesses. Consider the following restriction aspect:

```
var netRestriction = {
    pointcut: function(jp){ return jp.kind == NEW &&
                            jp.fun === XMLHttpRequest; },
    advice: function(jp){ throw "Cannot access the net."; }
};
```

This aspect forbids the access to the network. Its pointcut identifies instantiations of XMLHttpRequest objects, and the advice throws an exception with an informative message. Another possibility is not to throw an exception but to silently abort the sensitive resource access. For instance:

```
var alertRestriction = {
    pointcut: function(jp){ return jp.kind == EXEC && jp.fun === alert; },
    advice: function(jp){ /* do nothing */ }
};
```

This restriction aspect simply skips the execution of the alert method, in order to avoid popups.

### A scoping strategy for access control.

Restriction aspects are limited to see only illegal resource accesses by means of scope control. However, scope control based on control flow only, as provided by AspectJ, is insufficient to directly support features like privileged execution and permission contexts [31]. For this reason, ModAC relies on a more expressive scoping control mechanism, *scoping strategies* [26, 27, 29].

A scoping strategy permits fine-grained control over the scope of a deployed aspect. A scoping strategy itself is specified by two *propagation functions*: a *call stack* propagation function c specifies how an aspect propagates along with method calls, and a *delayed evaluation* function d specifies whether or not an aspect is "captured" in objects when they are created.[3] Intuitively, the former allows controlling dynamic scoping of aspects, stopping propagation when a certain condition is met. The latter allows an aspect to follow an object: the aspect sees all join points occurring *lexically* within all methods of the object (and may potentially propagate further in method calls done by the object depending on the call stack propagation function). Propagation functions are predicates over join points: the call stack propagation function matches *call* join points for which the aspect should propagate, while the delayed evaluation propagation function matches *object creation* join points.

Scoping strategies in AspectScript are provided as an (optional) first argument to the aspect deployment constructs: deploy(s,asp,fun), which deploys aspect asp on the body of fun; and deployOn(s,asp,obj), which deploys asp on the object obj. In both cases, s is a scoping strategy, and asp can be a single aspect or an array of aspects.

The scoping strategy for access control that supports basic permission checking, privileged execution, and permission contexts is:

```
var acs = [ //access control strategy
    function(jp){ return !(jp.fun === doPrivileged && jp.target === jp.context);},
    function(jp){ return jp.target instanceof ACContext; }
];
```

The call stack propagation function expresses both basic permission checking and privileged execution. Essentially, it specifies that a restriction aspect always propagates on the call stack, except on privileged calls. A privileged call is a *self call* to doPrivileged. This way, a restriction aspect propagating through the stack stops

---

[1] In Java, a privileged action is started by calling the static method AccessController.doPrivileged.

[2] Aspects are standard objects in AspectScript. They have one pointcut and one advice, defined by the pointcut and advice attributes respectively. Both pointcuts and advices receive a join point as parameter. All advices are around advices.

[3] Scoping strategies also include a third component, called *activation function*. Activation is not used in this work, so we omit it.

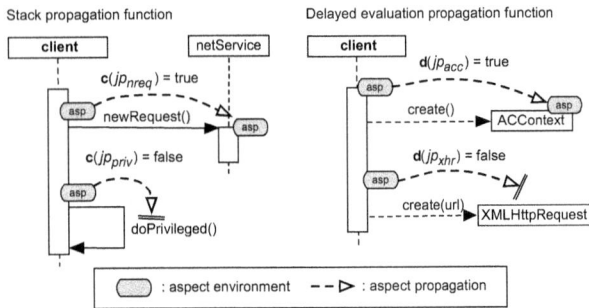

**Figure 1: Propagation of aspects with the access control strategy.**

```
1   var ACDeployer = {
2     acs: ...,  //access control strategy
3     pointcut: function(jp){ return jp.kind == NEW; }, // creation of objects
4     advice: function(jp){
5       var obj = jp.proceed();
6       var restrictions = getRestrictionsFor(obj);
7       deployOn(acs, restrictions, obj); //per−object deployment
8       return obj;
9   } };
10  deployOn([false,true],ACDeployer, function(){ /∗ main program ∗/ });
```

**Figure 2: Deployer aspect for deploying restriction aspects.**

its propagation upon a privileged call, and hence does not see resource accesses that occur in the control flow of that call. Considering self calls for privileged execution permits to maintain the aspects of the object initiating the action.

The delayed evaluation propagation function expresses the capture of permission contexts. It ensures that restriction aspects propagate to instances whose prototype is ACContext; therefore, creating such an object is a means to take a snapshot of the restriction aspects present at that point in time. Later on, it is enough to include these objects in the stack to restore the permission context. This is done by an overloaded version of doPrivileged that accepts an ACContext as extra parameter—more details can be found in [31].

Figure 1 depicts the propagation of an aspect asp deployed with the access control strategy. If asp is currently deployed (*i.e.* it is in the current aspect environment), it propagates on calls to newRequest ($jp_{nreq}$) but not on self-calls to doPrivileged ($jp_{priv}$). Therefore asp sees join points occurring during the execution of newRequest. Similarly, asp gets captured in new ACContext objects ($jp_{acc}$), and not in new XMLHttpRequest objects ($jp_{xhr}$). Hence, asp sees the subsequent activity of these ACContext objects.

ModAC fully modularizes aspect control, by relying only on the aspect language. As a matter of fact, scoping strategies replace the need for an ad-hoc, VM-supported mechanism specific to access control, as is the case of access control in the JVM and the CLR. For sure, the aspect language must support scoping strategies; however, scoping strategies are a general-purpose construct, with a range of applications beyond access control [26, 27, 29].

*Bootstrapping access control.*

Since access control is fully modularized, it is just one more aspect. In order for it to be effective in a given system, it has to be activated. In a language with dynamic aspect deployment, the only way is to do so explicitly in the program (*e.g.* around the main method, around the loading of a script, etc.). In a language with static deployment, access control must still be equivalently activated (*e.g.* command line or configuration file).

In the case of ModAC/AS, the activation of access control is performed by wrapping the main program in a deployment of the ACDeployer aspect (Figure 2). ACDeployer ensures that the relevant parts of the activity of all objects are under control of restriction aspects. It does so by deploying these restriction aspects on newly-created objects with the access control scoping strategy acs defined previously. Crucially, the deployment of restriction aspects must be done exactly in between the creation of an object and the beginning of its initialization. This way, when the object initiates computation, the necessary restriction aspects are already deployed on it.

The ACDeployer aspect deploys restriction aspects on objects when they are created. First, its pointcut matches all object creations (line

3). Then, the advice (lines 4-8) deploys the corresponding restriction aspects on the newly-created object (line 5), using deployOn (line 7) and specifying the access control scoping strategy (line 2). Finally, the object is returned (line 8). The set of restriction aspects that corresponds to a particular object is determined by the getRestrictionsFor method (line 6). This method abstracts the process of determining the needed restrictions. A possible implementation is to mimic the access control architecture of the JVM by returning the restriction aspects that correspond to the permissions declared in a policy file. Another implementation is to return restrictions based on dynamic conditions, such as the kind of user currently interacting with the application, as in role-based access control [14]. Line 10 deploys ACDeployer such that it propagates in all created objects (delayed evaluation is set to true); this ensures that it sees all object creations.

## 2.3 Threats to modular access control

ModAC seems to be a proof by existence that access control *can* be fully modularized using aspects, provided the aspect language supports a sufficiently expressive scoping mechanism. However, our previous work does not provide any formal guarantee in this respect. Most importantly, it does not consider threats posed by the presence of other, possibly untrusted, aspects.

We consider a simple attack model where the attacker can define an aspect whose purpose is to defeat access control. The attacker cannot alter the specifications of what entities are considered trusted or untrusted. These policies, the aspect weaver, and ModAC components themselves are part of the trusted computing base.

*Inhibition.*

Following the same attack model, De Borger *et al.* showed how easy it is to interfere with access control by means of aspects [8]. For instance, this AspectJ aspect completely inhibits access control in Java programs:

```
public aspect MaliciousAspect{
  void around(): execution(void SecurityManager+.check∗(..)){ }
}
```

As opposed to the JVM and the CLR, ModAC does not exhibit the previous vulnerability, simply because there are no explicit calls to a stack inspection algorithm. However, there are other alternatives for untrusted aspects to inhibit access control, to which ModAC is vulnerable: *i.e.* to prevent access control components—restriction aspects, the access control strategy, and the ACDeployer aspect—from actually controlling accesses.

We introduce the distinction between implicit and explicit inhibition. *Implicit inhibition* is based on using the aspect weaving mechanism to inhibit access control, such as in the above AspectJ example. *Explicit inhibition* consists of using other means provided by the base language (*e.g.* side effects) to prevent the different components of the access control system to fulfill their role.

*Explicit inhibition.*

There are many kinds of explicit inhibition, depending on the considered programming language. In a purely functional language, it is impossible to alter a function or mutate existing bindings and data structures. But in a stateful world, risks exist if the state of the access control components can be aliased and mutated. Such risks are exacerbated in languages like JavaScript, where it is possible to dynamically remove object members.

Fortunately, explicit inhibition requires the malicious entity to perform explicit actions, which can be observed and prevented by dedicated restriction aspects. For instance, the following restriction forbids any action on netRestriction (*e.g.* modification of its properties, invocation of its methods):

```
var metaNetRestriction = {
  pointcut: function(jp){ return jp.target === netRestriction; },
  advice: function(jp){ throw "Cannot manipulate the netRestriction aspect"; }
};
```

For any kind of explicit inhibition, a dedicated restriction must be defined. This shows how ModAC elegantly protects itself from explicit inhibition.

*Implicit inhibition.*

Because explicit inhibition can be prevented by means of restriction aspects, this paper focuses on implicit inhibition. Indeed, implicit inhibition is peculiar because it is directly enabled by the use of an aspect-oriented language; also, implicit inhibition can be achieved in any aspect language, regardless of whether or not the language allows arbitrary effects.

In the case of ModAC, there are three kinds of implicit inhibition: pointcut inhibition, advice inhibition, and scoping strategy inhibition. Pointcut inhibition consists in preventing the pointcut of an access control component from matching at relevant join points. For instance, the following malicious aspect inhibits the pointcut pc of a restriction aspect:

```
var maliciousAspect = {
  pointcut: function(jp){ return jp.kind == PCEXEC && jp.fun === pc; },
  advice : function(jp){ return false; }
};
```

The other kinds of inhibition follow a similar pattern: making a pointcut return false as above, making an advice do nothing by matching its execution but never proceeding, or impeding propagation of restriction aspects by making their propagation functions return always false, etc.

# 3. Ř: ONE ASPECT TO RULE THEM ALL

In this section we present Ř (pronounced "ring"), a self-protecting restriction aspect that prevents untrusted aspects from inhibiting access control in ModAC. We first introduce some terminology to discriminate different kinds of aspects (Section 3.1). We then describe and justify our design goals for secure modular access control (Section 3.2), and a general approach to control untrusted aspects (Section 3.3). We finally present Ř and explain how it prevents inhibition of both access control and itself (Section 3.4).

## 3.1 Aspect classification

First, we refer to all aspects that are part of ModAC—restriction aspects and the ACDeployer aspect—as *access control aspects*. We then make the distinction between *trusted aspects*, which should be given unrestricted freedom; and *untrusted aspects*, which are potentially trying to inhibit acces control. Classifying aspects as trusted or untrusted depends on the access control policy of a given

application. For example, a possible policy consists in considering all aspects defined in local code as trusted, whereas aspects defined in remote code are deemed untrusted. We do not commit to any specific means to express this classification.

In addition, we introduce a set of *protected aspects*. By definition, this set contains all aspects whose inhibition must be prevented. In order to secure ModAC, this set must include all access control aspects (but is not restricted to those aspects).

## 3.2 Securing ModAC: design goals

Our design goals for secure and modular access control are as follows:

**G1** *The base language must be completely oblivious to access control.*

**G2** *Untrusted aspects must not inhibit protected aspects, but are otherwise free to advise any join points.*

**G3** *Trusted aspects should be able to advise any join point.*

The first goal (G1) is the *raison d'être* of ModAC. Beyond being an important validation for AOP itself, fully modularizing access control with aspects allows access control to be added to other aspect languages, without requiring ad hoc support for it. The other two design goals are concerned with securing ModAC without overly restricting the programming model.

Design goal (G2) states that the non-inhibition property must be achieved without simply ruling out untrusted aspects. Untrusted aspects must be able to do whatever their access policies specify; the only strong requirement is that they do not inhibit protected aspects.

Design goal (G3) states that trusted aspects should be able to see any join point. This goal discards a restrictive approach that prohibits any kind of weaving (trusted or not) in certain core classes—thereby strongly coupling access control and weaving.

For instance, the Aspect-Oriented Permission System (AOPS) [8] ensures non-inhibition by disallowing any kind of weaving at join points lexically located in access control aspects and other sensitive components such as permission classes and the PermissionManager class. Doing so impedes even trusted aspects to advise these classes. In addition, it means that the weaver (and hence the aspect language semantics) is specifically tailored to take access control into account, something that we discard as of design goal (G1). Therefore, AOPS violates two of our design goals, (G1) and (G3).

## 3.3 Preventive inhibition

In order to reconcile goals (G2) and (G3)—*i.e.* preventing the inhibition of protected aspects by untrusted aspects, while allowing trusted aspects to see any join point—we introduce a simple technique: *preventive inhibition*. Preventive inhibition consists in inhibiting untrusted aspects *before* they get a chance to inhibit protected aspects.

To achieve preventive inhibition, it is sufficient to ensure that untrusted aspects do not apply at join points occurring in the control flow of protected aspects. For restriction aspects, this means that untrusted aspects cannot interfere with the identification of resource accesses nor with the process of aborting these accesses. For the ACDeployer aspect, this means that untrusted aspects cannot interfere with the identification of object creations nor with the calculation and deployment of restriction aspects.

## 3.4 The Ř restriction

How can preventive inhibition be achieved while maintaining (G1), *i.e.* without requiring modifications to the aspect language

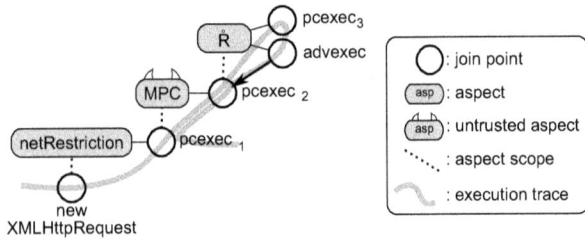

**Figure 3: Pointcut inhibition prevented by Ř.**

semantics? We describe a simple solution that is essentially just a programming pattern of ModAC. The approach relies on using a specific restriction aspect, called Ř. Ř is in charge of preventive inhibition for all protected aspects, including itself, thereby fulfilling goals (G2) and (G3). Because Ř is a restriction like any other, goal (G1) is fulfilled as well: there is no need to change the language semantics to support it.

## Inhibition with Ř.

Ř is deployed on untrusted objects at creation time, just like other restriction aspects. Its definition is:

```
var Ř = {
  pointcut: function(jp){
    return jp.kind == PCEXEC &&
      cflow(function(jp){ return protectedAspects.contains(jp.target); });
  },
  advice: function(jp){ return false; }
};
```

Ř inhibits every pointcut execution it sees, provided that the execution is in the control flow of a join point whose target is in the protected aspects set. In consequence, all aspects in the protected aspects set cannot be inhibited by untrusted aspects, simply because untrusted aspects do not even get a chance to see the join points they would potentially advise. Note that Ř is the first-class equivalent of the pointcut conjunction discussed in the previous section. Making it a restriction aspect like any other is the key to enforce this inhibition check without affecting the language semantics.

## Illustration.

Figure 3 illustrates how Ř avoids pointcut inhibition by an untrusted aspect MPC on the netRestriction aspect presented before. When a new XMLHttpRequest instance is created, a join point is generated. The netRestriction aspect sees this creation, and therefore, its pointcut is evaluated. This generates a pointcut execution join point (pcexec₁), which is observed by MPC. Consequently, the MPC pointcut is evaluated, which generates another pointcut execution join point (pcexec₂). Since MPC is untrusted, Ř was deployed on it. Hence, the pointcut of Ř sees pcexec₂, and matches it (it is a pointcut execution join point and a protected aspect, netRestriction, is in the control flow). In consequence, Ř inhibits the pointcut of MPC. Advice and scoping strategies inhibitions are prevented in a similar way.

## Self-protection.

Crucially, Ř can protect *itself* from inhibition by untrusted aspects, following the exact same principle. To do so, Ř is added to the set of protected aspects. Self-protection of Ř can be observed in the same Figure 3, by replacing the reference to netRestriction on the figure with Ř. An untrusted aspect can try to inhibit Ř as many times as it wants in the same flow of execution. If the interaction

is infinite, the program does not terminate[4]. If the interaction is finite, Ř eventually rules the untrusted aspect. Self-protection of Ř elegantly secures ModAC by not introducing any additional mechanism; Ř is just a restriction aspect protecting access control aspects, including itself, and other protected aspects, from inhibition by untrusted aspects.

## Bootstrapping.

Ř uses the protectedAspects set to identify the aspects it must protect from implicit inhibition. Naturally, untrusted entities must not be allowed to interfere with this data structure. Inhibiting access control by interfering with the protectedAspects set can either be done implicitly via weaving, or through explicit manipulation. Implicit inhibition is already prevented by Ř itself (because the protectedAspects is manipulated only by entities pertaining to the set itself, like Ř). Explicit inhibition is avoided by using a dedicated restriction aspect:

```
var paRestriction = {
  pointcut: function(jp){ //any action on protectedAspects
    return jp.target === protectedAspects; },
  advice: function(jp){ throw "Cannot manipulate the protected aspects set"; }
};
```

This restriction follows the same pattern as the metaNetRestriction presented before; it forbids *any* action over protectedAspects. This restriction must be deployed on all untrusted entities at creation time. Note that this restriction is just another restriction, and therefore (G1) is still fulfilled.

## 4. A CALCULUS FOR ASPECTSCRIPT

The previous section has informally explained how ModAC can be made secure thanks to the Ř restriction aspect. ZAC, a JavaScript library based on ModAC [33], has been extended to include Ř. However, ModAC itself has never been proven to be effective, even in the absence of untrusted aspects; and it remains to be proven that Ř is effectively securing ModAC.

In order to do so, we focus on AspectScript and establish a formal basis for it: the $\lambda_{AS}$ calculus. Section 5 then states formally that ModAC/$\lambda_{AS}$, the implementation of ModAC in the $\lambda_{AS}$ calculus, is correct and secure. Proofs and executable semantics are provided online [32]. Note that Section 5 is also accessible to readers who prefer not to dive into the formal semantics, as it includes a precise description of the argumentation line.

$\lambda_{AS}$ is a core calculus for AspectScript, and as such has to be faithful to JavaScript. We use the $\lambda_{JS}$ calculus [17] as a starting point. The aspect-oriented part of AspectScript is based on the calculus of LAScheme [28], which models first-class aspects with dynamic deployment and execution levels in a higher-order procedural language. ModAC however also requires (a form of) scoping strategies, which are not part of the existing LAScheme formalization. As a result, even though we simplify the treatment of execution levels in the calculus, $\lambda_{AS}$ is more complex. But this complexity is directly drawn from the required characteristics of the language (JavaScript-based, first-class aspects with dynamic deployment, execution levels, and scoping strategies).

We now first give a brief overview of $\lambda_{JS}$, and then describe its extension to support aspect weaving with dynamic aspect deployment and scoping strategies.

---

[4] Any untrusted piece of code is (a priori) given the power of the base language (which is Turing-complete) and can therefore always provoke non-termination. Different mechanisms (including restriction aspects!) can be used to avoid this misbehavior (*e.g.* timeout, limit on the number of produced join points), but this is out of the scope of this work [33].

$$
\begin{array}{llcl}
Value & v & ::= & c \mid \mathbf{fun}(x\cdots)\{e\} \mid o \mid l \\
Bool & b & ::= & \mathbf{true} \mid \mathbf{false} \\
Const & c & ::= & n \mid str \mid b \mid \mathbf{undefined} \mid \mathbf{null} \\
Object & o & ::= & \{str : v \cdots\} \\
Expr & e & ::= & x \mid v \mid \mathbf{let}\ (x = e)\ e \mid e(e\cdots) \mid e[e] \mid \\
& & & e[e] = e \mid e = e \mid \mathbf{ref}\ e \mid \mathbf{deref}\ e \\
Store & \mu & ::= & \epsilon \mid \mu + (l \mapsto o)
\end{array}
$$

$n \in \mathcal{N}$, the set of numbers; $str \in \mathcal{S}$, the set of strings; $x \in \mathcal{X}$, the set of variable names; $l \in \mathcal{L}$, the set of locations.

**Figure 4: Syntax of the $\lambda_{JS}$ language (excerpt; slightly modified).**

## 4.1 Core JavaScript: $\lambda_{JS}$

Guha *et al.* designed $\lambda_{JS}$ as a core subset of JavaScript to which JavaScript programs are desugared. The interest of $\lambda_{JS}$ is its compactness. We briefly describe the syntax of $\lambda_{JS}$, the desugaring process, and a few reduction rules.

*Syntax.*

Figure 4 shows part of the syntax of $\lambda_{JS}$. The language has primitive values such as numbers, strings, booleans, and two special values **null** and **undefined**, in addition to functions (**fun**) and objects $o$. Objects are a series of attribute-value pairs enclosed in curly braces. Expressions include identifiers, values, a **let** construct, function application, property access, and property write. In order to support first-class mutable references, values are augmented with store locations. Objects in the store are explicitly referenced and dereferenced using **ref** and **deref**, respectively. $\lambda_{JS}$ also includes typical control operators and primitive n-ary operators; we omit these for brevity.

*Desugaring.*

Several JavaScript constructs are specified via translation (called "desugaring") to $\lambda_{JS}$ [17]. For example, the desugaring of function creation is:

```
desugar[[ function(x···){e} ]] = ref {
    "code": fun(this, fthis, x···) { desugar[[ e ]] },
    "prototype": ref {"__proto__": (deref Object)["prototype"]}}
```

A function is desugared into an object (using the {...} notation) with two attributes: code and prototype. The code attribute is the actual function (note that function is a JavaScript term, and **fun** is a $\lambda_{JS}$ term). Also, this is an ordinary identifier: it is the first formal parameter of a desugared function. In JavaScript, a method is a function, which is a value, and can be shared between objects; this refers to the currently-executing object. For the sake of properly dealing with aspect environments in $\lambda_{AS}$, we slightly extend $\lambda_{JS}$ and pass a second parameter to every desugared function; the parameter, named fthis, is bound to the function object thus created by the desugaring process. Note that desugaring reveals some of JavaScript peculiarities: the prototype attribute of a function object is an object whose prototype is the prototype attribute of Object.

The semantics of $\lambda_{JS}$ is defined as a small-step reduction relation $\hookrightarrow$. A program configuration $\langle \mu, e \rangle$ consists of a store and an expression. The reduction relation is standard. Evaluation contexts [36] are used to specify a call-by-value, left-to-right evaluation semantics. E.g., the reduction rule for object creation is:

$$\langle \mu, E[\mathbf{ref}\ \{str : v \cdots\}]\rangle \hookrightarrow \langle \mu', E[l]\rangle \qquad \text{NEW}$$
$$\text{where } l \notin dom(\mu) \text{ and } \mu' = \mu + (l \mapsto \{str : v \cdots\})$$

**ref** simply allocates a new location in the store and returns it.

$$
\begin{array}{lcl}
J & ::= & \epsilon \mid j + J \\
j & ::= & \lceil k, l_o, l_f, p \rceil \\
k & ::= & \texttt{new} \mid \texttt{call} \mid \texttt{exec} \mid \texttt{pc-exec} \mid \texttt{adv-exec}
\end{array}
$$

$p \in \mathcal{T}$, the set of thunks $J \in \mathcal{J}$, the set of join point stacks

$$
\begin{array}{lcl}
Expr \quad e & ::= & \ldots \mid \texttt{jp}\,(j,\alpha) \mid \texttt{in-jp}\,(e) \mid \\
& & \texttt{c/asp}\ k\ e\ e \cdots \mid \\
EvalCtx \quad E & ::= & \ldots \mid \texttt{in-jp}\,(E) \mid \\
& & \texttt{c/asp}\ k\ v \cdots E\ e \cdots \\
v & ::= & \ldots \mid \boxed{J}
\end{array}
$$

**Figure 5: Join points**

| $k$ | new | call / exec / pc-exec / adv-exec |
|---|---|---|
| $l_o$ | object prototype | target object |
| $l_f$ | **null** | target function |
| $p$ | | primitive operation |

**Figure 6: Join point abstraction attributes per kind.**

The function application rule is the standard $\beta_v$ reduction:

$$\langle \mu, E[\mathbf{fun}(x\cdots)\{e\}(v\cdots)]\rangle \hookrightarrow \langle \mu, E[e[v\cdots/x\cdots]]\rangle \quad \text{CALL}$$

## 4.2 AspectScript Semantics

We now describe the syntax and operational semantics of $\lambda_{AS}$, a core calculus for AspectScript based on $\lambda_{JS}$. Its operational semantics is defined via the reduction relation $\hookrightarrow: \mathcal{M} \times \mathcal{A} \times \mathcal{J} \times \mathcal{E} \to \mathcal{M} \times \mathcal{A} \times \mathcal{J} \times \mathcal{E}$.

We extend the $\lambda_{JS}$ configuration with two additional elements: a $\lambda_{AS}$ program configuration $\langle \mu, \alpha, J, e \rangle$ consists of a store $\mu \in \mathcal{M}$, an aspect environment $\alpha \in \mathcal{A}$, a join point stack $J \in \mathcal{J}$, and an expression $e \in \mathcal{E}$. The *stack aspect environment* $\alpha$ is used to maintain the aspects propagated through the stack by means of the call stack propagation function.[5]

In the following we describe the semantics of join points, aspects and their deployment, as well as the weaving semantics. The formalism is based on the semantics of LAScheme [28], an aspect-oriented Scheme-like language with execution levels, itself based on a combination of Clifton and Leavens's work [6] (modeling of the join point stack) and Dutchyn *et al.* [10] (weaving semantics). By convention, when we introduce new user-visible syntax (*e.g.* the aspect deployment expression), we use **bold** font. Internal terms are written in `typewriter` font.

### 4.2.1 Join Points

The join point stack $J$ is a list of *join point abstractions* $j$, which are tuples $\lceil k, l_o, l_f, p \rceil$ (Figure 5). We introduce five kinds of join points: new for object creation, call for function application and method invocation, and exec, pc-exec, adv-exec for function, pointcut, and advice execution, respectively. Figure 6 describes the different values for the components of join point abstractions, depending on their kind. For instance, $p$ is always the primitive operation (used to perform the original computation); $l_o$ denotes the prototype of the object being created in a new join point, and the target object for call and the three execution join points.

In order to keep track of the join point stack in the semantics we introduce two internal expression forms. $\texttt{jp}\,(j,\alpha)$ introduces

---

[5] We also maintain the currently-executing object/function in the program configuration, omitted here for simplicity. The online Redex model includes the full configuration.

163

$$Expr \quad e \quad ::= \quad \ldots \mid \textbf{deployOn}[e,e](e,e)$$
$$EvalCtx \quad E \quad ::= \quad \ldots \mid \textbf{deployOn}[E,e](e,e) \mid$$
$$\textbf{deployOn}[b,E](e,e) \mid$$
$$\textbf{deployOn}[b,b](E,e) \mid$$
$$\textbf{deployOn}[b,b](v,E)$$

$$AspectEnv \quad \alpha \quad ::= \quad \alpha + (b_c, b_d, l) \mid \epsilon$$
$$Store \quad \mu \quad ::= \quad \epsilon \mid \mu + (l \mapsto o^\alpha)$$
$$\mathsf{asps}(l) \quad = \quad \alpha, \text{ where } \mu(l) = o^\alpha$$

$$\langle \mu, \alpha, J, E[\textbf{deployOn}[b_c, b_d](l_{asp}, l_{obj})] \rangle \qquad \text{DEPLOYON}$$
$$\hookrightarrow \langle \mu', \alpha, J, E[l_{obj}] \rangle$$
$$\text{where} \quad \mu(l_{obj}) = o^\alpha \text{ and } \mu' = \mu(l_{obj} \mapsto o^{\alpha' + (b_c, b_d, l_{asp})})$$

**Figure 7: Aspects and deployment.**

$$\langle \mu, \alpha, J, E[\textbf{ref} \{str : v \cdots\}] \rangle \qquad \text{NEW}$$
$$\hookrightarrow \langle \mu, \alpha, J, E[\mathsf{jp}(\lceil \mathtt{new}, proto, \mathbf{null}, p \rceil, \alpha)] \rangle$$
where
$$proto = v_i \text{ if } str_i = \texttt{"\_\_proto\_\_"}$$
$$\alpha' = (\mathsf{asps}(\mathtt{cobj}()) \oplus \mathsf{asps}(\mathtt{cfun}()) \oplus \alpha)|_d$$
$$p = \textbf{fun}()\{ \text{ new/prim } \{str : v \cdots\}^{\alpha'} \}$$

$$\langle \mu, \alpha, J, E[\textbf{fun}(x\cdots)\{e\}(l_0\, l_1\, v \cdots)] \rangle \qquad \text{CALL}$$
$$\hookrightarrow \langle \mu, \alpha, J, E[\mathsf{jp}(\lceil \mathtt{call}, l_0, l_1, p_c \rceil, \alpha')] \rangle$$
where
$$\alpha' = (\mathsf{asps}(\mathtt{cobj}()) \oplus \mathsf{asps}(\mathtt{cfun}()) \oplus \alpha)|_c$$
$$p_e = \textbf{fun}()\{ \text{ app/prim } \textbf{fun}(x\cdots)\{e\}\, l_0\, l_1\, v \cdots \}$$
$$p_c = \textbf{fun}()\{ \text{ app/prim } \textbf{fun}()\{\mathsf{jp}(\lceil \mathtt{exec}, l_0, l_1, p_e \rceil, \alpha')\}\}$$

$$\langle \mu, \alpha, J, E[\texttt{c/asp}\, k\, \textbf{fun}(x\cdots)\{e\}\, l_0\, l_1\, v \cdots] \rangle \qquad \text{C/ASP}$$
$$\hookrightarrow \langle \mu, \alpha, J, E[\mathsf{jp}(\lceil k, l_0, l_1, p \rceil, \alpha)] \rangle$$
$$\text{where} \quad p = \textbf{fun}()\{ \text{ app/prim } \textbf{fun}(x\cdots)\{e\}\, l_0\, l_1\, v \cdots \}$$

**Figure 8: Join point creation.**

a join point $j$ whose underlying computation via proceed will be executed with aspect environment $\alpha$. $\texttt{in-jp}(e)$ keeps track of the fact that execution of $e$ is proceeding under a dynamic join point. We extend the definition of evaluation contexts accordingly (Figure 5). The expression $\texttt{c/asp}$ (which stands for "call/aspect") is used later to treat pointcut and advice execution join points similarly. It is a function application annotated with the kind of join point $k$ that needs to be created; this expression form is generated by the weaver, discussed later on.

A join point abstraction captures the minimum context information necessary for ModAC to work (target object and function), as well as to trigger its corresponding computation when necessary (the $p$ function). We write $\boxed{J}$ to denote the reification of the join point stack $J$ as a $\lambda_{AS}$ value. A number of introspection primitives are provided; for instance, $\textbf{kind}(\boxed{J})$ is the $\lambda_{AS}$ equivalent of jp.kind in AspectScript. Similarly, $\textbf{tobj}$ (resp. $\textbf{tfun}$) can be used to retrieve the (location of the) target object (resp. function).

### 4.2.2 Aspects and Deployment

For the sake of conciseness and simplicity, we make the three following simplifications to $\lambda_{JS}$ in this paper: *i)* scoping strategies have constant boolean components (instead of join point predicates); *ii)* only per-object deployment (**deployOn**) is described; *iii)* we do not account for context exposure (*i.e.* pointcuts simply return **true** if they match, instead of an environment). These simplifications do not affect the validity of our results: constant propagation functions are enough to state and prove the desired properties of ModAC, **deployOn** is strictly more expressive than **deploy** [29], and context exposure is an orthogonal feature for this work.

As described on Figure 7, an aspect environment $\alpha$ is a list of tuples $(b_c, b_d, l)$ where $l$ denotes the reference to the aspect, and the two boolean values corresponds to the c and d components of the scoping strategy specified at deployment time. An aspect can be any object whose pointcut attribute is a function that takes a join point stack as input and produces either **true** or **false**. To compensate for the absence of context exposure from pointcuts, an advice function also receives as first argument the current join point stack. An advice proceeds using the $\textbf{proceed}(\boxed{J})$ primitive.

An aspect is deployed with **deployOn**. Because **deployOn** embeds an aspect within an object, the stack aspect environment of the program configuration is not enough; each object needs to have its own aspect environment as well. To do so, we annotate an object

$o$ with its aspect environment $\alpha$ as $o^\alpha$. By construction, an object is annotated with its aspect environment as soon as it is allocated in the store (with **ref**). We therefore extend the definition of the store, and introduce an internal function $\mathsf{asps}$ in order to access the aspects of an object in the store.

The DEPLOYON rule shows the semantics of per-object deployment: the aspect (at location) $l_{asp}$ is added at the end of the aspect environment of the object (at location) $l_{obj}$, along with the specified scoping strategy components.

### 4.2.3 Join Point Creation & Disposal

We change the NEW rule of $\lambda_{JS}$ to account for the creation of new join points (Figure 8). The join point abstraction components are filled according to Figure 6. The primitive operation $p$ is a thunk that returns a fresh reference to the newly-created object. Actual object creation is done using **new/prim**, an internal expression that performs creation without generating any join point. Note that the object value passed to **new/prim** is annotated with its initial aspect environment, $\alpha'$. This environment is calculated as the order-preserving union ($\oplus$) of three aspect environments: the ones deployed on the currently-executing object and function (obtained with $\mathtt{cobj}()$ and $\mathtt{cfun}()$, respectively); and the stack aspect environment. Only aspects that propagate in newly-created objects are included in $\alpha'$. The notation $\alpha|_d$ refers to the aspects in $\alpha$ whose d component is true.

To account for the creation of $\texttt{call}$ and $\texttt{exec}$ join points, we change the $\lambda_{JS}$ evaluation rule for function application/method invocation as well. The new CALL rule generates a $\texttt{call}$ join point whose components are filled according to Figure 6. The primitive operation $p_c$ is a thunk that generates an $\texttt{exec}$ join point when applied. The primitive operation of this $\texttt{exec}$ join point, $p_e$, performs the actual function execution by means of **app/prim**, another internal expression that does not generate join points. Note that the jp expressions associated to both join points specify that the stack aspect environment must change to $\alpha'$ when $p_c$ or $p_e$ are applied in order to reflect the propagation of aspects through the stack. This aspect environment is determined by taking the order-preserving union of three aspects environments: the ones deployed on the currently-executing object and function; and the stack aspect environment; and filtering the resulting environment along the c

$\langle \mu, \alpha, j + J, E[\texttt{in-jp}\,(v)] \rangle \hookrightarrow \langle \mu, \alpha, J, E[v] \rangle \qquad \textsc{OutJp}$

$\langle \mu, \alpha, j + J, E[\texttt{in-jp}\,(\mathbf{err}\,v)] \rangle \hookrightarrow \langle \mu, \alpha, J, E[\mathbf{err}\,v] \rangle \ \textsc{OutJp-Err}$

**Figure 9: Join point disposal.**

$\langle \mu, \alpha, J, E[\texttt{jp}(\lceil k, l_o, l_f, p \rceil, \alpha_p)] \rangle \qquad\qquad \textsc{Weave}$

$\hookrightarrow \langle \mu, \alpha, J', E[\texttt{in-jp}(\texttt{swap}(\texttt{app/prim}\,W[\![\alpha']\!]_{\alpha_p, J'}, \epsilon))] \rangle$

where

$\qquad J' = \lceil k, l_o, l_f, p \rceil + J$

$\qquad \alpha_s = \epsilon$ if $k \in \{\texttt{pc-exec}, \texttt{adv-exec}\}, \alpha$ otherwise

$\qquad \alpha' = \texttt{asps}(\texttt{cobj}()) \oplus \texttt{asps}(\texttt{cfun}()) \oplus \alpha_s$

$W[\![\epsilon]\!]_{\alpha, \lceil k, l_o, l_f, p \rceil + J} = \mathbf{fun}()\{\texttt{swap}(\texttt{app/prim}\,p, \alpha)\}$

$W[\![\alpha_w + (b_c, b_d, l_{asp})]\!]_{\alpha, \lceil k, l_o, l_f, p \rceil + J} =$

$\quad \texttt{app/prim}$

$\quad \mathbf{fun}(next)\{$

$\qquad \mathbf{let}(pc = (\mathbf{deref}\,l_{asp})["pc"])$

$\qquad \mathbf{if}(\texttt{c/asp pc-exec}\,(\mathbf{deref}\,pc)["code"]\,l_{asp}\,pc\,\boxed{j_p + J})\{$

$\qquad\quad \mathbf{let}(adv = (\mathbf{deref}\,l_{asp})["adv"])$

$\qquad\quad \mathbf{fun}()\{$

$\qquad\qquad \texttt{c/asp adv-exec}\,(\mathbf{deref}\,adv)["code"]\,l_{asp}\,adv\,\boxed{j_a + J}\ \}$

$\qquad \}\mathbf{else}\{\ next\ \}$

$\quad \}$

$\quad W[\![\alpha_w]\!]_{\alpha, \lceil k, l_o, l_f, p \rceil + J, p}$

where $\quad j_a = \lceil k, l_o, l_f, \mathbf{fun}()\{\texttt{app/prim}\,next\}\rceil$

$\qquad\quad j_p = \lceil k, l_o, l_f, \mathbf{fun}()\{\mathbf{err}\ \texttt{"pc cannot proceed"}\}\rceil$

**Figure 10: Aspect weaving.**

component (written $\alpha|_c$), which determines the aspects that should propagate on the call stack.

Rule C/Asp accounts for the creation of pc-exec and adv-exec join points. This rule matches a function application/method invocation, but receives a first argument ($k$) that specifies which join point must be generated. Because invocations of pointcuts and advices are implicit, C/Asp does not generate call join points. Join point attributes are filled according to Figure 6; the primitive operation $p$ performs the pointcut/advice execution by means of app/prim, just like in the case of exec join points.

Once the computation underlying a join point is reduced to a value, the OutJp rule gets rid of the join point and the in-jp expression (Figure 9). OutJp-Err does the same in the case of an error.

### 4.2.4 Weaving

We now turn to the semantics of aspect weaving, specified by the Weave rule (Figure 10). A jp expression reduces to an in-jp expression (to signal the fact that the upcoming computation is associated to a join point), and the join point is pushed onto the stack (we discuss the use of swap and $\alpha_s$ later below). The list of aspects in scope $\alpha'$ is calculated as the order-preserving union of the aspect environments of the object and function in context, and the aspects propagated through the stack.

The weaving process is based on evaluating the function returned by the $W$ metafunction. $W$ recurs on $\alpha'$ and returns a composed procedure whose structure reflects the way advice is going to be dispatched. The base case, $W[\![\epsilon]\!]$, corresponds to the execution of the primitive operation. Otherwise, for each aspect $(b_c, b_d, l_{asp})$

$Expr \quad e \ ::= \ \dots \mid \texttt{app/prim}\,e\,e \cdots \mid \texttt{new/prim}\,e$

$EvalCtx \quad E \ ::= \ \dots \mid \texttt{app/prim}\,v \cdots E\,e \cdots$

$\qquad\qquad\qquad \mid \texttt{new/prim}\,E$

$\langle \mu, \alpha, J, E[\texttt{app/prim}\,\mathbf{fun}(x \cdots)\{e\}\,v \cdots] \rangle \qquad \textsc{AppPrim}$

$\hookrightarrow \langle \mu, \alpha, J, E[e[v \cdots/x \cdots]] \rangle$

$\langle \mu, \alpha', J, E[\texttt{new/prim}\,o^\alpha] \rangle \hookrightarrow \langle \mu', \alpha', J, E[l] \rangle \quad \textsc{NewPrim}$

where $\quad l \notin dom(\mu)$ and $\mu' = \mu + (l \mapsto o^\alpha)$

**Figure 11: Primitive function application and object allocation.**

in the environment, $W$ first applies its pointcut to the current join point stack (which generates a pc-exec join point using the c/asp construct). If the pointcut matches, then $W$ returns a function that applies the advice of $l_{adv}$ (and generates an adv-exec join point). All this process is parameterized by the function to proceed with, $next$. In order to allow an advice to call **proceed** to trigger either the base computation or the next advice in the chain, rule Weave creates an auxiliary join point $j_a$ whose $p$ component is a thunk that applies $next$. To be complete, an auxiliary join point $j_p$ is also created and passed to the pointcut; its $p$ component triggers an error if **proceed** is called. Finally, If an aspect does not apply, then $W$ simply returns $next$.

### Primitive forms.

The semantics of $\lambda_{AS}$ use internal primitive forms app/prim and new/prim, described in Figure 11. app/prim is an application that does not trigger a join point: rule AppPrim simply performs the classical $\beta_v$ reduction. app/prim is used to perform the actual application of a function, as well as to hide "administrative" application, *i.e.* the initial application of the composed aspect chain, and its recursive applications. Similarly, new/prim allocates an object in the store and reduces to the corresponding location without producing a join point.[6]

### Execution levels.

The weaving semantics explained previously is insufficient, because any aspect language must take precautions with infinite regression. Indeed, if we omitted the use of swap and $\alpha_s$ in Figure 10, a $\lambda_{AS}$ program would never terminate. Tanter addressed this issue with execution levels [28], which ensure that pointcut and advice computation by default always happen at a higher level than base computation, avoiding infinite loops such as those due to pointcuts matching against themselves. Recall that in $\lambda_{AS}$, pointcuts and advices are standard functions. With execution levels, pointcuts and advices are always evaluated at the level above the expression that generates a join point. When the last advice in the chain proceeds, execution shifts back to the original level in order to run the base computation.[7]

---

[6] These primitive forms are necessary for the semantics to allow actual computation to happen. The fact that they are *internal* means that it is not necessary to protect them from untrusted aspects: they cannot be used by any user code, and cannot be advised since they do not produce join points. Recall that in a higher-order aspect language, the use of execution levels is key to supporting these primitive forms as internal only [28].

[7] Full-fledged execution levels include the possibility to explicitly shift execution up and down if needed, as well as to define level-capturing functions [28]. We do not include these advanced facilities in this work.

$$\begin{array}{lcl} Expr & e & ::= \ \dots \ | \ \mathtt{swap}(e,\alpha) \ | \ \mathtt{in\text{-}swap}(e,\alpha) \\ EvalCtx & E & ::= \ \dots \ | \ \mathtt{in\text{-}swap}(E,\alpha) \end{array}$$

$$\langle \mu, \alpha, J, E[\mathtt{swap}(e,\alpha')]\rangle \qquad\qquad \textsc{In-Swap}$$
$$\hookrightarrow \langle \mu, \alpha', J, E[\mathtt{in\text{-}swap}(e,\alpha)]\rangle$$
$$\langle \mu, \alpha', J, E[\mathtt{in\text{-}swap}(v,\alpha)]\rangle \qquad\qquad \textsc{Out-Swap}$$
$$\hookrightarrow \langle \mu, \alpha, J, E[v]\rangle$$
$$\langle \mu, \alpha', J, E[\mathtt{in\text{-}swap}((\mathbf{err}\ v),\alpha)]\rangle \qquad \textsc{Out-Swap-Err}$$
$$\hookrightarrow \langle \mu, \alpha, J, E[\mathbf{err}\ v]\rangle$$

**Figure 12: Swapping aspect environments.**

We introduce a simple modeling of execution levels, that does not require having to explicitly track the current execution level in the program configuration. Instead, we use the call stack with internal expressions so as to *swap* aspect environments and restore them when appropriate (Figure 12). Swapping per se is a very simple process: given an expression $e$ and an aspect environment $\alpha'$, swap installs the aspect environment, and evaluates the expression (IN-SWAP). in-swap is used to restore the swapped aspect environment $\alpha$ when the expression is fully reduced (OUT-SWAP). Additionally, $\alpha_s$ is used to remove aspects in the stack aspect environment from scope when weaving pc-exec and adv-exec join points. This prevents the aspects deployed on the currently executing object/function from seeing their own activity. Note that this approach does support multiple levels of execution.

Weaving (Figure 10) uses swap exactly where the original levels semantics [28] uses **up** and **down** shifting. The whole weaving process is wrapped by a swap, so that the current aspect environment is swapped with an empty environment $\epsilon$ that represents the upper level environment . This environment is used to evaluate pointcuts and advices. Of course, the fact that the stack aspect environment starts empty does not prevent aspects that have been deployed in objects and functions from taking effect. If the last advice proceeds (the base case of $W$), aspect environments are swapped again, in order to restore the original environment to evaluate the base computation. The environment in which weaving is carried out is restored once the base computation has completed. Finally, when the whole weaving is complete, the original aspect environment is restored.

## 5. PROPERTIES OF MODULAR ACCESS CONTROL

In this section we state two theorems corresponding to the following properties of ModAC:

**Basic effectiveness.** ModAC is effective in absence of untrusted aspects. This means that restriction aspects are actually deployed on untrusted objects, see illegal resource accesses, and effectively prevent them.

**Non-inhibition.** ModAC with $\mathring{R}$ is effective even in presence of untrusted aspects. This means that $\mathring{R}$ effectively prevents untrusted aspects from inhibiting protected aspects.

More precisely, we show the results for ModAC/$\lambda_{AS}$. The extension of these results to ModAC/AS (and other aspect languages) in Section 6. This section sketches the formal argument by describing the main intermediate steps. Each step includes an informal explanation, and a formal statement. The actual proofs, which rely on the operational semantics above, are provided online [32].

First, we describe the properties that define basic effectiveness.

DEFINITION 1 (BASIC EFFECTIVENESS). *An implementation of ModAC is said to comply with basic effectiveness if the following properties are fulfilled:*

- Restrictions deployment. *Restrictions are deployed on all the corresponding objects before these objects can be used.*

- Restrictions scope. *A restriction aspect sees all the computation produced by the objects it is deployed on.*

- Restrictions effectiveness. *A restriction aspect always prevents the resource accesses it identifies.*

THEOREM 1 (MODAC/$\lambda_{AS}$ BASIC EFFECTIVENESS). *ModAC/$\lambda_{AS}$ complies with basic effectiveness.*

This theorem is a direct consequence of Lemmas 1, 2, and 3, exposed below, which address each property of basic effectiveness separately.

Lemma 1 states that any aspect (referenced by $l_{depl}$), in particular ACDeployer, deployed with scoping strategy (**false**,**true**) propagates to every new object in the store, and does so *in the first position* in the aspect environment of these objects. This ensures that $l_{depl}$ sees all object creations in the application and gets a reference to these objects before any other entity. The only prerequisite is that $l_{depl}$ is already deployed in the first position on all objects at a given point. This can be straightforwardly achieved in the bootstrapping process by exhaustively deploying ACDeployer on every object.[8]

LEMMA 1 (RESTRICTIONS DEPLOYMENT). *Let $C = \langle \mu, \cdot, \cdot, \cdot\rangle$ be a program configuration where $\forall\, (l \mapsto o^\alpha) \in \mu$, $\alpha = (\textbf{false},\textbf{true}, l_{depl}) + \alpha'$, for some $\alpha'$, and $l_{depl} \in dom(\mu)$. If $C \hookrightarrow \langle \mu', \cdot, \cdot, \cdot\rangle$, then $\forall\, (l \mapsto o^\alpha) \in \mu'$, $\alpha = (\textbf{false},\textbf{true}, l_{depl}) + \alpha''$, for some $\alpha''$.*

Lemma 2 states that all aspects in the stack aspect environment deployed with c = **true**, propagate through the stack if the same level of execution is considered; *i.e.* the stack inspection algorithm is correctly implemented by means of scoping strategies.

LEMMA 2 (RESTRICTIONS SCOPE). *Let $C = \langle \cdot, \alpha, \cdot, \cdot\rangle$ be a program configuration and $\alpha_s = \alpha|_c$. If $C \hookrightarrow^* \langle \cdot, \alpha', \cdot, \cdot\rangle$, and the sequence of reductions starts and ends at the same execution level, then $\alpha_s \subseteq \alpha'$.*

Lemma 3 states that if a restriction aspect R matches a join point $j$ and does not proceed, then the primitive operation associated to $j$ is not evaluated. Consequently a restriction aspect fulfills its role no matter in which position it is woven at the illegal resource access join point.

LEMMA 3 (RESTRICTIONS EFFECTIVENESS). *Let $C = \langle \mu, \cdot, J, E[e]\rangle$ be a program configuration where $J = \lceil \cdot, \cdot, \cdot, p\rceil + J'$, for some $J'$, $e = \mathtt{app/prim}\ W[\![\alpha]\!]_{\cdot, J}$, $(\cdot, \cdot, l_R) \in \alpha$, and $l_R$ is a valid aspect reference in $\mu$ to a restriction aspect that matches $J$ and does not proceed for $J$. If $C \hookrightarrow^* \langle \cdot, \cdot, J, E[v]\rangle$, for some $v$, then $p$ is not applied in these reductions.*

Finally, we present the non-inhibition theorem. This theorem states that if the evaluation of a pointcut whose aspect has $\mathring{R}$ deployed on it reduces to a value, this value is either **false** or (**err** $\cdot$). This holds whenever the join point stack contains a join point whose target is in the set of protected aspects $PA$. Notice that the theorem implicitly permits the existence of other untrusted aspects trying to inhibit $\mathring{R}$ itself.

---

[8]Whenever an element of an entity (program configuration, join point, tuple, etc.) is not required, we use $\cdot$ as a wildcard.

THEOREM 2 (NON-INHIBITION). *If $l_{asp}$ is a valid aspect reference in $\mu$, $\mathring{R} \in \mathsf{asps}(l_{asp})$, and $\lceil \cdot, s, \cdot, \cdot \rceil \in J$; where $s \in PA$, then:*
*If $\langle \mu, \alpha, J, E[\mathsf{jp}(\lceil \mathsf{pc\text{-}exec}, l_{asp}, \cdot, \cdot \rceil, \cdot)] \rangle \hookrightarrow^* \langle \cdot, \alpha, J, E[v] \rangle$, then $v = \mathbf{false}$ or $v = (\mathbf{err} \cdot)$.*

# 6. DISCUSSION

We discuss how to extend our results from $\lambda_{AS}$ to full-fledged AspectScript, and the requirements for a general-purpose aspect language to securely support ModAC.

## From $\lambda_{AS}$ to AspectScript.

Due to desugaring, results obtained in $\lambda_{JS}$ do not immediately apply to JavaScript [17]. This is because desugaring introduces new behavior that was not present in the original code. When going from $\lambda_{AS}$ to AspectScript, the theorems remain valid because they are based on the aspect-oriented features of the language, which have no relation to the desugaring process. However, access control aspects can be led to behave incorrectly if they use "exploitable" features that introduce holes upon desugaring. For example, consider a slight modification of the pointcut of the netRestriction aspect in order to allow communication with safe.cl:

```
function(jp){ return /* same as before */ && !(jp.args[0] == "safe.cl"); }
```

The equality operator == forces both operands to be of the same type [11]. For this reason, jp.args[0] is transformed to a string by an invocation of toString. The problem is that this extra method call opens the opportunity for bypassing access control:

```
var req = netService.newRequest(
        { t: 0, toString: function(){return ["safe.cl", "evil.com"][this.t++]}});
```

The toString method of the argument to newRequest returns "safe.cl" the first time it is invoked (in the pointcut of netRestriction) and "evil.com" the second time (in the body of newRequest).

In order to avoid such holes, the first possibility is to simply avoid using exploitable features in the definition of restriction aspects. For instance, it is safe to use reference equality === because it does not perform any kind of type conversion [11] (notice that all restriction aspects defined in this work follow this guideline). A less drastic solution is to permit the use of exploitable features, but to carefully examine access control aspects in order to check if their *particular usage* of the feature is safe. For example, the equality operator == is safe if both operands are of the same runtime type! As detailed by Guha *et al.*, this checking can be automated by a specialized type system [17].

Finally, AspectScript uses a scoping strategy acs, which supports privileged execution and capturable permission contexts; acs is expressed with propagation functions, c and d. We made a simplification in $\lambda_{AS}$ by supporting only constant boolean propagation values. As we said in Section 4.2.2, this simplification does not affect our results. In fact, supporting propagation functions only requires that $\mathring{R}$ prevents inhibition of these functions; this is achieved by extending the pointcut of $\mathring{R}$:

```
function(jp){return ... || cflow(function(jp){return acs.contains(jp.fun);});}
```

This way, $\mathring{R}$ also inhibits untrusted aspects in the control flow of acs components. Theorem 2 and its proof must be reformulated accordingly; but this is direct.

## Aspect languages for secure ModAC.

This paper focuses on AspectScript to be as close as possible to our practical implementation, ZAC [33]. Still, both ModAC and the approach for securing it using $\mathring{R}$ are independent of AspectScript.

They can be realized in any aspect language, provided it meets certain key conditions. First of all, the language must support scoping strategies, or an equivalently expressive scoping mechanism. Per-object aspects are only necessary if one wants to provide per-object access control. Execution levels are necessary to avoid infinite loops whenever pointcut and/or advice execution join points are exposed to weaving; in order to control implicit inhibition, $\mathring{R}$ relies on matching pointcut execution join points.

A crucial point in ModAC that is directly informed by the formal framework and explicitly used in Lemma 1 is related to aspect precedence: ACDeployer must always be the aspect with least precedence in the aspect environment to be woven at a new join point. This allows ACDeployer to deploy restriction aspects on objects before they get a chance to execute any piece of code. The semantics of $\lambda_{AS}$ ensures this premise because per-object aspects are "engrained" within the object following the semantics of dynamically-deployed aspects in AspectScheme [10]. In AspectScheme this is a design decision; here it is not—it is a *requirement*. If an aspect language uses a different approach to ordering aspects, or permits to undeploy aspects, then it must provide a mechanism to guarantee the above invariant related to the presence and position of ACDeployer. For example, in AspectJ [20], aspects cannot be undeployed, but manual ordering is provided. Therefore, some mechanism must be added, as in AOPS [8]. On a related note, it is necessary that ACDeployer can deploy the restrictions on a newly-created object before any code is run on behalf of this object. In $\lambda_{AS}$, this is obtained thanks to the desugaring, which creates an empty object and then calls an initializer. In AspectJ, this can be achieved thanks to the pre-initialization join points. If an aspect language does not exhibit this specific event of an object life time, then it is not possible to guarantee that restrictions see all the computation of untrusted objects.

# 7. RELATED WORK

The relation between aspects and security has a long history. We now discuss a number of related approaches. To the best of our knowledge, AOPS is the only approach that supports untrusted aspects while preventing inhibitions of access control aspects.

## Modularization of access control.

There are several proposals that modularize (part of) access control into aspects, particularly in Java [9, 19, 21, 23, 25, 34, 35]. However, these solutions implicitly assume that no other entities can affect the behavior access control aspects. This implies that access control is vulnerable to inhibition. Work on *inlined reference monitors* [12] is also related. These monitors are used to maintain access control state in the application, executing security actions whenever certain events occur. Monitors are inlined in the application code at appropriate places. Here again, it is assumed that no further code transformations can change the semantics of the security policies.

## Limiting the effects of advice.

A number of static reasoning approaches deal with ensuring that advice cannot have unwanted effect on the base program (*e.g.* Harmless Advice [7], EffectiveAdvice [22], Translucid Contracts [3]). These proposals focus on control flow and side effects, and can therefore express the fact some aspects cannot skip proceed. However, inhibition of access control as dealt with in this work requires more fine-grained control, since it limits untrusted aspects only on well-defined join points, leaving them otherwise unrestricted.

*Treatment of permission contexts.*

Caromel and Vayssière addressed the issue of correctly handling permission contexts in the presence of metaobjects [5]. The issue is to ensure that the permission context at the base level does not affect that of the metalevel, and vice-versa. The proposed solution relies on capturing the permission context when jumping to the metalevel, and restoring it when going back to the base level. Because permission contexts are part of aspect environments in ModAC, we generalize this approach to deal with aspect environments (using `swap` and `in-swap`); also, our execution-level based approach properly deals with `proceed`.

*Preventing access control inhibition.*

The aspect-oriented permission system (AOPS) [8] is the most related approach; it uses history-based access control (HBAC) [1], in which the decision of allowing access to a sensitive resource is taken based on *all* the entities that have participated in the execution trace. This characteristic makes HBAC a good alternative for discovering interferences produced by untrusted aspects. As discussed in Section 3.2, AOPS sacrifices two of our design goals: the aspect language semantics is customized to prevent weaving of crucial elements of the access control architecture (G1), thereby impeding even trusted aspects to apply at these points (G3). This being said, history-based access control is more expressive than stack-based access control. Extending or adapting our approach to HBAC is a potentially fruitful perspective.

# 8. CONCLUSION

Access control has been a recurrent target for aspect-oriented programming, mainly because of the obvious crosscutting nature of basic permission checking. However, security is a delicate concern, and therefore a correct aspectization cannot ignore potentially malicious aspects.

We have shown that access control, including privileged execution and first-class permission contexts, can be fully modularized as an aspect: the aspect language is oblivious to access control, untrusted aspects cannot inhibit access control, and trusted aspects are able to see any join point. The approach relies on defining $\mathring{R}$, a self-protecting restriction aspect in ModAC. $\mathring{R}$ is in charge of ensuring *non-inhibition* of access control. We define $\lambda_{AS}$, a core calculus for AspectScript, and use it for stating and proving the properties of ModAC. Crucially, the language must provide some guarantee with respect to aspect precedence.

The ZAC library for access control in JavaScript, implemented in AspectScript and based on ModAC and $\mathring{R}$, is the first practical realization of the proposed approach. Optimizing this implementation, and porting the approach to different languages are valuable perspectives for further validating the benefits of modular and secure access control with aspects.

# 9. REFERENCES

[1] M. Abadi and C. Fournet. Access control based on execution history. *In Proceedings of the 10th annual Network and Distributed System Security Symposium*, pages 107–121, 2003.

[2] *Proceedings of the 9th ACM International Conference on Aspect-Oriented Software Development (AOSD 2010)*, Rennes and Saint Malo, France, Mar. 2010. ACM Press.

[3] M. Bagherzadeh, H. Rajan, G. T. Leavens, and S. Mooney. Translucid contracts: Expressive specification and modular verification for aspect-oriented interfaces. In *Proceedings of the 10th ACM International Conference on Aspect-Oriented*

*Software Development (AOSD 2011)*, Porto de Galinhas, Brazil, Mar. 2011. ACM Press.

[4] D. Box and C. Sells. *Essential .NET: The common language runtime*, volume 1. Addison-Wesley, Nov. 2002.

[5] D. Caromel and J. Vayssière. A security framework for reflective Java applications. *Software: Practice and Experience*, 33(9):821–846, 2003.

[6] C. Clifton and G. T. Leavens. MiniMAO$_1$: An imperative core language for studying aspect-oriented reasoning. *Science of Computer Programming*, 63:312–374, 2006.

[7] D. S. Dantas and D. Walker. Harmless advice. In *Proceedings of the 33rd ACM SIGPLAN-SIGACT Symposium on Principles of Programming Languages (POPL 2006)*, pages 383–396, Charleston, South Carolina, USA, Jan. 2006. ACM Press.

[8] W. De Borger, B. De Win, B. Lagaisse, and W. Joosen. A permission system for secure AOP. In AOSD 2010 [2], pages 205–216.

[9] B. De Win, W. Joosen, and F. Piessens. Developing secure applications through Aspect-Oriented programming. In *Aspect-Oriented Software Development*, pages 633—650. Addison-Wesley Professional, Oct. 2004.

[10] C. Dutchyn, D. B. Tucker, and S. Krishnamurthi. Semantics and scoping of aspects in higher-order languages. *Science of Computer Programming*, 63(3):207–239, Dec. 2006.

[11] ECMA International. *ECMAScript Language Specification. ECMA-262*. 5th edition, Apr. 2009.

[12] U. Erlingsson and F. Schneider. IRM enforcement of Java stack inspection. In *Proceedings of the IEEE Symposium on Security and Privacy*, pages 246–255, 2000.

[13] M. Felleisen, R. B. Findler, and M. Flatt. *Semantics Engineering with PLT Redex*. MIT Press, 2009.

[14] D. Ferraiolo and R. Kuhn. Role-Based access control. *15th NIST-NCSC National Computer Security Conference*, pages 554–563, 1992.

[15] C. Fournet and A. D. Gordon. Stack inspection: theory and variants. *ACM Transactions on Programming Languages and Systems (TOPLAS)*, 25(3):360 – 399, 2003.

[16] J. Gosling, B. Joy, G. Steele, and G. Bracha. *The Java Language Specification, 3rd edition*. Addison-Wesley, 2005.

[17] A. Guha, C. Saftoiu, and S. Krishnamurthi. The essence of JavaScript. In T. D'Hondt, editor, *Proceedings of the 24th European Conference on Object-oriented Programming (ECOOP 2010)*, number 6183 in Lecture Notes in Computer Science, pages 126–150, Maribor, Slovenia, June 2010. Springer-Verlag.

[18] N. Hardy. The confused deputy. *SIGOPS Operating Systems Review*, 22(4):36–38, 1988.

[19] M. Huang, C. Wang, and L. Zhang. Toward a reusable and generic security aspect library. In *AOSD Technologies for Application-Level Security*, 2004.

[20] G. Kiczales, E. Hilsdale, J. Hugunin, M. Kersten, J. Palm, and W. Griswold. An overview of AspectJ. In J. L. Knudsen, editor, *Proceedings of the 15th European Conference on Object-Oriented Programming (ECOOP 2001)*, number 2072 in Lecture Notes in Computer Science, pages 327–353, Budapest, Hungary, June 2001. Springer-Verlag.

[21] A. Mourad, M. LaverdiÃÍre, and M. Debbabi. An aspect-oriented approach for the systematic security hardening of code. *Computers & Security*, 27(3-4):101–114, June 2008.

[22] B. C. d. S. Oliveira, T. Schrijvers, and W. R. Cook. EffectiveAdvice: discplined advice with explicit effects. In AOSD 2010 [2], pages 109–120.

[23] R. Ramachandran. *AspectJ for Multilevel Security*. Master Thesis, Victoria University of Wellington, 2006.

[24] P. Samarati and S. D. C. di Vimercati. Access control: Policies, models, and mechanisms. In *Foundations of Security Analysis and Design*, volume 2171 of *Lecture Notes in Computer Science*, pages 137–196. Springer Berlin / Heidelberg, London, UK, 2001.

[25] P. Słowikowski and K. Zieliński. Comparison study of aspect-oriented and container managed security. In *Proceedings of the Workshop on Analysis of Aspect Oriented Software*, Germany, 2003.

[26] É. Tanter. Expressive scoping of dynamically-deployed aspects. In *Proceedings of the 7th ACM International Conference on Aspect-Oriented Software Development (AOSD 2008)*, pages 168–179, Brussels, Belgium, Apr. 2008. ACM Press.

[27] É. Tanter. Beyond static and dynamic scope. In *Proceedings of the 5th ACM Dynamic Languages Symposium (DLS 2009)*, pages 3–14, Orlando, FL, USA, Oct. 2009. ACM Press.

[28] É. Tanter. Execution levels for aspect-oriented programming. In AOSD 2010 [2], pages 37–48.

[29] É. Tanter, J. Fabry, R. Douence, J. Noyé, and M. Südholt. Scoping strategies for distributed aspects. *Science of Computer Programming*, 75(12):1235–1261, Dec. 2010.

[30] R. Toledo, P. Leger, and É. Tanter. AspectScript: Expressive aspects for the Web. In AOSD 2010 [2], pages 13–24.

[31] R. Toledo, A. Núñez, É. Tanter, and J. Noyé. Aspectizing Java access control. *IEEE Transactions on Software Engineering*, 38(1):101–117, Jan./Feb. 2012.

[32] R. Toledo and É. Tanter. Secure and modular access control with aspects—supplementary material. http://users.dcc.uchile.cl/~rtoledo/modac-aosd/.

[33] R. Toledo and É. Tanter. Access control in JavaScript. *IEEE Software*, 28(5):76–84, Sept./Oct. 2011.

[34] B. Vanhaute, B. De Decker, and B. De Win. Building frameworks in AspectJ. *Workshop on Advanced Separation of Concerns (ECOOP)*, pages 1–6, 2001.

[35] J. Viega, J. Bloch, and P. Chandra. Applying Aspect-Oriented programming to security. *Cutter IT Journal*, 14(2):31–39, Feb. 2001.

[36] A. K. Wright and M. Felleisen. A syntactic approach to type soundness. *Journal of Information and Computation*, 115(1):38–94, Nov. 1994.

# A Typed Monadic Embedding of Aspects

### Nicolas Tabareau
ASCOLA Group
INRIA, France
nicolas.tabareau@inria.fr

### Ismael Figueroa
PLEIAD Lab & ASCOLA
DCC U. of Chile & INRIA
ifiguero@dcc.uchile.cl

### Éric Tanter[*]
PLEIAD Laboratory
CS Department (DCC)
University of Chile
etanter@dcc.uchile.cl

## ABSTRACT

We describe a novel approach to embed pointcut/advice aspects in a typed functional programming language like Haskell. Aspects are first-class, can be deployed dynamically, and the pointcut language is extensible. Type soundness is guaranteed by exploiting the underlying type system, in particular phantom types and a new anti-unification type class. The use of monads brings type-based reasoning about effects for the first time in the pointcut/advice setting, thereby practically combining Open Modules and EffectiveAdvice, and enables modular extensions of the aspect language.

## Categories and Subject Descriptors

D.3.3 [**Programming Languages**]: Language Constructs and Features; D.3.2 [**Language Classifications**]: Applicative (functional) languages

## General Terms

Languages, Design

## Keywords

Monads; aspect-oriented programming; type-based reasoning; modular language extensions.

## 1. INTRODUCTION

Aspect-oriented programming languages support the modular definition of crosscutting concerns through a join point model [15]. In the pointcut/advice mechanism, crosscutting is supported by means of pointcuts, which quantify over join points, in order to implicitly trigger advice [35]. Such a mechanism is typically integrated in an existing programming language by modifying the language processor, may it be the compiler (either directly or through macros), or the virtual machine. In a typed language, introducing pointcuts and advices also means extending the type system, if type soundness

---

[*]É. Tanter is partially funded by FONDECYT project 1110051.
I. Figueroa is funded by a CONICYT-Chile Doctoral Scholarship.

is to be preserved. For instance, AspectML [6] is based on a specific type system in order to safely apply advice. AspectJ [14] does not substantially extend the type system of Java and suffers from soundness issues. StrongAspectJ [7] addresses these issues with an extended type system. In both cases, proving type soundness is rather involved because a whole new type system has to be dealt with.

In functional programming, the traditional way to tackle language extensions, mostly for embedded languages, is to use monads [22]. Early work on AOP suggests a strong connection to monads. De Meuter proposed to use them to lay down the foundations of AOP [21], and Wand *et al.* used monads in their denotational semantics of pointcuts and advice [35]. Recently, Tabareau proposed a weaving algorithm that supports monads in the pointcut and advice model, which yields benefits in terms of extensibility of the aspect weaver [30], although in this work the weaver itself was not monadic but integrated internally in the system. This connection was exploited in recent preliminary work by the authors to construct an extensible monadic aspect weaver, in the context of Typed Racket [11], but the proposed monadic weaver was not fully typed because of limitations in the type system of Typed Racket.

This work proposes a lightweight, full-fledged embedding of aspects in Haskell, that is typed and monadic. By *lightweight*, we mean that aspects are provided as a small standard Haskell library[1]. The embedding is *full-fledged* because it supports dynamic deployment of first-class aspects with an extensible pointcut language—as is usually found only in dynamically-typed aspect languages like AspectScheme [9] and AspectScript [33] (Section 2).

By *typed*, we mean that in the embedding, pointcuts, advices, and aspects are all statically typed (Section 3), and pointcut/advice bindings are proven to be safe (Section 4). Type soundness is directly derived by relying on the existing type system of Haskell (type classes [34], phantom types [16], and some recent extensions of the Glasgow Haskell Compiler). Specifically, we define a novel type class for anti-unification [25, 26], key to define safe aspects.

Finally, because the embedding is *monadic*, we derive two notable advantages over ad-hoc approaches to introducing aspects in an existing language. First, we can directly reason about aspects and effects using traditional monadic techniques. In short, we can generalize the interference combinators of EffectiveAdvice [23] in the context of pointcuts and advice (Section 5). Second, because we embed a monadic weaver, we can modularly extend the aspect language semantics. We illustrate this with several extensions and show how type-based reasoning can be applied to language extensions (Section 6). Section 7 discusses several issues related to our approach, Section 8 reviews related work, and Section 9 concludes.

---

[1]Available, with examples, at http://pleiad.cl/haskellaop

## 2. INTRODUCING ASPECTS

A premise for aspect-oriented programming in functional languages is that function applications are subject to aspect weaving. We introduce the term *open application* to refer to a function application that generates a join point, and consequently, can be woven.

### Open function applications.

Opening all function applications in a program or only a few selected ones is both a language design question and an implementation question. At the design level, this is the grand debate about *obliviousness* in aspect-oriented programming. Opening all applications is more flexible, but can lead to fragile aspects and unwanted encapsulation breaches. At the implementation level, opening all function applications requires either a preprocessor or runtime support.

For now, we focus on *quantification*—through pointcuts—and opt for a conservative design in which open applications are realized *explicitly* using the # operator: `f # 2` is the same as `f 2`, except that the application generates a join point that is subject to aspect weaving. We will come back to obliviousness in Section 7.3, showing how different answers can be provided within the context of our proposal.

### Monadic setting.

Our approach to introduce aspects in a pure functional programming language like Haskell can be realized without considering effects. Nevertheless, most interesting applications of aspects rely on computational effects (*e.g.* tracing, memoization, exception handling, etc.). We therefore adopt a monadic setting from the start. Also, as we show in Section 5, this allows us to exploit the approach of EffectiveAdvice [23] in order to do type-based reasoning about effects in presence of aspects.

### Illustration.

As a basic example, consider the following:

```
advice:
ensurePos proceed n = proceed (abs n)
```

```
monadic version of sqrt:
sqrtM n = return (sqrt n)
```

```
using an aspect:
program n = do deploy (aspect (pcCall sqrtM) ensurePos)
               sqrtM # n
```

The advice `ensurePos` enforces that the argument of a function application is a positive number, by replacing the original argument with its absolute value. We then deploy an aspect that reacts to applications of `sqrtM`, the monadic version of `sqrt`, by executing this advice. This is specified using the pointcut (`pcCall sqrtM`). Evaluating `program -4` results in `sqrtM` to be eventually applied with argument 4. Aspects are created with `aspect` and deployed with `deploy`.

Our introduction of AOP therefore simply relies on defining aspects (pointcuts, advices), the underlying aspect environment together with the operations to deploy and undeploy aspects, and open function application.

The remainder of this section briefly presents these elements, and the following section concentrates on the main challenge: properly typing pointcuts and ensuring type soundness of pointcut/advice bindings.

## 2.1 Join Point Model

The support for crosscutting provided by a programming language lies in its *join point model* (JPM) [19]. A JPM is composed by three elements: *join points* that represents the points of a program that aspects can affect, a *means of identifying* join points—here, pointcuts—and a *means of effecting* at join points—here, advices.

### Join points.

Join points are function applications. A join point JP contains a function of type `a → m b`, and an argument of type `a`. `m` is a monad denoting the underlying computational effect stack. Note that this means that only functions that are properly lifted to a monadic context can be advised. In addition, in order for pointcuts to be able to reason about the type of advised functions, we require the functions to be `PolyTypeable`[2].

```
data JP m a b = (Monad m, PolyTypeable (a → m b)) ⇒
                JP (a → m b) a
```

From now on, we omit the type constraints related to `PolyTypeable` (the `PolyTypeable` constraint on a type is required each time the type has to be inspected dynamically; exact occurrences of this constraint can be found in the implementation).

### Pointcuts.

A pointcut is a predicate on the current join point. It is used to identify join points of interests. A pointcut simply returns a boolean to indicate whether it matches the given join point.

```
data PC m a b = Monad m ⇒ PC (∀ a' b'. m (JP m a' b' → m Bool))
```

A pointcut is represented as a value of type `PC m a b`. (`a` and `b` are used to ensure type safety, as discussed in Section 3.1.) The predicate itself is a function `∀ a' b'. m (JP m a' b' → m Bool)`, meaning it has access to the monadic stack. The ∀ declaration quantifies on type variables `a'` and `b'` (using rank-2 types) because a pointcut should be able to match against any join point, regardless of the specific types involved (we come back to this in Section 3.1).

We provide two basic pointcut designators, `pcCall` and `pcType`, as well as logical pointcut combinators, `pcOr`, `pcAnd`, and `pcNot`.

```
pcType f = let t = polyTypeOf f in PC (_type t)
           where _type t = return (\jp →
                 return (compareType t jp))

pcCall f = let t = polyTypeOf f in PC (_call f t)
           where _call f t = return (\jp →
                 return (compareFun f jp &&
                         compareType t jp))
```

`pcType f` matches all calls to functions that have a type compatible with `f` (see Section 3.1 for a detailed definition) while `pcCall f` matches all calls to `f`. In both cases, `f` is constrained to allow using the `PolyTypeable` introspection mechanism, which provides the `polyTypeOf` function to obtain the type representation of a value. This is used to compare types with `compareType`.

To implement `pcCall` we require a notion of function equality[3]. This is used in `compareFun` to compare the function in the join point to the given function. Note that we also need to perform a type comparison, using `compareType`. This is because a polymorphic function whose type variables are instantiated in one way is equal to the same function but with type variables instantiated in some other way (*e.g.* `id :: Int → Int` is equal to `id :: Float → Float`).

---

[2] Haskell has a mechanism to introspect types called `Typeable`, but it is limited only to monomorphic types. `PolyTypeable` is an extension that supports polymorphic types and thus can be defined for any type.

[3] For this notion of function equality, we use the `StableNames` API, which relies on pointer comparison. See Section 7.1 for discussion on the issues of this approach.

Users can define their own pointcut designators. For instance, we can define control-flow pointcuts like AspectJ's `cflow` (discussed briefly in Section 6), data flow pointcuts [18], pointcuts that rely on the trace of execution [8], etc.

### Advice.

An advice is a function that executes in place of a join point matched by a pointcut. This replacement is similar to open recursion in EffectiveAdvice [23]. An advice receives a function (known as the `proceed` function) and returns a new function of the same type (which may or may not apply the original `proceed` function internally). We introduce a type alias for advice:

```
type Advice m a b = (a → m b) → a → m b
```

For instance, the type `Monad m ⇒ Advice m Int Int` is a synonym for the type `Monad m ⇒ (Int → m Int) → Int → m Int`. For a given advice of type `Advice m a b`, we call `a → m b` the *advised type* of the advice.

### Aspect.

An aspect is a first-class value binding together a pointcut and an advice. Supporting first-class aspects is important: it makes it possible to support aspect factories, separate creation and deployment/undeployment of aspects, exporting opaque, self-contained aspects as single units, etc. We introduce a data definition for aspects, parameterized by a monad `m` (which has to be the same in the pointcut and advice):

```
data Aspect m a b c d = Aspect (PC m a b) (Advice m c d)
```

We defer the detailed definition of `Aspect` with its type class constraints to Section 3.2, when we address the issue of safe pointcut/advice binding.

## 2.2 Aspect Deployment

The list of aspects that are deployed at a given point in time is known as the *aspect environment*. To be able to define an heterogenous list of aspects, we use an existentially-quantified data `EAspect` that hides the type parameters of `Aspect`:[4]

```
data EAspect m = ∀ a b c d. EAspect (Aspect m a b c d)

type AspectEnv m = [EAspect m]
```

This environment can be either fixed initially and used globally [19], as in AspectJ, or it can be handled dynamically, as in AspectScheme [9]. Different scoping strategies are possible when dealing with dynamic deployment [31]. Because we are in a monadic setting, we can pass the aspect environment implicitly using a monad. An open function application can then trigger the set of currently-deployed aspects by retrieving these aspects from the underlying monad.

There are a number of design options for the aspect environment, depending on the kind of aspect deployment that is desired. Following the `Reader` monad, we can provide a fixed aspect environment, and add the ability to deploy an aspect for the dynamic extent of an expression, similarly to the `local` method of the `Reader` monad. We can also adopt a `State`-like monad, in order to support dynamic aspect deployment and undeployment with global scope. In this paper, without loss of generality, we go for the latter.

Since we are interested in using arbitrary computational effects in programs, we define the aspect environment through a *monad*

---

[4] Since existential quantification requires type parameters to be free of type class constraints, aspects with ad-hoc polymorphism have to be instantiated before deployment to statically solve each remaining type class constraint (see Section 7.2 for more details).

*transformer*, which allows the programmer to construct a monadic stack of effects [17]. A monad transformer is a type constructor that is applied to an underlying monad to construct a new monad enhanced with the effect introduced by the transformer, while retaining access to all the underlying effects. The `AOT` monad transformer is defined as follows:

```
data AOT m a = AOT {run :: AspectEnv (AOT m) →
                            m (a, AspectEnv (AOT m))}
```

Similar to the state transformer, we use a **data** declaration to define the type `AOT`. This type wraps a `run` function, which takes an initial aspect environment and returns a computation in the underlying monad `m` with a value of type `a`, and a potentially modified aspect environment.

The monadic `bind` and `return` functions of the composed `AOT m` monad are the same as in the state monad transformer. Note that the aspect environment is bound to the same monad `AOT m`. This provides aspects with access to open applications[5].

We now define the functions for dynamic deployment, which simply add and remove an aspect from the aspect environment (note the use of `$` to avoid extra parentheses):

```
deploy, undeploy :: EAspect (AOT m) → AOT m ()
deploy asp   = AOT $ \aenv → return ((), asp:aenv)
undeploy asp = AOT $ \aenv → return ((), deleteAsp asp aenv)
```

In order to extract the computation of the underlying monad from an `AOT` computation we define the `runAOT` function, with type `Monad m ⇒ AOT m a → m a` (similar to `evalStateT` in the state monad transformer), that runs a computation in an empty initial aspect environment. For instance, in the example of the `sqrt` function, we can define a `client` as follows:

```
client n = runIdentity (runAOT (program n))
```

## 2.3 Aspect Weaving

Aspect weaving is triggered through open applications, *i.e.* applications performed with the `#` operator, *e.g.* `f # x`.

### Open applications.

We introduce a type class `OpenApp` that declares the `#` operator. This makes it possible to overload `#` in certain contexts, and it can be used to declare constraints on monads to ensure that the operation is available in a given context.

```
class Monad m ⇒ OpenApp m where
    (#) :: (a → m b) → a → m b
```

The `#` operator takes a function of type `a → m b` and returns a (woven) function with the same type. Any monad composed with the `AOT` transformer has open application defined:

```
instance Monad m ⇒ OpenApp (AOT m) where
    f # a = AOT $ \aenv →
        do (woven_f, aenv') ← weave f aenv aenv (newjp f a)
           run (woven_f a) aenv'
```

An open application results in the creation of a join point (`newjp`) that represents the application of `f` to `a`. The join point is then used to determine which aspects in the environment match, produce a new function that combines all the applicable advices, and apply that function to the original argument.

---

[5] We could have defined `AOT` using the state monad transformer. However this would cause conflicts with existing monad transformer libraries when composing several effects. For instance, deploying `AOT` on a monadic stack that already contains a state component would imply using explicit lifting. We integrate `AOT` as a monad transformer that implicitly lifts operations for standard effects such as state, errors, IO, etc.

*Weaving.*

The function to use at a given point is produced by the `weave` function, defined below:

```
weave :: Monad m ⇒ (a → AOT m b) → AspectEnv (AOT m)
                 → AspectEnv (AOT m) → JP (AOT m) a b
                 → m (a → AOT m b, AspectEnv (AOT m))
weave f [] fenv _ = return (f, fenv)
weave f (asp:asps) fenv jp =
    case asp of EAspect (Aspect pc adv) →
       do (match,fenv') ← apply_pc pc jp fenv
          weave (if match
                    then apply_adv adv f
                    else f)
                asps fenv' jp
```

The `weave` function is defined recursively on the aspect environment. For each aspect, it applies the pointcut to the join point. It then uses either the partial application of the advice to `f` if the pointcut matches, or `f` otherwise[6], to keep on weaving on the rest of the aspect list. This definition is a direct adaptation of AspectScheme's weaving function [9].

*Applying advice.*

As we have seen, the aspect environment has type `AspectEnv m`, meaning that the type of the advice function is hidden. Therefore, advice application requires *coercing* the advice to the proper type in order to apply it to the function of the join point:

```
apply_adv :: Advice m a b → t → t
apply_adv adv f = (unsafeCoerce adv) f
```

The operation `unsafeCoerce` of Haskell is (unsurprisingly) unsafe and can yield to segmentation faults or arbitrary results. To recover safety, we could insert a runtime type check with `compareType` just before the coercion. We instead make aspects type safe such that we can prove that the use of `unsafeCoerce` in `apply_adv` is *always* safe. The following section describes how we achieve type soundness of aspects; Section 4 formally proves it.

## 3. TYPING ASPECTS

Ensuring type soundness in the presence of aspects consists in ensuring that an advice is always applied at a join point of the proper type. Note that by "the type of the join point", we refer to the type of the function being applied at the considered join point.

### 3.1 Typing Pointcuts

The intermediary between a join point and an advice is the pointcut, whose proper typing is therefore crucial. The type of a pointcut as a predicate over join points does not convey any information about the types of join points it matches. To keep this information, we use *phantom type variables* a and b in the definition of PC:

```
data PC m a b = Monad m ⇒ PC (∀ a' b'. m (JP m a' b' → m Bool))
```

A phantom type variable is a type variable that is not used on the right hand-side of the data type definition. The use of phantom type variables to type embedded languages was first introduced by Leijen and Meijer to type an embedding of SQL in Haskell [16]; it makes it possible to "tag" extra type information on data. In our context, we use it to add the information about the type of the join points matched by a pointcut: `PC m a b` means that a pointcut can match applications of functions of type a → m b. We call this type the *matched type* of the pointcut. Pointcut designators are in charge of specifying the matched type of the pointcuts they produce.

*Least general types.*

Because a pointcut potentially matches many join points of different types, the associated type must be a *more general type*. For instance, consider a pointcut that matches applications of functions of type Int → m Int and Float → m Int. Its matched type is the parametric type a → m Int. Note that this is in fact the *least general type* of both types.[7] Another more general candidate is a → m b, but the least general type conveys more precise information.

As a concrete example, below is the type signature of the `pcCall` pointcut designator:

```
pcCall :: Monad m ⇒ (a → m b) → PC m a b
```

*Comparing types.*

The type signature of the `pcType` pointcut designator is the same as that of `pcCall`:

```
pcType :: Monad m ⇒ (a → m b) → PC m a b
```

However, suppose that `f` is a function of type Int → m a. We want the pointcut (`pcType f`) to match applications of functions of more specific types, such as Int → m Int. This means that `compareType` actually checks that the matched type of the pointcut is *more general* than the type of the join point.

*Logical combinators.*

We use type constraints in order to properly specify the matched type of logical combinations of pointcuts. The intersection of two pointcuts matches join points that are most precisely described by the *principal unifier* of both matched types. Since Haskell supports this unification when the same type variable is used, we can simply define pcAnd as follows:

```
pcAnd :: Monad m ⇒ PC m a b → PC m a b → PC m a b
```

For instance, a control flow pointcut matches any type of join point, so its matched type is a → m b. Consequently, if `f` is of type Int → m a, the matched type of pcAnd (pcCall f) (pcCflow g) is Int → m a.

Dually, the union of two pointcuts relies on *anti-unification* [25, 26], that is, the computation of the least general type of two types. Haskell does not natively support anti-unification. We exploit the fact that multi-parameter type classes can be used to define relations over types, and develop a novel type class `LeastGen` (for *least general*) that can be used as a constraint to compute the least general type t of two types t1 and t2 (defined in Section 4):

```
pcOr :: (Monad m, LeastGen (a → b) (c → d) (e → f)) ⇒
        PC m a b → PC m c d → PC m e f
```

For instance, if `f` is of type Int → m a and `g` is of type Int → m Float, the matched type of pcOr (pcCall f) (pcCall g) is Int → m a.

The negation of a pointcut can match join points of any type because no assumption can be made on the matched join points:

```
pcNot :: Monad m ⇒ PC m a b → PC m a' b'
```

*User-defined pointcut designators.*

The set of pointcut designators in our language is open. User-defined pointcut designators are however responsible for properly specifying their matched types. If the matched type is incorrect or too specific, soundness is lost.

A pointcut cannot make any type assumption about the type of the join point it receives as argument. The reason for this is again the homogeneity of the aspect environment: when deploying an

---

[6]`apply_pc` checks whether the pointcut matches the join point and returns a boolean and a potentially modified aspect environment. Note that `apply_pc` is evaluated in the full aspect environment `fenv`, instead of the decreasing (`asp:asps`) argument.

[7]The term *most specific generalization* is also valid, but we stick here to Plotkin's original terminology [25].

aspect, the type of its pointcut is hidden. At runtime, then, a pointcut is expected to be applicable to any join point. The general approach to make a pointcut safe is therefore to perform a runtime type check, as was illustrated in the definition of `pcCall` and `pcType` in Section 2.1. However, certain pointcuts are meant to be conjuncted with others pointcuts that will first apply a sufficient type condition.

In order to support the definition of pointcuts that *require* join points to be of a given type, we provide the `RequirePC` type:

```
data RequirePC m a b = Monad m ⇒
                RequirePC (∀ a' b'. m (JP m a' b' → m Bool))
```

The definition of `RequirePC` is similar to that of `PC`, with two important differences. First, the matched type of a `RequirePC` is interpreted as a type *requirement*. Second, a `RequirePC` is not a valid stand-alone pointcut: it has to be combined with a standard `PC` that enforces the proper type upfront. To safely achieve this, we overload `pcAnd`[8]:

```
pcAnd :: (Monad m, LessGen (a → b) (c → d)) ⇒
         PC m a b → RequirePC m c d → PC m a b
```

`pcAnd` yields a standard `PC` pointcut and checks that the matched type of the `PC` pointcut is *less general* than the type expected by the `RequirePC` pointcut. This is expressed using the constraint `LessGen`, which, as we will see in Section 4, is based on `LeastGen`.

To illustrate, let us define a poincut designator `pcArgGT` for specifying pointcuts that match when the argument at the join point is greater than a given n (of type a instance of the `Ord` type class):

```
pcArgGT :: (Monad m, Ord a) ⇒ a → RequirePC m a b
pcArgGT n = RequirePC $ return (\jp →
                return (unsafeCoerce (getJpArg jp) >= n))
```

The use of `unsafeCoerce` to coerce the join point argument to the type a forces us to declare the `Ord` constraint on a when typing the returned pointcut as `RequirePC m a b` (with a fresh type variable b). To get a proper pointcut, we use `pcAnd`, for instance to match all calls to `sqrtM` where the argument is greater than 10:

```
pcCall sqrtM `pcAnd` pcArgGT 10
```

The `pcAnd` combinator guarantees that a `pcArgGT` pointcut is always applied to a join point with an argument that is indeed of a proper type: no runtime type check is necessary within `pcArgGT`, because the coercion is always safe.

## 3.2  Typing Aspects

The main typing issue we have to address consists in ensuring that a pointcut/advice binding is type safe, so that the advice application does not fail. A first idea to ensure that the pointcut/advice binding is type safe is to require the matched type of the pointcut and the advised type of the advice to be the same (or rather, unifiable):

```
wrong!
data Aspect m a b = Aspect (PC m a b) (Advice m a b)
```

This approach can however yield unexpected behavior. Consider the following example:

```
idM x = return x

adv :: Monad m ⇒ Advice (Char → m Char)
adv proceed c = proceed (toUpper c)

program = do deploy (aspect (pcCall idM) adv)
             x ← idM # 'a'
             y ← idM # [True,False,True]
             return (x, y)
```

The matched type of the pointcut `pcCall idM` is `Monad m ⇒ a → m a`. With the above definition of `Aspect`, `program` passes the typechecker because it is possible to unify a and Char to Char. However, when evaluated, the behavior of `program` is undefined because the advice is unsafely applied with an argument of type `[Bool]`, for which `toUpper` is undefined.

The problem is that during typechecking, the matched type of the pointcut and the advised type of the advice can be unified. Because unification is symmetric, this succeeds even if the advised type is more specific than the matched type. In order to address this, we again use the type class `LessGen` to ensure that the matched type is less general than the advice type:

```
data Aspect m a b c d = (Monad m, LessGen (a → m b) (c → m d))
                      ⇒ Aspect (PC m a b) (Advice m c d)
```

This constraint ensures that pointcut/advice bindings are type safe: the coercion performed in `apply_adv` always succeeds. We formally prove this in the following section.

## 4.  TYPING ASPECTS, FORMALLY

We now formally prove the safety of our approach. We start briefly summarizing the notion of type substitutions and the *is less general* relation between types. Note that we do not consider type class constraints in the definition. Then we describe a novel anti-unification algorithm implemented with type classes, on which the type classes `LessGen` and `LeastGen` are based. We finally prove pointcut and aspect safety, and state our main safety theorem.

### 4.1  Type Substitutions

In this section we summarize the definition of type substitutions, which form the basis of our argument for safety. We consider a typing environment $\Gamma = (x_i : T_i)_{i \in \mathbb{N}}$ that binds variables to types.

DEFINITION 1    (TYPE SUBSTITUTION, FROM [24]).
*A type substitution $\sigma$ is a finite mapping from type variables to types. It is denoted $[x_i \mapsto T_i]_{i \in \mathbb{N}}$, where $dom(\sigma)$ and $range(\sigma)$ are the sets of types appearing in the left-hand and right-hand sides of the mapping, respectively. It is possible for type variables to appear in $range(\sigma)$.*

*Substitutions are always applied simultaneously on a type. If $\sigma$ and $\gamma$ are substitutions, and $T$ is a type, then $\sigma \circ \gamma$ is the composed substitution, where $(\sigma \circ \gamma)T = \sigma(\gamma T)$. Application of substitution on a type is defined inductively on the structure of the type.*

*Substitution is extended pointwise for typing environments in the following way: $\sigma(x_i : T_i)_{i \in \mathbb{N}} = (x_i : \sigma T_i)_{i \in \mathbb{N}}$. Also, applying a substitution to a term t means to apply the substitution to all type annotations appearing in t.*

DEFINITION 2    (LESS GENERAL TYPE).
*We say type $T_1$ is less general than type $T_2$, denoted $T_1 \preceq T_2$, if there exists a substitution $\sigma$ such that $\sigma T_2 = T_1$. Observe that $\preceq$ defines a partial order on types (modulo $\alpha$-renaming).*

DEFINITION 3    (LEAST GENERAL TYPE).
*Given types $T_1$ and $T_2$, we say type $T$ is the least general type iff $T$ is the supremum of $T_1$ and $T_2$ with respect to $\preceq$.*

---

[8]The constraint is different from the previous constraint on `pcAnd`. This is possible thanks to the recent `ConstraintKinds` extension of `ghc`.

```
1  class LeastGen' a b c σ_in σ_out | a b c σ_in → σ_out
2
3  Inductive case: The two type constructors match,
4  recursively compute the substitution for type arguments a_i, b_i.
5  instance (LeastGen' a_1 b_1 c_1 σ_0 σ_1, ...,
6           LeastGen' a_n b_n c_n σ_{n-1} σ_n,
7           T c_1 ... c_n ~ c)
8      ⇒ LeastGen' (T a_1 ... a_n) (T b_1 ... b_n) c σ_0 σ_n
9
10 Default case: The two type constructors don't match, c has to be a variable,
11 either unify c with c' if c' ↦ (a, b) or extend the substitution with c ↦ (a, b)
12 instance (Analyze c (TVar c),
13          MapsTo σ_in c' (a, b),
14          VarCase c' (a, b) c σ_in σ_out)
15     ⇒ LeastGen' a b c σ_in σ_out
```

**Figure 1:** LeastGen'

## 4.2 Statically Computing Least General Types

In an aspect declaration, we statically check the type of the point-cut and the type of the advice to ensure a safe binding. To do this we encode an anti-unification algorithm at the type level, exploiting the type class mechanism of Haskell. A multi-parameter type class R $t_1 \ldots t_n$ can be seen as a *relation* R on types $t_1 \ldots t_n$, and **instance** declarations as ways to (inductively) define this relation, in a manner very similar to logic programming.

The type classes LessGen and LeastGen used in Section 3 are defined as particular cases of the more general type class LeastGen', shown in Figure 1. This class is defined in line 1 and is parameterized by types $a$, $b$, $c$, $\sigma_{in}$ and $\sigma_{out}$. $\sigma_{in}$ and $\sigma_{out}$ denote substitutions encoded at the type level as a list of mappings from type variables to *pairs* of types. We use pairs of types in substitutions because we have to simultaneously compute substitutions from $c$ to $a$ and from $c$ to $b$[9]. To be concise, lines 5-8 present a single definition parametrized by the type constructor arity but in practice, there needs to be a different instance declaration for each type constructor arity.

PROPOSITION 1. *If* LeastGen' $a$ $b$ $c$ $\sigma_{in}$ $\sigma_{out}$ *holds, then substitution* $\sigma_{out}$ *extends* $\sigma_{in}$ *and* $\sigma_{out}c = (a, b)$.

PROOF 1. *By induction on the type representation of $a$ and $b$.*

*A type can either be a type variable, represented as* TVar a, *or an $n$-ary type constructor* T *applied to $n$ type arguments[10]. The rule to be applied depends on whether the type constructors of $a$ and $b$ are the same or not.*

*(i) If the constructors are the same, the rule defined in lines 5-8 computes* (T $c_1$ ... $c_n$) *using the induction hypothesis that $\sigma_i c_i = (a_i, b_i)$, for $i = 1 \ldots n$. The component-wise application of constraints is done from left to right, starting from substitution $\sigma_0$ and extending it to the resulting substitution $\sigma_n$. The type equality constraint* (T $c_1$ ...) ~ $c$ *checks that $c$ is unifiable with* (T $c_1$ ...) *and, if so, unifies them. Then, we can check that $\sigma_n c = (a, b)$.*

*(ii) If the type constructors are not the same the only possible generalization is a type variable. In the rule defined in lines 12-15 the goal is to extend $\sigma_{in}$ with the mapping $c \mapsto (a, b)$ such that $\sigma_{out}c = (a, b)$, while preserving the injectivity of the substitution (see next proposition).* □

---

[9]The $a$ $b$ $c$ $\sigma_{in}$ → $\sigma_{out}$ expression means that $\sigma_{out}$ is functionally dependent on the other parameters. Functional dependencies were proposed by Jones [13] as a mechanism to more precisely control type inference in Haskell. An expression c e | c → e means that fixing the type c should fix the type e.

[10]We use the Analyze type class from PolyTypeable to get a type representation at the type level. For simplicity we omit the rules for analyzing type representations.

PROPOSITION 2. *If $\sigma_{in}$ is an injective function, and* LeastGen' $a$ $b$ $c$ $\sigma_{in}$ $\sigma_{out}$ *holds, then $\sigma_{out}$ is an injective function.*

PROOF 2. *By construction* LeastGen' *introduces a binding from a fresh type variable to $(a, b)$, in the rule defined in lines 12-15, only if there is no type variable already mapping to $(a, b)$—in which case $\sigma_{in}$ is not modified.*

*To do this, we first check that $c$ is actually a type variable (*TVar c*) by checking its representation using* Analyze. *Then in relation* MapsTo *we bind $c'$ to the (possibly inexistent) type variable that maps to $(a, b)$ in $\sigma_{in}$. In case there is no such mapping $c'$ is* None.

*Finally, relation* VarCase *binds $\sigma_{out}$ to $\sigma_{in}$ extended with $\{c \mapsto (a, b)\}$ in case $c'$ is* None, *otherwise $\sigma_{out} = \sigma_{in}$. It then unifies $c$ with $c'$. In all cases $c$ is bound to the variable that maps to $(a, b)$ in $\sigma_{out}$, because it was either unified in rule* MapsTo *or in rule* VarCase.

*The hypothesis that $\sigma_{in}$ is injective ensures that any preexisting mapping is unique.* □

PROPOSITION 3. *If $\sigma_{in}$ is an injective function, and* LeastGen' $a$ $b$ $c$ $\sigma_{in}$ $\sigma_{out}$ *holds, then $c$ is the least general type of $a$ and $b$.*

PROOF 3. *By induction on the type representation of $a$ and $b$.*

*(i) If the type constructors are different the only generalization possible is a type variable $c$.*

*(ii) If the type constructors are the same, then $a = T a_1 \ldots a_n$ and $b = T b_1 \ldots b_n$. By Proposition 1, $c = T c_1 \ldots c_n$ generalizes $a$ and $b$ with the substitution $\sigma_{out}$. By induction hypothesis $c_i$ is the least general type of $(a_i, b_i)$.*

*Now consider a type $d$ that also generalizes $a$ and $b$, i.e. $a \preceq d$ and $b \preceq d$, with associated substitution $\alpha$. We prove $c$ is less general than $d$ by constructing a substitution $\tau$ such that $\tau d = c$.*

*Again, there are two cases, either $d$ is a type variable, in which case we set $\tau = \{d \mapsto c\}$, or it has the same outermost type constructor, i.e. $d = T d_1 \ldots d_n$. Thus $a_i \preceq d_i$ and $b_i \preceq d_i$; and since $c_i$ is the least general type of $a_i$ and $b_i$, there exists a substitution $\tau_i$ such that $\tau_i d_i = c_i$, for $i = 1 \ldots n$.*

*Now consider a type variable $x \in dom(\tau_i) \cap dom(\tau_j)$. By definition of $\alpha$, we know that $\sigma_{out}(\tau_i(x)) = \alpha(x)$ and $\sigma_{out}(\tau_j(x)) = \alpha(x)$. Because $\sigma_{out}$ is injective (by Proposition 2), we deduce that $\tau_i(x) = \tau_j(x)$ so there are no conflicting mappings between $\tau_i$ and $\tau_j$, for any $i$ and $j$. Thus we can define $\tau = \bigcup \tau_i$ and check that $\tau d = c$.* □

DEFINITION 4 (LeastGen TYPE CLASS). *To compute the least general type* c *for* a *and* b, *we define:*

LeastGen a b c $\triangleq$ LeastGen' a b c $\sigma_{empty}$ $\sigma_{out}$, *where $\sigma_{empty}$ is the empty substitution and $\sigma_{out}$ is the resulting substitution.*

DEFINITION 5 (LessGen TYPE CLASS). *To establish that type* a *is less general than type* b, *we define:*

LessGen a b $\triangleq$ LeastGen a b b

## 4.3 Pointcut Safety

We now establish the safety of pointcuts with relation to join points.

DEFINITION 6 (POINTCUT MATCH).
*We define the relation* matches(pc, jp), *which holds iff applying pointcut* pc *to join point* jp *in the context of a monad* m *yields a computation* m True.

Now we prove that the matched type of a given pointcut is more general than the join points matched by that pointcut.

PROPOSITION 4. *Given a join point term* jp *and a pointcut term* pc, *and type environment* $\Gamma$, *if*

$\Gamma \vdash$ pc: PC m a b
$\Gamma \vdash$ jp: JP m a' b'
$\Gamma \vdash$ matches(pc, jp)
*then* a' $\rightarrow$ m b' $\preceq$ a $\rightarrow$ m b.

PROOF 4. *By induction on the matched type of the pointcut.*

- *Case* pcCall: *By construction the matched type of a* pcCall f *pointcut is the type of* f. *Such a pointcut matches a join point with function* g *if and only if:* f *is equal to* g, *and the type of* f *is less general than the type of* g. *(On both* pcCall *and* pcType *this type comparison is performed by* compareType *on the type representations of its arguments.)*

- *Case* pcType: *By construction the matched type of a* pcType f *pointcut is the type of* f. *Such a pointcut only matches a join point with function* g *whose type is less general than the matched type.*

- *Case* pcAnd *on* PC PC: *Consider* pc1 `pcAnd` pc2. *The matched type of the combined pointcut is the* principal unifier *of the matched types of the arguments—which represents the intersection of the two sets of join points. The property holds by induction hypothesis on* pc1 *and* pc2.

- *Case* pcAnd *on* PC RequirePC: *Consider* pc1 `pcAnd` pc2. *The matched type of the combined pointcut is the type of* pc1 *and it is checked that the type required by* pc2 *is* more general *so the application of* pc2 *will not yield an error. The property holds by induction hypothesis on* pc1.

- *Case* pcOr: *Consider* pc1 `pcOr` pc2. *The matched type of the combined pointcut is the* least general type *of the matched types of the argument, computed by the* LeastGen *constraint—which represents the union of the two sets of join points. The property holds by induction hypothesis on* pc1 *and* pc2.

- *Case* pcNot: *The matched type of a pointcut constructed with* pcNot *is a fresh type variable, which by definition is more general than the type of any join point.*

- *User-defined pointcuts must maintain this property, otherwise safety is lost.* □

## 4.4 Advice Type Safety

If an aspect is well-typed, the advice is more general than the matched type of the pointcut:

PROPOSITION 5. *Given a pointcut term* pc, *an advice term* adv, *and a type environment* $\Gamma$, *if*

$\Gamma \vdash$ pc: PC m a b
$\Gamma \vdash$ adv: Advice m c d
$\Gamma \vdash$ (aspect pc adv): Aspect m a b c d
*then* a $\rightarrow$ m b $\preceq$ c $\rightarrow$ m d.

PROOF 5. *Using the definition of* Aspect *(Section 3.2) and because* $\Gamma \vdash$ (aspect pc adv): Aspect m a b c d, *we know that the constraint* LessGen *is satisfied, so by Definitions 4 and 5, and Proposition 1,* a $\rightarrow$ m b $\preceq$ c $\rightarrow$ m d. □

## 4.5 Safe Aspects

We now show that if an aspect is well-typed, the advice is more general than the advised join point:

THEOREM 1 (SAFE ASPECTS). *Given the terms* jp, pc *and* adv *representing a join point, a pointcut and an advice respectively, given a type environment* $\Gamma$, *if*

$\Gamma \vdash$ pc: PC m a b
$\Gamma \vdash$ adv: Advice m c d

```
module Fib (fib, pcFib) where
import AOP

fibBase n = return 1

pcFib = pcCall fibBase `pcAnd` pcArgGT 2

fibAdv proceed n = do f1 ← fibBase # (n-1)
                      f2 ← fibBase # (n-2)
                      return (f1 + f2)

fib :: Monad m ⇒ m (Int → m Int)
fib = do { deploy (aspect pcFib fibAdv); return $ fibBase # }
```

**Figure 2: Fibonacci module.**

$\Gamma \vdash$ (aspect pc adv): Aspect m a b c d
*and*
$\Gamma \vdash$ jp: JP m a' b'
$\Gamma \vdash$ matches(pc, jp)
*then* a' $\rightarrow$ m b' $\preceq$ c $\rightarrow$ m d.

PROOF 6. *By Proposition 4 and 5 and the transitivity of* $\preceq$. □

COROLLARY 1 (SAFE ADVICE APPLICATION).
*The coercion of the advice in* apply_adv *is safe.*

PROOF 7. *Recall* apply_adv *(Section 2.3):*

apply_adv :: Advice m a b → t → t
apply_adv adv f = (unsafeCoerce adv) f

*By construction,* apply_adv *is used only with a function* f *that comes from a join point that is matched by a pointcut associated to* adv. *Using Theorem 1, we know that the join point has type* JP m a' b' *and that* a' $\rightarrow$ m b' $\preceq$ a $\rightarrow$ m b. *We note* $\sigma$ *the associated substitution. Then, by compatibility of substitutions with the typing judgement [24], we deduce* $\sigma\Gamma \vdash \sigma$adv: Advice m a' b'. *Therefore* (unsafeCoerce adv) *corresponds exactly to* $\sigma$adv, *and is safe.* □

# 5. TYPE-BASED REASONING ABOUT ASPECTS

This section illustrates how we can exploit the monadic embedding for type-based reasoning about aspects, regarding both control flow properties and computational effects. In essence, this section shows how the approach of EffectiveAdvice [23] can be used in the context of pointcut/advice AOP, or dually, how Open Modules [1] can be extended with effects.

## 5.1 A Simple Example

We first describe a simple example that serves as the starting point. Figure 2 describes a Fibonacci module, following the canonical example of Open Modules. The module uses an internal aspect to implement the recursive definition of Fibonacci: the base function, fibBase simply implements the base case, and the fibAdv advice implements recursion when the pointcut pcFib matches. Note that pcFib uses the user-defined poincut pcArgGT (defined in Section 3.1) to check that the call to fibBase is done with an argument greater than 2. The fib function is defined by first deploying the internal aspect, and then partially applying # to fibBase. This transparently ensures that an application of fib is open. The fib function is exported, together with the pcFib pointcut, which can be used by an external module to advise applications of the internal fibBase function. Figure 3 presents a Haskell module that provides a more efficient implementation of fib by using a memoization advice. To

177

```
module MemoizedFib (fib) where
import qualified Fib
import AOP

memo proceed n =
 do m ← get
    if member n m then return (m ! n)
    else do { y  ← proceed n ; m' ← get; put (insert n y m');
              return y }

fib = do { deploy (aspect Fib.pcFib memo); Fib.fib }
```

**Figure 3: Memoized Fibonacci module.**

benefit from memoization, a client only has to import `fib` from the `MemoizedFib` module instead of directly from the `Fib` module.

Note that, if we consider that the aspect language only supports the `pcCall` pointcut designator, this implementation actually represents an open module proper. Preserving the properties of open modules, in particular protecting from external advising of internal functions, in presence of arbitrary quantification (*e.g.* `pcType`, or an always-matching pointcut) is left for future work. Importantly, just like Open Modules, the approach described here does not ensure anything about the advice beyond type safety. In particular, it is possible to create an aspect that incorrectly calls `proceed` several times, or an aspect that has undesired computational effects. The type system can assist us in expressing and enforcing specific interference properties.

## 5.2 Protected Pointcuts

In order to extend Open Modules with effect-related enforcement, we introduce the notion of *protected pointcuts*, which are pointcuts enriched with restrictions on the effects that associated advice can exhibit. Simply put, a protected pointcut embeds a *typed combinator* that is applied to the advice in order to build an aspect. If the advice does not respect the (type) restrictions expressed by the combinator, the aspect creation expression simply does not typecheck and hence the aspect cannot be built.

A typed combinator is any function that can produce an advice:

```
type Combinator t m a b = Monad m ⇒ t → Advice m a b
```

The `protectPC` function packs together a pointcut and a combinator:

```
protectPC :: (Monad m, LessGen (a → m b) (c → m d)) ⇒
  PC m a b → Combinator t m c d → ProtectedPC m a b t c d
```

A protected pointcut, of type `ProtectedPC`, cannot be used with the standard aspect creation function `aspect`. The following `pAspect` function is the only way to get an aspect from a protected pointcut (the constructor `PPC` is not exposed):

```
pAspect :: Monad m ⇒ ProtectedPC m a b t c d → t
                    → Aspect m a b c d
pAspect (PPC pc comb) adv = aspect pc (comb adv)
```

The key point here is that when building an aspect using a protected pointcut, the combinator `comb` is applied to the advice `adv`.

We now show how to exploit this extension of Open Modules to control both control flow and computational effects, using the proper type combinators.

## 5.3 Control Flow Properties

Rinard *et al.* present a classification of advice in four categories depending on how they affect the control flow of programs [27]:

- **Combination:** The advice can call `proceed` any number of times.

```
type Narrow m a b c = Monad m ⇒
                      (a → m Bool, Augment m a b c, Replace m a b)
narrow :: Monad m ⇒ Narrow m a b c → Advice m a b
narrow (p, aug, rep) proceed x =
   do b ← p x
      if b then replace rep proceed x
           else augment aug proceed x
```

**Figure 4: Narrowing advice combinator (adapted from [23]).**

- **Replacement:** There are no calls to `proceed` in the advice.

- **Augmentation:** The advice calls `proceed` exactly once, and it does not modify the arguments to or the return value of `proceed`.

- **Narrowing:** The advice calls `proceed` at most once, and does not modify the arguments to or the return value of `proceed`.

Memoization is a typical example of a narrowing advice: the combination of a replacement advice ("return memoized value without proceeding") and an augmentation advice ("proceed and memoize return value"), where the choice between both is driven by a runtime predicate ("is there a memoized value for this argument?").

Oliveira *et al.* [23] show a type-based enforcement of these categories, through advice combinators. Figure 4 shows the definition of the `narrow` combinator, which takes a triple consisting of the predicate, the augmentation advice and the replacement advice, and builds a proper advice with the narrowing logic. For brevity, we do not detail all combinators and advice types, taken from [23]. These combinators fit the general `Combinator` type we described in Section 5.2, and can therefore be embedded in protected pointcuts. It is now straightforward for the `Fib` module to expose a protected pointcut that restricts valid advice to narrowing advice:

```
module Fib (fib, ppcFib) where
...
ppcFib = protectPC pcFib narrow
```

The protected pointcut `ppcFib` embeds the `narrow` type combinator. Therefore, only advice that can be statically typed as narrowing advice can be bound to that pointcut.

## 5.4 Effect Interference

The typed monadic embedding of aspects also allows to reason about computational effects. For instance, consider a modification of the Fibonacci module where the `fibErr` function can throw an error message when called with a negative integer (Figure 5). In that situation, it is interesting to ensure that the advice bound to the exposed pointcut cannot throw or catch in the same Error monad.

This can be done as in EffectiveAdvice [23], by enforcing advices to be parametric with respect to the monad used by the base computation. To distinguish between base and aspect parts of the monadic stack, we first have to introduce a modified `AOT` monad transformer that manages a splitting of the monadic stack into a monad transformer `t` that collects effects available to aspect computations and a monad `m` that collects effects available to the base computation. Thus, we define the data type `NIAOT` as follows (`NI` stands for non-interference):

```
data NIAOT t m a = NIAOT {runNI :: AspectEnv (NIAOT t m)
                          → t m (a, AspectEnv (NIAOT t m)) }
```

and extend other definitions (`weave`, `deploy`, ...) accordingly.

## Effect interference and pointcuts.

The novelty compare to EffectiveAdvice is that we also have to deal with interferences for pointcuts. But to allow effect-based reasoning on pointcuts, we need to distinguish between the monad used by the base computation and the monad used by pointcuts. Indeed, in the interpretation of the type `PC m a b`, `m` stands for both monads, which forbids to reason separately about them. To address this issue, we need to interpret `PC m a b` differently, by saying that the matched type is `a → b` instead of `a → m b`. In this way, the monad for the base computation (which is implicitly bound by `b`) does not have to be `m` at the time the pointcut is defined. To accommodate this new interpretation with the rest of the code, very little changes have to be made[11]. The main changes are in the type of `pcCall`, `pcType` and in the definition of `Aspect`

```
pcCall, pcType :: Monad m ⇒ (a → b) → PC m a b
```

```
data Aspect m a b c d = (Monad m, LessGen (a → b) (c → m d)) ⇒
                          Aspect (PC m a b) (Advice m c d)
```

Note how the definition of `Aspect` forces the monad of the pointcut computation to be unified with that of the advice, and with that of the base code. The result of Section 4 can straightforwardly be rephrased with these new definitions.

## Typing non-interfering poincuts and advices.

It now becomes possible to restrict the type of pointcuts and advices by using rank-2 types. The following type synonyms guarantee that an aspect of type `NIPC t a b` only uses effects available in `t` (and similarly for `NIAdvice t a b`).

```
type NIPC t a b = ∀ m. (Monad (t m), Monad m, ...) ⇒
                     PC (NIAOT t m) a b
type NIAdvice t a b = ∀ m. (Monad (t m), Monad m, ...) ⇒
                     Advice (NIAOT t m) a b
```

By universally quantifying over the type `m` of the effects used in the base computation, these types enforce, through the properties of parametricity, that pointcuts (or advices) cannot refer to specific effects in the base program.

We can define aspect construction functions that enforce different (non-)interference patterns, such as non-interfering pointcut `NIPC` with unrestricted advice `Advice`, unrestricted pointcut `PC` with non-interfering advice `NIAdvice`, etc. Symmetrically, we can check that a part of the base code cannot interfere with effects available to aspects by using the type synonym `NIBase`, which universally quantifies over the type `t` of effects available to the advice:

```
type NIBase m a b = ∀ t. (Monad (t m), MonadTrans t, ...) ⇒
                     a → NIAOT t m b
```

Coming back to Open Modules and protected pointcuts, in order to enforce non-interfering advice, we need to define a typed combinator that requires an advice of type `NIAdvice`:

```
niAdvice :: (Monad (t m), Monad m) ⇒
            NIAdvice t a b → Advice (NIAOT t m) a b
niAdvice adv = adv
```

The `niAdvice` combinator is computationally the identity function, but does impose a type requirement on its argument. Using this combinator, the `Fib` module can expose a protected pointcut that enforces non-interference with base effects:

```
module Fib (fib, ppcFib) where
...
ppcFib = protectPC pcFib niAdvice
```

EffectiveAdvice is restricted with respect to AOP in that it does not support quantification at all [23]. We have shown that in our

---

[11]The implementation available online uses this interpretation of `PC m a b`.

```
fibErr :: (..., MonadError String (t m)) ⇒
          NIAOT t m (Int → NIAOT t m Int)
fibErr = do deploy (niAspect pcFib fibAdv)
            return errorFib
       where errorFib n = if n < 0
                          then throwError "Not defined"
                          else fibBase # n
```

**Figure 5: Fibonacci with error.**

```
type Level = Int
data ELT m a = ELT {run :: Level -> m (a, Level)}
```
*primitive operations*
```
inc = ELT $ \l -> return ((), l + 1)
dec = ELT $ \l -> return ((), l - 1)
at l = ELT $ \_ -> return ((), l)
```
*user visible operations*
```
current = ELT $ \l -> return (l, l)
up c   = do {inc; result <- c; dec; return result}
down c = do {dec; result <- c; inc; return result}
lambda_at f l = \arg -> do
{n <- current; at l; result <- f arg; at n; return result}
```

**Figure 6: Execution levels monad transformer and level-shifting operations**

---

embedding, protected pointcuts can be used to extend EffectiveAdvice to support quantification, thereby combining it with Open Modules.

# 6. LANGUAGE EXTENSIONS

The typed monadic embedding of aspects supports modular extensions of the aspect language. More precisely, we can modularly implement new semantics for aspect scoping and weaving. We briefly discuss in this section three possible extensions (available in the online distribution): i) aspect weaving with execution levels [32]; ii) secure weaving in which a set of join points can be hidden from advising; iii) privileged aspects that can see hidden join points. At the end of this section, we discuss how the types allow us to reason about these extensions.

## 6.1 Execution Levels

Execution levels avoid unwanted *computational interference* between aspects, *i.e.* when an aspect execution produces join points that are visible to others, including itself [32]. Execution levels give structure to execution by establishing a tower in which the flow of control navigates. Aspects are deployed at a given level and can only affect the execution of the underlying level. The execution of an aspect (both pointcuts and advices) is therefore not visible to itself and to other aspects deployed at the same level, only to aspects standing one level above. The original computation triggered with `proceed` is always executed at the level at which the join point was emitted. If needed, the programmer can use level-shifting operators to move execution up and down in the tower.

The monadic semantics of execution levels are implemented in the `ELT` monad transformer (Figure 6). The `Level` type synonym represents the level of execution as an integer. `ELT` wraps a `run` function that takes an initial level and returns a computation in the underlying monad `m`, with a value of type `a` and a potentially-modified level. As in the `AOT` transformer, the monadic `bind` and `return` functions are the same as in the state monad transformer. The private operations `inc`, `dec`, and `at` are used to define `current`, `up`, `down`, and `lambda_at`. In addition to level shifting with `up` and `down`, `current` reifies the current level, and `lambda_at` creates a *level-capturing function* bound at

```
deployInEnv (Aspect (pc::PC (AOT (ELT m)) tpc) adv) aenv =
  let pcEL ldep = (PC $ return (\jp → do
                     lapp ← current
                     if lapp == ldep
                       then up $ runPC pc jp
                       else return False)) :: PC (AOT (ELT m)) tpc
      advEL ldep proceed arg =
                     up $ adv (lambda_at proceed ldep) arg
  in do l ← current
        return EAspect (Aspect (pcEL l) (advEL l)) : aenv
```

**Figure 7: Redefining aspect deployment for execution levels semantics. An aspect is made level-aware by transforming its pointcut and advice.**

level 1. When such a function is applied, execution jumps to level 1 and then goes back to the level prior to the application [32].

The semantics of execution levels can be embedded in the definition of aspects themselves, by transforming the pointcut and advice of an aspect at deployment time[12]. This is done through the two functions, pcEL and advEL (Figure 7). pcEL first ensures that the current execution level lapp matches ldep, the level at which the aspect is deployed. If so it then runs the pointcut one level above. Similarly, advEL ensures that the advice is run one level above, with a proceed function that captures the deployment level.

## 6.2 Secure Weaving

For security reasons it can be interesting to protect certain join points from being advised. To support such a secure weaving, we define a new monad transformer AOT_s, which embeds an (existentially quantified) pointcut that specifies the hidden join points, and we modify the weaving process accordingly (not shown here).

```
data EPC m = forall a b. EPC (PC m a b)

data AOT_s m a = AOT_s { runAOT_s ::
       AspectEnv (AOT_s m) → EPC (AOT_s m) →
       m (a, (AspectEnv (AOT_s m), EPC (AOT_s m)))}
```

This can be particularly useful when used with the cflow pointcut (described below) to protect the computation that occurs in the control flow of critical function applications. For instance, we can ensure that the whole control flow of function f is protected from advising during the execution of program p:

```
runAOT_s (EPC (pcCflow f)) p
```

### cflow *pointcut.*

The pcCflow pointcut is defined using a (join point) stack monad and one aspect that matches every join point and stores it in the stack; then the pcCflow pointcut is just a test on this stack. Using effect non-interference enforcement (Section 5.4), we can guarantee that this stack is private to pcCflow. Alternative optimizations can be defined, for example putting in the stack only relevant join points, or a per-flow deployment that allows to use a boolean instead of a stack.

## 6.3 Privileged Aspects

Hiding some join points to *all* aspects may be too restrictive. For instance, certain "system" aspects like access control should be treated as privileged and view all join points. Another example is the aspect in charge of maintaining the join point stack for the sake of control flow reasoning (used by pcCflow). In such cases, it

is important to be able to define a set of privileged aspects, which can advise all join points, even those that are normally hidden. The implementation of a privileged aspects list is a straightforward extension to the secure weaving mechanism described above.

## 6.4 Reasoning about Language Extensions

The above extensions can be implemented in an untyped language such as LAScheme [32]. However, it is not possible in an untyped setting to statically reason about effects provided by a language extension or enforce that a piece of code is used with a particular weaving semantics.

### *Enforcing non-interference in language extensions.*

We can combine the monadic interpretation of execution levels with the management of effect interference (Section 5.4) in order to reason about level-shifting operations performed by base and aspect computations. For instance, it becomes possible to prevent aspect and/or base computation to use effects provided by the ELT monad transformer, thus ensuring that the default semantics of execution levels is preserved (and therefore that the program is free of aspect loops [32]). If more advanced use of execution levels is required, this contraint can be explicitly relaxed in the AOT monad transformer, thus stressing in the type that it is the responsability of the programmer to avoid infinite regression.

### *Using types to enforce weaving semantics.*

The type system makes it possible to specify functions that can be woven, but only within a specific aspect monad. For instance, suppose that we want to define a critical computation, which must only be run with secure weaving for access control. The computation must therefore be run within the AOT_s monad transformer with a given pointcut pc_ac (ac stands for access control).

To enforce the use of AOT_s with a specific pointcut value would require the use of a dependent type, which is not possible in Haskell. This said, we can use the newtype data constructor together with its ability to derive automatically type class instances, to define a new type AOT_ac that encapsulates the AOT_s monad transformer and forces it to be run with the pc_ac pointcut:

```
newtype AOT_ac m a = AOT_ac (AOT_s m a)
       deriving (Monad, OpenApp, ...)

runSafe (AOT_ac c) = runAOT_s (EPC pc_ac) c
```

Therefore, we can export the critical computation by typing it appropriately:

```
critical :: Monad m ⇒ AOT_ac m a
```

Because the AOT_ac constructor is hidden in a module, the only way to run such a computation typed as AOT_ac is to use runSafe. The critical computation is therefore only advisable with secure weaving for access control. Type based reasoning about aspect language extensions is, to the best of our knowledge, a novel contribution of this work.

## 7. DISCUSSION

We now discuss a number of issues related to our approach: how to define a proper notion of function equality, how to deal with overloaded functions, how to enhance the handling of the monadic stack, and finally, we analyze the issue of obliviousness.

## 7.1 Supporting Equality on Functions

Pointcuts quantify about join points, and a major element of the join point is the function being applied. The pcType designator relies on type comparison, implemented using the PolyTypeable type

---

[12]For simplicity, in Section 2.2 we only described the default semantics of aspect deployment; aspect (un)deployement is actually defined using overloaded (un)deployInEnv functions.

class in order to obtain representations for polymorphic types. The `pcCall` is more problematic, as it relies on function equality, but Haskell does not provide an operator like `eq?` in Scheme.

A first workaround is to use the `StableNames` API that allows to compare functions using pointer equality. Unfortunately, this notion of equality is fragile. `StableNames` equality is safe in the sense that it does not equate two functions that are not the same, but two functions that are equal can be seen as different.

The problem becomes even more systematic when it comes to bounded polymorphism. Indeed, each time a function with constraints is used, a new closure is created by passing the current method dictionary of type class instances. Even with optimized compilation (*e.g.* `ghc -O`), this (duplicated) closure creation is unavoidable and so `StableNames` will consider different any two constrained functions, even if the passed dictionary is the same.

To overcome this issue, we have overloaded our equality on functions with a special case for functions that have been explicitly tagged with a unique identifier at creation (using `Data.Unique`). This allows to have a robust notion of function equality but it has to be used explicitly at each function definition site.

## 7.2  Advising Overloaded Functions

From a programmer point of view, it can be interesting to advise an overloaded function (that is, the application of all the possible implementations) with a single aspect. However, deploying aspects in the general case of bounded polymorphism is problematic because of the resolution of class constraints. Recall that in order to be able to type the aspect environment, we existentially hide the matched and advised types of an aspect. This means that all type class constraints must be solved statically at the point an aspect is deployed. If the matched and advised types are both bounded polymorphic types, type inference cannot gather enough information to statically solve the constraints. So advising all possible implementations requires repeating deployment of the same aspect with different type annotations, one for each instance of the involved type classes.

To alleviate this problem, we developed a macro using Template-Haskell [28]. The macro extracts all the constrained variables in the matched type of the pointcut, and generates an annotated deployment for every possible combination of instances that satisfy all constraints. In order to retain safety, the advised type of an aspect must be less constrained than its matched type. This is statically enforced by the Haskell type system after macro expansion.

## 7.3  Obliviousness

The embedding of aspects we have presented thus far supports quantification through pointcuts, but is not oblivious: open applications are explicit in the code. A first way to introduce more obliviousness without requiring non-local macros or, equivalently, a preprocessor or ad hoc runtime semantics, is to use *partial applications* of #. For instance, the `sqrtM` function can be turned into an implicitly woven function by defining `sqrtM'= sqrtM #`. This approach was used in Figure 2 for the definition of `fib`. It can be sufficient in similar scenarios where quantification is under control. Otherwise, it can yield to issues in the definition of pointcuts that rely on function identity, because `sqrtM'` and `sqrtM` are different functions. Also, this approach is not entirely satisfactory with respect to obliviousness because it has to be applied specifically for each function.

In [21], De Meuter proposes to use the binder of a monad to redefine function application. His approach focuses on defining one monad per aspect, but can be generalized to a list of dynamically-deployed aspects as presented in Section 2.2. For this, we can re-

define the monad transformer `AOT` to make all monadic applications open transparently:

```
instance Monad m ⇒ Monad (AOT m) where
    return a = AOT (\aenv → return (a, aenv))
    k >>= f  = do { x ← k; f # x }
```

This presentation improves obliviousness because any monadic application is now an open application, but it suffers from a major drawback: it breaks the *monadic laws*. Indeed, left identity and associativity can be invalidated, depending on the current list of deployed aspects. This is not surprising as AOP allows to redefine the behavior of a function and even to redefine the behavior of a function depending on its context of execution. Breaking monadic laws is not prohibited by Haskell, but it is very dangerous and fragile; for instance, some compilers exploit the laws to perform optimizations, so breaking them can yield to incorrect optimizations.

## 8.  RELATED WORK

The earliest connection between aspects and monads was established by De Meuter in 1997 [21]. In that work, he proposes to describe the weaving of a given aspect directly in the binder of a monad. As we have just described above (Section 7.3), doing so breaks the monad laws, and is therefore undesirable.

Wand *et al.* [35] formalize pointcuts and advice and use monads to structure the denotational semantics; a monad is used to pass the join point stack and the store around evaluation steps. The specific flavor of AOP that is described is similar to AspectJ, but with only pure pointcuts. The calculus is untyped. The reader may have noticed that we do not model the join point stack in this paper. This is because it is not *required* for a given model of AOP to work. In fact, the join point stack is useful only to express control flow pointcuts. In our approach, this is achieved by specifying a user-defined pointcut designator for control flow, which uses a monad to thread the join point stack (or, depending on the desired level of dynamicity, a simple control flow state [19]). Support for the join point stack does not have to be included as a primitive in the core language. This is in fact how AspectJ is implemented.

Hofer and Osterman [12] shed some light on the modularity benefits of monads and aspects, clarifying that they are different mechanisms with quite different features: monads do not support declarative quantification, and aspects do not provide any support for encapsulating computational effects. In this regard, our work does not attempt at unifying monads and aspects, contrary to what De Meuter suggested. Instead, we exploit monads in a given language to build a flexible embedding of aspects that can be modularly extended. In addition, the fully-typed setting provides the basis for reasoning about monadic effects.

The notion of *monadic weaving* was described by Tabareau [30], where he shows that writing the aspect weaver in a monadic style paves the way for modular language extensions. He illustrated the extensibility approach with execution levels [32] and level-aware exception handling [10]. The authors then worked on a practical monadic aspect weaver in Typed Racket [11]. However, the type system of Typed Racket turned out to be insufficiently expressive, and the top type `Any` had to be used to describe pointcuts and advices. This was the original motivation to study monadic weaving in Haskell. Also in contrast to this work, prior work on monadic aspect weaving does not consider a base language with monads. In this paper, both the base language and the aspect weaver are monadic, combining the benefits of type-based reasoning about effects (Section 5) and modular language extensions (Section 6)—including type-based reasoning about language extensions.

Haskell has already been the subject of AOP investigations using the type class system as a way to perform static weaving [29]. AOP idioms are translated to type class instances, and type class resolution is used to perform static weaving. This work only supports simple pointcuts, pure aspects and static weaving, and is furthermore very opaque to modular changes as the translation of AOP idioms is done internally at compile time.

The specific flavor of pointcut/advice AOP that we developed is directly inspired by AspectScheme [9] and AspectScript [33]: dynamic aspect deployment, first-class aspects, and extensible set of pointcut designators. While we have not yet developed the more advanced scoping mechanisms found in these languages [31], we believe there are no specific challenges in this regard. The key difference here is that these languages are both dynamically typed, while we have managed to reconcile this high level of flexibility with static typing.

In terms of statically-typed functional aspect languages, the closest proposal to ours is AspectML [6]. In AspectML, pointcuts are first-class, but advice is not. The set of pointcut designators is fixed, as in AspectJ. AspectML does not support: advising anonymous functions, aspects of aspects, separate aspect deployment, and undeployment. AspectML was the first language in which first-class pointcuts were statically typed. The typing rules rely on anti-unification, just like we do in this paper. The major difference, though, is that AspectML is defined as a completely new language, with a specific type system and a specific core calculus. Proving type soundness is therefore very involved [6]. In contrast, we do not need to define a new type system and a new core calculus. Type soundness in our approach is derived straightforwardly from the type class that establishes the anti-unification relation. Half of section 4 is dedicated to proving that this type class is correct. Once this is done (and it is a result that is independent from AOP), proving aspect safety is direct. Another way to see this work is as a new illustration of the expressive power of the type system of Haskell, in particular how phantom types and type classes can be used in concert to statically type embedded languages.

Aspectual Caml [20] is another polymorphic aspect language. Interestingly, Aspectual Caml uses type information to influence matching, rather than for reporting type errors. More precisely, the type of pointcuts is inferred from the associated advices, and pointcuts only match join points that are valid according to these inferred types. We believe this approach can be difficult for programmers to understand, because it combines the complexities of quantification with those of type inference. Aspectual Caml is implemented by modifying the Objective Caml compiler, including modifications to the type inference mechanism. There is no proof of type soundness.

The advantages of our typed embedding do not only lie within the simplicity of the soundness proof. They can also be observed at the level of the implementation. The AspectML implementation is over 15,000 lines of ML code [6], and the Aspectual Caml implementation is around 24,000 lines of Objective Caml code [20]. In contrast, our implementation, including the execution levels extension (Section 6), is only 1,600 lines of Haskell code. Also, embedding an AOP extension entirely inside a mainstream language has a number of practical advantages, especially when it comes to efficiency and maintainability of the extension.

Finally, reasoning about advice effects has been studied from different angles. For instance, harmless advice can change termination behavior and use I/O, but no more [5]. A type and effect system is used to ensure conformance. Translucid contracts use grey box specifications and structural refinement in verification to reason about control effects [4]. In this work, we rather follow the type-based approach of EffectiveAdvice (EA) [23], which also accounts for various control effects and arbitrary computational effects. A limitation of EA is its lack of support quantification, relying instead on open recursive functions. A contribution of this work is to show how to extend this approach to the pointcut/advice mechanism. The subtlety lies in properly typing pointcuts. An interesting difference between both approaches is that in EA, it is not possible to talk about "the effects of all applied advices". Once an advice is composed with a base function, the result is seen as a base function for the following advice. In contrast, our approach, thanks to the aspect environment and dynamic weaving, makes it possible to keep aspects separate and ensure base/aspect separation at the effect level even in presence of multiple aspects. We believe that this splitting of the monadic stack is more consistent with programmers expectations.

## 9. CONCLUSION

We develop a novel approach to embed aspects in an existing language. We exploit monads and the Haskell type system to define a typed monadic embedding that supports both modular language extensions and reasoning about effects with pointcut/advice aspects. We show and prove how to ensure type soundness by design, including in presence of user-extensible pointcut designators, relying on a novel type class for establishing anti-unification. Compared to other approaches to statically-typed polymorphic aspect languages, the proposed embedding is more lightweight, expressive, extensible, and amenable to interference analysis. The approach can combine Open Modules and EffectiveAdvice, and supports type-based reasoning about modular language extensions.

**Acknowledgments** This work was supported by the INRIA Associated team RAPIDS.

## 10. REFERENCES

[1] J. Aldrich. Open modules: Modular reasoning about advice. In A. P. Black, editor, *Proceedings of the 19th European Conference on Object-Oriented Programming (ECOOP 2005)*, number 3586 in Lecture Notes in Computer Science, pages 144–168, Glasgow, UK, July 2005. Springer-Verlag.

[2] *Proceedings of the 7th ACM International Conference on Aspect-Oriented Software Development (AOSD 2008)*, Brussels, Belgium, Apr. 2008. ACM Press.

[3] *Proceedings of the 9th ACM International Conference on Aspect-Oriented Software Development (AOSD 2010)*, Rennes and Saint Malo, France, Mar. 2010. ACM Press.

[4] M. Bagherzadeh, H. Rajan, G. T. Leavens, and S. Mooney. Translucid contracts: Expressive specification and modular verification for aspect-oriented interfaces. In *Proceedings of the 10th ACM International Conference on Aspect-Oriented Software Development (AOSD 2011)*, Porto de Galinhas, Brazil, Mar. 2011. ACM Press.

[5] D. S. Dantas and D. Walker. Harmless advice. In *Proceedings of the 33rd ACM SIGPLAN-SIGACT Symposium on Principles of Programming Languages (POPL 2006)*, pages 383–396, Charleston, South Carolina, USA, Jan. 2006. ACM Press.

[6] D. S. Dantas, D. Walker, G. Washburn, and S. Weirich. AspectML: A polymorphic aspect-oriented functional programming language. *ACM Transactions on Programming Languages and Systems*, 30(3):Article No. 14, May 2008.

[7] B. De Fraine, M. Südholt, and V. Jonckers. StrongAspectJ: flexible and safe pointcut/advice bindings. In AOSD 2008 [2], pages 60–71.

[8] R. Douence, P. Fradet, and M. Südholt. Trace-based aspects. In R. E. Filman, T. Elrad, S. Clarke, and M. Akşit, editors, *Aspect-Oriented Software Development*, pages 201–217. Addison-Wesley, Boston, 2005.

[9] C. Dutchyn, D. B. Tucker, and S. Krishnamurthi. Semantics and scoping of aspects in higher-order languages. *Science of Computer Programming*, 63(3):207–239, Dec. 2006.

[10] I. Figueroa and É. Tanter. A semantics for execution levels with exceptions. In *Proceedings of the 10th Workshop on Foundations of Aspect-Oriented Languages (FOAL 2011)*, pages 7–11, Porto de Galinhas, Brazil, Mar. 2011. ACM Press.

[11] I. Figueroa, É. Tanter, and N. Tabareau. A practical monadic aspect weaver. In *Proceedings of the 11th Workshop on Foundations of Aspect-Oriented Languages (FOAL 2012)*, pages 21–26, Potsdam, Germany, Mar. 2012. ACM Press.

[12] C. Hofer and K. Ostermann. On the relation of aspects and monads. In *Proceedings of AOSD Workshop on Foundations of Aspect-Oriented Languages (FOAL 2007)*, pages 27–33, 2007.

[13] M. P. Jones. Type classes with functional dependencies. In *Proceedings of the 9th European Symposium on Programming Languages and Systems*, ESOP '00, pages 230–244, London, UK, UK, 2000. Springer-Verlag.

[14] G. Kiczales, E. Hilsdale, J. Hugunin, M. Kersten, J. Palm, and W. Griswold. An overview of AspectJ. In J. L. Knudsen, editor, *Proceedings of the 15th European Conference on Object-Oriented Programming (ECOOP 2001)*, number 2072 in Lecture Notes in Computer Science, pages 327–353, Budapest, Hungary, June 2001. Springer-Verlag.

[15] G. Kiczales, J. Lamping, A. Mendhekar, C. Maeda, C. V. Lopes, J.-M. Loingtier, and J. Irwin. Aspect-oriented programming. In M. Akşit and S. Matsuoka, editors, *Proceedings of the 11th European Conference on Object-Oriented Programming (ECOOP 97)*, volume 1241 of *Lecture Notes in Computer Science*, pages 220–242, Jyväskylä, Finland, June 1997. Springer-Verlag.

[16] D. Leijen and E. Meijer. Domain specific embedded compilers. In T. Ball, editor, *Proceedings of the 2nd USENIX Conference on Domain-Specific Languages*, pages 109–122, 1999.

[17] S. Liang, P. Hudak, and M. Jones. Monad transformers and modular interpreters. In *Proceedings of the 22nd ACM SIGPLAN-SIGACT symposium on Principles of programming languages*, POPL '95, pages 333–343, New York, NY, USA, 1995. ACM.

[18] H. Masuhara and K. Kawauchi. Dataflow pointcut in aspect-oriented programming. In *Proceedings of the First Asian Symposium on Programming Languages and Systems (APLAS'03)*, volume 2895 of *Lecture Notes in Computer Science*, pages 105–121, Nov. 2003.

[19] H. Masuhara, G. Kiczales, and C. Dutchyn. A compilation and optimization model for aspect-oriented programs. In G. Hedin, editor, *Proceedings of Compiler Construction (CC2003)*, volume 2622 of *Lecture Notes in Computer Science*, pages 46–60. Springer-Verlag, 2003.

[20] H. Masuhara, H. Tatsuzawa, and A. Yonezawa. Aspectual Caml: an aspect-oriented functional language. In *Proceedings of the 10th ACM SIGPLAN Conference on Functional Programming (ICFP 2005)*, pages 320–330, Tallin, Estonia, Sept. 2005. ACM Press.

[21] W. D. Meuter. Monads as a theoretical foundation for aop. In *In International Workshop on Aspect-Oriented Programming at ECOOP*, page 25. Springer-Verlag, 1997.

[22] E. Moggi. Notions of computation and monads. *Inf. Comput.*, 93(1):55–92, July 1991.

[23] B. C. d. S. Oliveira, T. Schrijvers, and W. R. Cook. EffectiveAdvice: discplined advice with explicit effects. In AOSD 2010 [3], pages 109–120.

[24] B. C. Pierce. *Types and programming languages*. MIT Press, Cambridge, MA, USA, 2002.

[25] G. D. Plotkin. A note on inductive generalization. *Machine Intelligence*, 5:153–163, 1970.

[26] J. C. Reynolds. Transformational systems and the algebraic structure of atomic formulas. *Machine Intelligence*, 5:135–151, 1970.

[27] M. Rinard, A. Salcianu, and S. Bugrara. A classification system and analysis for aspect-oriented programs. In *Proceedings of the 12th ACM Symposium on Foundations of Software Engineering (FSE 12)*, pages 147–158. ACM Press, 2004.

[28] T. Sheard and S. P. Jones. Template meta-programming for haskell. *SIGPLAN Not.*, 37(12):60–75, Dec. 2002.

[29] M. Sulzmann and M. Wang. Aspect-oriented programming with type classes. In *Proceedings of the 6th workshop on Foundations of aspect-oriented languages*, FOAL '07, pages 65–74, New York, NY, USA, 2007. ACM.

[30] N. Tabareau. A monadic interpretation of execution levels and exceptions for aop. In É. Tanter and K. J. Sullivan, editors, *Proceedings of the 11th International Conference on Aspect-Oriented Software Development (AOSD 2012)*, Potsdam, Germany, Mar. 2012. ACM Press.

[31] É. Tanter. Expressive scoping of dynamically-deployed aspects. In AOSD 2008 [2], pages 168–179.

[32] É. Tanter. Execution levels for aspect-oriented programming. In AOSD 2010 [3], pages 37–48.

[33] R. Toledo, P. Leger, and É. Tanter. AspectScript: Expressive aspects for the Web. In AOSD 2010 [3], pages 13–24.

[34] P. Wadler and S. Blott. How to make ad-hoc polymorphism less ad hoc. In *Proceedings of the 16th ACM Symposium on Principles of Programming Languages (POPL 89)*, pages 60–76, Austin, TX, USA, Jan. 1989. ACM Press.

[35] M. Wand, G. Kiczales, and C. Dutchyn. A semantics for advice and dynamic join points in aspect-oriented programming. *ACM Transactions on Programming Languages and Systems*, 26(5):890–910, Sept. 2004.

# On Exceptions, Events and Observer Chains

Mehdi Bagherzadeh   Hridesh Rajan   Ali Darvish
Dept. of Computer Science, Iowa State University
Ames, IA, 50011, USA
{mbagherz,hridesh,ali2}@iastate.edu

## ABSTRACT

Modular understanding of behaviors and flows of exceptions may help in their better use and handling. Such reasoning tasks about exceptions face unique challenges in event-based implicit invocation (II) languages that allow subjects to implicitly invoke observers, and run the observers in a chain. In this work, we illustrate these challenge in Ptolemy and propose $Ptolemy_\chi$ that enables modular reasoning about behaviors and flows of exceptions for event announcement and handling. $Ptolemy_\chi$'s *exception-aware specification expressions* and *boundary exceptions* limit the set of (un)checked exceptions of subjects and observers of an event. *Exceptional postconditions* specify the behaviors of these exceptions. Greybox specifications specify the flows of these exceptions among the observers in the chain. $Ptolemy_\chi$'s type system and refinement rules enforce these specifications and thus enable its modular reasoning. We evaluate the utility of $Ptolemy_\chi$'s exception flow reasoning by applying it to understand a set of aspect-oriented (AO) bug patterns. We also present $Ptolemy_\chi$'s semantics including its *sound* static semantics.

***Categories and Subject Descriptors*** D.3.3 [**Programming Languages**]: Language Constructs and Features—*Exceptions*

***General Terms***   Languages, Theory

***Keywords***   event, exceptional behavior, exception flow

## 1.  INTRODUCTION

Exceptions are useful for structured separation of normal and error handling code. However, their improper use and handling could put a system in undetermined risky states or even crash it [29]. Understanding exceptions, especially their behaviors and flows may help with their better use and handling [28, 35]. Previous work, such as JML [25], ESC/Java [12], the work of Jacobs *et al.* [16] anchored exceptions [39], Jex [33] and EJFlow [5] enable reasoning about behaviors or flows of exceptions. However, they are tailored for systems in which invocation relations among the modules are explicit and known, i.e. explicit invocation (EI) [7]. With EI, in languages such as Java, a module explicitly invokes another mod-

ule with a method call E.m() in which both the static type of the invoked module E and the name of the method m are known. EI reasoning techniques use this knowledge to incorporate the behaviors and flows of the exceptions of the method m into its invoking module [7]. Supertype abstraction enables reasoning independent of the dynamic type of E. However, the invocation relations among modules of a system are not limited to explicit invocations. This is true in languages such as Java or C# when using events and delegates or in languages such as AspectJ [21,22] or Ptolemy [31] when a module invokes another module implicitly and without knowing about it, i.e. implicit invocation (II) [15]. Ptolemy is an event-based II extension of Java.

Modular reasoning about behaviors and flows of exceptions faces unique challenges in II languages such as Ptolemy or AspectJ that allow a (subject) module to invoke other (observer) modules without knowing about them and run them in a chain. In Ptolemy, a subject announces an event and zero or more observers register for the event and handle it. The observers form a chain based on their dynamic registration order and may invoke each other. The observers are not explicitly mentioned in the subject and are invoked implicitly by an II mechanism, upon announcement of the event. The following code snippet in Ptolemy illustrates a subject module that announces an event Ev causing its unknown observers to be invoked in a chain.

```
announce Ev(){}
```

To reason about this subject, especially behaviors and flows of exceptions during announcement and handling of Ev, behaviors of its observers when throwing the exceptions and flows of these exceptions in the chain of the observers should be understood. Depending on the order of the observers in the chain, an exception thrown by one observer may be caught by *another* observer or propagated down the chain. However, the observers, their orders in the chain, and the behaviors and flows of their exceptions are not known to the subject. Even if the such information were available for individual observers, a naive use of EI reasoning techniques may require all possible orders of execution of the observers to be considered, i.e. *n!* for *n* observers. Such dependency of reasoning on system configuration, i.e. individual observers, their order and behaviors and flows of their exceptions, threatens its modularity since any changes in the system configuration could invalidate any previous reasoning. The main difference from Java's Event Model or the Observer pattern [24] is that observers form a chain based on their dynamic registration order and may invoke each other.

Previous work on modular reasoning about Ptolemy-like II languages, such as translucid contracts [2], join point interfaces (JPI) [36], crosscutting programming interfaces (XPI) and the work of Khatchadourian *et al.* [20], provides the subject with the knowledge about behaviors of its unknown observers via rely-guarantee-like [7] techniques. The rely-guarantee allows the subject to rely on the provided knowledge about its observers independent of the system configuration [19]. However, the previous work is mostly focused on normal behaviors of the observers and is less concerned about their exceptions, and behaviors and flows of these exceptions. Previous work on Join Point Interfaces (JPI) [3] provides the subject with the knowledge about the types of the exceptions of its observers but not their behaviors or flows.

In summary, the following problems exist when understanding a subject in Ptolemy, especially modular reasoning about behaviors and flows of exceptions in announcement and handling of its event:

- *problem (1)*: subject does not know about the exceptions that its observers may throw;

- *problem (2)*: subject does not know about the behaviors of the exceptions of its observers;

- *problem (3)*: subject does not know about the flows of the exceptions among its observers in the chain of the observers.

We propose *Ptolemy$_\chi$*, as an exception-aware extension of Ptolemy, to solve these problems. To address the problem *(1)* for checked exceptions, similar to JPI, we limit the set of checked exceptions of a subject and its observers. These exceptions are referred to as *boundary exceptions*. To address the problem *(1)* for unchecked exceptions we propose *exception-aware specification expressions* to limit the set of unchecked exceptions of the subject and its observers. For the problem *(2)* we use exceptional postconditions to specify the state of the subject and observers upon throwing the exceptions. And finally to solve the problem *(3)*, we use greybox specifications to specify the flow of the exceptions among the observers in their chain, by limiting their implementation structures. *Ptolemy$_\chi$*'s type system, refinement rules and runtime assertion checks ensure that the subject and its observers satisfy these specifications and thus enable modular reasoning about the behaviors and flows of the exceptions for announcement and handling of the event, independent of system configuration, i.e. individual observers and their execution order in their chain.

## 1.1 Contributions

In summary the contributions of this paper are the following:

- Enabling modular reasoning about behaviors and flows of exceptions for event announcement and handling in *Ptolemy$_\chi$*;

- Evaluating *Ptolemy$_\chi$*'s exception flow reasoning in understanding Coelho *et al.*'s aspect-oriented bug patterns [6];

- Presenting *Ptolemy$_\chi$*'s sound static and dynamic semantics.

The rest of the paper continues as the following. Section 2 illustrates the problems *(1)–(3)* for modular reasoning about behaviors and flows of exceptions in announcement and handling of events in Ptolemy. Section 3 describes our proposed language design of *Ptolemy$_\chi$* and how it solves the example modular reasoning problems of Section 2. Section 4 discusses *Ptolemy$_\chi$*'s sound type system and its refinement rules that enable its modular reasoning. Section 5 illustrates the usability of *Ptolemy$_\chi$*'s exception flow reasoning in understanding Coelho *et al.*'s AO bug patterns [6]. This section also discusses the overhead of applying *Ptolemy$_\chi$* to Ptolemy programs. Section 6 compares our proposal with existing work. Section 7 concludes and discusses future work.

## 2. PROBLEM

In this section we illustrate the problems *(1)–(3)* for modular reasoning about *(i)* behaviors and *(ii)* flows of exceptions during announcement and handling of events in Ptolemy [31].

## 2.1 Modular Reasoning about Behaviors of Exceptions

As an example of modular reasoning about behaviors of exceptions during event announcement and handling, consider verification of the JML-like assertion $\Phi_b$ on line 11 of Figure 1. Figure 1 and Figure 2 illustrate a savings bank account with a withdraw functionality. The assertion $\Phi_b$ says that: *an account is not withdrawn if an exception RegDExc is thrown during the announcement and handling of the event WithdrawEv, lines 5–8.* In other words the balance of the account is not changed if the announcement and handling of WithdrawEv terminates abnormally by throwing a RegDExc exception. The expression **old** refers to the values of variables at the beginning of the method withdraw.

To verify the assertion $\Phi_b$ one should understand the exceptional behavior of the announcement and handling of the event WithdrawEv, lines 5–8, for the exception RegDExc. This involves understanding the exceptional behaviors of the unknown observers of WithdrawEv, which are invoked in a chain upon its announcement, and also the exceptional behavior of the body of the announce expression itself, lines 6–7. The exceptional behavior of an observer for an exception is the state of the observer right before throwing the exception [25]. For such understanding to be modular [23], one may only use the implementation and the interface of the subject module Savings, lines 1–14, and the interfaces it references, i.e. event type WithdrawEv, lines 29–38 of Figure 2. The reasoning must be independent of system configuration, i.e. unknown observers of WithdrawEv or their execution order in their chain. However, neither the implementation of the subject nor the event type declaration provides any knowledge about the exceptions that the observers of WithdrawEv may throw, i.e. problem *(1)*, or their exceptional behaviors if they terminate abnormally by throwing RegDExc, i.e. problem *(2)*. To better understand these problems we provide a short background on Ptolemy.

### 2.1.1 Ptolemy Language in a Nutshell

The language Ptolemy is an II extension of Java with support for explicit announcement and handling of typed events [31]. In Figure 1, written in Ptolemy, the subject module Savings explicitly announces the event WithdrawEv using the **announce** expression, lines 5–8, with the event body on lines 6–7. The observer module Check, lines 15–27 shows interest in the event using the **when** − **do** binding declaration, line 24, and dynamically registers itself to handle it using the **register** expression, line 26. The observer Check runs the observer handler method check upon announcement of WithdrawEv. The observer Check checks for undesired behaviors of withdraw, such as overdrawing or violation of maximum number of withdrawals of accounts, etc. For brevity, Check only checks for maximum number of monthly withdrawals of savings accounts that is limited to 6 according to *U.S. Federal Reserve board Regulation D*. The field numWithdrawals keeps track of the number of withdrawals. The observer Check throws an exception RegDExc if Regulation D is violated, line 20. In Ptolemy, the subject Savings and the observer Check know about the event WithdrawEv, however, they do not know about each other which in turn means that the subject does not know about its observers or their exceptions, i.e. problem *(1)*.

In Ptolemy, zero or more observers could register for the same event. Similar to AspectJ [21, 22], these observers form a chain

```
1  class Savings {                              15  class Check {
2   int bal; int numWithdrawals;               16   void check(WithdrawEv next) throws Throwable{
3   void withdraw(int amt) throws Throwable{    17    refining
4    try{                                       18     establishes next.acc().bal==old(next.acc().bal){
5     announce WithdrawEv(this,amt){            19     if(next.acc().numWithdrawals>=6)
6       bal-= amt;                              20      throw new RegDExc();
7       numWithdrawals++;                       21     }
8     }                                         22    next.invoke();
9    }                                          23   }
10   catch(RegDExc e){                          24   when WithdrawEv do check;
11    //@ assert this.bal==old(this.bal);       25
12    throw e; }                                26   Check(){ register(this);  }
13   catch(RuntimeException e){ e.printStackTrace(); } 27 }
14  } .. }                                      28  class RegDExc extends Exception {}
```

**Figure 1: Savings bank account example in Ptolemy [31] with the subject `Savings` and the observer `Check`.**

based on their dynamic registration order, with the event body at the end of the chain. Upon announcement of the event the control is transferred from the subject to the first observer in the chain that at some point during its execution may decide to invoke the next observer in the chain, using the `invoke` expression, line 22. The `invoke` expression is the equivalent of AspectJ's `proceed`. The chain of the observers for `WithdrawEv` is stored in an event closure `next`, that is passed as a parameter to its observer handler method `check`, line 16. The subject `Savings` does not statically know about its observers or their execution order in the chain.

The event `WithdrawEv` should be declared before it is announced or handled. The declaration of `WithdrawEv` in Figure 2 lists a set of context variables, lines 30–31, and has a greybox translucid contract, lines 32–37. The context variables are shared information between the subject and its observers and their values are provided by the subject when the event is announced, line 5. Translucid contracts [2] are discussed later in this section. The event `WithdrawEv` does not give the subject `Savings` any information about the exceptions of its observers, i.e. problem *(1)*.

```
29  void event WithdrawEv{
30   Savings acc;
31   int amt;
32   requires acc!=null
33   assumes{
34    establishes next.acc().bal==old(next.acc().bal);
35    next.invoke();
36   }
37   ensures acc.bal<=old(acc.bal)
38  }
```

**Figure 2: Event `WithdrawEv` with a translucid contract, lines 32–37.**

***Exception Handling in Ptolemy*** To cope with the unknown exceptions of its observers, and especially their checked exceptions, the subject `Savings` adds a general throws clause **`throws Throwable`** to the signature of the method `withdraw` which announces `WithdrawEv`, line 3. The same is true for the observer `Check` and its observer handler method `check`, line 16, since it invokes other unknown observers of `WithdrawEv` with their unknown exceptions using the invoke expression. However, these general catch clauses hardly provide any information about the exceptions of the observers of `WithdrawEv` since they basically allow any checked exceptions. As in Java, in Ptolemy checked exceptions are propagated explicitly by declaring them in method headers but unchecked exceptions are propagated implicitly.

***Translucid Contracts in Ptolemy*** The translucid contract of `WithdrawEv`, lines 32–37 of Figure 2, is a greybox specifica-

tion [4] that enables modular reasoning about the *normal* behavior and control effects of its announcement and handling. The normal behavior of the announcement and handling of `WithdrawEv` is specified using the precondition **`requires`**, line 32, and the postcondition **`ensures`**, line 37. The control effects for the normal behavior are specified by the **`assumes`** block, lines 33–36, that limits the implementation structure of the observers of the event. The assumes block is a combination of program and specification expressions. The program expression, line 35, exposes the control effects of interest in the implementations of the observers such as `Check`, whereas the specification expression, line 34, hides the rest of the implementations of the observers and allows them to vary as long as they respect the specification. The assumes block on lines 33–36 says that observers of `WithdrawEv` can do anything as long as they do not change the balance of the account `acc`, i.e. **`establishes`** next.$acc()$.$bal ==$ **`old`**(next.$acc()$.$bal$), and then invoke the next observer in the chain using **`next.invoke()`**. The expression **`next`**.$acc()$ returns the context variable `acc`.

***Refinement of Translucid Contracts*** Refinement rules for the translucid contract of `WithdrawEv` enable modular reasoning about normal behavior and control effects of its announcement and handling, independent of its unknown observers and their execution order [2]. The refinement rules require each subject and observer of `WithdrawEv` to satisfy the pre- and postconditions of its translucid contract. They also require the observers to structurally refine the assumes block of the translucid contract. For the observer handler method `check`, lines 16–23 in Figure 1, to structurally refine the assume block of `WithdrawEv`, lines 33–36 in Figure 2, the following conditions should hold: for each program expression in the assumes block, line 35, the refining observer `check` must have a textually matching program expression in its implementation at the same place, line 22; and for each specification expression in the assumes block, line 34, the observer must have a **`refining`** expression with the same specification in its implementation at the same place, lines 17–21. Using the contract of `WithdrawEv` one may conclude that upon announcement and normal termination of `WithdrawEv` all of its observers in the chain are invoked. This is valid since each observer of `WithdrawEv` has to refine its contract and have the invoke expression in its implementation. However the translucid contract only works for normal execution of announcement and handling of `WithdrawEv`, i.e. no exceptions thrown, and doe not provide any information about the exceptions of its observers, their behaviors or flows, i.e. problems *(1)–(3)*.

### 2.1.2  Boundary Exceptions in Ptolemy

As illustrated, in Ptolemy the subject `Savings` does not know about exceptions of its observers, i.e. problem *(1)*, and the throws clauses **throws Throwable** in the signature of the observer handler methods do not address the problem. Previous work on join point interfaces (JPI) [3] solves this problem for checked exceptions by specifying the set of checked exceptions that the observers may throw. Following JPI, we limit the set of checked exceptions of the observers of an event and call these exceptions boundary exceptions. Figure 3 illustrates the boundary exception RegDExc for the observers of WithdrawEv in its declaration, line 29. The boundary exception RegDExc also limits the set of the checked exceptions of the subject Savings. With the boundary exception RegDExc, the throws clauses of the methods withdraw and check should be changed to **throws** RegDExc since it is the only checked exception they can throw.

```
29  void event WithdrawEv throws RegDExc{
30  .. /* the same as before */
31  }
```

**Figure 3: boundary exception RegDExc, line 29.**

However, the boundary exception of WithdrawEv does not say anything about the behaviors and flows of RegDExc or the unchecked exceptions of its observers, their behaviors or flows in the chain of the observers, i.e. problems *(2)–(3)*.

## 2.2 Modular Reasoning about Flows of Exceptions

*Flows of Checked Exceptions* As an example of modular reasoning about flows of checked exceptions during event announcement and handling, consider verification of a requirement $\Phi_{f1}$ that says: *the checked boundary exception RegDExc must be propagated back to the subject Savings, if thrown by any observer of WithdrawEv during its announcement and handling.* In other words, an exception RegDExc thrown by one observer of WithdrawEv must not be caught by another one of its observers in the chain of the observers. The checked exception RegDExc is used by the observer Check to tell the subject that the withdrawal operation terminated unsuccessfully. The subject in turn propagates the exception back to its clients, line 12, which supposedly have more contextual information to handle it. If the exception RegDExc is thrown but does not reach the subject, then the subject and consequently its clients may wrongly assume the successful termination of the withdraw, which is an undesired behavior.

```
1  class Logger{
2  void log(WithdrawEv next) throws RegDExc{
3   try{
4    refining
5     establishes next.acc().bal==old(next.acc().bal){
6      Log.logTrans(next.acc(),"withdraw");
7     }
8    next.invoke();
9   }
10   catch(Exception e){} /* swallow */
11  }
12  when WithdrawEv do log;
13 }
```

**Figure 4: Observer Logger could swallow RegDExc depending on its execution order in the chain of the observers.**

This could happen if another observer, say Logger [1] in Figure 4, registers and runs before Check in the chain of the ob-

servers. Now when WithdrawEv is announced, the observer Logger runs and invokes the observer Check using its invoke expression, line 8, which throws a RegDExc. However, the **try – catch** expression around the invoke expression of Logger, lines 3 and 10, catches the exception and swallows it, line 10.

To reason about and verify $\Phi_{f1}$, one should understand the flow of RegDExc among the observers of WithdrawEv in the chain of the observers. For the reasoning to be modular, one may only use the implementation and interface of the subject Savings and the declaration of the event WithdrawEv, independent of system configuration. However, neither the subject nor the event provides any knowledge about the flows of the exceptions among the observers, i.e. problem *(3)*.

The execution order of the observers in their chain is also important in reasoning about $\Phi_{f1}$. If Logger runs before Check then $\Phi_{f1}$ is violated because Logger swallows RegDExc thrown by Check. However, if the execution order is reversed, i.e. Check runs before Logger, then RegDExc propagates back to the subject and $\Phi_{f1}$ holds.

*Flows of Unchecked Exceptions* The boundary exception of WithdrawEv tells the subject Savings that its observers may only throw the checked exception RegDExc, however, it does not say anything about their unchecked exceptions. To handle the unknown unchecked exceptions of the observers the subject Savings catches these exceptions using **catch** (**RuntimeException** $e$) {..} and prints out their backtraces, line 13. However, this catch clause could be unused especially if no unchecked exceptions of the observers reaches the subject. We refer to this property as $\Phi_{f2}$. Such an unused catch clause, which is trying to catch exceptions that do not reach it, is an example of a bad programming practice. Coelho *et al.* [6] categorizes such unused catch clauses as residual catch bugs. Bug patterns are discussed in more details in Section 5.

Again, different orders of the execution of the observers in their chain entails different results regarding $\Phi_{f2}$. The catch clause on line 13 of Figure 1 is unused, if the observer Logger runs before other observers of WithdrawEv in the chain. Recall that in Ptolemy the body of the announce expression is at the end of the chain. The try-catch expression around the invoke expression in Logger, lines 3 and 10, catches and swallows not only the checked boundary exception RegDExc but also all unchecked exceptions of the observers which run after it in the chain, before these exceptions reach the subject Savings. However, if an observers runs before Logger in the chain and throws an unchecked exception then this exception could reach the subject Savings and thus its catch clause will actually be used and necessary.

## 3. *Ptolemy$_\chi$*

In this section we describe *Ptolemy$_\chi$*'s core syntax and specification features for stating behaviors and flows of exceptions during event announcement and handling and address the problems *(1)–(3)*.

### 3.1 Program Syntax

*Ptolemy$_\chi$* is an exception-aware extension of Ptolemy [31] with support for specification and modular reasoning about behaviors and flows of exceptions during event announcement and handling. Similar to Ptolemy, *Ptolemy$_\chi$* is an implicit invocation (II) object-oriented (OO) language with support for explicit announcement and handling of typed events. The formal definition of *Ptolemy$_\chi$* is given as an expression language.

---

[1] Translucid contracts in Ptolemy [2] do not support throwing and handling of exceptions and only work for normal execution. One way of treating try-catch expressions in observers such as Logger

is to discard the catch part and only keep the body of the try part, which represents the normal execution. This allows Logger to structurally refine its contract in Figure 2.

The syntax of *Ptolemy*$_\chi$'s executable programs is shown in Figure 5. The syntax supports (typed) events, exceptions, classes, objects, inheritance and subtyping for classes. Similar to Java, exceptions are treated like objects [8] and are divided into checked and unchecked exceptions. For simplicity, *Ptolemy*$_\chi$ does not have packages, privacy modifiers, abstract classes and methods, static members, interfaces or constructors. The superscript $\overline{term}$ shows a sequence of zero or more *term* whereas [*term*] means zero or one, i.e. optional. A *Ptolemy*$_\chi$'s program (*prog*) is a collection of declarations ($\overline{decl}$) followed by an expression (*e*), which is like a main method in Java. There are two types of declarations in *Ptolemy*$_\chi$: class and event type declarations.

$$
\begin{array}{rcl}
prog & ::= & \overline{decl}\ e \\
decl & ::= & \texttt{class}\ c\ \texttt{extends}\ d\ \{\ \overline{form}\ \overline{meth}\ \overline{binding}\ \} \\
 & | & c\ \texttt{event}\ p\ \texttt{throws}\ \overline{x}\ \{\ \overline{form}\ [\ contract\ ]\ \} \\
t & ::= & c\ |\ \texttt{thunk}\ c\ p \\
meth & ::= & t\ m\ (\overline{form})\ \{\ e\ \}\ \texttt{throws}\ \overline{x} \\
form & ::= & t\ var, \quad \text{where}\ var \neq \texttt{this} \\
binding & ::= & \texttt{when}\ p\ \texttt{do}\ m \\
ep & ::= & var\ |\ ep.f\ |\ ep == ep|\ ep < ep\ |\ !\ ep\ |\ ep\ \&\&\ ep|\ \texttt{old}(ep) \\
e, se & ::= & n\ |\ \texttt{null}\ |\ var\ |\ \texttt{new}\ c()\ |\ e.m(\overline{e})\ |\ e.f\ |\ e.f = e \\
 & | & \texttt{if}\ (ep)\ \{\ e\ \}\ \texttt{else}\ \{\ e\ \}\ |\ \texttt{cast}\ c\ e\ |\ form = e;\ e \\
 & | & \texttt{announce}\ p\ (\overline{e})\ \{\ e\ \}\ |\ e.\texttt{invoke}()\ |\ \texttt{register}(e) \\
 & | & \texttt{refining}\ spec\ \{\ e\ \}\ |\ spec\ |\ \texttt{either}\ \{\ e\ \}\ \texttt{or}\ \{\ e\ \} \\
 & | & \texttt{throw}\ e\ |\ \texttt{try}\ \{e\}\ \texttt{catch}\ (x\ var)\ \{e\} \\
contract & ::= & \texttt{requires}\ ep\ [\ \texttt{assumes}\ \{\ se\ \}\ ]\ \texttt{ensures}\ ep\ \overline{excp} \\
excp & ::= & \texttt{signals}\ (\overline{x})\ ep \\
spec & ::= & \texttt{requires}\ ep\ \texttt{ensures}\ ep\ \texttt{throws}\ \overline{x}
\end{array}
$$

**where**

$$
\begin{array}{rcl}
c & \in & \mathcal{C}, \text{set of class names} \\
d & \in & \mathcal{C} \cup \{\texttt{Object}\}, \text{a set of superclass names} \\
p & \in & \mathcal{P}, \text{set of event type names} \\
f & \in & \mathcal{F}, \text{set of field names} \\
var & \in & \{\texttt{this}, \texttt{next}\} \cup \mathcal{V}, \mathcal{V}\ \text{is a set of variable names} \\
x & \in & \mathcal{X} \cup \{\texttt{NoRuntimeExc}, \texttt{Exception}, \texttt{RuntimeException} \\
 & & \texttt{ClassCastException}, \texttt{NullPointerException}\}, \\
 & & \mathcal{X}\ \text{is a set of exception names}
\end{array}
$$

**Figure 5: *Ptolemy*$_\chi$'s syntax, based on [2, 31]**

### 3.1.1 Declarations

In a class declaration, the class has a name (*c*), a super class (*d*), a set of fields ($\overline{form}$), methods ($\overline{meth}$) and binding declarations ($\overline{binding}$). A binding declaration associates an event type (*p*) with an observer handler method (*m*). In *Ptolemy*$_\chi$ observers are normal classes with handler methods which take a handler chain as their parameter. Both observer handler methods and non-handler regular methods (*meth*) have to declare their checked exceptions ($\overline{x}$) in their **throws** clauses. *However, similar to Java, the declaration of unchecked exceptions is not necessary neither for the observer handler methods nor for the regular methods.*

In an event type declaration, the event has a name (*p*), a return type (*c*), a set of checked boundary exceptions ($\overline{x}$), a set of context variables ($\overline{form}$) and an optional translucid contract (*contract*). The boundary exceptions of the event are checked exceptions, named in its **throws** clause, that limit the set of checked exceptions that subjects and observers of the event can throw. All the other checked exceptions must be handled locally by the subjects and observers. The boundary exceptions address the problem *(1)* for checked exceptions. For example, Figure 3 lists a boundary exception RegDExc in the declaration of WithdrawEv, line 29. Declaration of the context variables specify types and names of information shared between the subjects and observers.

### 3.1.2 Specifications

The idea is to use the translucid contract of an event to reason about behaviors and flows of the exceptions of its observers independent of the unknown observers or their execution order in their chain. The translucid contract, (*contract*), is a greybox specification that specifies behaviors of both checked and unchecked exceptions of subject and observers of an event, during its announcement and handling. The contract also specifies the set of unchecked exceptions that subject and observers of the event can throw and the flows of exceptions among the observers of the event in their chain.

***Behaviors of (Un)checked Exceptions*** In a translucid contract, behaviors of checked or unchecked exceptions thrown during announcement and handling of an event is specified using the signals clauses. A signals clause **signals** $(\overline{x})\ ep$ lists checked or unchecked exceptions ($\overline{x}$) of the subjects and observers of the event and associates them with a side effect free postcondition (*ep*). The signals clause says if announcement or handling of the event terminates abnormally by throwing any of the exceptions $\overline{x}$ then it terminates in a state that satisfies the exceptional postcondition *ep*. In other words, if an observer or a subject throws an exception ($\overline{x}$) then it must satisfy the exceptional postcondition (*ep*). Postcondition of an exception is the state right before the exception is thrown [25]. Exceptions with no exceptional postconditions, have the default postcondition of **true**. Exceptional postconditions address the problem *(2)*. Figure 6 illustrates the exceptional postconditions for the checked exception RegDExc, line 9, and the unchecked exceptions RuntimeException, line 10.

The translucid contract also specifies normal behavior of the announcement and handling of an event [2]. The normal behavior **requires** $ep_1$ **ensures** $ep_2$ says that if the event is announced in a state that satisfies the precondition $ep_1$ and its announcement and handling terminates normally, i.e. throws no exceptions, then it terminates in a state that satisfies the postcondition $ep_2$. The specification **establishes** $ep$ is sugar for **requires** $true$ **ensures** $ep$.

An **assumes**$\{se\}$ block in a translucid contract limits the implementation structure of the observers of its event. The body (*se*) of the assumes block is a combination of program expressions and exception-aware specification expressions *spec*. Program expressions reveal interesting exceptional control flows, e.g. try-catch or invoke expressions in the implementation of the refining observers, whereas exception-aware specification expressions hide the rest of the implementation under **refining** expressions. Figure 12 illustrates an assumes block in the translucid contract of WithdrawEv, lines 2–9 with the program expressions on lines 3, 6 and 7 and the exception-aware specification expressions on lines 4–5 and 8.

***Limiting Unchecked Exceptions*** Exception-aware specification expressions in an assumes block limit the set of unchecked exceptions of the subjects and observers of an event, in combination with program expressions. The exception-aware specification expression **requires** $ep_1$ **ensures** $ep_2$ **throws** $\overline{x}$ says that in a refining implementation of an observer, if the execution starts in a state satisfying the precondition $ep_1$ and terminates normally, it terminates in a state satisfying the postcondition $ep_2$. However, if the execution terminates abnormally by throwing an unchecked exception, then the thrown exception must an exception in $\overline{x}$ otherwise the specification expression is violated. The exception-aware specification expression **requires** $ep_1$ **ensures** $ep_2$ is sugar for **requires** $ep_1$ **ensures** $ep_2$ **throws RuntimeException**. Exception-aware specification expressions address the problem *(1)* for unchecked exceptions.

For example, Figure 6 illustrates an exception-aware specification expression on lines 4–5 that allows all unchecked exceptions RuntimeException to be thrown by the observers of WithdrawEv. The program expression **next.invoke**(), line 6 can throw any unchecked exception since it invokes the next observer of WithdrawEv. Note that the event closure **next** is

189

non-null by construction. Figure 11 illustrates another exception-aware specification expression, line 3–4, that allows no unchecked exceptions, i.e. `NoRuntimeExc`. The program expressions `next.invoke()` in this figure, line 5, cannot throw any unchecked exceptions since it invokes the next observer. Finally, Figure 12 illustrates two exception-aware specification expressions, lines 4–5 and 8. The first one allows all unchecked exceptions by a refining implementation in an observer of `WithdrawEv`, however, the program expression try-catch which surrounds it, lines 3 and 7, catches all such unchecked exceptions and handles them in a way that itself may not throw any unchecked exceptions, line 8.

*Flows of (Un)checked Exceptions* An invoke expression causes the invocation of the next observers in a chain of observers. Revealing the invoke expression and its enclosing expression in an assumes block specifies the flows of exceptions of the observers in the chain of the observers. For example, in Figure 6 an observer of `WithdrawEv` is not allowed to catch any exception thrown by other observers in the chain because there is no try-catch expression around its invoke expression, line 6. This means any exception thrown by an observer in the chain propagates back all the way to its subject without being caught by another observer in the chain. In Figure 12 an observer can catch an unchecked exception thrown by another observer in the chain and swallow it because its invoke expression, line 6, is enclosed by a try-catch expression, lines 3 and 7, however, it cannot catch checked exceptions such as `RegDExc`. Because invoke expressions plays a critical role in understanding the flows of exceptions in the chain of the observers, revealing them in the the assumes block of the translucid contract of the event is *mandatory*. Revealing invoke expressions in the assumes blocks addresses the problem *(3)*.

In summary, the translucid contract and boundary exceptions of an event limit the set of checked and unchecked exceptions of the subjects and unknown observers of an event. The contract specifies the behaviors of the subjects and observers of the event upon throwing these exceptions and the flows of these exceptions among the observers in their chain. The idea is to use the translucid contract of the event to reason about its announcement and handling independent of its unknown observers or their execution order in the chain. *Ptolemy*$_\chi$'s typing and refinement rules, in Section 4, ensure that the subjects and observers satisfy these specifications.

### 3.1.3 Expressions

*Ptolemy*$_\chi$ has expressions for throwing and handling of exceptions and explicit announcement and handling of events. The expressions `throw` and `try − catch` are standard and similar to Java [8]. The expressions `finally` and try-catch expressions with multiple catch clauses are sugars and are omitted from the syntax. *Ptolemy*$_\chi$ supports the built-in exceptions `Throwable`, `Exception`, `RuntimeException`, `NullPointerException` and `ClassCastException`, similar to exceptions in Java with the same subtyping relations. Unchecked exceptions are subtypes of `RuntimeException`. The exception `NoRuntimeExc` stands for no unchecked exception.

In *Ptolemy*$_\chi$, a subject explicitly announces an event (*p*) using an `announce` expression with parameters ($\overline{e}$) as its context variables and the event body (*e*). The announce expression starts the execution of the chain of observers for the event (*p*). An observer is registered using a `register` expression that evaluates its parameter (*e*) to an object and puts it into a chain of observers. Expression `invoke` evaluates its receiver object (*e*) into an event closure and runs it, which in turn causes the next observer in the chain of observers to run. An event body is at the end of the chain of the observers. Event closures are also passed to observer handler

methods, e.g. line 16 of Figure 1. The type (`thunk` *c p*) ensures the value of the corresponding actual parameter is of type event closure with return type (*c*) for an event (*p*). The expression `next` is the placeholder for the event closure.

## 3.2 Modular Reasoning about Behaviors of Exceptions

To enable modular reasoning about behaviors of exceptions in event announcement and handling, e.g. reasoning about $\Phi_b$ in Section 2.1, *Ptolemy*$_\chi$ provides exceptional postconditions. Figure 6 illustrates the exceptional postconditions for the checked boundary exception `RegDExc`, line 9, and unchecked exceptions `RuntimeException`, line 10, The exceptional postcondition for `RegDExc` says that if during announcement and handling of `WithdrawEv` an exception `RegDExc` is thrown by its observers then they must guarantee that the balance of the account `acc` does not change, i.e. *acc.bal* == **old**(*acc.bal*).

```
1 void event WithdrawEv throws RegDExc{ ..
2   requires acc!=null
3   assumes{
4     establishes next.acc().bal==old(next.acc().bal)
5       throws RuntimeException;
6     next.invoke();
7   }
8   ensures acc.bal<=old(acc.bal)
9   signals (RegDExc) acc.bal==old(acc.bal)
10  signals (RuntimeException) true
11 }
```

**Figure 6: Exception-aware specification expression, lines 4–5 and exceptional postconditions, lines 9–10.**

Using this guarantee one is able to verify the assertion $\Phi_b$ in a modular fashion using only the implementation of the subject `Savings`, in Figure 1, and the contract of the event `WithdrawEv` especially its exceptional postcondition for `RegDExc`, line 9. Recall that the assertion $\Phi_b$ says if an exception `RegDExc` is thrown during announcement and handling of `WithdrawEv` the balance of the account is not changed. Upon announcement of `WithdrawEv` the control reaches line 10 of Figure 1 if an exception `RegDExc` is thrown by an observer and propagated back to the subject. According to the exceptional postcondition of `RegDExc` the observers of `WithdrawEv` ensure that the predicate *acc.bal* == **old**(*acc.bal*) is true upon throwing `RegDExc`. By replacing the variable **this** for `acc` in this exceptional postcondition we get the predicate **this**.*bal* == **old**(**this**.*bal*) which is the same as the assertion $\Phi_b$ that we wanted to verify. Recall that **this** is passed as the context variable `acc`, line 5 of Figure 1.

Such a reasoning is independent of system configuration, i.e. unknown observers of `WithdrawEv` or their execution order in their chain. This reasoning is valid because *Ptolemy*$_\chi$'s refinement rules require all the observers and subjects of `WithdrawEv` to satisfy the exceptional postcondition of `RegDExc` if they throw it. The refinement rules are discussed in Section 4.

## 3.3 Modular Reasoning about Flows of Exceptions

To enable modular reasoning about flows of exceptions of event announcement and handling, e.g. reasoning about $\Phi_{f1}$ and $\Phi_{f2}$ in Section 2.2, *Ptolemy*$_\chi$ provides exception-aware specification expressions and translucid contracts. The contract in Figure 6 says that a refining observer of `WithdrawEv` can throw any unchecked exception, lines 4–5, however it does not catch any exception thrown by *another* observers in the chain of observers, line 6.

The guarantee that an observer of `WithdrawEv` does not catch any exception thrown by another observer, means that the property $\Phi_{f1}$ holds. Recall that $\Phi_{f1}$ says if an exception `RegDExc` is thrown by an observer of `WithdrawEv` during its announcement and handling, the exception is propagated back to its subject. This guarantee also means that the catch clause `catch` (`RuntimeException` e) {..} in `Savings`, line 13 of Figure 1 is necessary, i.e. $\Phi_{f2}$, since an unchecked exception thrown by an observer propagates back to the subject. However, if the contract in Figure 11 is used instead of the contract in Figure 6 in reasoning, then such a catch clause will be unused, since the observers refining this contract cannot throw any unchecked exceptions, lines 3–4, and thus no unchecked exceptions reaches the subject.

Again only the implementation of the subject `Savings` and the contract of event `WithdrawEv` is enough for reasoning about $\Phi_{f1}$ and $\Phi_{f2}$, independent of system configuration, i.e. unknown observers or their execution order in their chain. This is only valid because $Ptolemy_\chi$'s type system and refinement rules ensure that subjects and observers of an event respect their contract.

# 4. MODULAR REASONING AND REFINEMENT IN $Ptolemy_\chi$

$Ptolemy_\chi$ enables reasoning about behaviors and flows of exceptions of observers using translucid contracts, as illustrated in Section 3.2 and Section 3.3. Such a reasoning is modular and independent of system configuration, i.e. unknown observers or their execution order in their chain, because of the following guarantees:

- each subject and observer of an event only throws the exceptions it is allowed to throw according;

- each subject and observer satisfies the exceptional postconditions of these exceptions upon throwing them; and

- each observer only throws and handles the exceptions at the places specified in its implementation.

These guarantees are provided by $Ptolemy_\chi$'s type checking rules, refinement rules for static structural refinement and runtime assertion checking of translucid contracts. These refinement rules restrict both subjects and observers of an event and are different from refinement rules in specification languages such as JML [25], although they may look similar syntactically.

## 4.1 Static Semantics

$Ptolemy_\chi$'s typing rules ensure that each subject and observer of an event only throws the *checked* boundary exception they are allowed to. The typing rules also check for structural refinement of the event's contract that ensures each observer only throws and handles the exceptions at the specified places in its implementation. Previous work on join point interfaces [3] informally discusses the typing rules for boundary-like exceptions, however, it does not provide any formalization or soundness proof.

### 4.1.1 Type Attributes

$Ptolemy_\chi$ tying rules use the type attributes of Figure 7 in which regular types are augmented with a set of checked exceptions $\overline{x}$.

The type attribute `exp` $t, \overline{x}$ denotes expressions of type $t$ that may throw the checked exceptions $\overline{x}$. Variables do not throw exceptions and are denoted by `var` $t$. The type attribute OK is used to type check top level declarations whereas OK in $c$ is used for type checking of lower level declarations in the context of an upper level declaration $c$. In $Ptolemy_\chi$, similar to Java, a checked

$$
\begin{array}{lll}
\theta ::= & & \text{``type attributes''} \\
\quad \text{OK} & & \text{``program/top-level declaration''} \\
\quad | \text{ OK in } c & & \text{``method, binding''} \\
\quad | \textbf{ var } t & & \text{``var/formal/field''} \\
\quad | \textbf{ exp } t, \overline{x} & & \text{``expression''} \\
\qquad \text{where } \forall x \in \overline{x}. \; isChecked(x) & & \\
\quad | \perp, \overline{x} & & \text{``bottom type''} \\
\Gamma ::= \{var : t\} & & \text{``type environment''} \\
\qquad \text{where } var \subseteq (\{\textbf{this}, \textbf{next}\} \cup \mathcal{V}), & & \\
\Pi ::= \{loc : t\} & & \text{``store typing''} \\
\qquad \text{where } loc \subseteq \mathcal{L}, & & \\
\qquad \mathcal{L} \text{ is set of location names} & & \\
\Gamma | \Pi \;\; \vdash e : \theta & & \text{``typing judgement''}
\end{array}
$$

**Figure 7: Type attributes, based on [31].**

exception is a subtype of `Exception` that is not a subtype of `RuntimeException`. The auxiliary function *isChecked* checks if an exception is a checked exception. The typing rules and auxiliary functions use a fixed class table $CT$ which is a list of event type and class declarations. We require distinct top level names and acyclic inheritance relations for classes in $CT$. The typing judgement $\Gamma | \Pi \;\; \vdash e : \theta$ says that in the typing environment $\Gamma$ and store typing $\Pi$ the expression $e$ has the type $\theta$. Figure 8 shows select typing rules for $Ptolemy_\chi$. The full set of typing rule could be found in our technical report [1].

### 4.1.2 Declaration Typing Rules

One may assume that subjects and observers of an event can throw any subtypes of its checked boundary exceptions. However as observed in previous work on JPIs [3], the observers can throw any subtype of the boundary exceptions and the subjects can only throw them invariantly. This is to avoid situations in which an observer may throw an exception that its subjects cannot handle, e.g. for a boundary exception $E_b$ if a subject throws $E_s$ and an observer throws $E_o$, such that $E_s <: E_o <: E_b$, then the subject cannot handle $E_o$ [3]. The observers can throw any subtypes of the boundary exceptions. Such a relation is denoted using $\subseteq$: which combines subtype and subset relations.

The rule (T-BINDING) type checks an observer of the event ($p$) and its binding declaration. The rule checks for the relation $\subseteq$: between the checked exceptions ($\overline{x'}$) of the observer handler method ($m$) of the event ($p$) and its boundary exceptions ($\overline{x}$), i.e. $\overline{x'} \subseteq: \overline{x}$. Similar to non handler methods, the observer handler method ($m$) must declare its checked exceptions ($\overline{x'}$). The rule (T-BINDING) also ensures that the body ($e$) of the observer handler method ($m$) structurally refines the body ($se$) of the assumes block of its translucid contract of ($contract$), i.e. $se \sqsubseteq e$. The structural refinement $\sqsubseteq$ is discussed in Section 4.2.

The rule (T-EVENTTYPE) type checks declaration of the event ($p$). The body ($se$) of the assumes block in the translucid contract of ($p$) specifies the implementation structure of its observers. This means, the same $\subseteq$: relation between checked exceptions of an observer and the boundary exceptions of its event must exist between the checked exceptions ($\overline{x'}$) of the assumes block ($se$) and the boundary exceptions ($\overline{x}$) of its event ($p$), i.e. $\overline{x'} \subseteq: \overline{x}$. The type ($t'$) of ($se$) should also be a subtype <: of the return type ($c$) of the event.

### 4.1.3 Expression Typing Rules

Similar to the rule (T-BINDING) that type checks observers of an event ($p$), the rule (T-ANNOUNCE) type checks its subjects and especially their announce expressions. The subjects can only throw exact boundary exceptions invariantly. This means that the set of the checked exceptions ($\overline{x''}$) of the subject must be in a subset relation $\subseteq$ with the boundary exceptions ($\overline{x}$) of the event ($p$), i.e. $\overline{x''} \subseteq \overline{x}$. The relation $\subseteq$ is different from $\subseteq$: since it does not allow the sub-

(T-BINDING)
$$t\,m(\textbf{thunk}\ t\ p\ var)\ \textbf{throws}\ \overline{x'}\ \{e\} = CT(c,m)$$
$$t\ \textbf{event}\ p\ \textbf{throws}\ \overline{x}\ \{\overline{form}\ contract\} \in CT$$
$$contract = \textbf{requires}\ ep_1\ \textbf{assumes}\ \{se\}\ \textbf{ensures}\ ep_2\ \overline{excp}$$
$$\overline{x'} \subseteq: \overline{x} \qquad se \sqsubseteq e$$
$$\overline{\vdash (\textbf{when}\ p\ \textbf{do}\ m)\ :\ OK\ in\ c}$$

(T-EVENTTYPE)
$$contract = \textbf{requires}\ ep_1\ \textbf{assumes}\ \{se\}\ \textbf{ensures}\ ep_2\ \overline{excp}$$
$$\forall x \in \overline{x}.\ isChecked(x)$$
$$\overline{var:t} \vdash se : \textbf{exp}\ t',\ \overline{x'} \qquad t' <: c \qquad \overline{x'} \subseteq: \overline{x}$$
$$\overline{\vdash c\ \textbf{event}\ p\ \textbf{throws}\ \overline{x}\ \{t\ var;\ contract\}\ :\ OK}$$

(T-ANNOUNCE)
$$c\ \textbf{event}\ p\ \textbf{throws}\ \overline{x}\ \{t'\ var';\ contract\} \in CT$$
$$\forall e_i \in \overline{e}.\ \Gamma|\Pi \vdash e_i : \textbf{exp}\ t_i, \overline{x_i} \qquad \Gamma|\Pi \vdash e : \textbf{exp}\ t', \overline{x'}$$
$$\forall t_i, t'_i \in \overline{t'}.\ t_i <: t'_i \qquad t' <: c \qquad \overline{x''} = \bigcup \overline{x_i} \cup \overline{x'} \qquad \overline{x''} \subseteq \overline{x}$$
$$\overline{\Gamma|\Pi \vdash \textbf{announce}\ p\ (\overline{e})\ \{e\} : \textbf{exp}\ c, \overline{x''}}$$

(T-INVOKE)
$$c\ \textbf{event}\ p\ \textbf{throws}\ \overline{x}\ \{\overline{form}\ contract\} \in CT$$
$$\Gamma|\Pi \vdash e : \textbf{exp}\ \textbf{thunk}\ c\ p, \overline{x'} \qquad \overline{x'} == \overline{x}$$
$$\overline{\Gamma|\Pi \vdash e.\textbf{invoke}\ ()\ :\ \textbf{exp}\ c, \overline{x}}$$

(T-REGISTER)
$$\Gamma|\Pi \vdash e : \textbf{exp}\ t, \overline{x}$$
$$\overline{\Gamma|\Pi \vdash \textbf{register}\ (e) : \textbf{exp}\ t, \overline{x}}$$

(T-SPEC)
$$\Gamma|\Pi \vdash ep_1 : \textbf{exp}\ t_1, \emptyset$$
$$\Gamma|\Pi \vdash ep_2 : \textbf{exp}\ t_2, \emptyset \qquad \forall x \in \overline{x}.\ x <: \texttt{RuntimeException}$$
$$\overline{\Gamma|\Pi \vdash \textbf{requires}\ ep_1\ \textbf{ensures}\ ep_2\ \textbf{throws}\ \overline{x}\ :\ \textbf{exp}\ \bot, \emptyset}$$

(T-REFINING)
$$spec = \textbf{requires}\ ep_1\ \textbf{ensures}\ ep_2\ \textbf{throws}\ \overline{x}$$
$$\Gamma|\Pi \vdash spec : \textbf{exp}\bot, \emptyset \qquad \Gamma|\Pi \vdash e : \textbf{exp}\ t, \overline{x}$$
$$\overline{\Gamma|\Pi \vdash \textbf{refining}\ spec\ \{e\} : \textbf{exp}\ t, \overline{x}}$$

(T-THROW)
$$\Gamma|\Pi \vdash e : \textbf{exp}\ t, \overline{x}$$
$$t <: \texttt{Throwable} \qquad isChecked(t)\ ?\ \overline{x'} = \{t\} : \overline{x'} = \emptyset$$
$$\overline{\Gamma|\Pi \vdash \textbf{throw}\ e : \textbf{exp}\ \bot, \overline{x'} \cup \overline{x}}$$

(T-TRYCATCH)
$$\Gamma|\Pi \vdash e_1 : \textbf{exp}\ t, \overline{x_1}$$
$$\Gamma, var : x|\Pi \vdash e_2 : \textbf{exp}\ t, \overline{x_2} \qquad \overline{x} = \{x_i\ |\ x_i \in \overline{x_1} \wedge\ !(x_i <: x)\}$$
$$\overline{\Gamma|\Pi \vdash \textbf{try}\ \{e_1\}\ \textbf{catch}\ (x\ var)\ \{e_2\} : \textbf{exp}\ t, \overline{x} \cup \overline{x_2}}$$

Auxiliary Functions:
$$isChecked(x) = (x <: \texttt{Exception})\ \wedge\ !(x <: \texttt{RuntimeException})$$
$$\overline{x_1} \subseteq: \overline{x_2} = \forall x_1 \in \overline{x_1}, \exists x_2 \in \overline{x_2}.\ x_1 <: x_2$$

**Figure 8:** *Ptolemy*$_\chi$'s select typing rules, based on [8,31].

ject to throw subtypes of the boundary exceptions. The set of the checked exception $(\overline{x''})$ of the subject is the union of the set of checked exceptions $(\overline{x'})$ for the body $(e)$ of its announce expression and the set of checked exceptions $(\overline{x_i})$ for its parameters $(e_i)$, i.e. $\overline{x''} = \bigcup \overline{x_i} \cup \overline{x'}$. The rule also checks that the types $(t_i)$ of the parameters $(e_i)$ of the announce expression are subtypes of the types $(t'_i)$ of the context variables of the event. The same should hold for the type $(t')$ of the body $(e)$ of the announce expression and the return type $(c)$ of the event $(p)$.

The rule (T-INVOKE) type checks an invoke expression. The invoke expression invokes the next observer of an event in the chain

of its observers. The chain of the observers is included in its event closure receiver object $(e)$. The event closure for an event $(p)$ with return type $(c)$ and the boundary exceptions $(\overline{x})$ is of type **thunk** $c\ p, \overline{x'}$. The rule checks for the equality of the set checked exceptions $(\overline{x'})$ of the invoke expression and the set of the boundary exceptions $(\overline{x})$ of its event $(p)$, i.e. $\overline{x'} == \overline{x}$. The rule (T-REGISTER) says that the set of checked exceptions $(\overline{x})$ of a register expression is set of checked exceptions of its parameter $(e)$.

The rule (T-SPEC) type checks an exception-aware specification expression. It checks that the side effect free pre- and postconditions $(ep_1)$ and $(ep_2)$ throw no checked exceptions and the exception set $(\overline{x})$ listed in its throws clause are unchecked exceptions, i.e. subtype of **RuntimeException**. The exception-aware specification expression throws no checked exceptions because it is a specification expression and has the bottom type $\bot$ that is a subtype of any other type. The rule (T-REFINING) type checks a refining expression that refines an exception-aware specification expression $(spec)$. The rule simply says that the set of checked exceptions $(\overline{x})$ of the refining expression is the same as the set of checked exceptions of its body $(e)$.

The rule (T-THROW) type checks a throw expression. It adds the thrown exception $(t)$ to the exception set $(\overline{x})$ of the expression, if the exception is a checked exception. The conditional $isChecked(t)\ ?\ \overline{x'} = \{t\} : \overline{x'} = \emptyset$ checks if $(t)$ is a checked exception. Recall that the typing rules only keep track of the checked exceptions and not unchecked exceptions. The rule (T-TRYCATCH) type checks a try-catch expression. It computes the set of checked exceptions $(\overline{x_1})$ of the body $(e_1)$ of the try part and then factors out checked exceptions that are subtypes of the exception $(x)$ handled by the catch part. The result is the set of the checked exceptions $(\overline{x})$ that are thrown by the try part and not handled in the catch part. The body of the catch part itself can throw the checked exceptions $(\overline{x_2})$. The set of the checked exceptions of the try-catch expression is the union of $(\overline{x})$ and $(\overline{x_2})$. The rest of the expression typing rules compose the set of checked exceptions thrown by subexpressions of an expression, based on the compositional rules of the language [1].

### 4.1.4 Soundness and Dynamic Semantics

*Ptolemy*$_\chi$'s type system is proven sound following the standard progress and preservation arguments. Treatment of features such as exception-aware specification expressions, translucid contracts and refining expressions are new in the dynamic semantics and the proof. The proof of soundness and the full set of *Ptolemy*$_\chi$'s dynamic and static semantics rules could be found in our report [1].

## 4.2 Structural Refinement

Structural refinement rules ensure that each observer of an event only throws and handles the exceptions at the places in its implementation as specified by the assumes block of its contract. Figure 9 shows select rules for *Ptolemy*$_\chi$'s structural refinement.

For an observer of the event $(p)$, the implementation $(e)$ of its observer handler method $(m)$ structurally refines the assumes block $(se)$ of its translucid contract, i.e. $se \sqsubseteq e$, if the following holds: for each program expression in $(se)$ there is a textually match program expression in $(e)$ at the same place, e.g. the rule *var* for the variable expressions. And for each exception-aware specification expression *spec* in $(se)$ there is a refining expression **refining** $spec\{e\}$ at the same place in the implementation which *claims* to refine *spec*. The rule for **either**$\{se_1\}$**or**$\{se_2\}$ allows the observer to either refine $(se_1)$ or $(se_2)$ and thus allow variability in the observers. Structural refinement for other expressions is based on the refinement of their subexpressions. For example, the try-catch expression **try** $\{se_1\}$ **catch** $(x\ var)$ $\{se_2\}$ is structurally refined by

| $c$ **event** $p$ **throws** $\overline{x}$ $\{\overline{form\ contract}\} \in CT$ | | |
|---|---|---|
| $contract = $ **requires** $ep_1$ **assumes** $\{se\}$ **ensures** $ep_2\ \overline{excp}$ | | |
| **when** $p$ **do** $m$ and $c\ m(\textbf{thunk}\ c\ p\ \textbf{next})\{e\} \in CT$ | | |
| $se$ is structurally refined by $e$, $se \sqsubseteq e$, as follows: | | |
| Cases of ($se$) | Refined by ($e$) | Side conditions |
| $var$ | $var$ | |
| $t\ var = se_1; se_2$ | $t\ var = e_1; e_2$ | $se_1 \sqsubseteq e_1, se_2 \sqsubseteq e_2$ |
| **if**$(sp)\{se_1\}$ **else**$\{se_2\}$ | **if**$(ep)\{e_1\}$ **else**$\{e_2\}$ | $sp \sqsubseteq ep,$ $se_1 \sqsubseteq e_1, se_2 \sqsubseteq e_2$ |
| **either** $\{se_1\}$ **or** $\{se_2\}$ | $e$ | $se_1 \sqsubseteq e \vee se_2 \sqsubseteq e$ |
| $se.$**invoke** () | $e.$**invoke** () | $se \sqsubseteq e$ |
| **announce** $p(\overline{se})\ \{se\}$ | **announce** $p(\overline{e})\ \{e\}$ | $\overline{se} \sqsubseteq \overline{e}, se \sqsubseteq e$ |
| **try** $\{se_1\}$ **catch** $(x\ var)\{se_2\}$ | **try** $\{e_1\}$ **catch** $(x\ var)\{e_2\}$ | $se_1 \sqsubseteq e_1, se_2 \sqsubseteq e_2$ |
| **throw** $se$ | **throw** $e$ | $se \sqsubseteq e$ |
| $spec$ | **refining** $spec\{e\}$ | |

**Figure 9: Select rules for structural refinement, based on [2,34].**

**try**$\{e_1\}$**catch**$(x\ var)\{e_2\}$ if its subexpressions ($se_1$) and ($se_2$) are refined by the subexpressions ($e_1$) and ($e_2$).

## 4.3 Runtime Assertion Checking

Structural refinement rules ensure that for each exception-aware specification expression ($spec$) of the form **requires** $ep_1$ **ensures** $ep_2$ **throws** $\overline{x}$ in a translucid contract of an event, there is a refining expression **refining** $spec\{e\}$ which claims to refine ($spec$), in the implementation of the observers of the event. However, they do not statically check this claim that the body ($e$) of the refining expression only throws the specified unchecked exceptions ($\overline{x}$). In $Ptolemy_\chi$, runtime assertion checking (RAC) does this check. RAC also ensures that the observers and subjects of an event satisfy the exceptional postconditions of their exceptions when they throw them. Figure 10 illustrates, in grey, the runtime assertion checking for the observer handler method check of Figure 1 with its translucid contract in Figure 6. The code for RAC is generated automatically by the compiler.

In Figure 10, a try-catch expression surrounds the original body of the check to catch the checked and unchecked exceptions that the observer is allowed to throw, that is RegDExc, lines 22–25, and RuntimeException, line 26–29, and to check for their exceptional postconditions, lines 24 and 28. After checking for the exceptional postconditions, the exceptions are rethrown to avoid changing of the exceptional control flow of the program, lines 25 and 29. The method Con.signal checks for the exceptional postcondition of the exception e and aborts the execution or raises an error of type **Error** in case of their violation.

There is another try-catch that surrounds the refining expression, lines 7–15. This try-catch expression catches all unchecked exceptions of the refining expression and checks if they are allowed to be thrown according to the exception-aware specification expression that the refining expression claims to refine, lines 8–12. The method Con.allowedExc checks if the thrown exception e is among the set of allowed exceptions and aborts the program if it is not, otherwise it rethrows the exception. In this example the exception-aware specification expression allows all unchecked exceptions to be thrown by its refining expression, i.e. **throws RuntimeException** on line 9. RAC also checks for the pre- and normal postcondition of the observer, lines 4 and 20 and the pre- and normal postcondition of the refining expression, lines 6 and 17, using the methods Con.require and Con.ensure.

## 5. EVALUATION

In this section, usability of $Ptolemy_\chi$'s exception flow reasoning is evaluated by using it to understand (non-) occurrence of a set of

```
1  void check(WithdrawEv next) throws RegDExc{
2    try{
3      /* observer's precondition */
4      Con.require(next.acc()!=null);
5      /* refining precondition */
6      Con.require(true);
7      try{
8        refining establishes next.acc().bal==
9          old(next.acc().bal) throws RuntimeException{
10         if(next.acc().numWithdrawal>=6)
11           throw new RegDExc();
12       }
13       /* allowed unchecked exceptions */
14     } catch(RuntimeException e){
15       Con.allowedExc(e, {RuntimeException.class}); }
16     /* refining postcondition */
17     Con.ensure(next.acc().bal==old(next.acc().bal));
18     next.invoke();
19     /* observer's normal postcondition */
20     Con.ensure(next.acc().bal<=old(next.acc().bal));
21   }
22   catch(RegDExc e){
23     /* exceptional postcondition */
24     Con.signal(next.acc().bal==old(next.acc().bal),e);
25     throw e; }
26   catch(RuntimeException e){
27     /* exceptional postcondition */
28     Con.signal(true, e);
29     throw e; }
30 }
```

**Figure 10: Runtime assertion checking in the observer Check.**

bug patterns [6] for aspect-oriented (AO) programs. The overhead of the application of $Ptolemy_\chi$ to a simple Ptolemy program is also discussed. Understanding the bug patterns requires knowledge about the set of checked and unchecked exceptions of unknown observers and flows of these exceptions in the chain of the observers.

### 5.1 AO Bug Patterns in $Ptolemy_\chi$

Despite their differences [31], $Ptolemy_\chi$ and AO languages such as AspectJ share some similarities. In these languages a subject (base code) announces an event implicitly or explicitly, observers (aspects) register for the event and are invoked implicitly and run in a chain upon announcement of the event. This in turn suggests that there may be similarities between bug patterns of these languages. Coelho *et al.* [6] introduces a set of AO bug patterns especially with regard to exception handling. In this section we adapt these bug patterns to $Ptolemy_\chi$ for our running bank account example and show how their occurrence or non occurrence could be understood using $Ptolemy_\chi$'s exception flow reasoning.

Coelho *et al.* [6] recognize 5 category of bug patterns in AO: *(i)* throw without catch *(ii)* residual catch *(iii)* exception stealer *(iv)* path dependent throw and *(v)* fragile catch. The last two bug patterns are not applicable to $Ptolemy_\chi$. A path dependent throw bug occurs when exceptions that a method throws vary based on different call chains leading to its invocation. This could happen because of AO scope designator constructs, such as **within** and **cflow** in AspectJ, that are not supported in $Ptolemy_\chi$. A fragile catch bug occurs when an aspect is supposed to catch an exception in a specific program point but misses it due to a problematic pointcut, i.e. pointcut fragility. Pointcut fragility does not happen in $Ptolemy_\chi$, because of its explicit event announcement [31].

193

## 5.2 Throw Without Catch

In *Ptolemy*$_\chi$, a throw without catch bug happens when an observer throws an exception during announcement and handling of an event and there is no handler to catch it, neither on the observer nor the subject side. This is especially true for unchecked exceptions since checked exceptions cannot go uncaught.

***Occurrence*** The subject `Savings` in Figure 1 and the contract in Figure 6 together illustrate a throw without catch bug, if the `catch (RuntimeException` $e$`){..}` on line 13 of Figure 1 is inadvertently forgotten by the developer. Similar to Java, *Ptolemy*$_\chi$ does not complain about the missing catch clause for unchecked exceptions. The contract for `WithdrawEv` in Figure 6 allows the observers of `WithdrawEv` to throw any unchecked exception, lines 4–5. It also does not allow an observer in the chain to catch any exceptions thrown by another observer of `WithdrawEv` since there is no try-catch surrounding the invoke expression, line 6. Thus if an observer of `WithdrawEv` throws an unchecked exception the exception propagates back to the subject `Savings` which also does not catch the exceptions, i.e. the exception goes uncaught during announcement and handling of `WithdrawEv`. One may argue that leaving the catch clause `catch (RuntimeException` $e$`){..}` in its place on line 13 of Figure 1 avoids the throw without catch bug, however, this catch clause itself could be unnecessary and unused and a residual bug if the observers of `WithdrawEv` do not throw any unchecked exceptions.

***Non Occurrence*** To ensure non occurrence of a throw without catch bug during announcement and handling of an event, one may want to force the observers of the event to not throw any unchecked exceptions. Figure 11 illustrates a variation of the contract of Figure 6 in which the observers of `WithdrawEv` cannot throw any unchecked exceptions. The exception-aware specification expression on lines 3–4 limits the set of unchecked exceptions of the observers of `Withdraw` to nothing, i.e. **throws NoRuntimeExc**.

```
1  void event WithdrawEv throws RegDExc{ ..
2   assumes{
3    establishes next.acc().bal==old(next.acc().bal)
4     throws NoRuntimeExc;
5     next.invoke();
6   } .. }
```

**Figure 11: No unchecked exceptions for observers, lines 3–4.**

Figure 12 illustrates another variation of the contract for `WithdrawEv` in which each observer has a try-catch expression in its body, lines 3 and 7–8. The try-catch expressions catches, line 7, any unchecked exception that might be thrown by the observers and handles them in a way that does not throw any unchecked exception itself, line 8.

```
1  void event WithdrawEv throws RegDExc{ ..
2   assumes{
3    try{
4     establishes next.acc().bal==old(next.acc().bal)
5      throws RuntimeException;
6     next.invoke();
7    } catch(RuntimeException e){
8     establishes true throws NoRuntimeExc; }
9   } .. }
```

**Figure 12: Observers handle their unchecked exceptions locally, lines 3 and 7 and 8.**

Unlike the contract in Figure 11, the contract in Figure 12 requires the implementation of the observers of `WithdrawEv` to be surrounded by the specified try-catch expression. It also allows an observer to catch and swallow the unchecked exceptions thrown by other observers in the chain which is not allowed in Figure 11.

## 5.3 Residual Catch

In *Ptolemy*$_\chi$, a residual catch bug happens when a subject tries to catch an exception that is not thrown or is already caught by its observers before reaching it.

***Occurrence*** The subject `Savings` in Figure 1 and the contract in Figure 11 together illustrate the occurrence of a residual catch bug. The subject `Savings` catches the unchecked exceptions thrown by its observers during announcement and handling of `WithdrawEv`, line 13. However, the contract for `WithdrawEv` in Figure 11 does not allow its observers to throw any unchecked exceptions, lines 3–4. This in turn means the catch clause on line 13 is not necessary since the subject is trying to catch exceptions that are not thrown by its observers. Recall that in *Ptolemy*$_\chi$, the contract limits the set of exceptions of both observers and the subject, 5–8 in Figure 1, of the event. Similarly, the subject `Savings` and the contract in Figure 12 illustrate a residual bug since the observers do not throw any unchecked exceptions, however the subject is still trying to catch them.

***Non Occurrence*** A residual bug does not occur during announcement and handling of `WithdrawEv` in the subject `Savings`, Figure 1, if its observers do not throw any unchecked exceptions and its catch clause, lines 13 is omitted. The subject `Savings` without the `catch (RuntimeException` $e$`) {..}` and the contract in Figure 11 illustrate non occurrence of a residual catch bug. The contract for `WithdrawEv` in Figure 11 does not allow its observer to throw any unchecked exception and the subject does not catch any unchecked exceptions and there is no unnecessary catch clause. The same is true for the contract in Figure 12 since it does not allow its observers to throw unchecked exceptions too. The subject `Savings` with the catch clause **catch** (`RuntimeException` $e$) {..} on line 13 and the contract in Figure 6 illustrate another example of the non occurrence of a residual bug. Here the contract allows the observers of `WithdrawEv` to throw any unchecked exceptions that could be propagated to the subject. This in turn means that the catch clause in the subject is actually necessary and used.

## 5.4 Exception Stealer

An exception stealer bug happens when an exception that was supposed to be caught by an exception handling observer is caught by a subject. This bug pattern is a special case of an unintended handler action where an exception is caught wrongly by an unintended handler. In *Ptolemy*$_\chi$, a variation of this bug happens when an exception thrown by an observer that was supposed to be caught by a subject is caught by another observer in the chain.

***Occurrence*** The subject `Savings` in Figure 1 and the contract for its `WithdrawEv` in Figure 13 illustrate the occurrence of an exception stealer bug. According to the property $\Phi_{f1}$ of the bank account example, Section 2.2, if an exception `RegDExc` is thrown during announcement and handling of `WithdrawEv` it is propagated back to the subject to be handled. However, the try-catch expression of the contract, lines 3–6 that surrounds the invoke expression allows an observers of `WithdrawEv` to catch any exception, including `RegDExc`, if thrown by another observer in the chain and swallow it, line 6, before it reaches the subject `Savings`.

***Non Occurrence*** The subject `Savings` in Figure 1 and the contract in Figure 6 illustrate non occurrence of the exception stealer bug for the exception `RegDExc`. The contract for `WithdrawEv` in Figure 6 does not allow any try-catch expression to surround the

```
1  void WithdrawEv throws RegDExc{ ..
2   assumes{
3    try{
4     establishes next.acc().bal==old(next.acc().bal);
5     next.invoke();
6    } catch(Exception e){ }/*swallow*/
7   } .. }
```

**Figure 13: Stealing and swallowing unchecked and checked exceptions through subsumption, line 6.**

invoke expression in the implementation of its observers. This in turn means if an exception RegDExc is thrown by an observer, it is not caught by other observers in the chain and is propagated back to the subject. Similarly the contract in Figure 12 does not allow the checked exception RegDExc thrown by one observer to be stolen by another observer and propagates it back to Savings. However, unchecked exceptions thrown by one observer could be caught by another observer in the chain, line 7.

## 5.5 Summary of Bug Patterns

Out of the three Coelho *et al.* 's bug patterns that are applicable to *Ptolemy*$_\chi$, (non) occurrence of all of them could be understood using *Ptolemy*$_\chi$'s exception reasoning technique. The key in reasoning about these patterns is to know about the set of checked and unchecked exceptions of subjects and observers of an event and especially flow of these exceptions in the chain of the observers. These patterns in our running bank account example are understood using only the translucid contract of WithdrawEv.

| Bug Pattern | Occurrence | Non Occurrence |
|---|---|---|
| Throw without catch | ✓ | ✓ |
| Residual catch | ✓ | ✓ |
| Exception stealer | ✓ | ✓ |
| Path dependent throw | Not applicable to *Ptolemy*$_\chi$ | |
| Fragile catch | Not applicable to *Ptolemy*$_\chi$ | |

**Figure 14: *Ptolemy*$_\chi$'s exception flow reasoning and understanding of AO bug patterns [6].**

## 5.6 Application to Ptolemy

In this section we discuss the overhead of the application of *Ptolemy*$_\chi$ to a simple figure editor Ptolemy program, that is shipped with its compiler distribution. The figure editor allows creation of simple figures such as points and lines. An event is announced when a figure moves and an observer updates the screen by invoking a mock method. The example has 1 event, 1 subject and 1 observer, 7 classes in total with 127 lines of code. One requirement for the figure editor could be that the observers of the event do not throw any checked exception and their unchecked exceptions must be propagated back to the subject for handling.

To specify this requirement a simple translucid contract **requires true assumes{** next.invoke(); **establishes true throws RuntimeException;** } **ensures true** should be added to the event declaration. Ptolemy's event declarations do not have any throws clause which is interpreted as no checked boundary exceptions in *Ptolemy*$_\chi$. The observer of the event should have a refining expression **refining establishes true throws RuntimeException**{..} in its implementation to structurally refine the contract. The contract and the refining expression adds 8 lines of the code to the program which increases its size by 6.2%. However, the compiler could be easily modified to not require the default pre- and postconditions of **requires true** and **ensures true** and insert the refining expression automatically

based on the structural refinement rules. Thus the real overhead in terms of lines of code written by the programmer is only 4 lines of code, i.e. about 3.1%. The programmer only needs to write the contract in the event type declaration. This overhead varies for different programs depending on number of events, observers, etc.

## 6. RELATED WORK

***Reasoning in Implicit Invocation (II)*** Garlan *et al.* [14] propose a compositional II reasoning technique to reason about normal behavior by breaking down a system into independent subsystems and relying on system configuration such as number of events, their consumption policy, etc. Dingel *et al.* [7] propose a rely-guarantee-like reasoning technique by relying on sound and complete announcement of semantic-carrying events, and invariants weakened by location predicates. Krishnaswami *et al.* [24] verify the structure and normal behavior of the Observer design pattern using separation logic by relying on system configuration such as number of observers and their individual invariants. Our work is focused on modular reasoning about behaviors and flows of exceptions for abnormal termination of event announcement with chained observers, independent of system configuration.

Join point interfaces (JPIs) [3] enable modular type checking for an aspect-oriented (AO) language in the presence of exceptions. Khatchadourian *et al.* [20] propose a rely-guarantee-like technique to reason about normal behaviors and traces of AO programs. Crosscutting programming interfaces (XPIs) [37] and Join Point Types (JPTs) [36] allow informal specification and reasoning about normal behaviors of aspects in AO. Pipa [40] enables global reasoning about normal and exceptional behaviors in AspectJ. Execution levels [10] use level shifting operations to specify the flow and interaction of exceptions of aspects and base code in AspectJ to avoid exception conflation. Our proposal is mostly focused on reasoning about both exceptional behaviors and flows of exceptions in a non-global and modular fashion.

***Reasoning in Explicit Invocation (EI)*** Supertype abstraction [26] and ESC/Java [12] allow modular reasoning about normal and exceptional behaviors using specifications in JML [25]. Jacobs *et al.* [16] propose a modular verification technique for invariants and locking patterns in the presence of unchecked exceptions in multithreaded programs. Failboxes [17] provide exceptional behavior guarantees in terms of dependency safety. Anchored exceptions [39] and the work of Malayeri and Aldrich [28] enable modular reasoning about flow of exceptions using more expressive exception interfaces. EJFlow [11] enables reasoning about exception flow using architectural level exception ducts. Jex [33], and the works of Jo *et al.* [18], Sinha *et al.* [35] and Fu *et al.* [13] propose global flow analyses for exceptions in Java. Leroy and Pessaux [27] propose a type based program analysis to estimate the set of uncaught exceptions in ML. Abnormal types [9] guarantee types of exceptions upon abnormal termination. As discussed in Section 1, these techniques use known invocation relations among modules of a system and are not concerned about II languages in which a subject can invoke unknown observers and run them in a chain.

***Reasoning Using Greybox Specifications*** Tyler and Soundarajan [38] and Shaner *et al.* [34] use greybox specifications for verification of mandatory calls. Rajan *et al.* [32] use greybox specification to reason about web service policies.

## 7. CONCLUSIONS AND FUTURE WORK

Modular reasoning about behaviors and flows of exceptions faces unique challenges in event-based implicit invocation (II) languages such as Ptolemy that allow subjects to invoke unknown observers and run them in a chain. In this work we have illustrated these challenges in Ptolemy and proposed *Ptolemy*$_\chi$ to address them. *Ptolemy*$_\chi$'s exception-aware specification expressions, boundary exceptions, exceptional postconditions along with greybox contracts allow limiting the set of exceptions that subjects and observers of an event may throw and specifying behaviors and flows of these exceptions. *Ptolemy*$_\chi$'s sound type system, static structural refinement and runtime assertion checks enable its modular reasoning independent of unknown observers of an event or their execution order in the chain of observers.

For future work, the first task would be a precise formalization of the relation between refinement [30] and structural refinement for abnormal termination. For normal termination structural refinement implies refinement [2,34]. The second task would be application of *Ptolemy*$_\chi$ to similar languages and evaluate its robustness.

*Acknowledgements* This work was supported in part by NSF grant CCF-10-17334. We thank Rex D. Fernando, Robert Dyer and the anonymous reviewers of *AOSD'13*.

# 8. REFERENCES

[1] M. Bagherzadeh, H. Rajan, and A. Darvish. On exceptions, events and observer chains. Technical Report 12-12, Iowa State U., 2012.

[2] M. Bagherzadeh, H. Rajan, G. T. Leavens, and S. Mooney. Translucid contracts: expressive specification and modular verification for aspect-oriented interfaces. In *AOSD'11*.

[3] E. Bodden, E. Tanter, and M. Inostroza. Safe and practical decoupling of aspects with join point interfaces. Technical Report TUD-CS-2012-0106, Technische U. Darmstadt.

[4] M. Büchi and W. Weck. The greybox approach: When blackbox specifications hide too much. Technical Report 297, Turku Center for Computer Science, 1999.

[5] N. Cacho, F. C. Filho, A. Garcia, and E. Figueiredo. EJFlow: taming exceptional control flows in aspect-oriented programming. In *AOSD'08*.

[6] R. Coelho, A. Rashid, A. von Staa, J. Noble, U. Kulesza, and C. Lucena. A catalogue of bug patterns for exception handling in aspect-oriented programs. In *PLoP'08*.

[7] J. Dingel, D. Garlan, S. Jha, and D. Notkin. Towards a formal treatment of implicit invocation using rely/guarantee reasoning. *Formal Asp. Comput.'98*, 10(3).

[8] S. Drossopoulou, S. Eisenbach, and T. Valkevych. Java type soundness revisited. Technical report, Imperial College.

[9] S. Drossopoulou and T. Valkevych. Java exceptions throw no surprises. Technical report, Imperial College London, 2000.

[10] I. Figueroa and E. Tanter. A semantics for execution levels with exceptions. In *FOAL'11*.

[11] F. Filho, P. da S. Brito, and C. Rubira. Reasoning about exception flow at the architectural level. In *Rigorous Development of Complex Fault-Tolerant Systems'06*.

[12] C. Flanagan, K. R. M. Leino, M. Lillibridge, G. Nelson, J. B. Saxe, and R. Stata. Extended static checking for Java. In *PLDI'02*.

[13] C. Fu, A. Milanova, B. G. Ryder, and D. G. Wonnacott. Robustness testing of Java server applications. *TSE'05*, 31(4).

[14] D. Garlan, S. Jha, D. Notkin, and J. Dingel. Reasoning about implicit invocation. In *FSE'98*.

[15] D. Garlan and D. Notkin. Formalizing design spaces: Implicit invocation mechanisms. In *VDM'91*.

[16] B. Jacobs, P. Muller, and F. Piessens. Sound reasoning about unchecked exceptions. In *SEFM'07*.

[17] B. Jacobs and F. Piessens. Failboxes: Provably safe exception handling. In *ECOOP 2000*.

[18] J.-W. Jo, B.-M. Chang, K. Yi, and K.-M. Choe. An uncaught exception analysis for Java. *J. Syst. Softw.'04*, 72(1).

[19] C. B. Jones. Tentative steps toward a development method for interfering programs. *TOPLAS'83*, 5(4).

[20] R. Khatchadourian, J. Dovland, and N. Soundarajan. Enforcing behavioral constraints in evolving aspect-oriented programs. In *FOAL'08*.

[21] G. Kiczales, E. Hilsdale, J. Hugunin, M. Kersten, J. Palm, and W. G. Griswold. An overview of AspectJ. In *ECOOP'01*.

[22] G. Kiczales, J. Lamping, A. Mendhekar, C. Maeda, C. V. Lopes, J.-M. Loingtier, and J. Irwin. Aspect-oriented programming. In *ECOOP'97*.

[23] G. Kiczales and M. Mezini. Aspect-oriented programming and modular reasoning. In *ICSE'05*.

[24] N. Krishnaswami, J. Aldrich, L. Birkedal, K. Svendsen, and A. Buisse. Design patterns in separation logic. In *TLDI'09*.

[25] G. T. Leavens, A. L. Baker, and C. Ruby. Preliminary design of JML: a behavioral interface specification language for Java. *Softw. Eng. Notes'06*, 31(3).

[26] G. T. Leavens and W. E. Weihl. Specification and verification of object-oriented programs using supertype abstraction. *Acta Informatica'95*, 32(8).

[27] X. Leroy and F. Pessaux. Type-based analysis of uncaught exceptions. *TOPLAS 2000*, 22(2).

[28] D. Malayeri and J. Aldrich. Practical exception specifications. In *Advanced Topics in Exception Handling Techniques'06*.

[29] R. A. Maxion and R. T. Olszewski. Improving software robustness with dependability cases. In *FTCS'98*.

[30] J. M. Morris. A theoretical basis for stepwise refinement and the programming calculus. *Sci. Com. Program.'87*, 9(3).

[31] H. Rajan and G. T. Leavens. Ptolemy: A language with quantified, typed events. In *ECOOP'08*.

[32] H. Rajan, J. Tao, S. M. Shaner, and G. T. Leavens. Tisa: A language design and modular verification technique for temporal policies in web services. In *ESOP'09*.

[33] M. P. Robillard and G. C. Murphy. Static analysis to support the evolution of exception structure in object-oriented systems. *TOSEM'03*, 12(2).

[34] S. M. Shaner, G. T. Leavens, and D. A. Naumann. Modular verification of higher-order methods with mandatory calls specified by model programs. In *OOPSLA '07*.

[35] S. Sinha, A. Orso, and M. J. Harrold. Automated support for development, maintenance, and testing in the presence of implicit control flow. In *ICSE'04*.

[36] F. Steimann, T. Pawlitzki, S. Apel, and C. Kastner. Types and modularity for implicit invocation with implicit announcement. *TOSEM'10*, 20(1).

[37] K. Sullivan, W. G. Griswold, H. Rajan, Y. Song, Y. Cai, M. Shonle, and N. Tewari. Modular aspect-oriented design with XPIs. *TOSEM'10*, 20(2).

[38] B. Tyler and N. Soundarajan. Black-box testing of grey-box behavior. In *FATES'03*.

[39] M. van Dooren and E. Steegmans. Combining the robustness of checked exceptions with the flexibility of unchecked exceptions using anchored exception declarations. In *OOPSLA'05*.

[40] J. Zhao and M. Rinard. Pipa: a behavioral interface specification language for AspectJ. In *FASE'03*.

# Method Slots: Supporting Methods, Events, and Advices by a Single Language Construct

YungYu Zhuang
The University of Tokyo
zhuang@csg.ci.i.u-tokyo.ac.jp

Shigeru Chiba
The University of Tokyo
chiba@acm.org

## ABSTRACT

To simplify the constructs that programmers have to learn for using paradigms, we extend methods to a new language construct, a *method slot*, to support both the event-handler paradigm and the aspect paradigm. A *method slot* is an object's property that can keep more than one function closure and be called like a method. We also propose a Java-based language, *DominoJ*, which replaces methods in Java with *method slots*, and explain the behavior of *method slots* and the operators. Then we evaluate the coverage of expressive ability of *method slots* by comparing *DominoJ* with other languages in detail. The feasibility of *method slots* is shown as well by implementing a prototype compiler and running a preliminary microbenchmark for it.

## Categories and Subject Descriptors

D.3.3 [**Programming Languages**]: Language Constructs and Features

## Keywords

aspect-oriented programming; event-driven programming

## 1. INTRODUCTION

The event-handler paradigm has been recognized as a useful mechanism in a number of domains such as user interface, embedded systems, databases [32], and distributed programming. The basic idea of the event-handler paradigm is to register an action that is automatically executed when something happens. At first it was introduced as techniques and libraries [7, 28, 26] rather than supported at language level. Recently supporting it at language level is a trend since a technique such as the Observer pattern [7] cannot satisfy programmers' need. The code for event triggers and observer management scatters everywhere. To address the issues, supporting events by a language construct is proposed in a number of languages [17, 3, 22, 6, 13, 9]. Implicit

invocation languages [8] might be classified into this category.

On the other hand, the aspect paradigm [14] is proposed to resolve crosscutting concerns, which cannot be modularized by existing paradigms such as object orientation. Although the aspect paradigm and the event-handler paradigm are designed for different scenarios, the constructs introduced for them are similar and can work as each other from a certain point of view.

In order to simplify the language constructs programmers have to learn, we borrow the idea of slots from Self [30] to extend the method paradigm in Java. In Self, everything is a slot that may contain either a value or a method. In other words, there is no difference between fields and methods since a method is also an object and thus can be kept in a field. We extend the slot and bring it to Java-like languages by proposing a new language construct named *method slot*. A *method slot* is an object's property which can keep more than one closure. We also present a Java-based language named *DominoJ*, where all methods in plain Java are replaced with *method slots*, to support both the event-handler paradigm and the aspect paradigm.

Our contributions presented in this paper are two-fold. First, we propose a new language construct, a *method slot*, to extend the method paradigm. Second, we introduce *method slots* to a Java-based language named *DominoJ*, and demonstrate how to use for the event-handler paradigm and the aspect paradigm.

## 2. MOTIVATION

With the evolution of software, more and more programming paradigms are developed for various situations. During programmers' life, they are always learning new paradigms and thinking about which ones are most suitable for the job at hand. For example, the event-handler paradigm is widely adopted by GUI frameworks [31, 18, 24]. When we write GUI programs with modern GUI libraries, we usually have to write a number of handlers for different types of events. The AWT [24] of Java is a typical example. If we want to do something for mouse events occurring on a button, we have to prepare a mouse listener that contains handler methods for those mouse events, and register the listener to the specified button object. A GUI program can be regarded as a composite of visual components, events, and handlers. The visual components and handlers are main logic, and events are used for connecting them. Indeed we have been familiar with using the event-handler paradigm for GUI programs, but it is far from our first "hello world" program. We are told to carefully consider the total execution order when users' input is read. If the event-handler paradigm is used, we can focus on the reaction to users' input rather than the order of users' input. Whether the mouse is clicked first or not does not matter. Another example is the aspect paradigm. Aspect-oriented programming is developed to modularize crosscutting concerns such as logging, which cannot be modularized by using only object-oriented programming. With the aspect paradigm, crosscutting concerns can be gathered up in an aspect by advices. At the same time, programmers cannot check only one place for understanding the behavior of a method call since advices in other places are possibly woven together. It also takes effort to get familiar with the aspect paradigm since it is quite different from our other programming experience.

To use a paradigm, just learning its concept is not enough. After programmers got the idea of a paradigm, they still have to learn new language constructs for the paradigm. Some paradigms like the aspect paradigm are supported with dedicated language constructs since the beginning because they cannot be represented well by existing syntax. On the other hand, although other paradigms like the event-handler paradigm have been introduced at library level for a long time, there are still good reasons for re-introducing them with direct support at language level [17, 22, 9]. Maybe one reason is that events are complicated in particular when we are not users but designers of a library. Besides GUI libraries, the event-handler paradigm is also implemented in a number of libraries for several domains such as simple API for XML [29] and asynchronous socket programming. Some techniques such as the Observer pattern [7] used in those libraries cannot satisfy the needs of defining events and tend to cause code scattering and tangling. Supporting paradigms by language constructs is a trend since it makes code more clear and reusable. Furthermore, a language supported paradigm may have associated static checks.

However, learning language constructs for a paradigm is never easy, especially for powerful paradigms like the aspect paradigm. Moreover, the syntax is usually hard to share with other paradigms. Even though programmers got familiar with the language constructs for a paradigm, they still have to learn new ones for another paradigm from the beginning. Given that all language constructs we need can be put into a language together, they look too complex and redundant. How to pick up the best language to implement a program with all required paradigms is always a difficult issue. This motivates us to find out an easy, simple, and generic language construct supporting multiple paradigms.

If we look into the language constructs for the event-handler paradigm and the aspect paradigm, there is a notable similarity between them. Both of them introduce a way to define the effect of calling specified methods. The differences are where the reactions are and what the reactions are targeted at. Listing 1 is a piece of code in EScala[1] [9], which is a typical event mechanism, showing how to define a moved event for the setPosition method in the Shape class. Here we specify that refresh method on a Display object should be executed after setPosition method is executed. As shown in Listing 2, the reaction can also be represented in AspectJ [25], the most well-known aspect-oriented language.

By comparing the two pieces of code, we can find that

---

[1]The syntax follows the example in EScala 0.3 distribution.

**Listing 1: Defining a reaction in EScala**

```
class Display() {
  def refresh() {
    System.out.println("display is refreshed.")
  }
}
class Shape(d: Display) {
  var left = 0; var top = 0
  def setPosition(x: Int, y: Int) {
    left = x; top = y
  }
  evt moved[Unit] = afterExec(setPosition)
  moved += d.refresh
}
object Test {
  def main(args: Array[String]) {
    val d = new Display()
    val s = new Shape(d)
    s.setPosition(0, 0)
  }
}
```

**Listing 2: Defining a reaction in AspectJ**

```
public class Display {
  public static void refresh() {
    System.out.println("display is refreshed.");
  }
}
public class Shape {
  private int left = 0; private int top = 0;
  public void setPosition(int x, int y) {
    left = x; top = y;
  }
}
public aspect UpdateDisplay {
  after() returning:
    execution(void Shape.setPosition(int, int)) {
      Display.refresh();
    }
}
public class Test {
  public static void main(String[] args) {
    Shape s = new Shape();
    s.setPosition(0, 0);
  }
}
```

pointcuts are close to events and advices can work as the += operator for handlers. They both refresh the display when the specified method is executed, but there is a significant difference between them. In EScala version, one Display object is mapped to one Shape object and the refresh action is performed within the Shape object. On the other hand, in AspectJ version there is only one Display object in the whole program and the refresh action is in UpdateDisplay, which is completely separated from Display and Shape. From the viewpoint of the event-handler paradigm, such behavior is an interaction between objects, so the reaction is defined inside the class and targeted at object instances; the encapsulation is preserved. From the viewpoint of the aspect paradigm, it is important to extract the reaction for the obliviousness since it is a different concern cutting across several classes. So the reactions are grouped into a separate construct and targeted at the class. Although the two paradigms are developed from different points of view, the language constructs used for them are quite similar. Furthermore, both the two paradigms depend on the most basic paradigm, the method paradigm, since both events and pointcuts cause the exe-

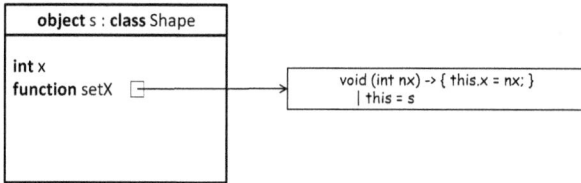

Figure 1: In JavaScript, both an integer and a function are fields on an object

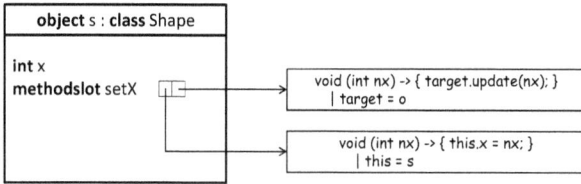

Figure 2: A method slot is an extended field that can keep more than one function closure

Listing 3: A sample code in DominoJ

```
1   public class Shape {
2     private int x;
3     public void setX(int nx) {
4       // default closure
5       this.x = nx;
6     }
7   }
8   public class Observer {
9     private int count;
10    public void update(int i) {
11      this.count++;
12    }
13    public static void main(String[] args) {
14      Shape s = new Shape();
15      Observer o = new Observer();
16      s.setX += o.update;
17      s.setX(10);
18    }
19  }
```

cution of a method-like construct. This observation led us to extend the method paradigm to support both the event-handler paradigm and the aspect paradigm. To a programmer, there are too many similar language constructs for different paradigms to learn, so we assume that the integration and simplification are always worth doing.

## 3. DOMINOJ

We extend methods to a new language construct named a *method slot*, to support methods, events, and advices. We also show our prototype language named *DominoJ*, which is a Java-based language supporting *method slots* and fully compatible with plain Java.

### 3.1 Method slots

Although methods and fields are different constructs in several languages such as C++ and Java, there is no difference between them in other languages like JavaScript. In JavaScript, a method on an object (strictly speaking, a function closure) is kept and used as other fields. Figure 1 shows a Shape object s, which has two fields: an integer field named x and a function field named setX. We use the following notation to represent a closure:

<return type> (<parameter list>) -> { <statements> }
| <variable binding list>

where the variable binding list binds non-local variables in the closure. The value stored in field setX is a function closure whose return type and parameter type are void and (int), respectively. The variable this used in the closure is bound to s given by the execution context. When we query the field by s.setX, the function closure is returned. When we call the field by s.setX(10), the function closure is executed.

We extend this field in JavaScript to keep an array of function closures rather than just one function closure. As shown in Figure 2, the extended field named a *method slot* can keep more than one function closure. DominoJ replaces a method with a method slot in plain Java. All method-like declarations and calls are referred to method slots. A method slot is a closure array and is an object's property

like a field. Like functions or other fields, method slots are typed and statically specified when they are declared. The type of method slot includes its return type and parameter types. All closures in it must be declared with the same type.

Listing 3 shows a piece of sample code in DominoJ. It looks like plain Java, but here setX is a method slot rather than a method. The syntax of method slot declaration is shown below:

<modifier>* <return type> <identifier>
"(" <parameter list>? ")" <throws>?
(<default closure> | ";")

The default closure is similar to the method body in Java except it is optional. The modifiers can be public, protected, or private for specifying the visibility of the method slot. This ensures that the access to the method slot can be controlled as the methods in plain Java. The modifier static can be specified as well. Such static method slots are kept on the class objects so can be referred using the class name like calling the static method in plain Java. The modifier abstract can also be used to specify that the method slot should be implemented by the subclasses. A method slot can be another kind of "abstract" by being declared without a default closure:

```
public void setX(int nx);
```

Unlike the modifier abstract, this declaration means that the method slot is an empty field and its behavior depends on the closures added to it later. In Listing 3, the method slot setX has a default closure, so the following function closure will be created and inserted into setX automatically when a Shape object, s, is instantiated:

```
void (int nx) -> { this.x = nx; }
| this = s
```

Now there is only one closure in the method slot setX. If we add another closure to setX, the object may look like the s object in Figure 2. How to add such a closure to a method slot will be demonstrated in the next subsection.

A method slot can also be declared with the modifier final to specify that it cannot be overridden in the subclasses. Although fields are never overridden in either prototype-based

199

```
1   ; call a methodslot
2   (define (call-methodslot object slotname args)
3     (let* ((methodslot (get-field object slotname
4                                   (get-type args)))
5            (return_type (get-return-type methodslot)))
6       (let execute-closures
7         ((closures (get-closures methodslot))
8          ($retval (cond ((boolean? return_type) #f)
9                         ((number? return_type) 0)
10                        (else '()))))
11        (if (null? closures)
12            $retval
13            (let
14              (($retval (execute-a-closure (car closures) args)))
15              (execute-closures (cdr closures) $retval)))))))
```

**Table 1: The four operators for method slots**

| Operator | Description |
|---|---|
| = | *add a new function closure and remove the others from the method slot.* |
| ^= | *insert a new function closure at the beginning of the array.* |
| += | *append a new function closure to the end of the array.* |
| -= | *remove function closures calling the method slot at the right-hand side.* |

languages like JavaScript or class-based languages like Java, method slots can be overridden in subclasses. Declaring a method slot with the same signature overrides but not hides the one in the superclass. When a method slot is queried or called on an object, the overriding method slot is selected according to the actual type of the object. It is also possible to access the overridden method slot in the superclass through the keyword super. Note that method slots must be declared within a class and cannot be declared as local variables. Thus the usage of this and super in the default closure are the same as in a Java method, which refer to the owning class and its superclass, respectively. Constructors are method slots as well, and super() is allowed since it calls the overridden constructor.

When a method slot is called by () operator, the closures in it are executed in order. The arguments given to the method slot are also passed to its closures. The return value returned by the last closure is passed to the caller (if it is not the void type). A closure can use a keyword $retval to get the return value returned by the preceding closure in the method slot. If the closure is the first one in the method slot, $retval is given by a default value (0, false, or null). If the method slot is empty, the caller will get the default value and no exception is thrown. It is reasonable since the empty state is not abnormal for an array and just means that nothing should be done for the call at that time. The behavior of a method slot can be dynamically modified at runtime, while still statically typed and checked at compile time. How to call a method slot is described in Scheme as shown in Listing 4.

## 3.2 Operators for method slots

DominoJ provides four operators for manipulating the closures in a method slot: =, ^=, +=, and -=, as shown in Table 1. These operators are borrowed from C# and EScala, and are the only different syntax from Java. It is possible to add and remove a function closure to/from a method slot at runtime.

Their operands at both sides are method slots sharing the same type. Those operators except -= create a new function closure calling the method slot at the right-hand side, and add it to the method slot at the left-hand side. The method slot called by the function closure will get the same arguments which are given to the method slot owning the function closure. In other words, a reference to the method slot at the right-hand side is created and added to the method

slot at the left-hand side. The syntax of using the operators to bind two method slots is shown below:

$<expr>$".".$<methodslot>$
    $<operator>$ $<expr>$".".$<methodslot>$;

where $<expr>$ can be any Java expression returning an object, or a class name if the following $<methodslot>$ is static. When the binding statement is executed at runtime, the $<expr>$ at both sides will be evaluated according to current execution context and then given to the operator. In other words, the $<expr>$ at the right-hand side is also determined at the time of binding rather than the time of calling. The object returned by the $<expr>$ at the left-hand side helps to find out the method slot at the left-hand side, where we want to add or remove the new function closure. The object got by evaluating the $<expr>$ at the right-hand side is attached to the new function closure as a variable target, which is given to the new function closure along with the execution context at the time of calling. For example, the binding statement in Line 16 of Listing 3 creates a new function closure calling the method slot update on the object o by giving target = o, and appends it to the method slot setX on the object s.

```
void (int nx) -> { target.update(nx); }
  | target = o
```

Then the status of the s object will be the same as the one shown in Figure 2. When the slot setX on the object s is called as Line 17 in Listing 3, the default closure and the slot update on the object o are sequentially called with the same argument: 10. Note that all closures in a method slot get the same execution context, where this refers to the object owning the method slot, and therefore, the callee method slot in target must be accessible from the caller method slot in this. With proper modifiers, a method slot cannot call and be called without any limitation. The behavior avoids breaking the encapsulation in object-oriented programming.

The -= operator removes function closures calling the method slot at the right-hand side from the method slot at the left-hand side. It is also possible to remove the default closure from a slot by specifying the same method slots at both sides:

```
s.setX -= s.setX;
```

Operators manipulate the default closure only when the method slots at both sides are the same one, otherwise operators regard the right-hand side as a closure calling that method slot. Note that the default closure is never destroyed even when it is removed. The algorithms of the four operators are described in Scheme in Listing 5.

## Listing 5: The algorithms of the four operators

```
1  ; operator =
2  (define (assign-closure methodslot object slotname)
3    (let ((closure
4            `(call-methodslot ,object ,slotname args)))
5      (set-closures methodslot closure)))
6
7  ; operator ^=
8  (define (insert-closure methodslot object slotname)
9    (let ((closure
10           `(call-methodslot ,object ,slotname args)))
11     (set-closures
12       methodslot
13       (append closure (get-closures methodslot)))))
14
15 ; operator +=
16 (define (append-closure methodslot object slotname)
17   (let ((closure
18           `(call-methodslot ,object ,slotname args)))
19     (set-closures
20       methodslot
21       (append (get-closures methodslot) closure))))
22
23 ; operator -=
24 (define (remove-closure methodslot object slotname)
25   (let ((closure
26           `(call-methodslot ,object ,slotname args)))
27     (set-closures
28       methodslot
29       (remove (lambda (x) (equal? x closure))
30               (get-closures methodslot)))))
```

## Listing 7: The algorithm of binding method slots

```
1  ; bind methodslots by operators
2  (define (bind-methodslots
3           operator l_object l_slotname r_object r_slotname)
4    (let ((l_methodslots (get-fields l_object l_slotname))
5          (r_methodslots (get-fields r_object r_slotname)))
6      (for-each
7        (lambda (l_methodslot)
8          (for-each
9            (lambda (r_methodslot)
10             (if (or (same-type? l_methodslot
11                                 r_methodslot)
12                     (generic-type? l_methodslot
13                                    r_methodslot))
14               (cond ((equal? operator "=")
15                      (assign-closure l_methodslot
16                                      r_object
17                                      r_slotname))
18                     ((equal? operator "^=")
19                      (insert-closure l_methodslot
20                                      r_object
21                                      r_slotname))
22                     ((equal? operator "+=")
23                      (append-closure l_methodslot
24                                      r_object
25                                      r_slotname))
26                     ((equal? operator "-=")
27                      (remove-closure l_methodslot
28                                      r_object
29                                      r_slotname)))))
30           r_methodslots))
31       l_methodslots)))
```

## Listing 6: The algorithm of checking the types

```
1  ; is same type
2  (define (same-type? l_methodslot r_methodslot)
3    (and (equal? (get-return-type l_methodslot)
4                 (get-return-type r_methodslot))
5         (equal? (get-parameter-types l_methodslot)
6                 (get-parameter-types r_methodslot))))
7
8  ; is generic type
9  (define (generic-type? l_methodslot r_methodslot)
10   (and (equal? (get-parameter-types r_methodslot)
11                "Object[]")
12     (if (equal? (get-return-type l_methodslot)
13                 "void")
14       (equal? (get-return-type r_methodslot)
15               "void")
16       (equal? (get-return-type r_methodslot)
17               "Object"))))
```

Although a method slot at the right operand of the operators such as += must have the same type that the left operand has, there is an exception. If a method slot takes only one parameter of the Object[] type and its return type is Object or void, then it can be used as the right operand whatever the type of the method slot at the left operand is. Such a method slot can be used as a generic method slot. The type conversion when arguments are passed is implicitly performed. Listing 6 shows how to check the type of two method slots in Scheme.

DominoJ allows binding method slots to constructors by specifying class name instead of the object reference and giving the keyword constructor as the method slot at the left-hand side. For example,

```
Shape.constructor += Observer.init;
```

means that creating a closure calling the static method slot init on the class object Observer and appending to the con-

structor of Shape. Here the return type of init should be void, and the parameter types must be the same as the constructor. Note that the closures appended to the constructor cannot block the object creation. This design ensures that the clients will not get an unexpected object, but additional objects can be created and bound to the new object. For example, in the default closure of init, an instance of Observer can be created and its update can be bound to the method slot setX of the new Shape object. Using constructor at the right-hand side is not allowed.

Since Java supports method overloading, some readers might think the syntax of method slots have ambiguity but that is not true. For example, the following expression does not specify parameter types:

```
s.setX += o.update;
```

If setX and/or update are overloaded, += operator is applied to all possible combinations of setX and update. Suppose that there are setX(int), setX(String), update(int), and update(String). += operator adds update(int) to setX(int), update(String) to setX(String). If there is update(Object[]), it is added to both setX(int) and setX(String) since it is generic. It is possible to introduce additional syntax for selecting method slots by parameters, but the syntax will be more complicated. Listing 7 is the algorithm in Scheme for picking up and binding two method slots by operators.

Since a method slot may contain closures calling other method slots, a call to the method slot can be regarded as a sequence of method slot calls among objects. When a method slot is explicitly called by a statement in a certain default closure, the method slots bound to it by operators are implicitly called by DominoJ. In order to make method slots more flexible, DominoJ provides keywords to get the preceding objects in the call sequence. In the default clo-

201

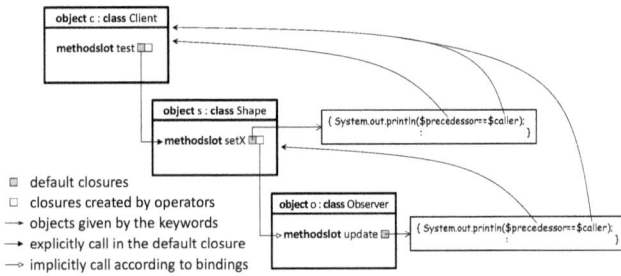

Figure 3: The keywords $caller and $predecessor

sure, i.e. the body of method slot declaration, the caller object can be got by keyword $caller. It refers to the object where we start the call sequence by a statement. The predecessor object, in other words, the object owning the preceding method slot in the call sequence, can also be got by the keyword $predecessor. It refers to the object owning the closure calling the current method slot whether explicitly or implicitly. Taking the example of Figure 2, suppose that we have a statement calling s.setX in the default closure of the method slot test in another class Client:

```
public class Client {
  public void test(Shape s) {
    s.setX(10);
  }
}
```

If test on an object instance of this class, for example c, is executed, the relationship between the objects c, s, and o can be described as shown in Figure 3. Note that calling other method slots explicitly by statements in the default closure of test, setX, or update will start separate call sequences. In Figure 3, using $caller in the default closure of setX and update both returns the object c since there is only one caller in a call sequence. However, the predecessor objects of s and o are different. Using $predecessor in the default closure of setX returns the object c, but using $predecessor in the default closure of update returns the object s. Note that both the apparent types of $caller and $predecessor are Object because the caller and the predecessor are determined at runtime. If current method slot is called in a static method slot, $caller or $predecessor will return the class object properly. The special method call proceed in AspectJ is introduced in DominoJ as well. It calls the default closure of the preceding method slot. In Figure 3, calling proceed in the default closure of update on o will execute the default closure of setX on s since s.setX is the preceding method slot of o.update. If there is no preceding method slot for current one, calling proceed will raise an exception.

## 4. EVALUATION

To show the feasibility of DominoJ and measure the overheads caused by method slots, we implemented a prototype compiler[2] of DominoJ built on top of JastAddJ [27]. The source code in DominoJ can be compiled into Java bytecode and run by Java virtual machine. In the following microbenchmark, the standard library is directly used without

---

[2]The prototype compiler of DominoJ is available from http://www.csg.ci.i.u-tokyo.ac.jp/projects/dominoj/

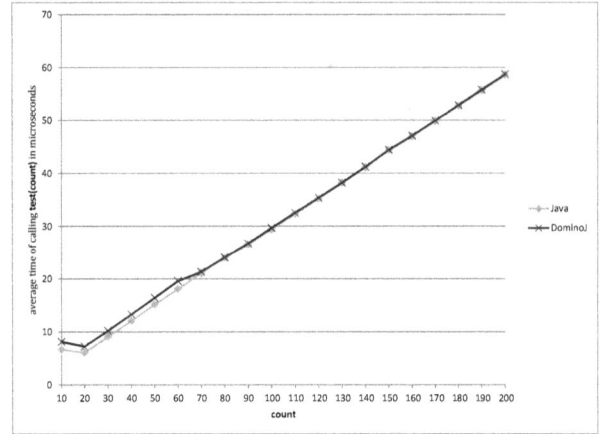

Figure 4: The average time per method call in Java and DominoJ

recompilation due to the performance concern. All methods in the standard library can be called as method slots which have only the default closure, but cannot be modified by the operators.

### 4.1 Microbenchmark

In order to measure the overheads of method slots, we executed a simple program and compared the average time per method call in DominoJ and in plain Java. The program calls a method that calculates $sin(\pi/6)$ count times by expanding Taylor series up to 100th order. In the program we call the method with count = 10, 20, 30, ..., 200 sequentially and observe the execution time. In order to ensure the execution time can be measured accurately, for different count we call the method one thousand times and calculate the average. The program was compiled by our prototype compiler and run on the JVM of OpenJDK 1.6.0_24 and Intel Core i7 (2.67GHz, 4 cores) with 8GB memory.

Figure 4 shows the result of the experiment. At first the execution time in DominoJ was 22% slower than in Java ($8.14\mu s$ against $6.67\mu s$), but the gap was getting narrow and almost the same as in Java after count = 70. These numbers on execution overheads are not serious in practice since they are not very large and can be optimized by JVM at runtime.

### 4.2 Enhance the method paradigm

Method slots extend the method paradigm to support the event-handler paradigm and the aspect paradigm, while still reserving the original behavior in the method paradigm. In DominoJ, if the operators for method slots are not used, the code works as in plain Java. In other words, the method in Java can be regarded as a method slot that has only the default closure.

We also analyze how method slots can be applied to "GoF" design patterns [7], and classify the patterns into four groups as shown in Table 2. The key idea of the patterns in group I can be considered event propagation—from the outer object to the inner object, or among colleague objects. Using method slots can avoid code scattering caused by the pattern code since event implementation is eliminated. Code tangling caused by combining multiple patterns can be eased as well. The patterns in group II use the inheritance to al-

**Table 2: Method slots can be applied to design patterns**

| | Pattern Name | Description |
|---|---|---|
| I | Adapter, Chain of Responsibility, Composite, Decorator, Facade, Mediator, Observer, Proxy | *Propagate events among objects* |
| II | Abstract Factory, Bridge, Builder, Factory Method, State, Strategy, Template Method, Visitor | *Change class behavior at runtime without inheritance* |
| III | Command, Flyweight, Interpreter, Iterator, Prototype | *Replace inheritance part in the logic* |
| IV | Memento, Singleton | *Not applicable* |

**Table 3: The roles and bindings of the event-handler paradigm in EScala and DominoJ**

| | Type | EScala | DominoJ |
|---|---|---|---|
| *role* | *Event* | field (evt) | method slot |
| | *Handler* | method | |
| *binding* | *Event-to-Handler* | += | += |
| | | -= | -= |
| | *Event-to-Event* | \|\| | +=, ^= |
| | | && | use Java expression in the default closure of method slots |
| | | \ | |
| | | filter | |
| | | map | |
| | | empty | |
| | | any | |
| | *Handler-to-Event* | afterExec | += |
| | | beforeExec | ^= |
| | | imperative | explicit trigger is possible |

ter the class behavior at runtime. Different implementation can be added to the method slot instead of overriding in subclasses. In that sense, method slots can be used as an alternative to the polymorphism. Although method slots are not perfect replacement for the inheritance, it is convenient in particular when programmers are forced to choose between two superclasses due to single inheritance limitation. The patterns classified under group III also use the inheritance as a part of their pattern code, so programmers may use method slots or not depending on the situation. As to the patterns in group IV, DominoJ is not beneficial to object creation as what AspectJ can do in [10]. The reason is that DominoJ does not support the inter-type declaration and cannot stop the object creation. Details of this analysis is available in [33].

## 4.3 The event-handler paradigm

To evaluate how DominoJ works for the event-handler paradigm, first we analyze the bindings between the event and the handler in a typical event mechanism like EScala, and compare them with DominoJ. In languages directly supporting the event-handler paradigm, events are usually introduced as fields, which are separate from methods. In order to associate fields with methods, there are three types of binding between events (fields) and handlers (methods). The ways used for each type of binding are usually different in an event mechanism, and also different between event mechanisms. Table 3 shows the ways provided by EScala. The corresponding DominoJ syntax for the three types of binding is also listed, but actually there is only slot-to-slot binding in DominoJ since only method slots are involved in the event-handler paradigm. Every method slot can play an event role and a handler role at the same time. Listing 8 shows how to use DominoJ for the event-handler paradigm for the shape example mentioned in Section 2. Below we will discuss what the three types of binding are, and explain how DominoJ provides the same advantages with the simplified model.

The event-to-handler binding is the most trivial one since it means what action reacts to a noteworthy change. Whether supporting the event-handler paradigm by languages or not, in general the event-to-handler binding is dynamic and provided by a clear manner. For example, in the Observer pattern an observer object can call a method on the subject to register itself; in C# and EScala, += operator and -= operator are used to bind/unbind a method to a special field named event. In addition to the two operators, Domi-

**Listing 8: Using DominoJ for the event-handler paradigm**

```
1  public class Display {
2    public void refresh(int x, int y) {
3      System.out.println("display is refreshed.");
4    }
5  }
6  public class Shape {
7    private int left = 0; private int top = 0;
8    public void setPosition(int x, int y) {
9      left = x; top = y;
10   }
11   public Shape(Display d) {
12     this.setPosition += d.refresh;
13   }
14 }
15 public class Test {
16   public static void main(String[] args) {
17     Display d = new Display();
18     Shape s = new Shape(d);
19     s.setPosition(0, 0);
20   }
21 }
```

noJ provides ^= operator and = operator to make it easier to manipulate the array of handlers. In C# and EScala the handlers for an event can be only appended sequentially and removed individually, but in DominoJ programmers can use = operator to empty the array directly without deducing the state at runtime. Using ^= operator along with += operator also makes design intentions more clear since a closure can be inserted at the beginning without popping and pushing back.

The second one is the event-to-event binding that enables event composition and is not always necessary but greatly improves the abstraction. In a modern event mechanism, event composition should be supported. EScala allows programmers to define such higher-level events to make code more readable. An event-to-event binding can be simulated by an event-to-handler binding and a handler-to-event binding, but it is annoying and error-prone. In DominoJ, it is also possible to define a higher-level event by declaring a method slot without a default closure. Then operators += and ^= can be used to attach other events like what the operator \|\| in EScala does. Other operators in EScala such as

&& and map are not provided in DominoJ, but the same logic can be represented by statements in another handlers and attached by += operator. For example, in Listing 1 we can declare a new event adjusted that checks if left and top are the same as the arguments given to setPosition using the operator && in EScala:

```
evt adjusted[Unit] = afterExec(setPosition)
                     && ((left,top) != _._1)
adjusted += onAdjusted
```

where _._1 refers to the arguments given to setPosition and onAdjusted is the reaction. In DominoJ, we can declare a higher-level event adjusted and perform the check in another method slot checkAdjusted:

```
public void adjusted(int x, int y);
public void checkAdjusted(int x, int y) {
  if(!(x==left && y==top))  adjusted(x, y);
}
```

and then bind them as follows:

```
setPosition += checkAdjusted;
adjusted += onAdjusted;
```

Although the expression in DominoJ is not rich and declarative as in EScala, they can be used to express the same logic. In addition, the event-to-event binding in EScala is static, so that the definition of a higher-level event in EScala cannot be changed at runtime. On the other hand, it is possible in DominoJ since the slot-to-slot binding is totally dynamic.

The last one is handler-to-event binding, which is also called an event trigger or an event definition. It decides whether an event trigger can be implicit or not. In the Observer pattern and C#, an event must be triggered explicitly, so that the trigger code is scattering and tangling. EScala provides two implicit ways and an explicit way: after the execution of a method, before the execution of a method, or triggering an event imperatively. In DominoJ, an event can be triggered either implicitly or explicitly. A method slot can not only follow the call to another method slot but also be imperatively called. More precisely, there is no clear distinction between the two triggering ways. In EScala afterExec and beforeExec are provided for statically binding an event to the execution of a method while DominoJ provides += operator and -= operator for dynamically binding a method slot to the execution of another method slot. This sounds like that a method slot has two pre-defined EScala-like events for the default closure, but it is not correct. In DominoJ's model the only one event is the call to a method slot, and the default closure is also a handler like the other closures calling other method slots. This feature makes the code more flexible since the execution order of all handlers can be taken into account together. As to the encapsulation, in EScala the visibility of explicit events follows its modifiers, and the implicit events are only visible within the object unless the methods they depend on are observable. On the other hand, the encapsulation in DominoJ relies on the visibility of method slots. The design is simpler but limits the usage because a public method slot is always visible as an event to other objects.

There is one more important difference between EScala and DominoJ. In DominoJ, a higher-level event can be declared or not according to programmers' design decision. In

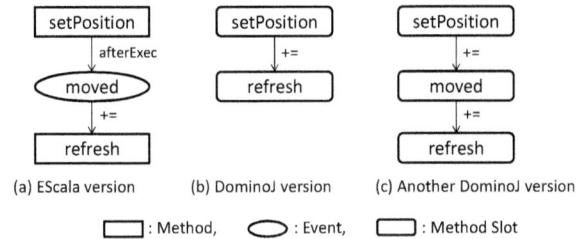

Figure 5: The execution order of the shape example in EScala and DominoJ

order to explain the difference, we use a tree graph to represent the execution order in the shape example by regarding setPosition as the root. As shown in Figure 5, we use rectangles, circles, and rounded rectangles to represent methods, events, and method slots, respectively. When a node is called, the children bound by beforeExec or ^= must be executed first, followed by the node itself and the children bound by afterExec or +=. Figure 5 (a) is the execution order of Listing 1, and Figure 5 (b) is the one of Listing 8. In the DominoJ version, the event moved is eliminated and its child refresh is bound to setPosition directly since we do not need additional events in such simple case. DominoJ is easier and simpler to apply the event-handler paradigm when events are not complicated but used everywhere. In EScala, events must be created since methods cannot be bound to each other directly. However, such events are still necessary if we want to keep the abstraction. In that case, method slots can be used as the events in EScala by declaring them without a default closure. For example, the event moved in Line 11 of Listing 1 can be translated into the following statements:

```
public void moved();
setPosition += moved;
```

Figure 5 (c) is another DominoJ version, which has the higher-level event as the EScala version. In DominoJ, programmers can choose between the simplified one and the original one depending on the situation.

Note that the number of lines of Listing 8 is one line longer than Listing 1 because the syntax of Scala looks more compact than Java. In Java the constructor and the fields used inside a class must be declared explicitly while they are omitted in Scala. In Listing 8 the constructor takes two more lines than Listing 1. If we do not take this into account, the EScala version is one line longer than the DominoJ version due to additional event declaration.

The line of code can also be analyzed according to Table 3. With regard to the roles, additional event declarations are necessary in EScala while they are combined into one declaration in DominoJ as we discussed above. For the event-to-handler binding, both the operators provided by EScala and DominoJ take one line. For the event-to-event binding, the operators provided by EScala can be written in the same line, but in DominoJ += operator and ^= operator cannot be merged into one line. In that case the code in DominoJ is longer than the EScala one. For example, a higher-level event changed can be defined by three events resized, moved, and clicked:

```
evt changed[Unit] = resized || moved || clicked
```

but in DominoJ they must be defined as follows:

```
resized += changed;
moved += changed;
clicked += changed;
```

That is why the expression in EScala is richer but complicated. Introducing appropriate syntax sugar to DominoJ to allow putting operators in one line is also possible, but we think it makes the design complicated. However, in this example we can also find passing the event value in EScala takes effort. In EScala, as far as we understand, only a value is kept in an event field. If we want to gather up the arguments x and y given to setPosition, and then pass to moved and changed, we need to declare additional classes such as Point and declare the events with the new type rather than Unit[3]. The additional classes increase the number of lines as well. For the handler-to-event binding, afterExec and beforeExec in EScala can define an anonymous event and share the same line of an event-to-handler binding. To sum up, in DominoJ the event declarations may be eliminated and thus the number of lines of source code can be reduced. On the other hand, the number of code of DominoJ version is longer when translating a complex EScala expression composed of a number of operators since DominoJ has less primitive syntax. DominoJ makes code clear because each method slot has a name explicitly, and each line for binding only defines the relation between two method slots.

## 4.4 The aspect paradigm

DominoJ can be used to express the aspect paradigm as well. In order to discuss language constructs concretely, we compare DominoJ with the most representative aspect-oriented language—AspectJ. The call to a method slot is a join point, and other method slots can be bound to it as advices. Note that aspect-oriented programming is broader as discussed in [14] and not restricted to the AspectJ style, which is the point-advice model. In this subsection first we analyze the necessary elements in the point-advice model, and compare the constructs provided by AspectJ and DominoJ. Then we use DominoJ to rewrite the shape example in Listing 2 and discuss the differences.

Since the purpose of the aspect paradigm is to modularize the crosscutting concerns, we need a method-like construct to contain the code piece, a way to attach the method-like construct to a method execution, and a class-like construct to group the method-like construct. In AspectJ, the class-like construct is the aspect construct, the method-like construct is the advice body, and the way of attaching is defined by the pointcut and advice declaration. In DominoJ, the method slot and the class construct in plain Java are used and only operators for method slots are introduced for attaching them. The method slots bound by += operator or ^= operator are similar to after/before advices, respectively. The method slots bound by = operator are similar to around advices and proceed can be used to execute the original method slot. It is expected that DominoJ cannot cover all expression in AspectJ since DominoJ's model is much simpler. For example, in DominoJ the inter-type declaration and the reflection are not provided. According to

---

[3]In EScala, declaring events with Unit type means that no data are passed [9].

Table 4: The mapping of language constructs for the aspect paradigm in AspectJ and DominoJ

| Construct | AspectJ | DominoJ |
|-----------|---------|---------|
| *grouping* | aspect | class |
| *code piece* | advice body | method slot body (default closure) |
| *pointcut and advice declaration* | after returning and execution | += and $retval |
| | before and execution | ^= |
| | around | = |
| | this | $caller |
| | target | $predecessor |
| | args | by parameters |

the three elements, Table 4 lists the mapping of language constructs in AspectJ and DominoJ.

In AspectJ programmers need to understand the special instance model for the aspect construct, but in DominoJ the class construct is reused. Although the instances of the construct for grouping need to be managed manually, there is no need to learn the new model and keywords like issingleton, pertarget, and percflow. In DominoJ programmers can create an instance of the aspect-like class and attach its method slots to specified objects according to the conditions at runtime. If the behavior of issingleton is preferred, programmers can declare all fields including method slots in the aspect-like class as static since static method slots are supported by DominoJ. The shape example of AspectJ in Section 2 can be rewritten by DominoJ as shown in Listing 9. Here the class UpdateDisplay is the aspect-like class. In Line 14, we attach the advice refresh in a static method slot init, so all Shape objects will share the class object of UpdateDisplay. Furthermore, we let init be executed after the constructor of Shape, so that we can avoid explicitly attaching refresh every time a Shape object is created. Moreover, we do not have to modify the constructor of Shape. If we need to count how many times setPosition is called for each Shape and thus pertarget is preferred, we can rewrite the class UpdateDisplay as shown in Listing 10. Every time a Shape object is created, a UpdateDisplay object is created for it implicitly. Note that the object ud will not be garbage-collected since its method slot count is attached to another method slot.

In DominoJ, there is no difference between methods and advices while in AspectJ they are different constructs. Although an advice in AspectJ can be regarded as a method body, it cannot be directly called. If the code of an advice is reusable, in AspectJ we must move it to another method but in DominoJ it is not necessary.

The pointcut and advice declaration in AspectJ and DominoJ are similar but not the same. First, what they target at is different. AspectJ is class-based while DominoJ is object-based. In other words, what AspectJ targets at are all object instances of a class and its subclasses but what DominoJ targets at are individual object instances. However, it is possible to emulate the class-based behavior in DominoJ by the code attaching to the constructor of a class as shown in Line 16 of Listing 9. Second, unlike AspectJ that has call and execution pointcut, in DominoJ only execution pointcut is supported. This limits the usage but reduces the complexity. In fact, the relation between advices is quite different in AspectJ and DominoJ. In AspectJ an advice is attached

## Listing 9: Using DominoJ as the aspect paradigm

```
1  public class Display {
2    public static void refresh(int x, int y) {
3      System.out.println("display is refreshed.");
4    }
5  }
6  public class Shape {
7    private int left = 0; private int top = 0;
8    public void setPosition(int x, int y) {
9      left = x; top = y;
10   }
11 }
12 public class UpdateDisplay {
13   public static void init() {
14     ((Shape)$predecessor).setPosition += Display.refresh;
15   }
16   static { Shape.constructor += UpdateDisplay.init; }
17 }
18 public class Test {
19   public static void main(String[] args) {
20     Shape s = new Shape();
21     s.setPosition(0, 0);
22   }
23 }
```

## Listing 10: Rewrite UpdateDisplay for pertarget

```
1  public class UpdateDisplay {
2    private int total = 0;
3    public void count(int x, int y) {
4      total++;
5    }
6    public static void init() {
7      UpdateDisplay ud = new UpdateDisplay();
8      ((Shape)$predecessor).setPosition += ud.count;
9    }
10   static { Shape.constructor += UpdateDisplay.init; }
11 }
```

to methods and cannot be directly attached to a specific advice, but in DominoJ a method slot is not only an advice but also a method. For example, if we need another advice for checking the dirty region in Listing 2, we may prepare an aspect CheckDirty containing this advice as shown in Figure 6 (a). However, the advice can only be attached to setPosition. In DominoJ, the advice can be attached to either setPosition or init as shown in Figure 6 (b).

The behavior of proceed in AspectJ and DominoJ is also a little different. The proceed in DominoJ should be used only along with = operator since it calls the default closure in the preceding method slot rather than the next closure. The root cause of the difference is the join point model: what DominoJ adopts is the point-in-time model while the one AspectJ adopts is the region-in-time model [15]. In other words, in AspectJ the arrays of the three types of advices are separate, but in DominoJ there is only one array. If += operator or ^= operator are used after using = operator to attach a method slot containing proceed, the behavior is not as expected as in AspectJ. Figure 7 shows an example of around advice in AspectJ and DominoJ. In AspectJ, the around advices localCache and memCache are attached to queryData in order. In DominoJ, we can do it similarly:

```
queryData = localCache;
localCache = memCache;
```

then using proceed in memCache and localCache will call the default closure of their preceding method slot, localCache

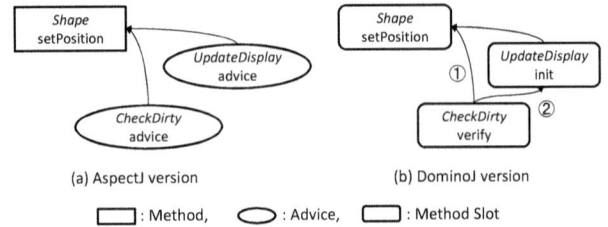

Figure 6: Adding another advice to the shape example in AspectJ and DominoJ

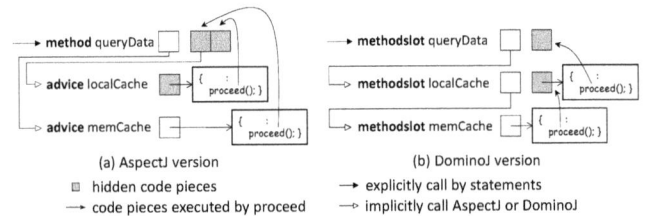

Figure 7: Calling proceed in AspectJ and DominoJ

and queryData, respectively. Another difference is that the args pointcut and the wildcard used in call and execution pointcuts in AspectJ are not supported in DominoJ. Method slots are simply matched by their parameters. If the overloading is not taken into account, the operators in DominoJ only select one method slot in one line statement.

As to the number of lines, the two versions are about the same. Comparing them line by line might not make much sense since there is no simple translation between DominoJ and AspectJ.

### 4.5 Summary of the coverage

For the event-handler paradigm, there are three significant properties: implicit events, dynamic binding, and event composition. DominoJ supports them all by method slots and only four operators. Rewriting a complex expression of event composition in EScala is also possible though it takes more lines. Introducing additional syntax may resolve the issue but it also complicates the model. As a result of regarding method slot calls as events, giving an event a different visibility from the methods it depends on is not supported by DominoJ.

The aspect paradigm of AspectJ has three important features that cannot be provided by the event-handler paradigm: around advice, obliviousness, and inter-type declaration. In DominoJ the around advice can be emulated by assigning a closure calling another method slot. DominoJ also supports the obliviousness in AspectJ by reusing the class construct as the aspect construct and attaching a method slot to a constructor of the target class. The inter-type declaration is not available in DominoJ. A possible solution is introducing a default method slot for undefined fields in a class like Smalltalk's doesNotUnderstand or what the no-applicable-method does in CLOS. In addition to being used for the event-handler paradigm and the aspect paradigm, DominoJ allows programmers to use both paradigms together.

# 5. RELATED WORK

The delegation introduced by C# [17] allows programmers to declare an event, define its delegate type, and bind a corresponding action to the event. Event composition is also supported by adding a delegate to two or more events. Although the delegate interface hides the executor from the caller, implicit events are not supported. The event must be triggered manually when the change happens. However, C# is able to emulate DominoJ using an unusual programming style: declaring an additional event for every method and always triggering the event rather than the method. From the point of view, a delegate is very similar to a method slot except the operator += in C# copies the handlers in the event but not creates a reference to the event. However, as in EScala, events and methods are still separate language constructs. Supporting by only one construct means that programmers do not need to decide between using such an unusual style or a normal style at the design stage whether newer modules might regard those methods as events or not. Furthermore, it is annoying that event fields and methods in C# cannot share the same name. Another disadvantage is that we have to ensure that there is at least one delegate for the event before triggering it. Otherwise it will raise an exception. This is not reasonable from the viewpoint of the event mechanism since it just means no one handles the event. In DominoJ no handlers for an event does not raise an exception and the one that triggers an event on a method slot is unaware of handlers.

There are a number of research activities on the integration of object-oriented programming and aspect-oriented programming. Those research use a single dispatch mechanism to unify OOP and AOP and reveal that the integration makes the model clearer, reusable, and compositable. Delegation-based AOP [11, 23] elegantly supports the core mechanisms in OOP and AOP by regarding join points as loci of late binding. The model proposed in [12] provides dedicated abstractions to express various object composition techniques such as inheritance, delegation, and aspects. The difference is that DominoJ integrates the event-handler paradigm and the aspect paradigm based on OOP. Another difference is that we propose a new language construct rather than a machine or language model, which makes it compatible with existing object-oriented languages such as Java. Other work such as FRED [21], composition filters [1], predicate dispatching [5], and GluonJ [2] can also be regarded as such integration work.

The method combination in Flavors and CLOS makes related methods easy to combine but not override. By default the combined method in Flavors first calls the before methods in the order that flavors are combined, following by the first primary method, then the after methods in the reverse order. The return value of the combined method is supplied by the primary method, while the return values of the before and after methods are ignored. Similarly, CLOS provides a standard method combination for generic functions. For a generic function call, all applicable methods are sorted before execution in the order the most specific one is first. Besides the primary, before, and after methods, CLOS provides the around methods and call-next-method for the primary and around methods. From the viewpoint of method combination, the default closure of a method slot looks like a primary method that can be dynamically added to other method slots as a before or after method, and even as an around method by assigning to the target method slot then using proceed as call-next-method. It is also easier to express the method combination as a hierarchy in DominoJ.

With regard to the event mechanism, several research activities are devoted to event declaration. Ptolemy [22] is a language with quantified and typed events, which allows a class to register handlers for events, and also allows a handler to be registered for a set of events declaratively. It has the ability to treat the execution of any expression as an event. The event model in Ptolemy solves the problems in implicit invocation languages and aspect-oriented languages. EventJava [6] extends Java to support event-based distributed programming by introducing the event method, which are a special kind of asynchronous method. Event methods can specify constraints and define the reaction in themselves. They can be invoked by an unicast or broadcast way. Events satisfying the predicate in event method headers are consumed by a reaction. Context-aware applications can be accommodated easily by the mechanism. Both the two research make events clear and expressive, but they do not support implicit events, which is one of the most significant properties as an event mechanism, whereas DominoJ supports it. Moreover, all events in their model are class-based, so that events for a specified object have to be filtered in the handlers. The binding in DominoJ is object-based, so it can describe the interaction between objects more properly.

On the other hand, several research support the event-handler paradigm upon the aspect paradigm. ECaesarJ [19] introduces events into aspect-oriented languages for context-handling. The events can be triggered explicitly by method calls or defined by pointcuts implicitly. EventCJ [13] is a context-oriented programming language that enables controlling layer activation modularly by introducing events. By declaring events, we can specify when and which instance layer is activated. It also provides layer transition rules to activate or deactivate layers according to events. EventCJ makes it possible to declaratively specify layer transitions in a separate manner. Comparing with DominoJ, using events in the two languages may beak modular reasoning since their event models rely on the pointcut-advice model. Furthermore, events are introduced as a separate construct from methods.

Flapjax [16] proposes a reactive model for Web applications by introducing behaviors and the event streams. Flapjax lets clients use the event-handler paradigm by setting data flows. The handlers for an event can be registered in an implicit way. However, unlike other event mechanisms, it requires programmers to use a slightly different event paradigm. The behavior of DominoJ is more similar to the typical event mechanism while it has the basic ability for the aspect paradigm as well.

Fickle [4] enables re-classification for objects at runtime. Programmers can define several state classes for a root class, create an object at a certain state, and change the membership of the object according to its state dynamically. With re-classification, repeatedly creating new objects between similar classes for an existing object can be avoided. Both Fickle and DominoJ allow changing the class membership of an object at runtime, so other objects holding the identity of the object can be unaware of the changes. The difference is that Fickle focuses on the changes between states while DominoJ focuses on the effect of calling specified methods. Fickle provides better structural ability such as declaring new fields in state classes. However, if the relation between states is not flat and cannot be separated clearly, programmers still have to maintain the same code between state classes. The common code to only part of states can be gathered up into one class in DominoJ. Furthermore, DominoJ is easier to use for the event-handler paradigm.

The lambda expressions [20] will be introduced in Java 8 as a new feature to support programming in a multicore environment. With the new expression, declaring anonymous classes for containing handlers can be eliminated. The lambda expression of Java 8 is a different construct from methods but method slots can be regarded as a superset of methods.

## 6. CONCLUSIONS

We discussed the similarity between the language constructs for the event-handler paradigm and the aspect paradigm, which motivates us to propose a new language construct, named *method slot*, to support both the two paradigms. We presented how a *method slot* is introduced as a language construct in a Java-based language, *DominoJ*. We then discussed how *method slots* can be used for the two paradigms and the coverage of expressive ability. Although the expression of *method slots* is not as rich as other languages, it is much simpler and able to express most functionality in the two paradigms. We also showed its feasibility by implementing a prototype compiler and running a preliminary microbenchmark.

## 7. REFERENCES

[1] L. Bergmans and M. Aksit. Composing crosscutting concerns using composition filters. *Commun. ACM*, 44(10):51–57, Oct. 2001.

[2] S. Chiba, A. Igarashi, and S. Zakirov. Mostly modular compilation of crosscutting concerns by contextual predicate dispatch. In *OOPSLA '10*, pages 539–554. ACM, 2010.

[3] J. Dedecker, T. V. Cutsem, S. Mostinckx, and W. D. Meuter. Ambient-oriented programming in AmbientTalk. In *ECOOP '06*, pages 230–254. Springer, 2006.

[4] S. Drossopoulou, F. Damiani, D. Dezani-Ciancaglini, and P. Giannini. Fickle: Dynamic object re-classification. In *ECOOP '01*, pages 130–149, 2001.

[5] M. Ernst, C. Kaplan, and C. Chambers. Predicate dispatching: A unified theory of dispatch. In *ECOOP '98*, pages 186–211. Springer-Verlag, 1998.

[6] P. T. Eugster and K. R. Jayaram. EventJava: An extension of Java for event correlation. In *ECOOP '09*, pages 570–594, 2009.

[7] E. Gamma, R. Helm, R. Johnson, and J. Vlissides. *Design Patterns*. Addison-Wesley, 1994.

[8] D. Garlan, S. Jha, D. Notkin, and J. Dingel. Reasoning about implicit invocation. In *SIGSOFT '98/FSE-6*, pages 209–221. ACM, 1998.

[9] V. Gasiunas, L. Satabin, M. Mezini, A. Núñez, and J. Noyé. EScala: modular event-driven object interactions in Scala. In *AOSD '11*, pages 227–240. ACM, 2011.

[10] J. Hannemann and G. Kiczales. Design pattern implementation in Java and AspectJ. In *OOPSLA '02*, pages 161–173. ACM, 2002.

[11] M. Haupt. A machine model for aspect-oriented programming. In *ECOOP '07*, pages 501–524, 2007.

[12] W. Havinga, L. Bergmans, and M. Aksit. A model for composable composition operators: expressing object and aspect compositions with first-class operators. In *AOSD '10*, pages 145–156. ACM, 2010.

[13] T. Kamina, T. Aotani, and H. Masuhara. EventCJ: a context-oriented programming language with declarative event-based context transition. In *AOSD '11*, pages 253–264. ACM, 2011.

[14] G. Kiczales, J. Lamping, A. Mendhekar, C. Maeda, C. Lopes, J. marc Loingtier, and J. Irwin. Aspect-oriented programming. In *ECOOP '97*. Springer-Verlag, 1997.

[15] H. Masuhara, Y. Endoh, and A. Yonezawa. A fine-grained join point model for more reusable aspects. In *APLAS '06*, pages 131–147, 2006.

[16] L. A. Meyerovich, A. Guha, J. Baskin, G. H. Cooper, M. Greenberg, A. Bromfield, and S. Krishnamurthi. Flapjax: a programming language for ajax applications. In *OOPSLA '09*, pages 1–20. ACM, 2009.

[17] Microsoft Corporation. C# language specification.

[18] Microsoft Corporation. Messages and message queues.

[19] A. Núñez, J. Noyé, and V. Gasiūnas. Declarative definition of contexts with polymorphic events. In *COP '09*, pages 2:1–2:6. ACM, 2009.

[20] Oracle Corporation. OpenJDK: Project Lambda. http://openjdk.java.net/projects/lambda/.

[21] D. Orleans. Incremental programming with extensible decisions. In *AOSD '02*, pages 56–64. ACM, 2002.

[22] H. Rajan and G. T. Leavens. Ptolemy: A language with quantified, typed events. In *ECOOP '08*, pages 155–179, 2008.

[23] H. Schippers, D. Janssens, M. Haupt, and R. Hirschfeld. Delegation-based semantics for modularizing crosscutting concerns. In *OOPSLA '08*, pages 525–542. ACM, 2008.

[24] Sun Microsystems. Abstract window toolkit. http://java.sun.com/products/jdk/awt/.

[25] The AspectJ Project. http://www.eclipse.org/aspectj/.

[26] The Boost Project. Boost.Signals. http://www.boost.org/libs/signals/.

[27] The JastAdd Project. JastAddJ: The JastAdd Extensible Java Compiler. http://jastadd.org/web/jastaddj/.

[28] The Qt Project. Signals & Slots. http://qt-project.org/doc/signalsandslots.

[29] The SAX project. Simple api for xml. http://www.saxproject.org/.

[30] The Self project. http://selflanguage.org/.

[31] The X.Org project. Xlib in x window system. http://www.x.org/.

[32] J. Widom and S. J. Finkelstein. Set-oriented production rules in relational database systems. In *SIGMOD '90*, pages 259–270. ACM Press, 1990.

[33] Y. Zhuang and S. Chiba. Applying DominoJ to GoF Design Patterns. Technical report, Dept. of Math. and Comp., Tokyo Institute of Technology, 2011.

# Context Traits

## Dynamic Behaviour Adaptation Through Run-Time Trait Recomposition

Sebastián González, Kim Mens
ICTEAM Institute
Université catholique de Louvain, Belgium
{s.gonzalez,kim.mens}@uclouvain.be

Marius Colăcioiu, Walter Cazzola
Department of Computer Science
Università degli Studi di Milano, Italy
{cazzola,colacioiu}@di.unimi.it

## ABSTRACT

Context-oriented programming emerged as a new paradigm to support fine-grained dynamic adaptation of software behaviour according to the context of execution. Though existing context-oriented approaches permit the adaptation of individual methods, in practice behavioural adaptations to specific contexts often require the modification of groups of interrelated methods. Furthermore, existing approaches impose a composition semantics that cannot be adjusted on a domain-specific basis. The mechanism of traits seems to provide a more appropriate level of granularity for defining adaptations, and brings along a flexible composition mechanism that can be exploited in a dynamic setting. This paper explores how to achieve context-oriented programming by using traits as units of adaptation, and trait composition as a mechanism to introduce behavioural adaptations at run time. First-class contexts reify relevant aspects of the environment in which the application executes, and they directly influence the trait composition of the objects that make up the application. To resolve conflicts arising from dynamic composition of behavioural adaptations, programmers can explicitly encode composition policies. With all this, the notion of context traits offers a promising approach to implementing dynamically adaptable systems. To validate the context traits model we implemented a JavaScript library and conducted case studies on context-driven adaptability.

## Categories and Subject Descriptors

D.3.3 [**Programming Languages**]: Language Constructs and Features

## General Terms

Design, Languages

## Keywords

Context-oriented programming; Dynamic composition; Run-time behaviour adaptation; Software traits

## 1. INTRODUCTION

Observing that traditional programming languages do not facilitate the construction of adaptable applications, Context-Oriented Programming (COP) [7] has emerged as a response by providing dedicated programming abstractions to define context-specific behaviour. Such behaviour can be recomposed dynamically when context changes are detected, so that applications can adapt to their varying circumstances of use. From a software engineering perspective, COP avoids both scattering and tangling of context-dependent behaviour throughout a program. Indeed, COP was born as a light-weight alternative to Aspect-Oriented Programming [6], albeit for a narrower family of languages.

Whereas existing COP approaches are fine grained in that they permit the adaptation of individual methods, methods seldom make sense on their own. Methods often come in groups of related functionality that should be adapted in unison. A typical example of this are the `push`, `pop` and `isEmpty` methods of stack data structures. Given that methods seldom work as isolated units, it would be more natural to adapt objects through slightly coarser-grained units of behaviour. Such units would make it easier to define *coherent* adaptations, because adaptations for methods that work together would also be bundled together.

When it comes to composing adaptations, typical COP approaches impose a layering order such that more recently activated layers overshadow the methods from less recently activated ones. This makes sense in most cases, though it precludes domain-specific compositions in which different orderings are needed. Traditional composition approaches such as mixins [3] and multiple inheritance [1] also provide insufficient control to reach a desired composition for every situation. In the case of mixins, mixins earlier in the inheritance chain are overshadowed by mixins appearing afterwards. With multiple inheritance, conflict resolution is fixed by the chosen method lookup mechanism.

The previous observations make of *traits* [8] a natural fit as behavioural units of adaptation. Traits are at the right level of abstraction for adaptability, because their intention is precisely to group methods that make sense together and are as cohesive as possible. Furthermore, traits have been conceived to expose composition conflicts, rather than automatically solving them through some fixed strategy imposed by inheritance or layering. Since any possible method combination can be obtained if needed, traits open up new possibilities for behaviour reconfiguration and thus foster adaptability.

The main contribution of this paper is a proposal for managed object recomposition at run time, aimed at supporting the adaptation of software to its context of use. In response to context changes, the composition of objects is modified to opportunistically improve or gracefully degrade the global behaviour of a system. To achieve this goal, we propose a novel *context traits* model, which as its name suggests is at the crossroads of COP and traits. Both of these composition mechanisms deal with important limitations of current-day programming languages: COP provides the necessary machinery to reify contexts and adapt behaviour dynamically, and traits provide a convenient unit of composition with well-defined composition semantics. To fully integrate the COP and trait worlds, we build a causal connection between contexts and object compositions, so that context changes detected in the environment are reflected directly as changes in the composition of objects. To manage such a dynamic system, we allow the definition of custom composition policies.

Since our approach proposes a highly dynamic object model, we chose JavaScript as implementation medium to prototype and test our ideas. The object core of JavaScript proved to be a good paradigmatic match to support the semantics we were after. However, to emphasise that the approach could be implemented in other object-based languages as well, the paper clearly separates the description of concepts from their implementation in JavaScript. For validation we implemented three case studies, each targeted at assessing different properties of the approach. From the experience we have gathered so far, context traits seem to bring advantages that cannot be observed in the COP and trait worlds alone.

The next section gives a conceptual description of our approach. Section 3 then outlines our current implementation in JavaScript. Section 4 explains how we validated the approach through case studies. Section 5 discusses design issues and future work. Finally, we present related approaches in Section 6, and conclude the paper in Section 7.

## 2. CONTEXT TRAITS

Today's software systems call for new computation models in which objects can adapt dynamically to the changing circumstances in which they run. In such a vision, objects can be regarded as evolving entities that are shaped by their environment —much like real life evolution shapes living entities in natural ecosystems. Under this metaphor, it is natural to think of objects as evolving compositions of *traits*, which are selected so that the object fits better in its changing logical and physical environment. This section explains such an adaptation mechanism, starting from core concepts and building up to the complete approach.

### 2.1 Contexts as Situation Reifiers

Standard object-oriented analysis prescribes that concepts found in the application domain that are relevant to the application should be modelled and represented explicitly as objects [16]. Following this principle, when some situation is of importance to an application and could potentially affect its behaviour, it should be modelled and given first-class status, just as any other relevant concept in the application domain. This is precisely what we do. We put forward the circumstances under which a software system executes as first-class entities that can be dealt with by the software. We thus define *context objects* as representations of the particular situations in which the system may execute [10]. This is in line with dictionary definitions of *context* such as "the situation within which something exists or happens, and

that can help explain it"[1] and "the interrelated conditions in which something exists or occurs".[2]

**Example:** A context BusyCPU can reify the situation that the CPU is highly loaded at the moment. A Connected context is for situations in which there is Internet connectivity. These are general contexts with respect to which a large class of applications can adapt. More specialised contexts are of course also possible. In the case of JavaScript, a Firefox context can model the situation that the program is running under such web browser, whereas a context NodeJS signifies that the script is running on the server-side using the node.js framework. ∎

A context becomes *active* when it is detected that the situation it reifies is taking place, and is otherwise *inactive*. The detection of such situations, and thus the triggering of activation and deactivation of contexts, is usually signalled through events. Events can come from a graphical user interface, network discovery service, sensor polling mechanism, and so forth. Context changes might also be triggered when receiving client requests from the network, or even internally by applications themselves when a logical situation changes within their own execution. However diverse these sources might be, the precise origin of context changes is irrelevant for the sake of this paper. All what matters for our purposes is that contexts are activated and deactivated to reflect as closely as possible the situation under which the system executes.

To manage the activation dynamics of contexts, every context has an *activation count* that is incremented for each activation request and correspondingly decremented for each deactivation request. The context becomes inactive only when the count reaches zero, and is otherwise active. A deactivation request of an inactive context makes no sense, since a situation cannot cease to exist if it was not occurring in the first place; for this reason, such a request is considered an error. However, it can very well be that a context is activated at some point and never deactivated again during the lifetime of a system. Hence, the number of activations is always greater or equal than that of deactivations. Activation counts help us deal with situations that can be caused by different occurrences of the same event. Such *multiply-occurring situations* hold until their last cause disappears from the environment.[3]

**Example:** A Connected context should remain active for as long as there is WiFi, 3G, Edge, or Ethernet connectivity. Connected is deactivated when none of the connectivity media are available. Examples of single-activation contexts are LowBattery (activated when the remaining battery charge goes below a certain threshold) and Firefox (activated when it is detected that the application is running in such browser). Firefox is also an example of a permanent context, which remains active for as long as the system runs. ∎

The *prototypical environment* of execution of the application is modelled through a Default context. The Default context is not just a straightforward application of the principle that relevant situations should be modelled explicitly, but also it brings formal and technical advantages discussed further in the paper. When in Default context, objects correspondingly exhibit their prototypical behaviour [14]. Beha-

---

[1] http://dictionary.cambridge.org/dictionary/british/context_1

[2] http://merriam-webster.com/dictionary/context

[3] Note that activation counts are not related to concurrency, despite their superficial resemblance to semaphores.

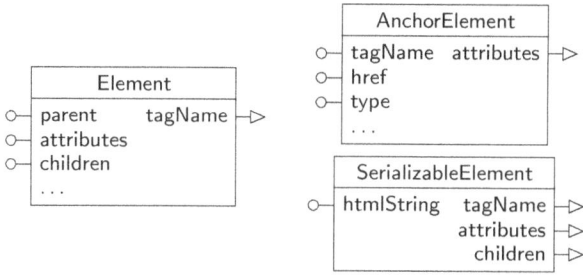

Figure 1: Traits for an HTML DOM library.

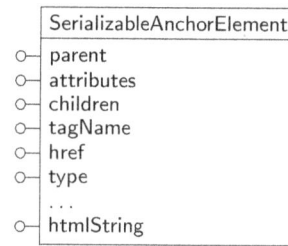

Figure 2: Result of **Element** + **SerializableElement** + **AnchorElement**, a plain trait composition.

vioural adaptations to other contexts are usually expressed as variations with respect to default behaviour. We explain next the terms in which such default and adapted behaviour can be defined.

## 2.2 Traits as Basic Behavioural Units

The kind of objects we propose for dynamic software adaptation are composed of traits [8]. Traits are behavioural units originally designed to facilitate software reuse. They are finer-grained than whole objects, because traits contain functionality that is as small and cohesive as possible. For instance, a Stack trait can contain the typical push, pop and isEmpty operations of any stack-like data structure —nothing more. Whole objects, in contrast, usually have broader functionality, some of which is shared with other objects. For instance, a full stack object normally features all base methods from its Collection hierarchy.[4]

The model of traits we use is a slight variation of the original [8], since our model is meant for classless objects, and more importantly, our emphasis is not on reuse but rather on dynamic recomposition. At the core, our notion of *trait* is defined just like in classical traits, including their formalisation [8]. Intuitively, a trait *provides* a minimal set of methods that semantically go together, and *requires* certain methods that it uses to fill in the remaining details of its internal implementation. Required methods are left open by trait developers for customisation by clients, akin to hook methods in traditional object-oriented frameworks.

**Example:** The depictions in Figure 1 are part of a trait-based design of a Document Object Model (DOM). For each trait, the methods on the left of the box are provided, and those on the right are required. The Element trait provides the bare minimum any DOM node must have, namely access to the parent, attributes and children of the node. However, since it is a generic node, the tagName is unspecified (i.e. it is required). The AnchorElement trait models a hypertext link, which knows its tagName (in this particular case, 'a') and anchor-specific attributes such as href and type. It reads the necessary information from a required attributes structure. Finally, the SerializableElement trait provides a method htmlString which produces a representation of the node <tagName attr$_1$="...", attr$_2$="...">...</tagName>, provided that it has access to the necessary properties. ■

Our model of traits benefits from simple object-based underpinnings. Unlike the classical model of traits, we have no notion of *dictionary*, *class* and super-sends. Hence, the relatively complex interaction between traits and classes [8]

needs not be considered (e.g. how methods of classes take precedence over methods of traits, and methods of traits over those of superclasses). This removed complexity makes our model simpler and homogeneous.

Formally, a *trait* is a function $T : \mathcal{N} \to \mathcal{B}^*$ that maps a set of method names $\mathcal{N}$ to a set of method bodies $\mathcal{B}^*$, just like in classical traits [8]. The set $\mathcal{B}^*$ includes an *undefined* method ($\bot$) and an *overspecified* method ($\top$) used to represent respectively required methods and conflicting methods. A trait is *free of conflicts* if none of its methods is $\top$.

Traits are composed by *summing* them ($+ : \mathcal{T} \times \mathcal{T} \to \mathcal{T}$), which amounts to taking the respective unions of all their provided and required methods. Since required methods specify only a signature and have no implementation, they cannot cause conflicts when composed. In contrast to required methods, conflicts between provided methods may arise if they share the same signature but have different implementations. Conflicts must be resolved explicitly, which avoids the limitations of the automatic linearisation mechanisms imposed by composition of mixins, and classes with multiple inheritance [8].

**Example:** The trait composition Element + SerializableElement + AnchorElement gives as result the trait SerializableAnchorElement illustrated in Figure 2. It constitutes a complete hypertext link that can be rendered as plain HTML, thanks to the provided method htmlString. ■

The main mechanism to avoid composition conflicts is *overriding* ($\rhd : \mathcal{T} \times \mathcal{T} \to \mathcal{T}$) of one trait over another. Since we don't have the notion of dictionaries, our definition differs slightly from that of classical traits:

$$(T_1 \rhd T_2)(a) = \begin{cases} T_2(a) & \text{if } T_1(a) = \bot \\ T_1(a) & \text{otherwise} \end{cases}$$

For cases in which overriding is not expressive enough, traits can be modified through *aliasing* ($\to : \mathcal{T} \times (\mathcal{N} \times \mathcal{N}) \to \mathcal{T}$), and *exclusion* ($- : \mathcal{T} \times \mathcal{N} \to \mathcal{T}$), which are defined as in classical traits [8].

Thanks to the composition operators, the final result of a trait composition may arbitrarily mix methods starting from a set of source traits $\{T_0, \ldots, T_n\}$. Such compositions preserve the *flattening property* of traits [8], though rephrased in object-based terms: for any object $o$ composed of traits $\{T_0, \ldots, T_n\}$, every non-overridden method in a trait $T_i$ has the same semantics as if it were implemented directly in the object $o$ in question.

## 2.3 Behaviour is State, and State is Behaviour

The distinction between state and behaviour in trait composition has been discussed in the past on a technical ground: state composition implies changing the memory layout of

---

[4]In the Pharo dialect of Smalltalk, this amounts to 160 methods.

objects, whereas behaviour composition does not [2]. Put this way, the distinction between state and behaviour seems to be the symptom of a *leaky abstraction*: added conceptual complexity caused by underlying implementation concerns. The reason why state composition generally changes memory layout, whereas behaviour does not, is that state is usually inlined in objects, whereas behaviour is usually stored in dictionaries —a more flexible data structure. It has been demonstrated long ago that the leak can be plugged while retaining efficiency [5].

Given the previous observations, we do not distinguish state and behaviour. In some sense, we could say that in our traits model *everything is state*, since methods can be regarded as plain object slots that happen to hold a method body instead of some other value. But we can also reverse this point of view and say that in our traits model *everything is behaviour*, because methods are obviously behaviour and any other value which is not a method can formally be regarded as a method whose only purpose is to return the value in question.[5] It is clear that the views of traits-as-state and traits-as-behaviour are not only reconcilable but are one and the same under different formal and technical guises. The distinction between state and behaviour is thus irrelevant at the formal and design levels of our traits model and its composition system.

## 2.4 Objects as Context-Driven Trait Compositions

In our model, an object $o \in \mathcal{O} \subset \mathcal{T}$ is a dynamically evolving composition of traits

$$o = T_{C_0} \cdot T_{C_1} \cdots T_{C_n}$$

where each $T_{C_i}$ is a trait containing the behaviour specific to a context $C_i$. We call this class of traits meant for context-driven adaptation *context traits*. Their composition operator $(\cdot)$ is introduced further on in this section. A trait $T_{C_i}$ becomes part of the composition of an object when the corresponding context $C_i$ becomes active, and is withdrawn when $C_i$ becomes inactive. Given that objects are obtained as pure compositions of traits, it follows that objects themselves are actually traits, albeit of a particular kind: objects must be *complete*, which means that they are free of conflicts and have no required methods.

According to the previous view of objects, the behaviour of an object $o$ in Default context must be contained in a trait $T_{\mathsf{Default}}$. Since the Default context is always active,[6] the composition of $o$ will always contain $T_{\mathsf{Default}}$. By convention we take $C_0 = \mathsf{Default}$, yielding

$$o = T_{\mathsf{Default}} \cdot T_{C_1} \cdots T_{C_n}$$

When Default is the only active context (i.e. in prototypical conditions), the object is identical to its default trait: $o = T_{\mathsf{Default}}$. For this reason, $T_{\mathsf{Default}}$ must be free of conflicts and have no required methods. The homogeneous treatment of default behaviour as just another trait renders the conceptual model purer and simplifies its implementation.

It should be noted that our model does not preclude the original use of traits for software modularisation and reuse,

since a context trait $T_C$ might be statically composed of other traits. Although perfectly admissible, for the sake of dynamic adaptation such static compositions need not be exposed and considered. The static structure of traits is opaque (irrelevant) to the run-time composition infrastructure, which needs to manage only those traits that participate in the dynamically evolving composition of objects.

The run-time system uses a total *object constitution* function constitution : $\mathcal{L} \times \mathcal{C} \to \mathcal{T}$, to retrieve the trait $T_{(l,C)} = $ constitution$(l, C)$ that adapts an object with identity $l$ to a certain context $C$. $T_{(l,C)} = \varnothing$ if the object designated by $l$ has no adaptation for context $C$. When the identity $l$ of an object is immaterial, we omit it from the notation $T_{(l,C)}$ and simply write $T_C$, like we did earlier in this section. Also for simplicity, we overload constitution when applied to a set of contexts, such that it is mapped over the given set: $T_{\{C_0,\ldots,C_n\}} = \{T_{C_0}, \ldots, T_{C_n}\}$.[7] Thanks to the constitution function, the system is able to retrieve the traits that make up an object for any given set of active contexts.

Note that the constitution function is not necessarily injective, meaning that two different contexts $C_1$ and $C_2$ can have a same associated trait $T_{(l,C_1)} = T_{(l,C_2)}$ for a given object $l$, and also that two different objects $l_1$ and $l_2$ can have the same trait $T_{(l_1,C)} = T_{(l_2,C)}$ for a given context $C$. In other words, a same trait can be reused as adaptation in different contexts and for different objects. Reusability of context traits can therefore leverage both the object-oriented and the context-oriented axes of our approach.

The *dynamic composition* operator $(\cdot)$ used previously is infix notation for the function composition : $2^{\mathcal{T}} \to \mathcal{O}$ which takes a set $\{T_{C_0}, \ldots, T_{C_n}\}$ of traits and returns a composed object $o$. Since composition takes a plain set as argument, the order in which traits are composed is irrelevant, and therefore the dynamic composition operator is commutative by definition. The dynamic composition function is defined as composition$(S) = \Sigma P_g(S)$, which is the summation[8] of the conflict-free set of traits obtained by applying the global composition policy $P_g$ to the given set of traits $S$. Composition policies are explained in Section 2.5.

The object constitution and object composition functions lead to the straightforward definition of the *object definition* function, object : $\mathcal{L} \times 2^{\mathcal{C}} \to \mathcal{O}$, as the function composition object = composition $\cdot$ constitution. Quite literally, an object is defined as the composition of its constitution. Applying the previous definitions this means

$$o_{(l,\{C_0,\ldots,C_n\})} = \Sigma P_g(\{T_{C_0}, \ldots, T_{C_n}\})$$

That is, an object is the controlled composition as per $P_g$ of its constituent traits for a certain set of active contexts. When the identity $l$ of the object is immaterial, and assuming that $A = \{C_0, \ldots, C_n\}$ is the set of currently active contexts, we denote $o_{(l,A)} = $ object$(l, A)$ simply as $o$, as was done earlier in this section.

The definition of object emphasises that all what is constant to an object is its identity, since its behaviour (i.e. trait composition) changes according to context. Also, the object function puts contexts in direct relationship with object compositions. The object function therefore plays a key role in establishing a *causal connection* [17] between the

---

[5]In this view of traits as pure behaviour, assigning a value to a method (e.g. point.x = 3) can be interpreted as replacing the method's body by another one (e.g. the body return 3).

[6]Our model does not forbid the deactivation of the Default context, but we have not found a use so far.

[7]Naturally, this overloaded definition of the function has signature constitution : $\mathcal{L} \times 2^{\mathcal{C}} \to 2^{\mathcal{T}}$.

[8]The $\Sigma$ operator is just a convenient notation for the $+$ operator introduced in Section 2.2.

composition of the computational system and the situation or circumstances in which the system executes.[9] Thanks to this causal connection, the physical and logical environment in which objects live literally shapes those objects, and as a result, the system as a whole.

## 2.5 Composition Policies

The setting for which we use trait-based composition is fundamentally different from the setting assumed by similar existing techniques. Whereas other approaches expect programmers to manually compose traits as a normal part of the application development cycle, we aim at composing traits automatically during system execution, so that objects are shaped according to their changing environment. In a system that is already deployed and running, trait composition cannot possibly be assisted, and thus any conflict arising from trait composition must be handled by the system itself. Hence, the knowledge on how to resolve composition conflicts must be somehow embedded into the system.

In our approach, developers can encode their strategies to conflict resolution as *composition policies* that the system can use during trait recomposition. Composition policies thus encapsulate domain-specific knowledge needed to resolve composition conflicts at run time. Through policies, developers can encode their conflict resolution strategies as first-class entities, and ship them together with the rest of the system or module they are developing.

A composition policy is a total function $P : 2^{\mathcal{T}} \to 2^{\mathcal{T}}$ that maps a possibly conflicting set of traits $S = \{T_1, \ldots, T_n\}$ to a partial resolution set $P(S) = \{R_1, \ldots, R_n\}$. Policies are free of side effects on their argument traits —that is, $P$ does not alter the definition of the given traits $T_1, \ldots, T_n$ in any way. The partial resolution traits $R_i$ are the result of composing the traits in $S$ in a way the policy deems appropriate. To this end, policies use regular trait composition operators exclusively (i.e. summation, overriding, aliasing and exclusion explained in Section 2.2). This means in particular that policies do not introduce foreign traits to the trait set $S$ being composed.[10]

A *(total) resolution* $R$ is a set containing exactly one conflict-free trait resulting from application of a policy. The trait in a resolution $R = P(S)$ can be used for composition instead of the original traits found in $S$. A policy $P$ that resolves every possible conflict set $S$ to a total resolution $R$ is said to be *general*. General policies can hardly be domain-specific, since they should be able to resolve arbitrarily diverse sets of traits. Section 2.6 gives an example of a general policy.

Composition of policies follows normal function composition: given two policies $P_1$ and $P_2$, $P_1 \cdot P_2$ is a composed policy. We call such compositions $P_c = P_1 \cdot P_2 \cdots P_{n-1} \cdot P_n$ *policy chains*. A policy chain $P_c$ processes a given conflict set in order, starting from $P_n$ and going through the chain until $P_1$ yields the final partial resolution of the original conflict set. Policy chains allow the aggregation of independently developed resolution strategies.

The context-driven composition machinery of our approach uses a global policy chain $P_g$ to resolve conflicts. Each partial policy $P_i$ that makes up $P_g$ encodes a particular (possibly domain-specific or object-specific) strategy to solving conflicts. To make sure that conflicts are always resolved, the system closes the chain with a general default policy $P_d$ that will solve any remaining conflict not handled by policies earlier in the chain. The global policy chain $P_g = P_d \cdot P_1 \cdots P_n$ is thus never empty, since it contains at least $P_d$.

## 2.6 Resolution by Activation Age

The general policy that we have found to make the most sense to be used as default policy $P_d$ relies on the intuition that contexts activated more recently in the system are more relevant than contexts activated long ago, since recently activated contexts are "hot" and correspond to situations that are just arising, whereas long-activated contexts are "cold" and correspond to background circumstances which are more stable. If a context has just been activated, the adapted behaviour it defines is probably better suited to the current situation than the behaviour of related though less recent adaptations.

To encode the previous observation as a policy, we define the *activation age* of an active context as a measure such that, if the last activation request for context $C_i$ happened more recently than the last activation request for context $C_k$, then the age of $C_i$ is less than the age of $C_k$, or equivalently, $C_i$ is *younger* than $C_k$. Note that the Default context is always the oldest context in the system, since it is activated at the start of execution. We make sure that two active contexts can never have the same age measure, so that the activation age defines a total order $<_a$ for any set of active contexts.

Given a set of conflicting traits $S = \{T_{C_1}, \ldots, T_{C_n}\}$ that adapt the behaviour of an object $o$, the *youngest trait composition policy* $P_a$ yields a resolved trait formed by favouring younger traits over older traits, as per activation age of their respective contexts. Assuming without loss of generality that $C_i <_a C_{i+1}$ for $1 \le i < n$, $P_a$ is defined as

$$P_a(\{T_{C_1}, \ldots, T_{C_n}\}) = \{T_{C_1} \triangleright T_{C_2} \triangleright \ldots \triangleright T_{C_{n-1}} \triangleright T_{C_n}\}$$

The resulting set contains a single conflict-free trait. Hence, the activation age policy $P_a$ is general, because it always returns a total resolution, and it can thus be used as default policy $P_d$.

The youngest trait composition policy $P_a$ produces a sort of layering between contexts, since all traits from a younger context $C_1$ will override all traits from an older context $C_2$. We therefore obtain the same semantics found in context-oriented systems based on layers [7] and delegation [11], though the ordering is not fixed: our approach allows to adjust it as needed on a domain-specific basis, by adding customised policies to the global policy chain $P_g$.

## 2.7 The Independent Behaviour Extensibility Challenge

Plain trait composition makes it possible to define adaptations that *replace* the original behaviour of an object through the override operator $\triangleright$ discussed previously. Such adaptations do not invoke the original behaviour, which is thus not manifested by the adapted object. In practice, however, many (if not most) behavioural adaptations need to be defined as *extensions* of the original behaviour of an object.

---

[9]The other key ingredient needed to establish this causal connection is the function that detects changes in the execution environment through events, and triggers context activations accordingly. As mentioned in Section 2.1 however, this paper does not discuss context detection.

[10]This might seem like a strong limitation, but our goal is to keep a simple semantics such that traits are introduced in a composition only because of active contexts, not policies.

Such extensions invoke the original behaviour of the object they adapt as a regular part of their adaptation logic.

Any approach to adaptation that allows behaviour extension thus needs to provide a mechanism to invoke overridden behaviour. In a plain trait-based approach, however, any attempt to compose an adaptation $T_2$ as an extension of another adaptation $T_1$ will result in conflicts for all extended methods, since the extended methods are precisely meant to have the same signature as the overridden methods. In the static composition setting of classical traits, such conflicts could be resolved by having $T_2$ override $T_1$, after having aliased each conflicting method of $T_1$ so that it can still be accessed from $T_2$, and —crucially— changing the code of each extension method in $T_2$ so that it uses the alias name instead of the original name to invoke overridden behaviour. In trait notation, this composition would be $T_2' \rhd (T_1[b_1 \to a_1][b_2 \to a_2] \ldots [b_n \to a_n])$, where $T_2'$ is the modified version of $T_2$, $a_i$ is the name of each extended method, and $b_i$ is the corresponding alias. Note that not only does the code of $T_2$ need to be modified by the programmer for the composition to work, but also the particular modifications needed in $T_2$ depend on the particular conflicts it has with $T_1$, and thus different modifications would be needed if $T_2$ were to extend some other trait $T_3$ with (for instance) one more or one less conflicting method. Plain trait composition was not designed to support *independent* behaviour extension, but it is hard to go without in a model where traits can be composed dynamically with other traits.

## 2.8 Extensibility Through proceed

Dynamic adaptation requires independent behaviour extensibility —once the software is deployed and running, programmers can no longer intervene to refactor their developed adaptations according to particular composition needs. Software extension by method aliasing as proposed originally for plain traits is not an option. Yet traits are all about software composition and reuse. We resolve this dilemma by plugging independent behaviour extensibility into traits in a way that leverages the existing traits composition mechanism, so that it does not need to be modified nor extended.

Any method $m_e$ designed to be a behaviour extension can send the message proceed to self to signal the invocation of overridden behaviour. This is akin to advice composition through proceed in AOP [12], method composition through call-next-method in CLOS [13], and super calls in typical object-oriented languages. Following regular trait semantics [8], such a self-send implies that proceed becomes required by the trait $T_e$ owning the method $m_e$:

Any such trait that requires proceed is called an *extension* trait. This makes sense conceptually: the trait is intended to be composed with the behaviour it extends, and thus is by itself incomplete —the trait requires original behaviour to work properly, and this is signalled through proceed. Also in keeping with the intention of required methods, it is not known in advance which concrete implementation of the overridden behaviour will be used for composition, since this decision depends on the context.

The composition of two extension traits is itself an extension trait, since the proceed requirement is propagated to

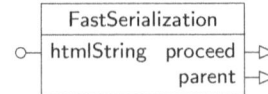

Figure 3: Fast DOM serialization as behavioural extension of plain DOM serialization.

the composed trait. To produce complete compositions, the proceed requirement should be fulfilled somehow. To comply with this requirement, we include as part of every dynamic trait composition an extensibility trait which provides such method:

The Extensible trait provides proceed, and given that it has no requirements, it will never render the composition in which it participates incomplete. Extensible is not only a technical solution: conceptually, any object that can be extended should feature this trait, much like an object that includes a Reflective trait to allow reflection, for example. Rather than having programmers add Extensible to every object they develop, the system does it automatically for them at dynamic composition time. Extensible is thus a system-defined part of the default trait $T_{\text{Default}}$ discussed in Section 2.4; in keeping with that discussion, Extensible is part of the prototypical behaviour of any object. To prevent that programmers inadvertently override the version of proceed provided by Extensible with behaviour that breaks the expected semantics, proceed needs to become a reserved method name.

**Example:** For efficiency, we would like to cache the serialisation of HTML nodes in a DOM. To this end, we introduce as illustrated in Figure 3 a FastSerialization trait extension. This extension features a specialised htmlString method that calls proceed to obtain a serialised representation of the current node, which it caches and uses upon the next invocation instead of calling proceed again. The serialisation cache invalidates itself when the structure of the DOM node changes (for instance, as a result of calling appendChild), and propagates invalidation to the parent node. Note that FastSerialization introduces parent as a new requirement that the extended trait SerializableElement does not have. Trait extensions are not restricted in this sense, and can introduce new requirements as needed. ∎

The behaviour that proceed invokes can be the original behaviour $m_o$ of the object, but it can also be yet another extension method that is part of the current composition. In principle it is unknown in advance which of the other available implementations of the current method will be invoked by proceed. Fortunately, we can rely on the conflict resolution machinery to perform method composition, as explained in the next section.

## 2.9 Resolution of proceed Through Policies

Consider two traits $T_1$, $T_2$ that conflict for a name $a$. They define implementations $T_1(a) = m_1$, $T_2(a) = m_2$ such that $m_1 \neq m_2$ (else it would not be a conflict). Apply the global policy $P_g$ to obtain a resolution $\{T_R\} = P(\{T_1, T_2\})$; $T_R$

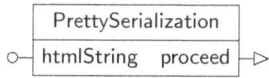

Figure 4: Definition of trait **PrettySerialization** that conflicts with trait **FastSerialization**.

```
1  var cop = require('context-traits');
2  var Untrusted = new cop.Context();
3  var LowBattery = new cop.Context({
4    name: 'low battery',
5    description: 'The remaining battery charge is low.'});
```

**Snippet 1: Context creation.**

will map $a$ to either $m_1$ or $m_2$.[11] Assume without loss of generality that $T_R(a) = m_1$. This means that $m_1$ is more specific than $m_2$ according to the policy $P_g$, which induces an order $m_1 < m_2$. For a more general set of conflicting traits $\{T_1, \ldots, T_n\}$, $P_g$ can be applied pairwise to induce a total order $m_1 < m_2 < \cdots < m_n$. Any general policy induces such an order on a set of methods in this way.[12]

For the case of the youngest trait composition policy $P_a$ explained in Section 2.6, the induced order will correspond to activation age. However, the global policy $P_g$ of which $P_a$ makes part can have small alterations or "tweaks" of this activation age order, according to the partial policies $P_i$ that currently make part of the chain.

The resulting total order induced by the global policy $P_g$ determines which method should be called next by proceed. The set of methods $\{m_1, \ldots, m_n\}$ that implement the same message as the currently executing method $m_c$ is ordered as a chain $m_1 < \cdots < m_c < m_{c+1} < \cdots < m_n$. Having found the position of $m_c$ in the chain, proceed can then invoke $m_{c+1}$. These chained invocations will end with the first method $m_i$ in the chain that does not invoke proceed. This could be a behaviour replacement introduced as an adaptation to the current context, but else it will be the default behaviour $m_d$, and it will be located at the end of the chain since the Default context is always the oldest.

**Example:** For readability, we want to format the HTML output produced by the htmlString method of trait SerializableElement. Figure 4 depicts a PrettySerialization trait that extends the original behaviour through the use of proceed. With this definition, traits PrettySerialization and FastSerialization have a conflict for method htmlString. Because the preferred order should be to cache the formatted output (as opposed to formatting the cached output every time again), the programmer can define a policy

$$P_{\text{Fast}}(\{\text{FastSerialization}, \text{PrettySerialization}\})$$
$$= \{\text{FastSerialization} \triangleright \text{PrettySerialization}\}$$

This domain-specific policy can be composed as part of the global policy $P_g$. ∎

This concludes the conceptual explanation of our approach to dynamic behaviour adaptation through run-time trait recomposition. In the next section we describe how this conceptual model has been implemented in JavaScript.

## 3. CONTEXT TRAITS IN JAVASCRIPT

In our context traits model, objects are purely composed out of traits, and their composition varies dynamically according to detected context changes. To lower the barrier

---

[11] Recall from Section 2.5 that policies do not introduce foreign traits into a composition, and thus a third method $m_3$ cannot appear.

[12] Note that the policy cannot be used more generally to induce an order on a set of traits, because the resolution $R = P(\{T_1, T_2\})$ does not necessarily return either $\{T_1\}$ or $\{T_2\}$; it can return a new composed trait $\{T_3\}$ that mixes methods from both.

for prototyping such an object-centric and dynamic model, a natural choice was JavaScript and the existing traits.js library [26]. Section 3.1 explains the implemented model from a programmer's perspective, and Section 3.2 discusses salient elements of the underlying implementation.

### 3.1 Main Mechanisms

We first present the main tools provided to JavaScript programmers to define, organise and discover contexts, to adapt behaviour to contexts, and to specify strategies for resolving composition conflicts. To make this functionality available, Snippet 1 (Line 1) shows how to import the context-traits.js module, thenceforth referenced through the cop variable.

#### 3.1.1 Context Creation

Snippet 1 shows how to create a minimal Untrusted context (Line 2), and how to create a LowBattery context (Line 3) with initialisation parameters. Note that, by convention, context variable names start with uppercase. This is to emphasize that these contexts are singleton objects —a very important property for interoperation that we discuss in Section 3.1.2. Since contexts are singletons, the classless object model of JavaScript is a better paradigmatic match than a class-based one, because we do not incur the creation of a class for the sole purpose of creating a context.

#### 3.1.2 Context Organisation

Contexts constitute a sort of shared vocabulary that allow different applications —or independently developed parts of the same application— to collaborate by defining adaptations with respect to the *same* specific situation [11]. It is crucial for context-oriented systems to be defined in terms of well-known contexts, since these contexts are common points of reference to define behavioural adaptations. Luckily, JavaScript —like any other object-based language— allows the use of objects as name spaces [23] to name and categorise contexts:

- cop.contexts.resources.memory.LowMemory
- cop.contexts.environment.acoustics.Noisy
- cop.contexts.programming.testing.FaultInjection
- cop.contexts.user.activities.Meeting

Name clashes are avoided by separating contexts in different domains, effectively constituting a shared vocabulary of prototypical contexts. This being said, local and vendor-specific contexts are also possible and useful for internal adaptations.

#### 3.1.3 Context Discovery

Context *detection* or *discovery* goes from straightforward to very sophisticated, depending on the nature of each context. For instance, whereas detecting contexts like Evening or Firefox is fairly simple, detecting UserSleepy may involve advanced image recognition and machine learning techniques. As mentioned in Section 2.1, context discovery matches well an event-based programming style, in which context activation and deactivation follow the occurrences of events. Snippet 2 illustrates the detection of LowBattery by registering a

215

```
1  window.addEventListener('batterystatus',
2    function (info) {
3      if (info.level <= 30) LowBattery.activate();
4      else LowBattery.deactivate(); })
```

**Snippet 2: Context detection.**

```
1  FastSerialization = Trait({
2    htmlString: function() {
3      var result = this.retrieve(); // probe cache
4      if (result == undefined) {    // cached?
5        result = this.proceed();     // get serialisation
6        this.store(result); }        // cache it
7      return result; }});            // return stored value
8
9  Production.adapt(HTMLElement, FastSerialization);
```

**Snippet 3: Object adaptation to context.**

listener to the event **batterystatus** of the **PhoneGap** library.[13]
A crucial point regarding context detection is that any associated code for detection, such as the conditional statement in Snippet 2, is written only *once*, rather than being scattered throughout the application at each point where behaviour depends on such detection.

### 3.1.4   Adaptation to Context

As explained in detail in Section 2.4, our approach is based on the definition of traits as behavioural adaptations to context. An object can specify a trait that contains context-dependent behaviour, so that the adapted behaviour becomes available when the context is activated, and the behaviour is withdrawn when the context deactivates. Snippet 3 shows the definition of the trait **FastSerialization** (lines 1–7), which was depicted in Figure 3. The call to method **adapt** in Line 9 suffices to register **FastSerialization** as adaptation of **HTMLElement** for a **Production** environment. The application thus applies such optimisations when running in production mode, but can run without if the **Production** context is inactive.

Since **Default** is a context like any other, it is possible to specify the default serialisation behaviour of the **HTMLElement** object using the same machinery for adaptation to other contexts:

```
cop.contexts.Default.adapt(HTMLElement, Serializable);
```

With this specification, HTML elements will be serialisable by default. Since the **Default** context is always active, the adaptation **SerializableElement** will be applied immediately to the object **HTMLElement**. Note the use of a name space in the adaptation expression: the **Default** context used here is one and the same used for every application, as discussed in Section 3.1.2.

Snippet 3 (Line 5) also illustrates the **proceed** mechanism in action, explained in detail in Section 2.8. The **proceed** method is used to invoke the original version of **htmlString** belonging to **HTMLElement**, which actually comes from the composition with **SerializableElement** shown previously.

### 3.1.5   Conflict Resolution

An adapted object can have different traits for different contexts. At run time, objects acquire traits depending on the set of currently active contexts. When those traits are composed, conflicts can arise if two or more traits provide a property with the same name but different values. To resolve conflicts, a composition policy can be specified as

---

[13]See http://phonegap.com.

```
1  cop.manager.addObjectPolicy(HTMLElement,
2    [FastSerialization, PrettySerialization]);
3
4  cop.manager.addObjectPolicy(HTMLElement,
5    [Production, Browser],
6    return function (FastSerialization, PrettySerialization) {
7      return Trait.override
8            (FastSerialization, PrettySerialization); });
```

**Snippet 4: Definition of a composition policy.**

illustrated in Snippet 4. Lines 1–2 show a shorthand invocation of the method **addObjectPolicy**, in which an array of traits [FastSerialization, PrettySerialization ] specifies a desired overriding order. This order is specific to object **HTMLElement**. Note that this policy is more specific than the policy given as example in Section 2.9, since the policy illustrated here is effective only for **HTMLElement** and not in general for every object.

Lines 4–8 in Snippet 4 show an invocation of **addObjectPolicy** in which an explicit function for conflict resolution is passed as argument. For illustration purposes, the passed function makes the policy functionally equivalent to the one of lines 1–2. The arguments passed to the resolution function (Line 6) are the traits corresponding to the contexts passed as arguments of **addObjectPolicy** (Line 5). For example, **Browser** is associated to **PrettySerialization**. The **Trait.override** method in Line 7 is provided by the traits.js library [26]. In general, any composition using summation, overriding, aliasing and exclusion is possible.

## 3.2   Implementation Notes

### 3.2.1   Context Activation Age

To implement the activation age measure described in Section 2.6, it is inconvenient to use a physical time stamp, because it depends on the resolution of the system clock. If the activation of two contexts happens so closely in time that they fall within the same clock tick, it will appear as if they were activated simultaneously. Rather than a physical measure, we use a logical one, defined with respect to the total number of activations that have taken place in the system. When any context is activated, the total activation count is incremented and the context is stamped with this new value. At any given time, the activation age of a context is the difference between the total activation count and the value stored in the context. If no other activations have taken place in the meantime, the age will be 0, meaning that this is the youngest context for the moment.

### 3.2.2   How To proceed

As explained in Section 2.8, the contract of **proceed** is that "it invokes the behaviour that has been overridden by the current extension method". The question remains as to how such contract can be implemented in practice. In languages that allow introspection of method invocations, such as Smalltalk, the **proceed** mechanism can be implemented straightforwardly.[14] For languages with less reflective power such as JavaScript, there is a less immediate though still

---

[14]In Smalltalk, the **proceed** method can have access to the **MethodContext** using the **thisContext** pseudo-variable. From the method context, it is possible to introspectively find the extension method that invoked **proceed**, and also recover the receiver and arguments used in this invocation. The **proceed**

216

```
1   traceable = function(name, method) {
2     wrapper = function() {
3       var invocations = cop.manager.invocations;
4       invocations.push([name, arguments, wrapper]);
5       try { return method.apply(this, arguments); }
6       finally { invocations.pop(); } };
7     return wrapper; };
8
9   proceed = function() {
10    var invocations = cop.manager.invocations;
11    if(invocations.length === 0) throw new Error
12      ('Proceed must be called from an adaptation');
13    var name, args, method;
14    [name, args, method] = invocations.top();
15    args = arguments.length === 0 ? args : arguments;
16    var alternatives = cop.manager.orderMethods(this, name);
17    var index = alternatives.indexOf(method);
18    if (index === -1) throw new Error
19      ('Cannot proceed from an inactive adaptation');
20    if (index+1 === alternatives.length) throw new Error
21      ('Cannot proceed further');
22    return alternatives[index+1].apply(this, args); }
```

**Snippet 5: Core of the proceed mechanism.**

valid solution, provided that the language features closures. Closures make it possible to build an auxiliary stack that mimics the method invocation stack. This ad-hoc stack can then be consulted by proceed.

The key parts of this technique are shown in Snippet 5. The traceable method is in charge of wrapping methods so that they become traceable —that is, their invocation is registered in the ad-hoc invocation stack. To achieve this, traceable takes a method name and implementation as arguments (Line 1), and creates an anonymous closure that wraps the given method (Line 2). The wrapper first fetches the ad-hoc stack (Line 3), and records the method invocation (Line 4).[15] The wrapper then invokes the wrapped method, preserving the receiver and arguments, and returns the result (Line 5). The invocation is removed from the stack after normal termination of the method, or even if it throws an exception (Line 6). The result of traceable is the wrapper method (Line 7). The net effect of traceable methods is simply that their invocation is reflected in the ad-hoc stack.

The traceable method just explained works hand in hand with the proceed method (Line 9). First of all, proceed fetches the ad-hoc invocation stack (Line 10), and checks whether it has been invoked from a method that makes part of an adaptation (lines 11–12). The necessary information corresponding to the current invocation is then read from the top of the ad-hoc stack (Line 14; recall that the stack is maintained by wrapper methods in lines 4 and 6). The assignment in Line 14 uses JavaScript's destructuring syntax to assign the three variables at once. If the user passes explicit arguments to proceed, these are used instead of those in the current method invocation (Line 15). Having obtained the method name corresponding to the current invocation, the run-time composition system finds out the methods that are applicable for the current receiver and method name (Line 16). Normally the arguments of the current invocation (held in the args variable) should also be passed to orderMethods along with the receiver and method

---

method can then invoke the overridden behaviour, passing the same receiver and arguments.

[15]The arguments local variable used in traceable is available within all JavaScript functions and needs not be declared.

```
1   Encrypter = Trait({
2     store: function(key, value) {
3       this.proceed(key, this.encrypt(value)); },
4     lookup: function(key) {
5       var encrypted = this.proceed();
6       if (encrypted)
7         return this.decrypt(encrypted); }});
8
9   Untrusted.adapt(cacheServer, Encrypter);
```

**Snippet 6: Adaptation of cache server to Untrusted.**

name, but in JavaScript arguments do not play any role in determining the applicability of methods, so they are ignored. The result is a list of method alternatives which is ordered according to the current composition policy, following the semantics explained in Section 2.9. In this list, the position of the currently executing method (the one which invoked proceed) is determined (Line 17), and unless there are any errors (lines 18–21), the next method in order of specificity is finally invoked and the result is returned (Line 22).

Methods need to be wrapped (i.e. made traceable) only when they override other methods —that is, when two or more traits provide the same method in a composition. This means the run-time overhead of wrappers is incurred only when necessary. Despite the added complexity of this solution as compared to a reflection-based approach, the technique just explained for implementing proceed can be ported to any language that supports closures, and is thus fairly generic.[16]

## 4. CASE STUDIES

Our proposal of context traits is at the crossroads of COP and traits. By conducting a number of small case studies, three of which are presented in this section, we tested the correctness of our implementation and explored some of the advantages this novel paradigmatic mix has to offer.

### 4.1 Cache Server

As a first fairly typical scenario for context-orientation, inspired by an example of Ghezzi et al. [9], we took the case study of an adaptive cache server. The server offers caching functionality for speeding up retrieval of information requested by other machines, and exposes a typical set of primitives used to store and retrieve generic resources, such as store(key, value), lookup(key) and remove(key). The cache server is adaptive: when in Untrusted context, the server activates an Encrypter trait that ciphers a resource before it is stored in the cache, and deciphers it upon retrieval; in Volatile context, the server backs up resources on persistent storage in order to be resilient to likely failures (Persistent trait); finally in context Auditing, the server logs each significant operation it performs (Logger trait). Snippet 6 illustrates the adaptation to Untrused mentioned previously. The code for the other adaptations is similar. For this case study, we identified 4 possible conflict scenarios:

1. Logger, Encrypter: store and lookup methods.
2. Logger, Persistent: store method.
3. Persistent, Encrypter: store method.
4. Logger, Persistent, Encrypter: store, and lookup methods.

As an example, Snippet 7 shows the definition of the composition policy for Logger and Encrypter (lines 2–6). The

---

[16]In particular, we have successfully applied the same technique to Ruby, which also lacks reflective call stack access.

```
1  cop.manager.addObjectPolicy(cacheServer,
2    [Auditing, Untrused],
3    function(Logger, Encrypter) {
4      return Trait.compose (Encrypter,
5        Trait.resolve({ store: undefined,
6                        lookup: undefined }, Logger)); });
```

**Snippet 7: Comp. policy for Logger vs. Encrypter.**

```
Android.adapt(ciclaMi, Trait({
  initHandlers: function() { ... /* Android specific */ },
  run: function(zoomLevel) { ... /* Android specific */ },
  gotoPosition: function(position, zoomLevel) {
    this.proceed();
    ... // Android specific
  },
  initScreen: function() {
    this.proceed();
    ... // Android specific
  },
  // New behaviour
  onCenterChanged: function() { ... },
  capturePhoto:    function() { ... },
  uploadData:      function() { ... },
  ... }));
```

**Snippet 8: Adaptation of CiclaMi for Android.**

policy solves the conflict by excluding store and lookup from trait Logger. As a result, the methods will be taken from the Encrypter trait. This simple cache server example illustrates that our proposal for composition and conflict resolution works fine in typical context-oriented scenarios.

## 4.2 CiclaMi

*CiclaMi* was conceived as an application for monitoring the state of bicycle paths in the city of Milan, taking advantage of the sensors typically embedded in smartphones. It serves as an example of how to write an application in such a way that it can either run as a web application, or as a mobile application, having different functionality depending on the platform. CiclaMi was originally designed and implemented as a traditional Web application, and two different code bases were needed for the mobile and web browser versions. Although the code was straightforward, it was duplicated (for the mobile and fixed versions) and suffered from tangling of environment sensing and base logic code. For instance, it contained several conditional statements invoking isAndroid to express Android-specific functionality. Similarly, code to detect whether the device had connectivity was tangled with the functionality that relied on having connectivity. To refactor CiclaMi as a COP application having a unique code base, we created a context Android, which can be activated by registering a PhoneGap event listener:

```
document.addEventListener
    ("deviceready", function() { Android.activate(); }, true);
```

Once active, context Android is never deactivated. To provide the behaviour of CiclaMi that is specific to Android, we defined the adaptation outlined in Snippet 8. This case study allowed us to improve the architecture of our application by separately defining the platform-dependent parts. It thereby illustrated a less publicised use of COP for static software modularisation and configuration.[17]

---

[17]This shows that COP is not a dynamic or static approach per se; the static or dynamic nature of behaviour goes with the *context* for which such behaviour has been defined.

```
1  $(document).ready(function() {
2    LowBattery.adapt($.mobile, Trait({
3      changePage: function(to, options) {
4        newOptions = jQuery.extend({}, options);
5        newOptions['transition'] = 'none';
6        this.proceed(to, newOptions); }})))});
```

**Snippet 9: Adaptation of jQuery core functionality.**

## 4.3 jQuery Mobile

The main goal of this third case study was to assess whether our approach could be used to dynamically adapt the source code of an existing system that was not written in a context-oriented style, *without* modifying its source code to introduce adaptations. To this end, we started from the jQuery Mobile library, which is a well-known library with one of the largest user bases in the JavaScript community. By default, jQuery Mobile uses animated transitions between pages. This "graphic candy" enhances the user experience, but consumes slightly more battery than not having intermediate animations for every modal transition in the user interface. For this reason, in this case study we wanted to adapt the library to low battery conditions by disabling animations. Snippet 9 shows the simple adaptation we introduced for the main $.mobile object in LowBattery context. The adaptation extends the changePage method of $.mobile. The changePage is used internally for the page loading and transitioning that occurs as a result of clicking a link or submitting a form. Our adapted version leaves the destination (to) untouched, but passes a new options object in which the transition has been disabled. In passing, Snippet 9 illustrates a call to proceed with explicit parameters (Line 6). This case study also illustrates *unanticipated* adaptation to context, since the code of jQuery Mobile was completely oblivious to our library, but could be adapted nevertheless. Finally, the case study also illustrates *non-invasive* definition of adaptations, since we did not need to touch the source code of jQuery Mobile to introduce our adaptation.

## 5. DISCUSSION AND FUTURE WORK

JavaScript naturally supports the view of traits-as-state discussed in Section 2.3, since both object state and behaviour are stored in assignable slots. However, JavaScript also supports the traits-as-behaviour view, thanks to the newly introduced *properties* abstraction:[18] in this case, a slot access such as point.x is actually a method call without arguments, as in Self [24]. Modern JavaScript thus aligns well with our traits model, which does not distinguish between state and behaviour. As to whether static languages such as Java can accommodate context traits, the question is still open. With recent progress on dynamic software updates [19], the technical barrier seems to have lowered.

Regarding the benefits our model brings to JavaScript, it could be argued that JavaScript by itself already allows —even invites— very flexible manipulation of objects. Programmers can change the implementation of any method at run time. What our approach adds is a conceptual framework on top of this low-level basis, a mindset that invites programmers to move from *thinking in isolation*, in a way that is mostly oblivious of the context in which the software will be used, towards *thinking in context* to build adaptive

---

[18]Properties are defined in the ECMAScript5 standard.

218

applications from the ground up. This new programming perspective comes with necessary language-level support, such as first-class contexts, traits, composition operators, and composition policies.

This paper focussed essentially on presenting the underlying concepts of our approach, backed by a prototype implementation in JavaScript. Although a first evaluation has already been performed through small case studies, we need to carry out a more thorough validation, including benchmarks to assess the performance overhead of our approach as compared to traditional approaches. Our prior experience with the design of context-oriented languages makes us confident, however, that performance loss will not be drastic and will be compensated by increased expressiveness [10]. Regarding correctness of context trait composition, it is an important topic that we are currently exploring generally for any context-based approach [4], and that we plan to tailor to the particular case of context traits.

In designing context traits we leveraged our previous experience and incorporated a minimal set of mechanisms that we have found to be most useful in practice, such as activation counters (Section 2.1) and composition policies based on activation age (Section 2.6). There remains a last generally useful mechanism, *context combinations* [11], that we will incorporate in the future. This should pose no problems, since context combinations are independent of the adaptation unit being used —whether methods or traits.

## 6. RELATED WORK

*Dynamic aspect weaving*, of which we take PROSE as iconic example [18], shares our motivation of streamlining dynamic behaviour adaptability, and also the approach of achieving it through run-time software recomposition. However, the means are quite different. Joinpoints in PROSE feature a rich set of events such as field accesses, method entry, catch and throw of exceptions, class loading and unloading, and breakpoints. Support of method invocations goes without saying. The interception of these events allows for detection of joinpoints at run time and execution of corresponding advice. Dynamic aspect weaving approaches such as PROSE are often more expressive than COP approaches like ours. PROSE can be seen as a sort of "composition assembler" that allows the execution of advise at practically any conceivable point during program execution. Our approach, on the other hand, proposes a high-level conceptual framework in terms of contexts, objects and adaptations. It should be noted that the purer the underlying object-oriented language, the closer COP and dynamic AOP come together, because with pure message-passing semantics, every action (including field accesses, caught and thrown exceptions, and so on) can be captured and adapted by COP. COP is conceptually simpler and works best with purer object-oriented models, whereas dynamic AOP is more complex but has better support for primitive constructs and mechanisms that bypass a strict object-oriented semantics.

*Chai$_3$* [22] is the first effort to introduce dynamic trait substitution. Since *Chai$_3$* has been conceived for class-based, statically typed languages, it allows the adaptation of existing methods, but the interface of adapted objects must be preserved. Objects are adapted by substituting one trait for a compatible one —one which provides and requires the same set of method signatures. Augmenting or reducing an object's protocol is not supported. *Chai$_3$* was described

formally, but the formalisation was not backed up by an implementation [22].

*Lexically-nested traits* [25] propose a novel trait composition mechanism based on prototype-based delegation. In this approach traits are actually closures, and are thus stateful. Nested scopes delegate to their parent scope. The state of a trait is visible only within the scope of the trait or its nested scopes. Like in our proposal, traits are regular objects, and any object can thus play the role of a trait. The idea of defining traits as closures and visibility control through lexical scoping inspired the development of the `traits.js` library [26], upon which we relied to build the dynamic adaptation machinery in our JavaScript implementation.

*Talents* [20] define behaviour that can be acquired by objects. Like classical traits, talents are stateless, compositional (i.e. new talents can be made from existing talents) and preserve the flattening property (i.e. objects containing talents cannot be distinguished semantically from objects that do not). Unlike traits, talents are added to objects, not to classes, leaving other instances of the same class unchanged. Further, talent composition is dynamic: objects can be recomposed at run time by adding and removing talents. Talents thus constitute a valid stepping stone towards software adaptation through dynamic object recomposition. Unfortunately, they do not feature a `proceed`-like mechanism for behaviour extension. Furthermore, since they were not specifically conceived for self-adaptive context-oriented systems, some of the necessary concepts and infrastructure are missing: they do not reify context nor build a causal connection between context and object composition, and therefore also lack a dynamic conflict resolution mechanism.

*ContextL* [7] is the seminal work on context orientation, and gave rise to a family of similar approaches that have come to be known as *layer-based* approaches [21]. Our approach rather belongs to the *context-based* branch of COP initiated by *Ambience* [11] —the first COP language to introduce first-class contexts, and in which the activation of adaptations was not dictated by base program control flow. Our approach improves on current approaches —whether layer based or context based— in that it offers more meaningful units of adaptation. Traits, as opposed to single methods, are geared towards flexible composition, liberating the programmer from the restrictions of inheritance mentioned in Section 2.2, and providing explicit abstractions to solve composition conflicts. Further, traits give first-class status to bundles of methods that have been designed to work together under a certain context. With single-method adaptation, the semantic connection between those methods remains implicit.

*ContextJS* [15], a JavaScript framework, does support the definition of adaptations as whole objects instead of individual methods, but when it comes to composition, such plain objects are far from having the status of full-blown traits with composition policies presented here. More importantly, there is no backing *conceptual mindset* such as the one presented in Section 2.4 that invites programmers to think in terms of meaningful units of adaptation.

## 7. CONCLUSION

Traits have been proposed to overcome some of the limitations of composition in object-oriented languages, by having better granularity and no imposed composition semantics. COP too has been proposed to overcome some of the limitations of composition in object-oriented languages, namely the lack of run-time behaviour adaptability. This paper proposes to marry the best of those two solutions. On the one hand, the paper presents a contribution to research on COP by proposing traits as units with convenient granularity for dynamic behaviour adaptation and a well-defined composition semantics. On the other hand, the paper proposes a contribution to research on traits by proposing a simpler traits model and a way of using it dynamically depending on detected context changes. At the crossroads of both approaches, we have found a convenient way of developing software applications that need to evolve according to their changing environment of execution.

# 8. ACKNOWLEDGEMENTS

This work has been supported by the ICT Impulse Programme of the Brussels Institute for Research and Innovation, and by the MIUR project CINA: Compositionality, Interaction, Negotiation, Autonomicity for the future ICT society. We thank the anonymous reviewers for their comments on earlier versions of this paper.

# References

[1] Kim Barrett, Bob Cassels, Paul Haahr, David A. Moon, Keith Playford and P. Tucker Withington. *A Monotonic Superclass Linearization for Dylan*. ACM SIGPLAN Notices 31.10 (1996).

[2] Alexandre Bergel, Stéphane Ducasse, Oscar Nierstrasz and Roel Wuyts. *Stateful Traits and Their Formalization*. Computer Languages, Systems and Structures 34.2–3 (2008).

[3] Gilad Bracha and William Cook. *Mixin-Based Inheritance*. ACM SIGPLAN Notices 25.10 (1990).

[4] Nicolás Cardozo, Jorge Vallejos, Sebastián González, Kim Mens and Theo D'Hondt. *Context Petri Nets: Enabling Consistent Composition of Context-Dependent Behavior*. PNSE Workshop Proceedings. CEUR-WS.org, 2012.

[5] Craig Chambers, David Ungar and Elgin Lee. *An Efficient Implementation of Self, a Dynamically-Typed Object-Oriented Language Based on Prototypes*. ACM SIGPLAN Notices 24.10 (1989).

[6] Pascal Costanza. *Dynamically Scoped Functions as the Essence of AOP*. ACM SIGPLAN Notices 38.8 (2003).

[7] Pascal Costanza and Robert Hirschfeld. *Language Constructs for Context-Oriented Programming: An Overview of ContextL*. DLS Symposium Proceedings. ACM Press, 2005.

[8] Stéphane Ducasse, Oscar Nierstrasz, Nathanael Schärli, Roel Wuyts and Andrew P. Black. *Traits: A Mechanism for Fine-Grained Reuse*. ACM Transactions on Programming Languages and Systems 28.2 (2006).

[9] Carlo Ghezzi, Matteo Pradella and Guido Salvaneschi. *An Evaluation of the Adaptation Capabilities in Programming Languages*. SEAMS Symposium Proceedings. ACM Press, 2011.

[10] Sebastián González, Nicolás Cardozo, Kim Mens, Alfredo Cádiz, Jean-Christophe Libbrecht and Julien Goffaux. *Subjective-C: Bringing Context to Mobile Platform Programming*. SLE Conference Proceedings. Springer-Verlag, 2011.

[11] Sebastián González, Kim Mens and Patrick Heymans. *Highly Dynamic Behaviour Adaptability Through Prototypes with Subjective Multimethods*. DLS Symposium Proceedings. ACM Press, 2007.

[12] Gregor Kiczales, Erik Hilsdale, Jim Hugunin, Mik Kersten, Jeffrey Palm and William Griswold. *An Overview of AspectJ*. ECOOP Conference Proceedings. Springer-Verlag, 2001.

[13] Gregor Kiczales, Jim des Rivières and Daniel G. Bobrow. *The Art of the Metaobject Protocol*. MIT Press, 1991.

[14] Henry Lieberman. *Using Prototypical Objects to Implement Shared Behavior in Object-Oriented Systems*. ACM SIGPLAN Notices 21.11 (1986).

[15] Jens Lincke, Malte Appeltauer, Bastian Steinert and Robert Hirschfeld. *An Open Implementation for Context-Oriented Layer Composition in ContextJS*. Science of Computer Programming 76.12 (2011).

[16] Ole Lehrmann Madsen and Birger Møller-Pedersen. *What Object-Oriented Programming May Be —and What It Does Not Have To Be*. ECOOP Conference Proceedings. Springer-Verlag, 1988.

[17] Pattie Maes. *Concepts and Experiments in Computational Reflection*. ACM SIGPLAN Notices 22.12 (1987).

[18] Andrei Popovici, Gustavo Alonso and Thomas Gross. *Just-in-Time Aspects: Efficient Dynamic Weaving for Java*. AOSD Conference Proceedings. ACM Press, 2003.

[19] Mario Pukall, Christian Kästner, Walter Cazzola, Sebastian Götz, Alexander Grebhahn, Reimar Schröter and Gunter Saake. *JavAdaptor —Flexible Runtime Updates of Java Applications*. Software: Practice and Experience (2012).

[20] Jorge Ressia, Tudor Gîrba, Oscar Nierstrasz, Fabrizio Perin and Lukas Renggli. *Talents: An Environment for Dynamically Composing Units of Reuse*. Software: Practice and Experience (2012).

[21] Guido Salvaneschi, Carlo Ghezzi and Matteo Pradella. *Context-Oriented Programming: A Software Engineering Perspective*. Journal of Systems and Software 85.8 (2012).

[22] Charles Smith and Sophia Drossopoulou. *Chai: Traits for Java-like Languages*. ECOOP Conference Proceedings. Springer-Verlag, 2005.

[23] David Ungar, Craig Chambers, Bay-Wei Chang and Urs Hölzle. *Organizing Programs Without Classes*. Lisp and Symbolic Computation 4.3 (1991).

[24] David Ungar and Randall B. Smith. *Self: The Power of Simplicity*. ACM SIGPLAN Notices 22.12 (1987).

[25] Tom Van Cutsem, Alexandre Bergel, Stéphane Ducasse and Wolfgang De Meuter. *Adding State and Visibility Control to Traits Using Lexical Nesting*. ECOOP Conference Proceedings. Springer-Verlag, 2009.

[26] Tom Van Cutsem and Mark S. Miller. *traits.js: Robust Object Composition and High-Integrity Objects for ECMAScript 5*. PLASTIC Workshop Proceedings. ACM Press, 2011.

# Author Index

www.ingramcontent.com/pod-product-compliance
Lightning Source LLC
Chambersburg PA
CBHW061413210326
41598CB00035B/6196